Progress in Botany

Volume 72

Series Editors
Ulrich Lüttge, TU Darmstadt, Institut für Botanik,
FB Biologie (10), Schnittspahnstraße 3–5,
64287 Darmstadt, Germany
Wolfram Beyschlag, Fakultät für Biologie, Lehrstuhl für
Experimentelle Ökologie und Ökosystembiologie,
Universität Bielefeld, Universitätsstraße 25, 33615 Bielefeld,
Germany
Burkhard Büdel, TU Kaiserslautern,
FB Biologie, Abt. Allgemeine Botanik,
Erwin-Schrödinger-Str., 67663 Kaiserslautern,
Gebäude 13/2, Germany
Dennis Francis, University of Cardiff, Cardiff School
of Biosciences, Cardiff, United Kingdom CF10 3TL

Ulrich Lüttge • Wolfram Beyschlag •
Burkhard Büdel • Dennis Francis
Editors

Progress in Botany 72

Springer

Editors
Prof. Dr. Ulrich Lüttge
TU Darmstadt
Inst. Botanik
Schnittspahnstr. 3-5
64287 Darmstadt
Germany
luettge@bio.tu-darmstadt.de

Prof. Dr. Wolfram Beyschlag
Universität Bielefeld
Fakultät für Biologie
LS für Experimentelle
Ökologie und Ökosystembiologie
Bielefeld
Germany
w.beyschlag@uni-bielefeld.de

Prof. Dr. Burkhard Büdel
TU Kaiserslautern
FB Biologie
Abt. Allgemeine Botanik
Erwin-Schrödinger-Str.
67663 Kaiserslautern
Gebäude 13/2
Germany
buedel@rhrk.uni-kl.de

Dr. Dennis Francis
University of Cardiff
Cardiff School of Biosciences
Cardiff
United Kingdom
francisd@cardiff.ac.uk

ISSN 0340-4773
ISBN 978-3-642-26528-0 ISBN 978-3-642-13145-5 (eBook)
DOI 10.1007/ 978-3-642-13145-5
Springer Heidelberg Dordrecht London New York

Softcover reprint of the hardcover 1st edition 2011

Cover design: WMXDesign GmbH, Heidelberg, Germany

Printed on acid-free paper

Springer is part of Springer Science+Business Media (www.springer.com)

Editorial

We regret to inform our readers that our enduring author **Prof. Dr. rer. nat. Dr. phil h.c. Burkhard Frenzel** passed away on February 6, 2010 in his 83rd year. Nearly every year since 1964, he has contributed valuable reviews in the field of paleobotany. His latest contribution was published last year in volume 71. The publisher and editors have lost a dedicated author who will be honourably remembered by his friends and colleagues throughout academia and beyond.

Contents

Contributors

Jayne Elisabeth Archer Department of English and Creative Writing, Penglais Campus, Aberystwyth University, Ceredigion SY23 3DY, UK

Irene Baroli Departamento de Biodiversidad y Biología Experimental, Facultad de Ciencias Exactas y Naturales, Universidad de Buenos Aires, Pabellón II, Ciudad Universitaria C1428EHA, Buenos Aires, Argentina, baroli@bg.fcen.uba.ar

M.B. Bitonti Department of Ecology, University of Calabria, Arcavacata Rende (CS), Italy, b.bitonti@unical.it

Anne M. Borland Molecular Plant Physiology, Newcastle Institute for Research on Sustainability, University of Newcastle Upon Tyne, Devonshire Building, NE1 7RU Newcastle, UK, a.m.borland@newcastle.ac.uk

Burkhard Büdel Molecular Phylogenetics, Department of Biology, University of Kaiserslautern, Postfach 3049, 67653 Kaiserslautern, Germany, buedel@rhrk.uni-kl.de

Johan Ceusters Afdeling Plantenbiotechniek, Departement Biosystemen, Faculteit Bio-Ingenieurswetenschappen, KULeuven, Willem Decroylaan 42, 3001 Heverlee, Belgium

A. Chiappetta Department of Ecology, University of Calabria, Arcavacata Rende (CS), Italy, a.chiappetta@unical.it

Arthur Gessler Institute for Landscape Biogeochemistry, Leibniz-Zentrum für Agrarlandschaftsforschung (ZALF) e.V., Eberswalderstr. 84, 15374 Müncheberg, Germany, gessler@zalf.de; Professorship for Landscape Biogeochemistry, Humboldt-University Berlin, Lentze-Allee 75, 14195 Berlin, Germany

Thorsten E.E. Grams Ecophysiology of Plants, Department of Ecology and Ecosystem Management, Technische Universität München, Am Hochanger 13, 85354 Freising, Germany, grams@tum.de

T.A. Jones USDA-Agricultural Research Service, Forage and Range Research Laboratory, Utah State University, Logan, UT 84322-6300, USA, Thomas.Jones@ars.usda.gov

Frank Kauff Molecular Phylogenetics, Department of Biology, University of Kaiserslautern, Postfach 3049, 67653 Kaiserslautern, Germany, fkauff@biologie.uni-kl.de

Franziska Kuhlmann Department of Chemical Ecology, Bielefeld University, Universitätsstr. 25, D-33615 Bielefeld, Germany

Tracy Lawson Department of Biological Sciences, University of Essex, Wivenhoe Park, Colchester CO4 3SQ, UK, tlawson@essex.ac.uk

Ulrich Lüttge Institut für Botanik, Technical University of Darmstadt, Schnittspahnstrasse 3-5, 64287 Darmstadt, Germany, luettge@bio.tu-darmstadt.de

Ulrike Mathesius ARC Centre of Excellence for Integrative Legume Research, Division of Plant Science, Research School of Biology, Australian National University, Canberra, ACT 0200, Australia, mathesius@anu.edu.au

Caroline Müller Department of Chemical Ecology, Bielefeld University, Universitätsstr. 25, D-33615 Bielefeld, Germany, caroline.mueller@uni-bielefeld.de

J.G. Robins USDA-Agricultural Research Service, Forage and Range Research Laboratory, Utah State University, Logan, UT 84322-6300, USA

Bernd Schneider Max-Planck-Institute for Chemical Ecology, Beutenberg Campus, Hans-Knöll-Str. 8, 07745 Jena, Germany, schneider@ice.mpg.de

U. Stahl Institute for Biotechnology, Berlin University of Technology, 13353 Berlin, Germany, Ulf.Stahl@lb.tu-berlin.de

L. Strack Institute for Biotechnology, Berlin University of Technology, 13353 Berlin, Germany, L.Strack@tu-berlin.de

Masashi Tazawa Professor Emeritus, The University of Tokyo, Chadom achi 6-15, Otsu 520-0066, Japan, mtazawa@kjd.biglobe.ne.jp

Howard Thomas IBERS, Edward Llwyd Building, Penglais Campus, Aberystwyth University, Ceredigion SY23 3DA, UK, hot@aber.ac.uk

Richard Marggraf Turley Department of English and Creative Writing, Penglais Campus, Aberystwyth University, Ceredigion SY23 3DY, UK

Susanne von Caemmerer Research School of Biology, Australian National University, GPO Box 475 Canberra, ACT2601, Australia, Susanne.Caemmerer@anu.edu.au

Michelle Watt CSIRO Plant Industry, Black Mountain Laboratories, GPO Box 1600, Canberra, ACT 2601, Australia

Part I
Review

Curriculum Vitae

Born in January 12, 1930 in Yokohama, Japan

Present address Chadomachi 6-15, Otsu 520-0066, Japan
E-mail mtazawa@kjd.biglobe.ne.jp

Education

1953	B.Sc. Faculty of Science, Osaka University
1960	Ph.D. Biology, Osaka University

Professional experience

1955–1968	Instructor, Department of Biology, Faculty of Science, Osaka University
1968–1977	Associate Professor, Department of Biology, Faculty of Science, Osaka University
1977–1990	Professor, Department of Botany, Faculty of Science, The University of Tokyo
1990–2002	Professor, Department of Applied Physics and Chemistry, Fukui University of Technology
1990	Professor emeritus, The University of Tokyo

Research Activities in Foreign Lands

1955–1956	DAAD student, Institute of Botany (head: Prof. Erwin Bünning), University of Tübingen
1956–1957	Research Assistant, ibid.
1967–1968	Invited Lecturer, Institute of Plant Physiology (head: Prof. Jacob Reinert), Free University of Berlin
1976	Senior Exchange Scholarship between Japan and Canada, Department of Biology (head: Prof. David Fensom), Mount Allison University

Honors

1985	Corresponding Member of The American Society of Plant Biologists
1990	Prize of The Japan Academy
1994	Honorary Member of The German Botanical Society
2004	Honorary Member of The Japanese Botanical Society

Sixty Years Research with Characean Cells: Fascinating Material for Plant Cell Biology

Masashi Tazawa

Contents

M. Tazawa
Professor emeritus, The University of Tokyo, Chadomachi, 6-15, Otsu 520-0066, Japan
e-mail: mtazawa@kjd.biglobe.ne.jp

U. Lüttge et al. (eds.), *Progress in Botany 72*,
DOI 10.1007/978-3-642-13145-5_1, © Springer-Verlag Berlin Heidelberg 2011

Abstract My nearly 60-year long and deeply rewarding research life can be divided into three periods, the first between 1952 and 1977 at Osaka University, the second between 1977 and 1990 at the University of Tokyo, and the third between 1990 and 2002 at the Fukui University of Technology. Throughout my research life, my main experimental material has been the internodal cells of the characean algae, *Nitella*, *Chara*, *Nitellopsis*, as well as the internodal cells of the brackish genus *Lamprothamnium*. The characean internodal cell is a giant cylindrical cell that is typically 50–100 mm in length and 0.5–1.0 mm in diameter (cf. my picture). The cell is characterized by a very active rotational cytoplasmic streaming with an unusually high speed amounting to 50–70 μm s^{-1} at 25°C as well as an exceptionally large action potential (150 mV), which can be generated by electrical, mechanical, and chemical stimuli. Due to these activities, the characean cell is often referred to the "green muscle" or the "green axon."

Professor Noburo Kamiya, my mentor throughout my research in the undergraduate (1952–1953) and graduate course (1953–1955) at the Biology Department of Osaka University, was already well-known for his famous work on cytoplasmic streaming in the plasmodium of the slime mold *Physarum polycephalum*. In 1940, he invented the so-called double-chamber method to measure the motive force of cytoplasmic streaming in *Physarum*. This simple yet elegant method allowed him to measure the motive force by controlling cytoplasmic streaming with air pressure (Kamiya 1940). Professor Kamiya's ingenious gift to develop new methods and instrumentation extended to the study of cytoplasmic streaming in characean cells (cf. Kamiya 1986). Under the influence of Professor Kamiya, my research on characean cells is characterized by the development of new methods. In 1964, I developed a perfusion technique to make cells that contain artificial cell saps so that the composition of the vacuole could be experimentally controlled. In 1976, I developed a method based on perfusion for removing the tonoplast of characean cells so that the composition of the cytosol could also be experimentally controlled. My daily exposure to cell surgeries in the Kamiya laboratory, such as the cutting and ligation of internodal cells, was inspirational in helping me to develop these intracellular perfusion techniques.

I learned from my mentor in Osaka that science is driven forward by the combination of a new idea with a new method. With this thought etched in my mind, I moved in 1977 from Osaka University to the University of Tokyo where I chaired the Laboratory of Plant Physiology in the Botany Department. Here we

Dedicated to Professor Noburo Kamiya (1913–1999).

took advantage of vacuolar perfusion and tonoplast-free cells to study the electrogenic ion pump, membrane excitation, cell motility, turgor regulation, salt tolerance, various vacuolar activities, including pH regulation, and protein degradation, as well as the intercellular transport of photoassimilates.

1 Biology as the Major

At noon of the 15th of August 1945, we, the pupils of the Preparatory Course of the Naval Payofficer's School, were assembled in the auditorium to hear the Emperor's (Tenno's) voice, which happened for the first time in our history. This was the Imperial message to the people conveying that he decided to accept the Potsdam Declaration issued on the 26th of July 1945. Japan was defeated after 3 years and 9 months' war with the Allied Forces, including USA, China, England, and other nations. I was 15 years old. After several days, we were forced to come back to our homes. My honest feeling at the moment of defeat was complex. On the one hand, I feared for the future of our nation, and on the other hand, I had a feeling that I was being saved from the death in battle. Most cities in Japan were almost completely destroyed by the indiscriminate bombing carried out by the US Air Force. However, my family lived in the suburbs of Osaka, the second biggest city in Japan, and although the City of Osaka was severely destroyed, the district where we lived was escaped from bombardment and my home survived without any damage. Soon, I returned to the former middle school.

In 1947, I entered the Third High School in Kyoto, which offered two courses, the liberal arts course and the science course. I selected the science course without any serious consideration about a future job. The fact that my father was an electric engineer might have helped me to select the science course. In the high school, I was a member of the rowing club of the school. The highlight for the school oarsmen was the eight boat race between the First High School in Tokyo and our school in Kyoto. To win the race, we trained hard in the Biwa-Lake, which is the largest lake in Japan. Exhausted from the training, we often cut classes at school, which resulted in getting poorer grades. When 3 years study in the High School was close to the end, I was forced to select the field for my future study in the university. First, I wished to study either agricultural science or the medical science, simply because the former might solve the food crisis under which the Japanese people suffered so much during the War. Also medical science attracted me, since it seemed directly linked to the relief of the people. I asked opinion of my intimate friend who was 2 years ahead of me. Hearing my wishes for the future course, he advised me to study biology, since biology constitutes the basis of agricultural and medical sciences. He further recommended me the Department of Biology in the Faculty of Science, Osaka University. I thought that his advice was rational and so I entered the Department of Biology of Osaka University in 1950 at the age of 20.

The Biology Department was founded in 1949 to advance the new trends in biology, which was to base biological reasoning and experimentation on chemical and physical principles. The first chairman of the Biology Department, Professor Shiro Akabori, who was a distinguished biochemist in the protein research, told the first undergraduate students at the entrance ceremony as follows: "Our new Biology Department places great emphasis on analyzing the biological phenomena on the basis of physics and chemistry." Along this line, Akabori invited excellent biologists to be Professors in the new Department. As already mentioned, Noburo Kamiya, who was already well known as a distinguished cell physiologist for the work I mentioned above, chaired the laboratory of cell physiology. Kazuo Okunuki, who had earned fame as a result of his discovery of cytochrome $c_{1,}$ chaired the laboratory of microbiology. Ichijiro Honjo, a forefront researcher in the field of sensory and behavioral physiology in animals, chaired the laboratory of comparative physiology. Hideo Kikkawa, who was well known for his discovery that the pigment found in *Bombyx* eggs is formed in the tryptophan – kynurenine – 3-hydroxykynurenine pathway, chaired the laboratory of genetics. Professor Kikkawa proposed the so-called one gene-one enzyme hypothesis in 1941, of George Beadle and Edward Tatum.

During the World War II, almost no scientific information from abroad came to Japan. Professor Kamiya who had stayed in William Seifriz's laboratory at the University of Pennsylvania from 1939 to 1942 (Kamiya 1989), and who revisited the USA in 1950 on the invitation of Seifriz told us after his return to Osaka that biology in Japan was at least 10 years behind that of USA. After the end of the War, we were very eager to catch up the level of the USA. At that time, many biology departments in old universities were divided into a Botany Department and a Zoology Department. In contrast, biology in Osaka, irrespective of whether the material was plant or animal, aimed at analyzing the biological phenomena on a cellular and physicochemical basis. We named this approach based on fundamentals as opposed to the traditional approach based on taxon, "modern biology," and we students were proud of studying modern biology.

In order to finish the undergraduate course and to get a Bachelor degree, I had to do research for 1 year. For this, I had to select the mentor. Through lectures and practices I had experienced, I thought that my talent might be suited to researches using physical means rather than chemical ones. Since the trend of research in Professor Kamiya's laboratory was physically oriented, I asked him to be my mentor.

2 Transcellular Osmosis and Polar Water Permeability

In April 1952, I started experiments to measure transcellular water transport in the internodal cells of *Nitella* by using the so-called double chamber osmometer (Kamiya and Tazawa 1956). To perform transcellular osmosis (TCO), an internodal cell is partitioned into two chambers A and B. When water in B is replaced with a sucrose solution, water moves transcellularly from chamber A to chamber B. The rate of the flow is proportional to the osmotic pressure (π_o) in B. The TCO constant (K) defined by Kamiya and Tazawa (1956) is obtained by dividing the initial rate of water flow (J_v) by the external osmotic pressure (π_o) that drives the flow. Namely,

$$J_v/\pi_o = K = (\mathrm{Lp_{en}Lp_{ex}}A_{en}A_{ex})/(\mathrm{Lp_{en}}A_{en} + \mathrm{Lp_{ex}}A_{ex}), \quad (1)$$

where Lp and A represent the hydraulic conductivity to either endosmosis (en) or exosmosis (ex) and the surface area of the cell part on either the endosmosis or the exosmosis side, respectively. If $\mathrm{Lp_{en}} = \mathrm{Lp_{ex}} = \mathrm{Lp}$ and $A_{en} = A_{ex} = A/2$, we get

$$\mathrm{Lp} = 4K/A, \quad (2)$$

where A represents the surface area of the whole cell and Lp in (2) represents the apparent hydraulic conductivity of the entire cell.

One day, Professor Kamiya suggested that I measure the hydraulic conductivity using an asymmetrical arrangement in order to test whether the water permeability might depend on the direction of osmosis. He certainly knew about the rectification property of the nerve fiber in terms of electric current. In axons, the electrical resistance to the depolarizing current is lower than it is to the hyperpolarizing one. Using an asymmetrical arrangement, to measure hydraulic resistance in characean cells, two reciprocal osmoses can be performed, one from the shorter half to the longer half and another in the reverse direction. These two reciprocal osmoses give two simultaneous equations corresponding to (1). Solving the equations, we get $\mathrm{Lp_{en}}$ and $\mathrm{Lp_{ex}}$.

We found that $\mathrm{Lp_{en}}$ is larger than $\mathrm{Lp_{ex}}$. Namely, water enters the cell easier than it escapes from the cell. The polar hydraulic conductivity was intensively studied for its cause by us in Osaka and by Dainty and coworkers in Edinburgh. In our first paper, the polar water permeability to water was attributed to an intrinsic characteristic of the plasma membrane (Kamiya and Tazawa 1956).

On the other hand, Dainty and Hope (1959) were of the view that the polar permeability to water was only apparent and was caused by a "sweeping away" of the solutes on the exosmosis side, thus lowering the effective concentration of the solute at the surface of the plasma membrane. In the asymmetrical arrangement, such a dilution-effect due to the sweeping away would be larger when the solution is given to the shorter cell part than when given to the longer cell part.

Later, Dainty and Ginzburg (1964a) found in *Chara* that hydraulic conductivity decreased markedly with an increase in the external sucrose concentration. They found that the inhibitory effect of the external osmolality cannot be attributed to the sweeping away effect. The inhibitory effect of the external osmotic pressure on hydraulic conductivity was reconfirmed by Kiyosawa and Tazawa (1972) in *Nitella flexilis*.

2.1 Hydraulic Conductivity (Lp) Is Affected by the Internal Osmotic Pressure

In view of the effect of the external osmotic pressure on hydraulic conductivity, I speculated that the internal osmotic pressure might also affect the hydraulic conductivity. I prepared cells that had a higher or a lower osmotic pressure using TCO and ligation (Kamiya and Kuroda 1956a) and then measured the hydraulic conductivity of these cells. We found that raising the internal osmotic pressure also decreased the hydraulic conductivity while lowering the internal osmotic pressure increased the hydraulic conductivity (Tazawa and Kamiya 1966). Later, we demonstrated that it is the osmotic pressure itself that affects the hydraulic conductivity and not the concentration of ions (Kiyosawa and Tazawa 1972).

We invented a new method, which enabled us to determine Lp_{en} and Lp_{ex} by inducing TCO in the symmetrical arrangement of the internodal cell of *Nitella*. (Tazawa and Kiyosawa 1973). Since the hydraulic conductivity of the tonoplast was much greater than the hydraulic conductivity of the plasma membrane (Kiyosawa and Tazawa 1977), the measured Lp_{en} and Lp_{ex} by the new method represent those of the plasma membrane.

Using the new method, we measured the specific resistance to endosmosis (Lp_{en}^{-1}) and that to exosmosis (Lp_{ex}^{-1}) at various external osmotic pressures (π_o) used as the driving force for the TCO (1). We found that Lp_{en}^{-1} decreases while Lp_{ex}^{-1} increases linearly with π_o. The polarity Lp_{en}/Lp_{ex} approaches unity when π_o approaches zero (Kiyosawa and Tazawa 1973).

The decrease in hydraulic conductivity caused by high external osmotic pressure was interpreted by Dainty and Ginzburg (1964b) to result from a reduction of water content of the membrane that would cause shrinkage of the membrane. The effect of internal osmotic pressure on hydraulic conductivity might be similar to the effect of the external osmotic pressure. The effect of both the internal and external osmotic pressure on hydraulic conductivity does not depend on the species of the solutes (Dainty and Ginzburg 1964a; Kiyosawa and Tazawa 1977).

Recently, Ye et al. (2004) proposed gating of water channels in *Chara* cells by high concentrations of solutes. They assert that the effects of intracellular and extracellular osmotic pressures on the hydraulic conductivity of characean cells may be explained in terms of the cohesion/tension model for water channels.

2.2 Water Channel

The idea of aqueous pores in plant cell membranes was proposed already in 1963 by Dainty. The apparent activation energy (E_a) for water flow across the plasma membrane was calculated by measuring the dependence of Lp on temperature. The average E_a value was 8.5 kcal mol^{-1} in *Nitella translucens* (Dainty and Ginzburg 1964a) and 5 kcal mol^{-1} in *Nitella flexilis* (Tazawa and Kamiya 1965). These values are higher than that for the self-diffusion of water, 4.2 kcal mol^{-1} (Tazawa and Kamiya 1965). The higher values of E_a are interpreted in terms of the penetration of water through very narrow pores of the cell membrane (Dainty and Ginzburg 1964a).

A breakthrough leading to an understanding of the nature of water channels was brought by Wayne and Tazawa (1990) who showed that hydraulic conductivity was significantly reduced by *p*-chloromercuriphenyl sulfonic acid, a sulfhydryl agent, which is known to inhibit the water permeability of erythrocytes and to bind specific proteins involved in water conduction (Benga et al. 1986). The inhibition was prevented by simultaneous treatment of cells with dithiothreitol. This was the first functional evidence for the existence of water channels in plant cells prior to the molecular identification of animal and plant aquaporins (cf. Maurel 1997).

3 Artificial Control of the Vacuolar Composition: Vacuolar Perfusion

While I was doing experiments in the Kamiya laboratory, my classmate Kiyoko Kuroda was studying the osmoregulation in *Nitella* cells. She used TCO to produce cells that had either a higher or a lower osmotic pressure than the internodal cell from which they were derived (Kamiya and Kuroda 1956a). The osmotic pressure of both the cells having a lower and a higher osmotic pressure return to the original osmotic pressure within several days. Kamiya and Kuroda demonstrated that osmoregulation in *Nitella* cells results from the regulation of the internal osmotic pressure and not the turgor pressure.

After finishing the undergraduate course in 1953, I entered the graduate course and continued my research in the same laboratory. Since osmoregulation attracted my interest, I asked Professor Kamiya if I could continue the work on the osmoregulation of *Nitella* and got his agreement.

Soon, I realized that replacement of the vacuolar sap with artificial solutions should be a powerful tool for analyzing the mechanism of osmoregulation.

3.1 Development of Vacuolar Perfusion Method

In 1955, the prototype of vacuolar perfusion was developed in *Nitella* by Kamiya and Kuroda. To perform the technique, an internodal cell is mounted in a chamber that has three compartments, each containing an isotonic artificial cell sap. After applying a negative pressure to the central compartment, both cell ends are cut off. Then, a small pressure difference is exerted between the two cell ends by gently raising the level of the artificial cell sap at one end of the chamber. This causes a flow of the artificial cell sap through the vacuole. After replacing the natural cell sap with an artificial one, the cell was ligated with threads. Kamiya and Kuroda (1955) reported that the perfused cells showed cytoplasmic streaming and survived for a while.

In order to get a perfused cell that can be used for the study of osmoregulation, the cell must be kept alive at least for several days – which is the time needed for *Nitella* cells to perform osmoregulation (Kamiya and Kuroda 1956a). I tried to prolong the survival time of the cells after replacing the cell sap with an artificial medium. I designed a device for vacuolar perfusion using a hand-made polyacrylate vessel composed of three pools (Tazawa 1964). An internodal cell whose vacuole had been stained previously with neutral red was mounted on the perfusion vessel. I filled all three pools with an artificial solution that was isotonic with respect to the internodal cell, and consequently, the cell lost its turgor. In this state, both cell ends were cut. When a slight difference in the water level between the side pools was given, the artificial solution flowed. The perfusion was stopped immediately after the red cell sap was pushed out from the cell. The cell was then ligated with silk thread at both cell ends. The perfused cell was observed under the microscope to see cytoplasmic streaming.

My attempts to obtain living cells after perfusion were unsuccessful. None of the perfused cells that were transferred to the hypotonic artificial pond water (APW) exhibited active cytoplasmic streaming. The source of the failure was a consequence of plasmolysis, which occurred during perfusion. The induced plasmolysis injured the plasma membrane. My trial continued for about 3 months, but I could not get a single cell that showed active cytoplasmic streaming. I almost gave up all hope of success. Then one day, I read in a book entitled "Physiology of Osmosis in Plant Cells" written by Toru Sakamura (1952) that "When a plant cell is exposed to dryness in the air, no plasmolysis occurs. The cell contracts forming folds, while the cell membrane and the protoplasm are stuck to the cell wall." Inspired by this sentence, I changed the condition of perfusion. Namely, the cell part in the central pool was exposed to the air instead of the isotonic medium. After the cell looked slightly contracted due to loss of cell water, both cell ends were cut and the vacuole was perfused with the isotonic medium. After perfusion, both cell ends were ligated. When I placed the ligated cell in APW and observed it under the microscope, I saw the most beautiful cytoplasmic streaming. I will never forget that moment. The cell perfused with the solution containing 150 mM KCl and 10 mM $CaCl_2$ survived for more than 1 month (Tazawa 1964).

3.2 Osmoregulation of Cells Having Artificial Cell Sap

Taking advantage of the vacuolar perfusion, we modified both the ionic composition and the osmotic pressure of the vacuole. Experiments on cells having variously modified vacuolar saps showed that the osmoregulation of the fresh water Characeae involved two mechanisms, one that regulated the osmotic pressure per se irrespective of the level of the vacuolar K^+ concentration and another that regulated the vacuolar K^+ concentration (Tazawa and Nagai 1966). We assumed that both mechanisms operated to keep the cytoplasmic K^+ concentration constant at a concentration around 80–100 mM (Nakagawa et al. 1974).

The assumption could be tested if we could modify the ionic composition of the cytoplasm independently of the osmotic pressure. I thought that the tonoplast-free cells would fit the purpose, since without the tonoplast, the ionic composition as well as the osmolality of the cytoplasm could be controlled by internal perfusion.

4 Artificial Control of the Cytoplasmic Composition: Tonoplast-Free Cell

Taylor et al. (1973) reported that the plasma membrane of an amoeba, *Chaos carolinensis*, could be ruptured by lowering the external concentration of free Ca^{2+} down to less than 10^{-7} M, and this gave me a hint of how to remove the tonoplast of characean cells. Following their recipe, I changed the perfusion medium from the one containing high Ca^{2+} (10 mM $CaCl_2$) to one containing EGTA, which would lower the free Ca^{2+} to less than 10^{-7} M (Tazawa et al. 1976). After perfusing the vacuole with an isotonic medium containing 5 mM EGTA and 1 mM ATP, the cell was ligated with polyester threads and placed under microscope. I observed that the clear boundary between the vacuole and the cytoplasm disappeared first at the ligated cell end where the streaming made a U-turn. It is assumed that the shearing stress upon the tonoplast caused the tonoplast to break; however, the break could not be repaired due to the very low Ca^{2+} concentration. The cytoplasmic streaming in tonoplast-free cells continues at the normal rate, suggesting that the tonoplast is not necessary for the streaming.

Just at the time when we succeeded in making the tonoplast-free cell, a shocking paper by Williamson (1975) appeared. He had perfused the vacuole with a highly concentrated (50 mM) EGTA solution at such a high rate that the streaming endoplasm was swept away. After about 60 s perfusion, the cytoplasmic organelles that still moved along the actin fibers abruptly stopped and became anchored to the actin bundles. He could restore the movement of these organelles by introducing ATP with Mg^{2+} into the perfusion fluid. The cell model of Williamson (1975) is valuable for studying the mechanism of cytoplasmic streaming, but it is not suitable for studying plasma membrane functions such as electrogenesis and excitability, since in his model, the outside of the cell was bathed in liquid paraffin. On the other hand, our tonoplast-free cells, which are bathed in APW, maintain not only cytoplasmic motility but also membrane activities. Unfortunately, because of the short survival time, tonoplast-free cells cannot currently be used for processes, including study of osmo- and ion-regulation that occur over a protracted time period. Tonoplast-free cells have found wide applications in various fields of cell physiology as mentioned in the following.

5 Turgor Regulation in *Lamprothamnium*, a Brackish Charophyte

A brackish water charophyte *Lamprothamnium* living in brackish water is exposed to daily fluctuation of the external osmotic pressure. Like marine algal cells, the internodal cell of *Lamprothamnium* keeps its turgor pressure even when challenged with a wide range of both external (Bisson and Kirst 1980) and internal osmotic pressures (Okazaki et al. 1984a). The turgor regulation is implemented mainly through the release of K^+ and Cl^- into a hypotonic medium and through the uptake of these ions from a hypertonic medium (Bisson and Kirst 1980; Okazaki et al. 1984b).

5.1 *Energetics of Movements of K^+ and Cl^- During Turgor Regulation*

In discussing the energetics of ion movements across the plasma membrane, it is essential to know the electrochemical potential gradient for each monovalent ion across the membrane. Consequently, we used the method of Kishimoto and Tazawa (1965a, b) to analyze the ionic concentrations of the cytoplasm separately from those of the vacuole in internodal cells of *Lamprothamnium*. To accomplish this, we perfused the vacuole with an isotonic medium containing $Ca(NO_3)_2$ and sorbitol. The K^+, Na^+, and Cl^- contents of the cell sap and the remaining cell were then analyzed separately. After calculating the electrochemical potential gradients of ions across the plasma membrane, we concluded that during the hypotonic turgor regulation, K^+ and Cl^- were released from the cell passively and that during the hypertonic regulation, K^+ was passively and Cl^- was actively absorbed. The passive movement of K^+ that occurred during hypertonic turgor regulation resulted from the hyperpolarization of the plasma membrane, which might have been caused by an activation of the electrogenic H^+ pump (Okazaki et al. 1984b).

5.2 *Ca^{2+} Signal as a Second Messenger in the Hypotonic Turgor Regulation*

The first membrane event that occurs with a lag of 1 min after hypotonic treatment is a depolarization of the plasma membrane potential from −60 to −100 mV, which is close to the equilibrium potential of K^+ (E_k). The plasma membrane conductance also begins to increase with a lag of 1 min and reaches a peak after 2–3 min. Within 1 h, the conductance of the plasma membrane returns to its resting level (Okazaki et al. 1984b).

The signal produced by the hypotonic treatment is assumed to be an error signal, which we defined as the difference between the increased turgor pressure P and the reference turgor P_{ref}. A transient rise in the cytoplasmic Ca^{2+} concentration ($[Ca^{2+}]_c$) serves as the second messenger in converting the error signal ($P - P_{ref}$) into the response. This increase in $[Ca^{2+}]_c$ required external Ca^{2+} (Okazaki and Tazawa 1986a). Nifedipine, a Ca^{2+} antagonist, abolished the hypotonic turgor regulation (Okazaki and Tazawa 1986b), suggesting that voltage-dependent Ca^{2+} channels are necessary for hypotonic turgor regulation. By measuring the light emission of aequorin, a Ca^{2+}-sensitive photoprotein that had been injected into the cytoplasm, we demonstrated that hypotonic treatment causes a transient increase in cytoplasmic Ca^{2+} with a lag of 1 min (Okazaki et al. 1987). What happens during the lag is still unknown, but it might include the process that senses the turgor change and a process that results in the activation of a voltage-dependent Ca^{2+} channel.

Intracellular free Ca^{2+} may activate monovalent ion channels as evidenced by experiments done on cytoplasmic droplets obtained from an internodal cell of *Lamprothamnium*. The droplets were shown to be covered with the tonoplast (Sakano and Tazawa 1986). The direct effect of Ca^{2+} on ion channels was studied using the cytoplasmic-side-out patches. In these cell models, an increase in the Ca^{2+} concentration from 10^{-8} to 10^{-5} M caused a dramatic increase in both the frequency and the duration of the K^+ channel opening (Katsuhara et al. 1989).

A calcium-dependent protein kinase (CDPK) may be the biochemical factor that recognizes the transient increase in $[Ca^{2+}]_c$. The CDPK may convert this transient increase in $[Ca^{2+}]_c$ into a sustainable response by activating monovalent ion channels (Yuasa et al. 1997). A 53-kDa CDPK was detected in cell extracts of *Lamprothamnium*, and this CDPK could be precipitated with the anti-*Dunaliella tertiolecta* CDPK antibody. Moreover, microinjection of the antibody into the cytoplasm of *Lamprothamnium* inhibits the hypotonic turgor regulation as does K-252a, an inhibitor of protein kinases.

For more information about turgor regulation in *Lamprothamnium*, please refer to the following review articles (Okazaki and Tazawa 1990; Okazaki 1996).

6 Mechanosensing in Fresh-Water Characean Cells

While brackish-water characean cells regulate their turgor pressure, fresh-water characean cells regulate their osmotic pressure instead (Kamiya and Kuroda 1956a). The transformation of the pressure signal into an electrical signal was studied further in *Chara corallina* by Shimmen (2003, 2006b). When the nodal cells are exposed to a sorbitol solution, their turgor pressure decreases quickly, while the turgor of the internode remains high. After a few seconds lag time, the membrane potential of the nodal region depolarizes. The pressure difference produced between the internodal cell and nodal cells causes a bending of the cell wall of the terminal end of the internodal cell (Shimmen 2003). The bending of the wall may activate the stretch-activated ion channels of the plasma membrane at the end of the internodal cell facing the node (Shimmen 2003).

Characeae algae provide a simple cell system for studying the wounding response in plants. A specimen consisting of two adjoining internodes with a node between them can be prepared. When one internode (victim cell) is cut, a very rapid depolarization is induced at the nodal end of the intact neighboring healthy cell (receptor cell). This is followed by a long-lasting depolarization, which is known as the receptor potential. The receptor potential often induces propagating action potentials that result from the activation of voltage-activated ion channels. The long-lasting receptor potential is induced by K^+ ions released from the victim cell. The amplitude of the receptor potential decreases when the cell turgor is experimentally reduced. This indicates that the stretching of the plasma membrane of the nodal part is involved in the depolarization (Shimmen 2006a). Here again, the nodal complex plays a central role in generation of the wounding signal (Shimmen 2006b).

The nodal region of the internodal cell is electrically distinct from the flank of the cell. This electrophysiological differentiation was demonstrated by showing that Ca^{2+} prevents a large depolarization induced by 100 mM KCl along the flank, while Ca^{2+} is ineffective in preventing the KCl-induced depolarization at the node (Shimmen 2008). When an osmotic shock (200 mM sorbitol) is given to the node, a large and long-lasting depolarization is induced. The depolarized state of the node continues after the removal of sorbitol. These characteristics of the node are suggested to be responsible for electrical responses to wounding in Characeae (Shimmen 2008).

Cytoplasmic streaming is also sensitive to the hydrostatic pressure. A positive hydrostatic pressure applied to one end of the internode of *Chara* induces an acceleration of the streaming that is away from the applied pressure and deceleration of the streaming that is toward the applied pressure. A negative pressure applied to one cell end causes the reverse responses of the streaming (Staves et al. 1992). The sensitivity to the hydrostatic pressure was lost by ligating the cell end, indicating that the site of pressure sensing is located at the nodal region.

7 Salt Tolerance and Ca^{2+}

When internodal cells of fresh-water charophyte *Nitellopsis obtusa* are subjected to 100 mM NaCl, they all die within a day. The salt treatment causes a significant increase in the cytoplasmic Na^+ concentration and a significant decrease in the cytoplasmic K^+ concentration. These complementary changes in the cytoplasmic concentrations of Na^+ and K^+ are accompanied by a large membrane depolarization and a great increase in the membrane conductance. By contrast, cells treated with 100 mM NaCl supplemented with 10 mM $CaCl_2$ survived for more than 2 weeks. In the internodal cells protected by the supplemental calcium, there are neither NaCl-induced changes in the cytoplasmic concentrations of Na^+ and K^+ nor

NaCl-induced changes in the membrane potential and the membrane conductance (Katsuhara and Tazawa 1986).

The protective effect of Ca^{2+} in providing salt tolerance also takes place in tonoplast-free cells, as long as the ATP concentration is greater than 0.1 mM. AMP and adenylyl-imidodiphosphate, a nonhydrolyzable analog of ATP, can replace ATP, indicating that ATP acts neither as an energy source nor as a substrate for protein phosphorylation. We assumed that ATP acts as a cofactor with Ca^{2+} to control the Na^+ permeability of the plasma membrane (Katsuhara and Tazawa 1987).

The hypothesis was tested directly by measuring the effect of adenosine phosphates on channel activity in cytoplasmic-side-out patches of the plasma membrane using a patch-pipette containing 100 mM Na^+ and 10 mM Ca^{2+}. The channel activity, which is high in the absence of ATP is depressed greatly by the addition of 1 mM ATP or AMP (Katsuhara et al. 1990).

The K^+ channel of the tonoplast also contributes to the salt tolerance. A transient treatment of cells with 100 mM NaCl for 30 min resulted in an increase in the cytoplasmic Na^+ concentration and a decrease in the cytoplasmic K^+ concentration. When these Na^+-loaded cells are transferred to 100 mM NaCl supplemented with 10 mM $CaCl_2$, the cytoplasmic Na^+ concentration decreases as a result of Na^+/K^+ exchange between the vacuole and the cytoplasm (Katsuhara and Tazawa 1988). A tonoplast K^+ channel was identified in the cytoplasmic drop-attached patches. Microinjection of Ca^{2+} into the drop greatly increases the opening probability of the K^+ channel (Katsuhara et al. 1991).

For more information on the salt tolerance of characean cells, please refer to the review by Katsuhara and Tazawa (1992).

8 Cytoplasmic Streaming and Ca^{2+}

Cytoplasmic streaming is one of the most beautiful and memorable processes that one can see under the microscope. If life can be defined, in part, by the ability to move in the absence of an exogenous force, then observing the rapid, rotational cytoplasmic streaming in characean internodal cells through a microscope is almost like seeing life itself. In characean cells, the moving endoplasm slides as a whole on the immobile cortical gel where motile fibrils are seated (Kamiya and Kuroda 1956b). The fibrils are composed of actin filaments, which interact with the myosin molecules attached to cell organelles (cf. Shimmen 2007).

8.1 Motive Force Measurement

The motive force of cytoplasmic streaming in characean cells was first measured by Kamiya and Kuroda (1958) using a centrifuge microscope. By varying the centrifugal force, which was directed antiparallel to the streaming endoplasm, they were able to stop cytoplasmic streaming. The magnitude of the endogenous motive force that drives cytoplasmic streaming is given by the centrifugal force required to stop cytoplasmic streaming.

While I was performing vacuolar perfusion, I noticed that the velocity of the cytoplasmic streaming antiparallel to the direction of the perfusion solution was retarded. This was due to the shearing force generated at the interface between the perfusion medium and the streaming endoplasm. An idea occurred to me that the motive force that drives cytoplasmic streaming could be obtained by determining the shearing force needed to just stop the streaming. Since the vacuole of characean cells is cylindrical, the shearing force (F in N m^{-2}) can be calculated by

$$F = Pr/2l, \tag{3}$$

where P is the difference in hydrostatic pressure between two ends of the cell, r and l represent the radius of the vacuole and the length of the cell, respectively (Tazawa 1968). In *Nitella* cells, the motive force of the cytoplasmic streaming determined by the perfusion method was on the average 0.17 N m^{-2} (1.7 dyn cm^{-2}), which is nearly equal to the value measured by the centrifugation method (0.16 N m^{-2} = 1.6 dyn cm^{-2}). Later, Kamiya and Kuroda (1973) determined the motive force in *Nitella* cells by a new method. The motive force with the new method also found good agreement with the data reported previously.

From the value of the motive force, we can calculate the work done per second by the 1 cm^2 (= 10^{-4} m^2) endoplasm that moves at a velocity (v) of 50×10^{-6} m s^{-1} using the following formula:

$$W = FAv, \tag{4}$$

where W represents the work done per second, F represents the motive force that drives the cytoplasmic streaming, and A represents the area of endoplasm that is moved by the driving force.

I calculate that the work performed by each square centimeter of endoplasm per second amounts to 8.5×10^{-10} J. We can estimate the amount of energy (E) consumed by each cm^2 (= 1×10^{-4} m^2) each day in order to power cytoplasmic streaming using the following formula:

$$E = 8.5 \times 10^{-10}\mathrm{J} \times 60 \times 60 \times 24. \tag{5}$$

I calculate that 7.3×10^{-5} J of energy is consumed by each square centimeter (10^{-4} m^{-2}) of endoplasm per day. Assuming that the thickness of the endoplasm is 10^{-5} m and the endoplasmic ATP concentration is 1.6 mM, (Mimura et al. 1984), I estimate the amount of ATP in the 1 cm^2 (= 10^{-4} m^2) to be 1.6×10^{-9} mol. If each molecule of ATP released 8×10^{-20} J of energy upon hydrolysis (Wayne 2009), then the complete hydrolysis of ATP pool would release a total of 7.3×10^{-5} J of energy – just enough to drive cytoplasmic streaming for a day. Luckily, the ATP pool is continually replenished through the activities of mitochondria and chloroplasts!

Using the perfusion method, I discovered novel features of the cytoplasmic streaming. I found that the decrease of the streaming velocity that occurs with a decrease in temperature is predominantly caused by an increase in the sliding resistance at the interface between the cortical gel and the endoplasm (Tazawa 1968). By contrast, the motive force is relatively unaffected by temperature in the range of 10–30°C.

I also found that lowering the internal osmotic pressure decreases both the motive force and the velocity, while raising the internal osmotic pressure increases both, suggesting that the compactness of the endoplasm is important for the interaction between myosin molecules attached to endoplasmic organelles and the actin filaments attached to the inner surface of the chloroplasts that are embedded in the cortical gel (Shimmen and Tazawa 1982b).

8.2 *Excitation–Cessation Coupling (E–C Coupling)*

Due to the spectacular motility of their cytoplasm and their excitability, characean internodal cells are often called "green muscle" or "green axon." Characean internodal cells can elicit action potential, and cytoplasmic streaming stops instantaneously upon the generation of an action potential. The phenomenon is called excitation–cessation coupling (E–C coupling) and it is comparable to the excitation–contraction coupling that occurs in muscle cells. How does electrical excitation stop cytoplasmic streaming? Two mechanisms for the E–C coupling can be postulated: excitation could result in a temporary disappearance of the motive force, or it could result in an increase in the sliding resistance due to an increase in the viscosity of the endoplasm. To solve the question we undertook the following experiment.

An electrical stimulus was given to the cell at the moment when the streaming cytoplasm, which was flowing antiparallel to the flow of the perfusion medium, was significantly retarded by the counter vacuolar perfusion. Simultaneous recording of the action potential and the cytoplasmic streaming showed that the decelerated streaming endoplasm started to flow in the direction of the flow of the perfusion medium within a second after the peak of the action potential (Tazawa and Kishimoto 1968). This fact suggests that the motive force vanishes as a result of the action potential. The streaming endoplasm once stopped by excitation begins to flow in its original direction within a minute after the action potential and regains its normal velocity after 5–10 min. During the recovery time, the motive force increased in parallel with the velocity, suggesting that the sliding resistance remained nearly constant during the action potential and the recovery period.

8.3 *Ca^{2+} as a Key Factor in E–C coupling*

Tonoplast-free cells, just like normal cells, generate an action potential. However, the cessation of cytoplasmic streaming, which accompanies the action potential in intact cells, had never been observed in tonoplast-free cells. Since the tonoplast-free cells contain EGTA, a Ca^{2+} -chelating agent, the lack of E–C coupling indicates that an increase in the Ca^{2+} concentration may be involved in the E–C coupling (Tazawa et al. 1976). The involvement of Ca^{2+} as a key factor in coupling the electrical stimulus to the motile response has been demonstrated in many ways. The presence of Ca^{2+} in the external medium is indispensable for E–C coupling (Barry 1969). Upon excitation, a significant increase in the influx of ^{45}Ca occurs (Hayama et al. 1979). An increase in cytoplasmic Ca^{2+} upon excitation has been documented by observing the burst of light emission that comes from aequorin that has been injected into the cytoplasm (Williamson and Ashley 1982). The rotation of chloroplasts in isolated cytoplasmic drops (Hayama and Tazawa 1980) as well as the cytoplasmic streaming (Kikuyama and Tazawa 1982) is reversibly inhibited by an iontophoretic injection of Ca^{2+}. Lastly, in the plasma membrane-permeabilized cell of *Nitella* (Shimmen and Tazawa 1982a, b), cytoplsamic streaming is reversibly inhibited by 1 μM Ca^{2+} (Tominaga et al. 1983).

Simultaneous recordings of cytoplasmic streaming, membrane potential, and light emission of aequorin demonstrated that an increase in the cytoplasmic Ca^{2+} concentration occurred simultaneously with the generation of the action potential. By contrast, velocity of the cytoplasmic streaming began to decrease only after the cytoplasmic Ca^{2+} concentration had reached a certain level (Kikuyama et al. 1993, 1996).

8.4 Nature of the Ca^{2+} Inhibition of Cytoplasmic Streaming

Upon studying the sensitivity of cytoplasmic streaming in tonoplast-free cells to Ca^{2+} by varying the intracellular Ca^{2+} concentration, we found that tonoplast-free cells were relatively insensitive to Ca^{2+} (Hayama et al. 1979). Consequently, we assumed that some Ca^{2+}-sensitizing component, which is present in both intact and plasma membrane-permeabilized cells, was lost or desensitized by unknown reasons in tonoplast-free cells (Tominaga et al. 1983).

Using permeabilized cells of *Chara*, Tominaga et al. (1985) found that calmodulin (CaM) inhibitors did not affect the Ca^{2+}-induced inhibition of cytoplasmic streaming but did inhibit the recovery resulting from the removal of Ca^{2+}. This indicated that Ca^{2+}-CaM is involved in the recovery process. Another Ca^{2+}-sensitizing factor was assumed that is responsible for the Ca^{2+}-induced inhibition (Tominaga et al. 1987).

Hints at the nature of the Ca^{2+}-sensitizing factor responsible for the Ca^{2+}-induced inhibition came from studies on the myosin isolated from a slime mold, *Physarum polycepharum*. The movement of *Physarum* myosin-coated latex beads along actin cables of *Chara* (Shimmen and Yano 1984) is also inhibited by micromolar Ca^{2+} concentrations (Kohama and Shimmen 1985). In the case of

Physarum myosin, the ATPase activity is only sensitive to Ca^{2+} when it is in the phosphorylated form (Kohama and Kendrick-Jones 1986).

In order to examine whether phosphorylation/dephosphorylation of myosin-like component in *Chara* also plays a role in the regulation of cytoplasmic streaming, we introduced protein phosphatase and its inhibitor that had been isolated from rabbit skeletal muscle directly to the cytoplasm of *Chara*. Assuming that the application of proteins to the cytoplasm of permeabilized cells would be hampered by presence of the cell wall, we improved upon the method we had used to make tonoplast-free cells in order to obtain tonoplast-free cells that were more sensitive to Ca^{2+} (Tominaga et al. 1987). Two important improvements were made for the perfusion procedure: one was the use of a slightly hypotonic perfusion medium instead of an isotonic one and the other was to perform a slow perfusion instead of rapid one (Tominaga et al. 1987). We may call the tonoplast-free cell prepared by the improved method "tonoplast-permeabilized" cell, since the integrity of the endoplasm is maintained after the loss of the tonoplast. Using both plasma membrane- and tonoplast-permeabilized cells, we tried to solve the secret of the Ca^{2+}-inhibition of the cytoplasmic streaming in *Chara*.

In our search to find the regulatory factors that sensitize myosin to Ca^{2+}, we discovered that α-naphthylphosphate, a synthetic inhibitor of protein phosphatase, inhibited the cytoplasmic streaming. In addition, we found that ATP-γ-S, applied to plasma membrane- permeabilized cells, strongly inhibited the recovery of Ca^{2+}-inhibited streaming. It is known that proteins that are thiophosphorylated with ATP-γ-S are not dephosphorylated by phosphataes. Moreover, vacuolar perfusion of the tonoplast-permeabilized cells with a medium containing protein phosphatase-1 and Ca^{2+} caused a gradual recovery of the streaming, which had been stopped by Ca^{2+}. Thus, we hypothesize that myosin is active when it is dephosphorylated and inactive when it is phosphorylated and that the cytoplasmic streaming in *Chara* is regulated through the phosphorylation/dephosphorylation of the putative myosin. When the action potential is induced, the elevated level of cytoplasmic Ca^{2+} stimulates a protein kinase to phosphorylate the myosin. This results in the cessation of the cytoplasmic streaming. The streaming recovers gradually when the phosphorylated myosin is dephosphorylated by a native protein phosphatase whose activity is stimulated by Ca^{2+}-calmodulin (Tominaga et al. 1987).

The scheme was supported later by the finding that the motility of *Chara* myosin is inhibited by okadaic acid, an inhibitor of protein phosphatase, and protein kinase C, but enhanced by staurosporin, an inhibitor of protein kinase (Morimatsu et al. 2002). McCurdy and Harmon (1992) have shown that a CDPK is associated with the subcortical actin bundles and to the surface of small organelles in *Chara*. The localization coincides well with localization of the proteins that reacted with the monoclonal antibody to the myosin heavy chain from mouse 3T3 cells (Grolig et al. 1988). It is to be noted that purified *Chara* myosin itself is not Ca^{2+}-sensitive as far as the actin-activated MgATPase activity and the in vitro motility are concerned (Awata et al. 2001).

8.5 Cell Models as Tools to Study the Structural and Molecular Basis of Cytoplasmic Streaming in Characean Cells

Using phase-contrast microscopy, Kamitsubo (1966) first visualized the immobile fibrils seated on the inner surface of chloroplasts, which are embedded in the cortical gel that served as the structural entity against which the endoplasm moved. Using electron microscopy, Nagai and Rebhun (1966) revealed that the fibrils were composed of mocrofilaments. Since heavy meromyosin (HMM) (Kersey et al. 1976) and subfragment one (S_1) (Williamson 1974) isolated from rabbit skeletal muscle myosin bound to the filaments and formed typical arrowhead structures directed opposite to the original cytoplasmic flow, the microfilaments were identified to be actin filaments.

The presence of myosin in the endoplasm of *Nitella* was first suggested by Chen and Kamiya (1975) who showed that when the endoplasm pretreated with an SH-reagent, *N*-ethylmaleimide (NEM), it is incapable of moving when it is brought into contact with the intact cortical gel. When the vacuole is rapidly perfused with a medium containing no ATP, the organelles remain attached to the cortical fibers. The organelles can be gradually freed from the fibers when either ATP or inorganic pyrophosphate is introduced into the cell. Williamson (1975) postulated that a myosin-like protein might be the component that links the organelles to the actin filaments. Higashi-Fujime (1980) provided more definitive evidence for the presence of myosin in the endoplasm by showing that when the endoplasm is squeezed out from a *Nitella* cell and used to coat a glass slide, in the presence of MgATP, actin cables slide along a glass surface at a rate similar to that of the cytoplasmic streaming in vivo. In the same functional assay, the exceptionally fast motor protein isolated from characean cells moved muscle F-actin at a rate of 60 $\mu m\ s^{-1}$ (Higashi-Fujime and Sumiyoshi 2001). This is ten times faster than myosin isolated from skeletal muscle that can move muscle F-actin.

Studies of another cell model derived from characean internodal cells provide evidence for the importance of the interaction between actin filaments and myosin for motility (Kuroda and Kamiya 1975). This movement is assumed to be caused by interaction of actin filaments attached to the surface of the chloroplast with myosin in the endoplasm. Evidence in support of this assumption comes from studies that show that chloroplast rotation is stopped by NEM. However, when the surface membrane of the drop bathed in a Ca^{2+}-free medium is ruptured with the tip of a fine glass needle in order to allow HMM to enter the droplet to replace the NEM-inactivated myosin, chloroplast rotation resumes.

Kuroda (1983, 1990) developed a “cut-open cell.” To make this type of cell model, an internodal cell of *Chara* was cut open parallel to its long axis. When the cut-open cell, whose cell wall is in contact with the glass slide, is bathed in Ca^{2+}-free medium, the tonoplast is ruptured spontaneously, and cytoplasmic streaming still takes place as in tonoplast-free cells. By contrast, cut-open cells prepared from cells pretreated with heat at 50°C lose their motile activity due to inactivation of heat-sensitive myosin (Kamitsubo 1981). However, streaming can be reconstituted following the application of HMM.

Shimmen and Tazawa (1982b) reconstituted cytoplasmic streaming by introducing an organelle suspension prepared from *Chara* cells into a *Nitella* cell whose endoplasm had been removed by intracellular perfusion. The *Chara* organelles immediately move along the *Nitella* actin cables, indicating that there is an association of "myosin" with the organelles. Sheez and Spudich (1983) succeeded in inducing the movement of HMM-coated fluorescent polymer beads that they had introduced onto the cut-open cell model.

The actin cables of characean internodal cells provided us with a functional assay for identifying myosin during isolation procedures. An alternative assay method was developed for the determination of isolated myosins. Yangida et al. (1984) observed the movement of single actin filaments stabilized with fluorescent phalloidin that interacted with soluble myosin fragments. Kron and Spudich (1986) observed that single actin filaments labeled with phalloidin exhibit ATP-dependent movement on a glass surface coated with skeletal myosin II. Since the rates of movement proved to be relatively independent of the type of actin, the system can be used as a quantitative myosin-movement assay with purified protein.

The method was applied to the isolation of myosin from lily pollen tubes (Kohno et al. 1991). A crude extract of lilly pollen tubes was brought into contact with the surface of a nitrocellulose-coated coverslip. When the assay solution containing rhodamine-phalloidin treated F-actin was introduced onto the coverslip, fluorescent images of actin filaments moved under a microscope, indicating that the crude extract contained myosin-like translocator. With aid of this motility assay, Yokota and Shimmen (1994) for the first time purified plant myosin. The molecular mass of the heavy chain of this myosin was 170 kDa and its ATPase activity was stimulated to 60 times by chicken breast actin. The velocity of fluorescent actin in the in vitro motility assay was similar to the in vivo velocity of the cytoplasmic streaming.

Again using the same motility assay, *Chara* myosin was isolated from the cytoplasm of cells whose vacuolar sap had been removed by vacuolar perfusion with a solution containing EGTA (Yamamoto et al. 1994). The molecular mass of *Chara* myosin is about 230 kDa, which is considerably larger than that of lily myosin. An antibody raised against this myosin did not recognize either smooth muscle myosin or myosin from lilly pollen tubes. Higashi-Fujime et al. (1995) also isolated myosin from *Chara* . Both myosins were soluble at low ionic strength, and their ATPase activities were stimulated 100–150 time by F-actin.

For more information about the progress of research on cytoplasmic streaming since the proposal of the "sliding theory" by Kamiya and Kuroda in 1956 (Kamiya and Kuroda 1956b), refer to the review article by Shimmen (2007).

9 Electrogenic H^+ Pump

9.1 *Light-Induced Potential Change*

During our study of osmoregulation in *Nitella* cells, we noticed that light markedly stimulated the uptake of K^+ ions into cells whose osmotic pressure had been lowered by TCO (Tazawa and Nagai 1960). We hypothesized that the membrane potential, a component of the electrochemical potential gradient, might be changed by light. To test this possibility, we measured the membrane potential of *Nitella* cell with a microelectrode. We found that light caused a large hyperpolarization of the membrane potential. It was the first intracellular recording of the light-induced potential change (LPC) in plant cells. Since blue and red light were found most effective, we assumed that the light-induced hyperpolarization was mediated through chlorophyll. We concluded that light enhances K^+ absorption by increasing the electrochemical potential gradient for K^+ (Nagai and Tazawa 1962).

Confirmative evidence for the involvement of chloroplasts in the LPC came from experiments done on *Spirogyra* (Fujii et al. 1978). We prepared a cell that lacked chloroplasts and cells that contained a small fragment of chloroplast. The LPC was not observed in the former cell but observed in the latter cell.

9.2 Direct Demonstration of the Electrogenic H^+-Pump (H^+-ATPase)

In search for the role of ATP in the LPC, we examined the effect of the cytoplasmic ATP concentration using the tonoplast-free cells of *Chara* . The LPC occurred independently of the ATP concentration in the range of 0.5–1.3 mM. However, at extremely low concentrations of ATP (1–2 μM), the LPC was completely abolished and the membrane potential stayed at a strongly depolarized level (Kikuyama et al. 1979).

Then, Shimmen, a graduate student who worked on the mechanism of membrane excitation using tonoplast-free cells (Shimmen et al. 1976), got an idea that the membrane potential of *Chara* cells might be controlled directly by ATP. He started the experiment using tonoplast-free cells in which both ends had been ligated with threads. It was the summer of 1976 where a part of the building of Faculty of Science of Osaka University was under construction. Because of vibrations caused by the pile-drivers it was impossible to carry out the electric measurements with a microelectrode inserted into the cell. Then he used the so-called "open-vacuole method" (Tazawa et al. 1975) that enabled us to measure the membrane potential and membrane conductance without using a microelectrode. The method had an additional merit in that the electric measurement can be done during intracellular perfusion. Shimmen found that the membrane potential is beautifully controlled by MgATP. Intracellular perfusion with a medium lacking either ATP or Mg^{2+} brought the membrane potential to the depolarized level. Perfusion with a medium containing 1 mM MgATP repolarized the plasma membrane by about 160 mV (Shimmen and Tazawa 1977). Later, we demonstrated that the electrogenic pump current calculated from the electric data could be explained by ATP-dependent net H^+ efflux (Takeshige et al. 1985).

Although we demonstrated a direct control of the membrane potential by ATP, a direct control of the LPC by ATP could not be demonstrated in intact *Chara* cells (Keifer and Spanswick 1979). However, in tonoplast-free cells of *Nitellopsis,* light caused a significant increase in the ATP concentration and a significant decrease in the concentrations of both ADP and AMP (Mimura and Tazawa 1986). These light-induced changes in adenine nucleotide concentrations were enough to account for the LPC in terms of enzyme kinetics, if we assume that K_m for ATP is 0.1 mM (Mimura et al. 1983) and the K_i values of ADP and AMP are 0.4 mM (Mimura et al. 1984). The discrepancy existing between intact and tonoplast-free cells may be due to the difference in local levels of adenine nucleotides between intact cells and tonoplast-free cells (Tazawa et al. 1987).

10 Membrane Excitation

10.1 Tonoplast Action Potential

The action potential in *Nitella* has two peaks, a rapid one and a slow one. When the vacuolar Cl^- concentration is reduced to less than 0.1 mM by vacuolar perfusion, the second component of the action potential changes its direction from a depolarizing potential to a hyperpolarizing one. Simultaneous recording of the plasma membrane potential and the tonoplast potential demonstrates that the Cl^--sensitive component of the action potential represents the tonoplast action potential (Kikuyama and Tazawa 1976). It was suggested that an increase in the cytoplasmic Ca^{2+} concentration caused by excitation of the plasma membrane induces an increase in the Cl^- permeability of the tonoplast (Shimmen and Nishikawa 1988).

10.2 Demonstration of the Voltage-Dependent Ca^{2+} Channel in the Plasma Membrane of Nitellopsis

My interest in membrane excitation was evoked by the action potential -induced cessation of cytoplasmic streaming. The development of the vacuolar perfusion method enabled us to survey the effect of the vacuolar ions on various electric parameters (Kishimoto et al. 1965). Taking advantage of vacuolar perfusion, we clarified the temporal relationship between the action potential and the cessation of cytoplasmic streaming (Tazawa and Kishimoto 1968). Meanwhile, data accumulated that showed that there was an increase in Ca^{2+} influx (Hayama et al. 1979) as well as an increase in the cytoplasmic Ca^{2+} concentration ($[Ca^{2+}]_c$) during membrane excitation (Williamson and Ashley 1982). In view of the fact that the Ca^{2+} influx is essential for the Ca^{2+}-induced streaming cessation, I leaped to the hypothesis that a sort of voltage-dependent Ca^{2+} channel might exist in the plasma membrane of characean cells as it does in animal cells. My idea was evoked also by electrophysiological data presented earlier by Australian workers. In 1961, Hope found that the peak value of the action potential is dependent on $[Ca^{2+}]_o$. Moreover, both the reversal potential (Findlay 1961) and the membrane conductance (Findlay 1964) at the time of the maximum inward current are dependent on $[Ca^{2+}]_o$.

As mentioned before, the action potential of intact characean cells consists of two components, the plasma membrane action potential and the tonoplast action potential. By contrast, the action potential of the tonoplast-free cell has only one component, the plasma membrane action potential. The large Cl^- efflux that is observed during excitation of intact cells is absent in tonoplast-free cells (Kikuyama et al. 1984). Moreover, in tonoplast-free cells, the peak value of the action potential does not depend on the intracellular concentration of Cl^- (Shimmen and Tazawa 1980). During the action potential in tonoplast-free cells, the inward current is carried by Ca^{2+}.

Next, we searched for evidence to demonstrate the existence of voltage-dependent Ca^{2+} channels using tonoplast-free cells of *Nitellopsis*. By measuring an excitation-induced inward current under voltage-clamp, we found that the maximum amplitude of the inward current at the membrane excitation was inhibited by the Ca^{2+} channel blockers, nifedipine, and La^{3+} (Shiina and Tazawa 1987a) but not by an anion channel inhibitor 9-anthracenecarboxylic acid (A-9-C) (Shiina and Tazawa 1987b).

Furthermore, we studied the mechanism of Cl^- channel activation that occurs during the action potential in intact cells of *Nitellopsis*. The Cl^- channel blocker, A-9-C inhibited Cl^- efflux and its associated transient inward current, but it did not inhibit E–C coupling. On the other hand, the Ca^{2+} channel blocker La^{3+} uncoupled the cessation of cytoplasmic streaming from the membrane excitation in addition to inhibiting the transient inward current and the Cl^- efflux. We also found that the Cl^- efflux was decreased by lowering the external Ca^{2+} concentration. These findings support the existence of the Ca^{2+}-activated Cl^- channel in the plasma membrane (Shiina and Tazawa 1987b). Later, we directly demonstrated that in tonoplast-free cells of *Nitellopsis,* the Cl^- channel in the plasma membrane is activated by raising the intracellular Ca^{2+} concentration (Shiina and Tazawa 1988).

10.3 Possible Involvement of Protein Phosphorylation/ Dephosphorylation in Regulation of Ca^{2+} Channel Activity

In our early work with tonoplast-free cells of *Chara*, we found that the inward current under voltage-clamp is abolished by removing intracellular ATP or Mg^{2+} (Shimmen and Tazawa 1977). Membrane excitability is maintained even when the MgATP-dependent electrogenic H^+ pump is inactivated by treatment of cells with vanadate (Shimmen and Tazawa 1982c) or with nitrogen (Mimura et al. 1984). Using tonoplast-free cells of *Nitellopsis*, however, we found that the amplitude of the inward current is decreased by agents that promote protein phosphorylation and increased by agents that promote phosphoprotein dephosphorylation (Shiina et al. 1988). Thus, it seems likely that the activity of the Ca^{2+} channel in characean cells is controlled through protein phosphoryolation and dephosphorylation. In higher plant cells, the Ca^{2+} current is regulated by Ca^{2+}-dependent protein kinases (cf. Ward et al. 2009).

11 Vacuolar Functions

The capacity of the vacuole to regulate its own pH can be observed by varying its pH by means of vacuolar perfusion and then observing its ability to return to the original pH (Moriyasu et al. 1984a). The recovery process was inhibited by dicyclohexylcarbodiimide (DCCD), a H^+-ATPase inhibitor, indicating that the pumping of H^+ from the cytoplasm to the vacuole is involved in the pH regulation. Later, we observed that specific inhibitors of the vacuolar type H^+-ATPase, bafilomycin A_1 (Okazaki et al. 1992) and concanamycin 4-B (Tazawa et al. 1995), inhibited the pH_v regulation. The dependence of the H^+-pumping on MgATP was demonstrated in vitro using tonoplast-vesicles prepared from *Chara* cells stained with neutral red (Moriyasu et al. 1984b). The red color of the vesicles, which indicated an acidic compartment, persists in the presence of MgATP but is lost in its absence. Later, Takeshige et al. (1988) prepared purified tonoplast vesicles by using the intracellular perfusion technique followed by the centrifugation of the tonoplast through a discontinuous Percoll gradient. The vesicles had both ATPase and PPase activities and both ATP and PPi supported H^+-pumping activity.

The protein-degrading activity of the vacuole was directly demonstrated by loading an exogenous protein into the vacuole (Moriyasu and Tazawa 1986). A Ca^{2+}-activated protease was found in the vacuole of *Chara* and was suggested to be a calpain (Moriyasu and Tazawa 1987), although it was later found to have properties that different from those of calpain (Moriyasu and Wayne 2004).

Loading of amino acids into the vacuole showed that alanine is completely metabolized and converted to glutamine, glycine, and ammonia (Sakano and Tazawa 1985). Thus, loading of metabolites by means of vacuolar perfusion is a powerful tool for analyzing not only the functions of the vacuole per se but also interaction between the vacuole and the cytoplasm.

12 Intercellular Transport of Ions and Photoassimilates

Chara provides a simple cell system for studying the intercellular transport of ions and photoassimilates. Moreover, in characean cells, the distribution of ions and photoassimilates between the vacuole, the streaming endoplasm, and the chloroplast-rich cortical gel layer can be determined by using the vacuolar perfusion. The cell sap pushed out by a Ca^{2+}-containing perfusion solution can be used to obtain the vacuolar sample and the remaining cell can be used as the cytoplasmic sample. The cytoplasm can be further separated into the sol endoplasm and the cortical gel. After the first perfusion, the vacuole is perfused with a solution containing EGTA, which disrupts the tonoplast. The extract pushed out by the second perfusion contains the sol endoplasm. The remaining cell without the vacuolar sap and the sol endoplasm can be used as the chloroplast-rich gel layer sample.

Using the one step perfusion, we analyzed the Rb^+ transport between two internodes that were connected by the node (Ding and Tazawa 1989). The rate of transnodal symplastic transport of Rb^+ was strictly dependent on the temperature. The rate of the transport was correlated with the rate of cytoplasmic streaming, which was also temperature-dependent. The Rb^+ transport was impeded by imposing a gradient of turgor pressure imposed between the two internodes. Thus, the plasmodesmata may be equipped with a valve mechanism that is sensitive to the pressure gradient (Ding and Tazawa 1989).

Using the two-step perfusion, we succeeded for the first time in the simultaneous measurement of the amino acids distribution in subcellular compartments under light and dark conditions (Mimura et al. 1990). Both the vacuole and the endoplasm showed the same distribution pattern and this pattern differs from that of the cortical gel. Upon illumination, the concentrations of the amino acids decreased in the vacuole and the cortical gel but remained nearly constant in the endoplasm.

In the isolated upper shoot composed of apex-internode-branchlet complex, we found that the transport of photoassimilates consisting of mainly sucrose and amino acids was polar, i.e., the rate of transport from the branchlet to the internode was five times greater than the rate of transport from the internode to the brachlet. The polar transport is supported by the gradient of cytoplasmic concentrations of photoassimilates between the branchlet and the internode (Ding et al. 1991a). The polarity of the transport vanished when the apex was removed (Ding et al. 1991b).

While closing my article, I would like to cite a sentence from a paper of my mentor, Noburo Kamiya, saying that "Characean cells give us an opportunity to perform various kinds of unusual cell manipulations and surgeries which would be impracticable with other material. As a matter of fact, the progress of research in this field since the middle of this century (twentieth century) owes much to the development of these novel methods which are not usual in plant physiology and plant cell biology" (Kamiya 1986).

Acknowledgments I would like to express my cordial thanks to my coworkers cited above. A wide variety of productive researches mentioned above could surely not have been achieved without their cooperation.

My thanks are also due to Professor Ulrich Lüttge, the Editor of "Progress in Botany," for his invitation to this review article, and Professor Randy Wayne, one of my former coworkers and now in the Department of Plant Physiology, Cornell University, for his kindness to read my manuscript and to give valuable suggestions in polishing the style of my English.

References

Awata J, Saitoh K, Shimada K, Kashiyama T, Yamamoto K (2001) Effect of Ca^{2+} and calmodulin on the motile activity of characean myosin in vitro. Plant Cell Physiol 42:828–834

Barry WH (1969) Coupling of excitation and cessation of cyclosis in *Nitella*: role of divalent cations. J Cell Physiol 72:153–160

Benga G, Popescu O, Borza V, Pop VI, Muresan A, Mocsy I, Brain A, Wrigglesworth JM (1986) Water permeability of human erythrocytes: identification of membrane proteins involved in water transport. Eur J Cell Biol 41:252–262
Bisson MA, Kirst GO (1980) *Lamprothamnium*, a euryhaline charophyte. II. Time course of turgor regulation. J Exp Bot 31:1237–1244
Chen JCW, Kamiya N (1975) Localization of myosin in the internodal cells of *Nitella* as suggested by differential treatment with N-ethylmaleimide. Cell Struct Funct 8:109–118
Dainty J (1963) Water relations of plant cells. Adv Bot Rev 1:279–326
Dainty J, Ginzburg BZ (1964a) The measurement of hydraulic conductivity (osmotic water permeability to water) of internodal cells by means of transcellular osmosis. Biochim Biophys Acta 79:102–111
Dainty J, Ginzburg BZ (1964b) The permeability of the cell membranes of *Nitella translucens* to urea and the effect of high concentrations of sucrose on this permeability. Biochim Biophys Acta 79:112–121
Dainty J, Hope AB (1959) The water permeability of cells of *Chara australis* R. BR. Aust J Biol Sci 12:136–145
Ding DQ, Tazawa M (1989) Influence of cytoplasmic streaming and turgor pressure gradient on the transnodal transport of rubidium and electrical conductance in *Chara corallina*. Plant Cell Physiol 30:739–748
Ding DQ, Mimura T, Amino S, Tazawa M (1991a) Intercellular transport and photosynthetic differentiation in *Chara corallina*. J Exp Bot 42:33–38
Ding DQ, Amino S, Mimura T, Nagata T, Tazawa M (1991b) Intercellular transport and subcellular distribution of photoassimilates in *Chara corallina*. J Exp Bot 42:1393–1398
Findlay GP (1961) Voltage-clamp experiments with *Nitella*. Nature 191:812–814
Findlay GP (1964) Ionic relations of cells of *Chara australis*. VIII. Membrane currents during a voltage clamp. Aust J Biol Sci 17:388–399
Fujii S, Shimmen T, Tazawa M (1978) Light-induced changes in membrane potential in *Spirogyra*. Plant Cell Physiol 19:573–590
Grolig F, Williamson RE, Parke J, Miller C, Anderton BH (1988) Myosin and Ca^{2+}-sensitive streaming in the alga *Chara*: detection of two polypeptides reacting with a monoclonal anti-myosin and their localization in the streaming endoplasm. Eur J Cell Biol 47:22–31
Hayama T, Tazawa M (1980) Ca^{2+} reversibly inhibits active rotation of chloroplasts in isolated cytoplasmic droplets of *Chara*. Protoplasma 102:1–9
Hayama T, Shimmen T, Tazwa M (1979) Participation of Ca^{2+} in cessation of cytoplasmic streaming induced by membrane excitation in Characeae internodal cells. Protoplasma 99:305–321
Higashi-Fujime S (1980) Active movement in vitro of bundles of microfilaments isolated from *Nitella* cells. J Cell Biol 87:569–578
Higashi-Fujime S, Sumiyoshi H (2001) Molecular sliding of the fastest *Chara* myosin from green algae. Curr Topics Biochem Res 4:61–70
Higashi-Fujime S, Ishikawa R, Iwasawa H, Kagami O, Kurimoto E, Kohama K, Hozumi T (1995) The fastest actin-based motor protein from the green algae, *Chara*, and its distinct mode of interaction with actin. FEBS Lett 375:151–154
Hope AB (1961) The action potential in cells of *Chara*. Nature 191:811–812
Kamitsubo E (1966) Motile protoplasmic fibrils in cells of characeae. II. Linear fibrillar structure and its bearing on protoplasmic streaming. Proc Jpn Acad 42:640–643
Kamitsubo E (1981) Effect of supraoptimal temperature on the function of the subcortical fibrils and an endoplasmic factor in *Nitella* internode. Protoplasma 109:3–12
Kamiya N (1940) The control of protoplasmic streaming. Science 92:462–463
Kamiya N (1986) Cytoplasmic streaming in giant algal cells: a historical survey of experimental approaches. Bot Mag Tokyo 99:441–467
Kamiya N (1989) My early career and the involvement of World War II. Annu Rev Plant Physiol Plant Mol Biol 40:1–18
Kamiya N, Kuroda K (1955) Some experiments on cell amputation. In 20th Ann Meet Bot Soc Jpn, pp 46–47

Kamiya N, Kuroda K (1956a) Artificial modification of the osmotic pressure of the plant cell. Protoplasma 46:423–436
Kamiya N, Kuroda K (1956b) Velocity distribution of the protoplasmic streaming in *Nitella* cells. Bot Mag (Tokyo) 69:544–554
Kamiya N, Kuroda K (1958) Measurement of the motive force of the protoplasmic rotation in *Nitella*. Protoplasma 50:144–148
Kamiya N, Kuroda K (1973) Dynamics of cytoplasmic streaming in a plant cell. Biorheology 10:179–187
Kamiya N, Tazawa M (1956) Studies on water permeability of a single cell by means of transcellular osmosis. Protoplasma 46:394–422
Katsuhara M, Tazawa M (1986) Salt tolerance in *Nitellopsis obtusa*. Protoplasma 135:155–161
Katsuhara M, Tazawa M (1987) ATP is essential for calcium-induced salt tolerance in *Nitellopsis obtusa*. Protoplasma 138:190–192
Katsuhara M, Tazawa M (1988) Changes in sodium and potassium in *Nitellopsis* cells treated with transient salt stress. Plant Cell Environ 11:71–74
Katsuhara M, Tazawa M (1992) Calcium-regulated channels and their bearing on physiological activities in characean cells. Phil Trans R Soc Lond B 338:19–29
Katsuhara M, Mimura T, Tazawa M (1989) Patch-clamp study on a Ca^{2+}-regulated K^+ channel in the tonoplast of the brackish Characeae *Lamprothamnium succinctum*. Plant Cell Physiol 30:549–555
Katsuhara M, Mimura T, Tazawa M (1990) ATP-regulated ion channels in the plasma membrane of a Characeae alga, *Nitellopsis obtusa*. Plant Physiol 93:343–346
Katsuhara M, Mimura T, Tazawa M (1991) Patch-clamp study on ion channels in the tonoplast of *Nitellopsis obtusa*. Plant Cell Physiol 32:179–184
Keifer DW, Spanswick RM (1979) Correlation adenosine triphosphate levels in *Chara corallina* with the activity of electrogenic pump. Plant Physiol 64:165–168
Kersey YM, Hepler PK, Palevitz BA, Wessels NK (1976) Polarity of actin filaments in characean algae. Proc Natl Acad Sci USA 73:165–167
Kikuyama M, Tazawa M (1976) Tonoplast action potential in *Nitella* in relation to vacuolar chloride concentration. J Membr Biol 29:95–110
Kikuyama M, Tazawa M (1982) Ca^{2+} reversibly inhibits the cytoplasmic streaming of *Nitella*. Protoplasma 113:241–243
Kikuyama M, Hayama T, Fujii S, Tazawa M (1979) Relationship between light-induced potential change and internal ATP concentration in tonoplast-free *Chara* cells. Plant Cell Physiol 20:993–1002
Kikuyama M, Oda K, Shimmen T, Hayama T, Tazawa M (1984) Potassium and chloride effluxes during excitation of Characeae cells. Plant Cell Physiol 25:965–974
Kikuyama M, Shimada K, Hiramoto Y (1993) Cessation of cytoplasmic streaming follows an increase of cytoplasmic Ca^{2+} during action potential in *Nitella*. Protoplasma 174:142–146
Kikuyama M, Tazawa M, Tominaga Y, Shimmen T (1996) Membrane control of cytoplasmic streaming in Characeae cells. J Plant Res 109:113–118
Kishimoto U, Tazawa M (1965a) Ionic composition of the cytoplasm of *Nitella flexilis*. Plant Cell Physiol 6:507–518
Kishimoto U, Tazawa M (1965b) Ionic composition and electric response of *Lamprothamnium succinctum*. Plant Cell Physiol 6:529–536
Kishimoto U, Nagai R, Tazawa M (1965) Plasmalemma potential in *Nitella*. Plant Cell Physiol 6:519–528
Kiyosawa K, Tazawa M (1972) Influence of intracellular and extracellular tonicities on water permeability in Characeae cells. Protoplasma 74:257–270
Kiyosawa K, Tazawa M (1973) Rectification characteristics of *Nitella* membrane in respect to water permeability. Protoplasma 78:203–214
Kiyosawa K, Tazawa M (1977) Hydraulic conductivity of tonoplast-free *Chara* cells. J Membr Biol 37:157–166

Kohama K, Kendrick-Jones J (1986) The inhibitory Ca^{2+}–regulation of the actin-activated Mg-ATPase activity of myosin from *Physarum polycephalum* plasmodia. J Biochem 99: 1433–1446
Kohama K, Shimmen T (1985) Inhibitory Ca^{2+}–control of movement of beads coated with *Physarum* myosin along actin-cables in *Chara*. Protoplasma 129:88–91
Kohno T, Okagaki T, Kohama K, Shimmen T (1991) Role of actin in the myosin-linked Ca^{2+}–regulation of ATP-dependent interaction between actin and myosin of a lower eukaryote, *Physarum polycephalum*. J Biochem 110:508–513
Kron SJ, Spudich JA (1986) Fluorescent actin filaments move on myosin fixed to a glass surface. Proc Natl Acad Sci USA 83:6272–6276
Kuroda K (1983) Cytoplasmic streaming in characean cells cut open by microsurgery. Proc Jpn Acad 59:126–130
Kuroda K (1990) Cytoplasmic streaming in plant cells. Int Rev Cytol 121:267–307
Kuroda K, Kamiya N (1975) Active movement of *Nitella* chloroplast in vitro. Proc Jpn Acad 51:774–777
Maurel C (1997) Aquaporins and water permeability of plant membranes. Annu Rev Plant Physiol Plant Mol Biol 48:399–429
McCurdy DW, Harmon AC (1992) Calcium-dependent protein kinase in the green alga *Chara*. Planta 188:54–61
Mimura T, Tazawa M (1986) Light-induced membrane hyperpolarization and adenine nucleotide levels in perfused characean cells. Plant Cell Physiol 27:319–330
Mimura T, Shimmen T, Tazawa M (1983) Dependence of the membrane potential on intracellular ATP concentration in tonoplast-free *Chara* cells. Planta 157:97–104
Mimura T, Simmen T, Tazawa M (1984) Adenine-nucleotide levels and metabolism-dependent membrane potential in cells of *Nitellopsis obtusa*. Planta 162:77–84
Mimura T, Sakano K, Tazawa M (1990) Changes in subcellular distribution of free amino acids in relation to light conditions in cells of *Chara corallina*. Bot Acta 103:42–47
Morimatsu M, Hasegawa S, Higashi-Fujime S (2002) Protein phosphorylation regulates actomyosin-driven vesicle movement in cell extracts isolated from the green algae, *Chara corallina*. Cell Motil Cytoskeleton 53:66–76
Moriyasu Y, Tazawa M (1986) Plant vacuole degrades exogenous proteins. Protoplasma 130:214–215
Moriyasu Y, Tazawa M (1987) Calcium-activated protease in the giant alga *Chara australis*. Protoplasma 140:72–74
Moriyasu Y, Wayne R (2004) A novel calcium-activated protease in *Chara corallina*. Eur J Phycol 39:57–66
Moriyasu Y, Shimmen T, Tazawa M (1984a) Vacuolar pH regulation in *Chara australis*. Cell Struct Funct 9:225–234
Moriyasu Y, Shimmen T, Tazawa M (1984b) Electric characteristics of the vacuolar membrane of *Chara* in relation to pH_v regulation. Cell Struct Funct 9:235–246
Nagai R, Rebhun L (1966) Cytoplasmic microfilaments in streaming *Nitella* cells. J Ultrastruct Res 14:571–589
Nagai R, Tazawa M (1962) Changes in resting potential and ion absorption by light in a single plant cell. Plant Cell Physiol 3:323–339
Nakagawa S, Kataoka H, Tazawa M (1974) Osmotic and ionic regulation in *Nitella*. Plant Cell Physiol 15:457–468
Okazaki Y (1996) Turgor regulation in a brackish water charophyte, *Lamprothamnium succinctum*. J Plant Res 109:107–112
Okazaki Y, Tazawa M (1986a) Involvement of calcium ion in turgor regulation upon hypotonic treatment in *Lamprothamnium succinctum*. Plant Cell Environ 9:185–190
Okazaki Y, Tazawa M (1986b) Ca^{2+} antagonist nifedipine inhibits turgor regulation upon hypotonic treatment in internodal cells of *Lamprothamnium*. Protoplasma 134:65–66
Okazaki Y, Tazawa M (1990) Calcium ion and turgor regulation in plant cells. J Membr Biol 114:189–194

Okazaki Y, Shimmen T, Tazawa M (1984a) Turgor regulation in a brackish charophyte. *Lamprothamnium succinctum*. I. Artificial modification of intracellular osmotic pressure. Plant Cell Physiol 25:565–571

Okazaki Y, Shimmen T, Tazawa M (1984b) Turgor regulation in a brackish charophyte, *Lamprothamnium succinctum*. II. Changes in K^+, Na^+ and Cl^- concentrations, membrane potential and membrane resistance during turgor regulation. Plant Cell Physiol 25:573–581

Okazaki Y, Yoshimoto Y, Hiramoto Y, Tazawa M (1987) Turgor regulation and cytoplasmic free Ca^{2+} in the alga *Lamprothamnium*. Protoplasma 140:67–71

Okazaki Y, Tazawa M, Moriyama Y, Iwasaki N (1992) Bafilomycin inhibits vacuolar pH regulation in a fresh water charophyte, *Chara corallina*. Bot Acta 105:421–426

Sakamura T (1952) Physiology of osmosis in plant cells (in Japanese). Yokendo, Tokyo

Sakano K, Tazawa M (1985) Metabolic conversion of amino acids loaded in the vacuole of *Chara australis* internodal cells. Plant Physiol 78:673–677

Sakano K, Tazawa M (1986) Tonoplast origin of the envelope membrane of cytoplasmic droplets prepared from *Chara* internodal cells. Protoplasma 131:247–249

Sheez MP, Spudich JA (1983) Myosin-coated fluorescent beads move on actin cables in vitro. Nature 303:31–35

Shiina T, Tazawa M (1987a) Demonstration and characterization of Ca^{2+} channel in tonoplast-free cells of *Nitellopsis obtusa*. J Membr Biol 96:263–276

Shiina T, Tazawa M (1987b) Ca^{2+}-activated Cl^- channel in plasmalemma of *Nitellopsis obtusa*. J Membr Biol 99:137–146

Shiina T, Tazawa M (1988) Ca^{2+}-dependent Cl^- efflux in tonoplast-free cells of *Nitellopsis obtusa*. J Membr Biol 106:135–139

Shiina T, Wayne R, Lim Tung HY, Tazawa M (1988) Possible involvement of protein phosphorylation/dephosphorylation in the modulation of Ca^{2+} channel in tonoplast-free cells of *Nitellopsis*. J Membr Biol 102:255–264

Shimmen T (2003) Studies on mechano-perception in the Characeae: transduction of pressure signals into electrical signals. Plant Cell Physiol 44:1215–1224

Shimmen T (2006a) Electrical perception of "death message" in *Chara*: characterization of K^+-induced depolarization. Plant Cell Physiol 47:559–562

Shimmen T (2006b) Electrophysiology in mechanosensing and wounding response. In: Volkov AG (ed) Plant electrophysiology – theory and methods. Springer, Berlin, pp 319–339

Shimmen T (2007) The sliding theory of cytoplasmic streaming: fifty years of progress. J Plant Res 120:31–43

Shimmen T (2008) Electrophysiological characterization of the node in *Chara corallina*: functional differentiation for wounding response. Plant Cell Physiol 49:264–272

Shimmen T, Nishikawa S (1988) Studies on the tonoplast action potential of *Nitella flexilis*. J Membr Biol 101:133–140

Shimmen T, Tazawa M (1977) Control of membrane potential and excitability of *Chara* cells with ATP and Mg^{2+}. J Membr Biol 37:167–192

Shimmen T, Tazawa M (1980) Intracellular chloride and potassium ions in relation to excitability of *Chara* membrane. J Membr Biol 55:223–232

Shimmen T, Tazawa M (1982a) Cytoplasmic streaming in the cell model of *Nitella*. Protoplasma 113:127–131

Shimmen T, Tazawa M (1982b) Reconstitution of cytoplasmic streaming in Characeae. Protoplasma 113:127–131

Shimmen T, Tazawa M (1982c) Effects of intracellular vanadate on electrogenesis, excitability and cytoplasmic streaming in *Nitellopsis obtusa*. Plant Cell Physiol 23:669–677

Shimmen T, Yano M (1984) Active sliding movement of latex beads coatred with skeletal muscle myosin on *Chara* actin bundles. Protoplasma 121:132–137

Shimmen T, Kikuyama M, Tazawa M (1976) Demonstration of two stable potential states of plasmalemma of *Chara* without tonoplast. J Membr Biol 30:225–247

Staves MP, Wayne R, Leopold AC (1992) Hydrostatic pressure mimics gravitational pressure in characean cells. Protoplasma 168:141–152
Takeshige K, Shimmen T, Tazawa M (1985) Electrogenic pump current and ATP-dependent H^+ efflux across the plasma membrane of *Nitellopsis obtusa*. Plant Cell Physiol 26:661–668
Takeshige K, Tazawa M, Hager A (1988) Characterization of the H^+ translocating adenosine triphosphatase and pyrophosphatase of vacuolar membranes isolated by means of a perfusion technique from *Chara corallina*. Plant Physiol 86:1168–1173
Taylor DL, Condeelis JS, Moore PL, Allen RD (1973) The contractile basis of amoeboid movement. The chemical control of motility in isolated cytoplasm. J Cell Biol 59:378–394
Tazawa M (1964) Studies on *Nitella* having artificial cell sap. I. Replacement of the cell sap with artificial solutions. Plant Cell Physiol 5:33–43
Tazawa M (1968) Motive force of the cytoplasmic streaming in *Nitella*. Protoplasma 65: 207–222
Tazawa M, Kamiya N (1965) Water permeability of a characean internodal cell. Ann Rep Biol Works Fac Sci Osaka Univ 13:123–157
Tazawa M, Kamiya N (1966) Water relations of characean internodal cell with special reference to its polarity. Aust J Biol Sci 19:399–419
Tazawa M, Kishimoto U (1968) Cessation of cytoplasmic streaming of *Chara* internodes during action potential. Plant Cell Physiol 9:361–368
Tazawa M, Kiyosawa K (1973) Analysis of transcellular water movement in *Nitella*. A new procedure to determine the inward and outward water permeabilities of membranes. Protoplasma 78:349–364
Tazawa M, Nagai R (1960) Die Mitwirkung von Ionen bei der Osmoregulation der *Nitella*zelle. Plant Cell Physiol 1:255–267
Tazawa M, Nagai R (1966) Studies on osmoregulation of *Nitella* internode with modified cell saps. Z Pflanzenphysiol 54:333–344
Tazawa M, Kikuyama M, Nakagawa S (1975) Open-vacuole method for measuring membrane potential and membrane resistance of Characeae cells. Plant Cell Physiol 16:611–621
Tazawa M, Kikuyama M, Shimmen T (1976) Electric characteristics and cytoplasmic streaming of Characeae cells lacking tonoplast. Cell Struct Funct 1:165–176
Tazawa M, Shimmen T, Mimura T (1987) Membrane control in the Characeae. Ann Rev Plant Physiol 38:95–117
Tazawa M, Okazaki Y, Moriyama Y, Iwasaki N (1995) Concanamycin 4-B: a potent inhibitor of vacuolar pH regulation in *Chara* cells. Bot Acta 108:67–73
Tominaga Y, Shimmen T, Tazawa M (1983) Control of cytoplasmic streaming by extracellular Ca^{2+} in permeabilized *Nitella* cells. Protoplasma 116:75–77
Tominaga Y, Muto S, Shimmen T, Tazawa M (1985) Calmodulin and Ca^{2+}-controlled cytoplasmic streaming in characean cells. Cell Struct Funct 10:315–325
Tominaga Y, Wayne R, Tung HYL, Tazawa M (1987) Phosphorylation-dephosphorylation is involved in Ca^{2+}-controlled cytoplasmic streaming of characean cells. Protoplasma 136: 161–169
Ward JM, Mäser P, Schroeder JI (2009) Plant ion channels: gene families, physiology, and functional genomics analyses. Annu Rev Physiol 71:59–82
Wayne R (2009) Plant cell biology: from astronomy to zoology. Elsevier, Amsterdam
Wayne R, Tazawa M (1990) Nature of the water channels in the internodal cells of *Nitellopsis*. J Membr Biol 116:31–39
Williamson RE (1974) Actin in the alga, *Chara corallina*. Nature 248:801–802
Williamson RE (1975) Cytoplasmic streaming in *Chara*: a cell model activated by ATP and inhibited by cytochalasin B. J Cell Sci 17:655–668
Williamson RE, Ashley CC (1982) Free Ca^{2+} and cytoplasmic streaming in the alga *Chara*. Nature 296:647–651
Yamamoto K, Kikuyama M, Sutoh-Yamamoto N, Kamitsubo E (1994) Purification of actin based motor protein from *Chara corallina*. Proc Jpn Acad 70:175–180

Yangida T, Nakase M, Nishiyama K, Oosawa F (1984) Direct observation of single actin filaments in the presence of myosin. Nature 307:58–60

Ye Q, Wiera B, Steudle E (2004) A cohesioin/tension mechanism explains the gating of water channels (aquaporins) in *Chara* internodes by high concentration. J Exp Bot 55:449–461

Yokota E, Shimmen T (1994) Isolation and characterization of plant myosin from pollen tubes of lily. Protoplasma 177:153–162

Yuasa T, Okazaki Y, Iwasaki N, Muto S (1997) Involvement of a calcium-dependent protein kinase in hypoosmotic turgor regulation in a brackish water Characeae *Lamprothamnium succinctum*. Plant Cell Physiol 38:586–594

Part II
Genetics

Root Apical Meristem Pattern: Hormone Circuitry and Transcriptional Networks

M.B. Bitonti and A. Chiappetta

Contents

Abstract In higher plants, growth and development relies on the spatiotemporal regulation of gene expression, which is under the control of both endogenous signals and external stimuli. In this chapter, recent advances in defining signaling machinery and genetic frameworks that underlie RAM patterning and maintenance are reviewed, with a focus on the interplay between different hormone classes. The evidence for an epigenetic control of the root developmental program is also briefly considered. Conceivably, many other aspects are still to be elucidated. Future challenges deal with, on the one hand, understanding how signaling and genetic programs are modulated to achieve adaptative traits under environmental pressure and, on the other, how cell fate is reprogrammed in vivo and in vitro. We conclude that knowledge from the plant model, *Arabidopsis thaliana*, could enhance our understanding of more complex species encountered in crop

M.B. Bitonti (✉) and A. Chiappetta
Department of Ecology, University of Calabria, Arcavacata Rende (CS), Italy
e-mail: b.bitonti@unical.it; a.chiappetta@unical.it

U. Lüttge et al. (eds.), *Progress in Botany 72*,
DOI 10.1007/978-3-642-13145-5_2,

systems and could provide relevant perspectives for both crop improvement and plant biotechnology.

1 Introduction

In higher plants, root growth is achieved by pronounced elongation of cells that are descendants of the root apical meristem (RAM), a specialized structure present at the root tip, which produces cells for virtually all of the root tissues formed during postembryonic development. In this localized microenvironment, meristematic cells can reside for potentially an indeterminate period of time and produce progeny cells while self-renewing, thus exhibiting key features of a stem-cell niche (Ohlstein et al. 2004; Li and Xie 2005; Dinneny and Benfey 2008; Morrison and Spradling 2008). This behavior relies on the capacity of RAM cells to undergo asymmetric cell division, which is another defining feature of stem-cells dictating that one daughter cell retains the meristematic fate and another daughter cell is programmed to differentiate into a specialized cell (McCulloch and Till 2005; Scheres 2007; Morrison and Spradling 2008). The acquisition of differential potential of the two daughter cells can result from an unequal partitioning of cell fate determinants through the asymmetric positioning of the cell division plate as well as from a different signaling from their surroundings.

The maintenance of the RAM is assured by a balance between the production of new meristematic cells and their displacement toward the differentiation process. However, in some cases, the RAM is genetically determined to become exhausted, and root growth shifts from an indeterminate to a determinate developmental pattern. Recently, an extensive review on the determinacy/indeterminacy of root growth has been provided by Shishkova et al. (2008), but it is outside the scope of this chapter.

Here, we review the recent advances in understanding transcriptional networks and hormone cross-talk underlying the establishment and regulation of the RAM. In particular, signals between the organizing region and the stem-cell population in the RAM as well as signals released by neighboring cells will be emphasized. Mainly, we focus on studies on the model plant, *Arabidopsis thaliana*.

2 An Overview of RAM Organization in Plants

Precociously determined during embryogenesis, and indeed before the establishment of the shoot apical meristem (SAM), the RAM exhibits a proliferative multicellular dome that includes apical initials and their derivative cells (Fig. 1) (Clowes 1976; Barlow 1997). The number of initials varies according to species (Webster and McLeod 1996), but they are almost permanent in position and divide continuously, producing at each division one cell that continues to act as an initial

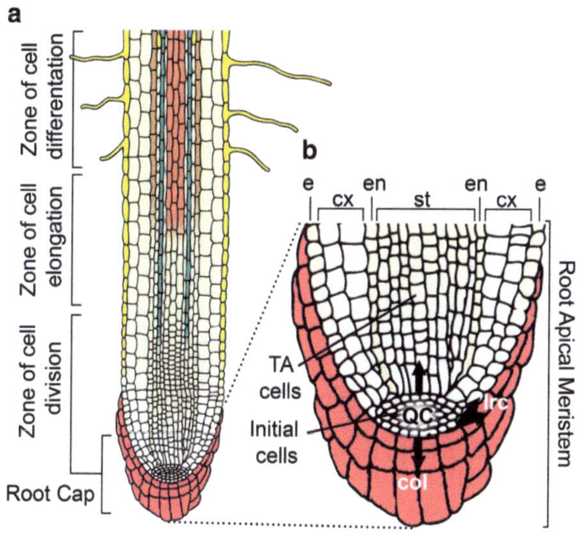

Fig. 1 Schematic of a higher plant primary root tip in longitudinal section (**a**) featuring zones of cell division, elongation, and differentiation. In (**b**), *arrows* indicate the direction of cell division; *col* columella; *cx* cortex; *e* epidermis; *en* endodermis; *lrc* lateral root cap; *QC* quiescent center; *st* stele; *TA* transit amplifying cells; adapted from Taiz and Zeiger (1998)

and one derivative cell. Derivative cells undergo numerous additional cell divisions producing a transit amplifying (TA) cell population, which is progressively displaced from initials and then allowed to differentiate (Fig. 1a). In the RAM, the initials divide proximally, laterally, or distally, thus giving rise to derivative cells for root cortex/endodermis/stele, epidermis/lateral rootcap, and columella, respectively (Fig. 1b). As a consequence, the RAM dome is located subterminally and is covered by the root cap, which protects the apical meristem, produces mucilage to facilitate a passage for the growing root, and serves as a gravity perceiving tissue. Behind the RAM is the root elongation zone where cells continue to elongate, followed by the differentiation zone, where cells acquire the characteristics of the mature differentiated primary tissues (Fig. 1a). Hence, the blue print for root function is laid down remarkably early during embryogenesis.

In angiosperms, the apical dome exhibits either an open or closed configuration (Clowes 1981; Heimsch and Seago 2008; and references cited therein). In the closed configuration, cell boundaries between cortical, epidermis, and root-cap regions are clearly distinguishable, while they are lacking in the open meristem (Fig. 2). Within both configurations, a large range of specific features exist in the Angiospermae (i.e., Basal angiosperms, Monocots, Eudicots) and in root ranks (i.e., primary vs. adventitious, primary vs. lateral, ephemeral vs. permanent), and have been extensively reviewed by Heimsch and Seago (2008). The specific organization of plant RAM is established early (i.e., the first day) after germination (Guttenberg 1968; Clowes 2000). However, some species exhibit RAMs that can change from

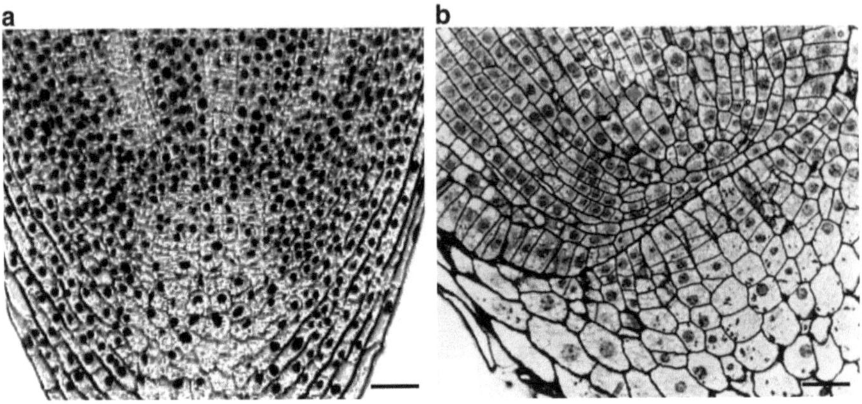

Fig. 2 Longitudinal sections through the root tip of (**a**) *Allium cepa* and (**b**) *Zea mays* [adapted from Webster and Mc Leod (1996)] exhibiting an open and closed organization, respectively. Bar scale (**a**) 70 μm; (**b**) 100 μm

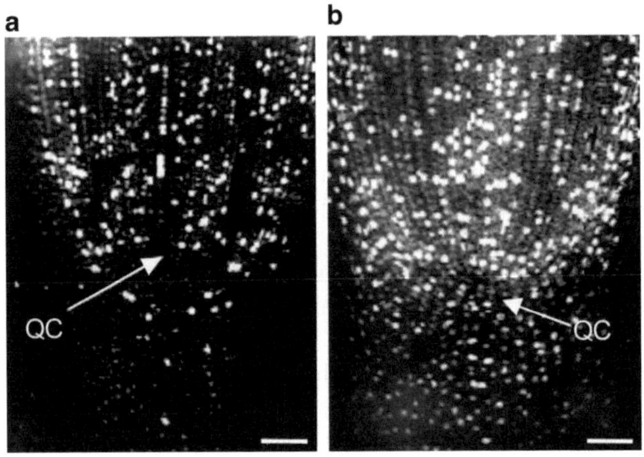

Fig. 3 Autoradiographs of longitudinal sections through the *Allium cepa* root apical meristem (RAM) of seedlings exposed to [Me-^{3}H]thymidine for 24 h: (**a**) root grown in water and (**b**) root supplied with ascorbic acid. Note the different size of the QC in (**a**) and (**b**), represented by area within the RAM exhibiting unlabeled nuclei. Bar scale = 120 μm (adapted from Liso et al. 1988)

closed to open configuration and vice-versa as the root grows (Seago and Heimsch 1969; Armstrong and Heimsch 1976).

Within the RAM, meristematic cells undergo division at differential rates. One major demarcation of differential rates of division can be observed in a group of cells located centrally (Fig. 3a), which rarely divide and therefore was named the quiescent center (QC) and bears some analogies to the central zone in the SAM (Clowes 1954, 1956). Since its discovery in roots of *Zea mays* (Clowes 1954), QC cells have long been considered a geometric necessity for root patterning as well as a reservoir of potentially meristematic cells that can be activated into division upon damage inflicted on other parts of the RAM (Clowes 1975; Barlow 1978). In line with this latter role, it has been demonstrated that QC cells are not permanently quiescent. Certainly, they can be induced to divide rapidly after irradiation (Clowes 1959, 1961), mechanical damage (Barlow 1974), carbohydrate starvation (Webster and Langenauer 1973), exposure to low temperature (Clowes and Stewart 1967), or exposure to ascorbic acid (Liso et al. 1988; Innocenti et al. 1990). However, and notably, in ascorbate-treated roots, a threshold number of QC cells were not induced to divide consistent with the existence of a minimal QC (Fig. 3b). Unlike the other treatments, the functions of meristem were not affected in these roots, thus suggesting a role for a minimal QC, maintaining and controlling RAM activity (Innocenti et al. 1990). As will be discussed later on, more recent evidence indicates that the QC represses cell differentiation of surrounding cells and therefore acts as the organizing center (OC) of the RAM stem niche (Van den berg et al. 1997).

3 Environmental Cues and RAM Patterning

In plants, both endogenous developmental factors and environmental cues can modulate cell differentiation and morphogenesis. Regarding root systems, an extensive literature exists about structural and functional modifications in a range of unrelated species, in response to water and nutrient availability, mechanical and gravitational stimuli, and responses to magnetic fields, as well as in relation to symbiotic or pathogenic interactions with heterologous organisms (Audus 1960; Behrens et al. 1982; Goodman and Ennos 1996; Fusconi et al. 1999; Stange et al. 2002; Waisel et al. 1997; Swarup et al. 2005; Potters et al. 2009).

However, with the exception of the effects on QC homeostasis, little attention has been paid to the effects induced by abiotic factors on RAM pattern as a whole (Feldman and Torrey 1975; Doncheva et al. 2005; Sanchez-Calderon et al. 2005; Sharma and Dubey 2007, De Tullio et al. 2009). Nevertheless, some data exist showing that exposure of *Zea mays* seedlings to a continuous electromagnetic field induced a stimulation in the rate of root elongation and resulted in a significant increase in cell expansion, in both the acropetal (proximal metaxylem cell lineage) (Fig. 4a, b) and basipetal (distal root cap cells) (Fig. 4e–h) directions. Concomitantly, a significant reduction in the size of the QC occurred (Fig. 4c, d) in roots exposed to this magnetic field together with an advanced differentiation of initials previously in surface contact with the QC (Bitonti et al. 2006). These results from a rather more complex system than *Arabidopsis* again underlie the importance of signals generated by cell-to-cell surface contact influencing differentiation and controlling root pattern.

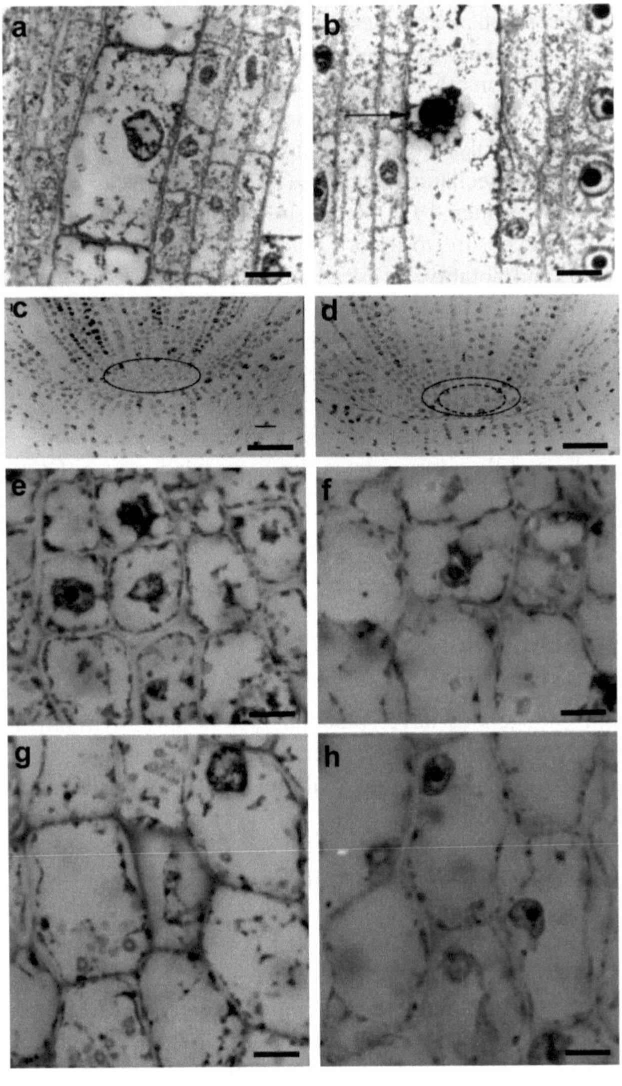

Fig. 4 Longitudinal sections through maize root tips in (**a**, **c**, **e**, **g**) control roots and (**b**, **d**, **f**, **h**) roots after 30-h exposure to a magnetic field. (**a**, **b**) Metaxylem cells at 800 mm from the cap-junction, scale bar = 40 μm; note differences in length between control (**a**) and exposed root (**b**). (**c**, **d**) Autoradiograph of root apical meristem of seedlings exposed to [Me-^{3}H]thymidine for 24 h; ring indicates quiescent centre (QC) characterized by absence of labeling, scale bar = 150 μm; note differences in QC size between control (**c**) and exposed root (**d**). (**e**, **f**) root cap cells at the junction with the RAM; (**g**, **h**) root cap cells in the distal zone of columella, scale bar = 15 μm; note differences in cell size and structure between control (**e**, **g**) and exposed root (**f**, **h**). Adapted from Bitonti et al. (2006)

4 Arabidopsis thaliana

4.1 Morphogenetic Establishment of the RAM During Embryogenesis

During embryogenesis, the morphogenetic organization of the future plant is defined through the establishment of two overlapping developmental patterns along the longitudinal (apical–basal) and radial axes. Along the longitudinal axis, the mature embryo exhibits the SAM, located between two cotyledons, the embryo axis or hypocotyl, and spatially separate from the RAM (Fig. 5a). The establishment of the radial pattern relies on the determination of three different tissues: protoderm, ground meristem, and procambium, which will generate cells that are fated to differentiate into epidermis, endodermis, and other cortical tissues and vascular tissues, respectively (Fig. 5b). In order to highlight the composite origin of the RAM, the morphogenetic events leading to its development are summarized in Fig. 6. Briefly, the stele, endodermis, cortex, epidermis, and lateral cap derive from the apical cell established at the bicellular embryo stage, while the QC and columella derive from the basal cell. It is evident that in plants very early during development (i.e., from the first zygotic division), cell fate determination relies on asymmetric cell division. Indeed, apical cell will give rise to most of the embryo proper while the basal cell will form the suspensor, a structure that physically positions the embryo within the seed and plays an active role in nutrient acquisition during early embryo development. In such a way, embryo polarity is established early through a mechanism that is common to zygotic embryo development in *Fucus* (Kropf et al. 1999). The importance of asymmetric cell division can be seen

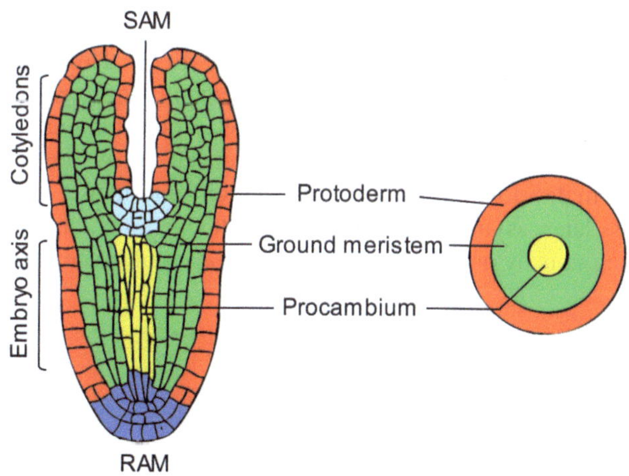

Fig. 5 Schematic of (**a**) longitudinal and (**b**) cross section of *Arabidopsis thaliana* mature embryo featuring the establishment of apical-basal and radial pattern, respectively. Adapted from Altamura et al. (2007)

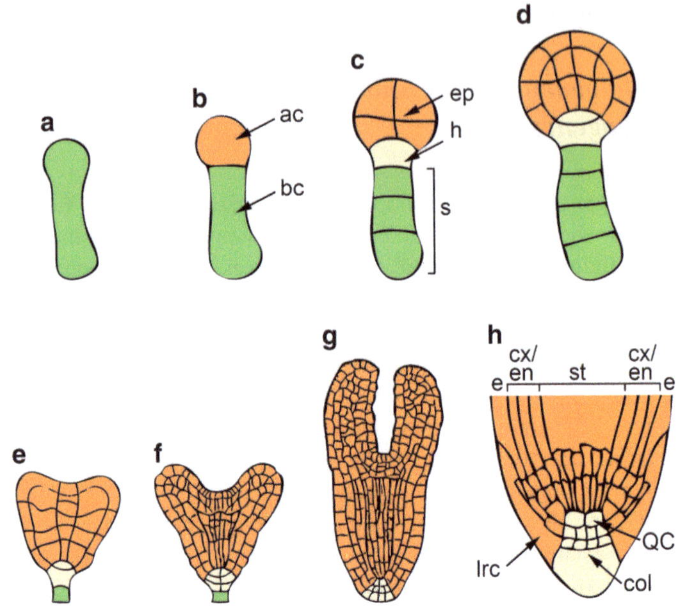

Fig. 6 Cartoon illustrating the main stages of *Arabidopsis thaliana* embryo development and the composite origin of the root pole. (**a**) zygotic cell, (**b**) bicellular stage, (**c**) octant stage, (**d**) globular stage, (**e**) triangular stage, (**f**) late-heart stage, (**g**) torpedo stage, (**h**) adult root. *ac* apical cell; *bc* basal cell; *col* columella; *e* epidermis; *en*/*cx* endodermis/cortex; *ep* embryo proper; *h* hypophysis; *lrc* lateral root cap; *QC* quiescent centre; *s* suspensor; *st* stele. Adapted from Altamura et al. (2007)

through phenotypic analysis of mutants that have lost this feature of cell partitioning. For example, the *gnome* (*gn*) mutant that does not exhibit asymmetric zygotic division is unable to organize an embryo with the normal polarity featuring spatially separated RAM and SAM (Mayer et al. 1993). However, it must be underlined that asymmetric division does not inherently block cell fate as demonstrated by *twin* mutants, which are able to produce viable additional embryos from the basal cell of the bicellular stage, suggesting that cell-to-cell communication is a relevant player in developmental processes (Vernon and Meinke 1994). Hence, as we will discuss later, asymmetric cell division, although a notable feature of cell fate determination, is by no means the sole determinant of tissue specification.

4.2 RAM Pattern

The cellular organization of the RAM in *Arabidopsis* seedlings was well defined by Dolan et al. (1993). The *Arabidopsis* RAM exhibits a dome typically organized as a three-tiered closed meristem, where distinct cell layers of initials give rise to the stele (S), endodermis/cortex (EC), and root-cap/epidermis (RCE) tissues (Fig. 7).

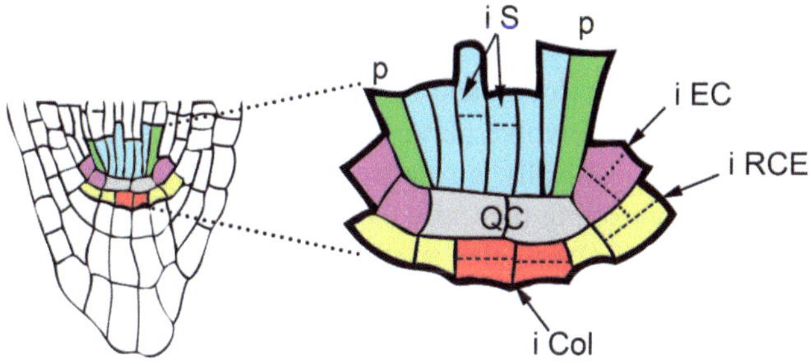

Fig. 7 Cartoon illustrating the organization of root apical meristem of *Arabidopsis thaliana* and the division planes of initial cells. *iCol* columella cell initial; *iEC* endodermis/cortex cell initial; *iRCE* lateral root cap/epidemis cell initial; *iS* stele cell initial

The QC is constituted by four founder cells surrounded by initials. Based on asymmetric division, the RCE initial firstly divides periclinally to give rise to one lateral root cap cell (the outer cell) and one inner that retains the role of an initial that divides anticlinally producing one epidermal cell (the upper cell) and one initial cell (the lower cell). In turn, the cortical and endodermal cells derive from two consecutive asymmetric cell divisions: the EC initial first undergoes an anticlinal division and the daughter cell in contact with the QC maintains the stem-cell fate, while the other derivate cell divides periclinally producing an inner and outer cell, which will differentiate into the endodermis, the innermost layer of the cortex and the next outer cortical layer, respectively. Finally, the S initial divides periclinally giving origin to xylem along only one central axis (i.e., 4/5 cell tiers in contact with QC) and phloem tissues. The most external S initials divide transversely to form the pericycle. Distally, columella initials below the QC divide periclinally producing central cells of the root cap (Fig. 7).

Cell ablation experiments have largely demonstrated that positional information more than clonal lineage is relevant in defining cell fate. Indeed, if an initial cell is ablated, an adjacent cell in contact with this cell divides, producing a new initial. Moreover, when a differentiated cell is ablated along a differentiating cell line, only an adjacent derivate or TA cell, which maintains the contact with a differentiated overlapping cell, does undergo asymmetric division to replace the ablated cell. Thus, a positional effect and a short-range cell-to-cell signal have a role in inducing asymmetric cell division (Van den Berg et al. 1995; Scheres 2001).

4.3 *Positional Signaling and Genetic Network Operating in Root Patterning*

Positional information established through both long-distance signals and cell-to-cell interactions as well as specific transcription factor activity are known to be important throughout plant development (Meyerowitz 2002; Vernoux and Benfey 2005; Qu and Zhu 2006; Veit 2006; Sablowski 2007; Scheres 2007; Benjamins and Scheres 2008; Kiefer 2009; Robert and Friml 2009; Stahl and Simon 2010).

Cell-to-cell interactions involve plasmodesmata and the cell wall. It is known that plasmodesmata dynamically guarantee or interrupt symplastic connection between cells, thus controlling the trafficking of molecules, which play key roles in developmental events. In particular, specific transcription factors at the protein or mRNA level but also micro (mi)RNAs can be translocated through plasmodesmata (Lucas and Lee 2004; Jackson 2005; Lee and Cui 2009). For example, the transcription factor, SHORTROOT (SHR), is specifically translocated into the QC and endodermis (Nakajima et al. 2001). More of this later (see Sect. 4.3.2).

The cell wall is also a relevant component of both long- and short-distance intercellular signaling. Apoplastic continuity along xylem vascular elements represents a route for long-distance signaling played by several types of molecules such as different hormones and sugars (Davies 1995; Francis and Halford 2006). Moreover, cell wall components can act as polarization markers. For example, LIPID TRANSFER PROTEIN (AtLTP1) is required at the protein level for cuticle formation. It is expressed in the globular stage of embryogenesis, at the level of the protoderm, but thereafter, its expression is more polar being confined to the developing hypocotyl and cotyledons but is excluded from the pole of the root where cuticle is neither formed nor required (Fig. 8) (Vroemen et al. 1996). Importantly, *ROOT-SHOOT-HYPOCOTYL-DEFECTIVE* (*RSH*), which encodes a cell wall hydroxyproline-rich glycoprotein, and *KNOLLE*, which encodes syntaxin, a specific cytokinesis protein, are necessary for the correct positioning of the cell plate during cytokinesis (Lauber et al. 1997; Hall and Cannon 2002). This is essential for defining the first asymmetric division of the zygote and subsequent normal embryo development. Indeed, *knolle* mutant embryos show highly irregular shapes comprising binucleate cells along with unusually shaped cells (Waizenegger et al. 2000).

An important role in both early embryogenesis and postembryonal root development that involves positional signaling is played by hormones. Polar auxin transport through specific alignment of plasma membrane-based influx and efflux carriers plays a major role in providing positional cues for developmental processes; localized auxin biosynthesis also contributes to create hormone gradients

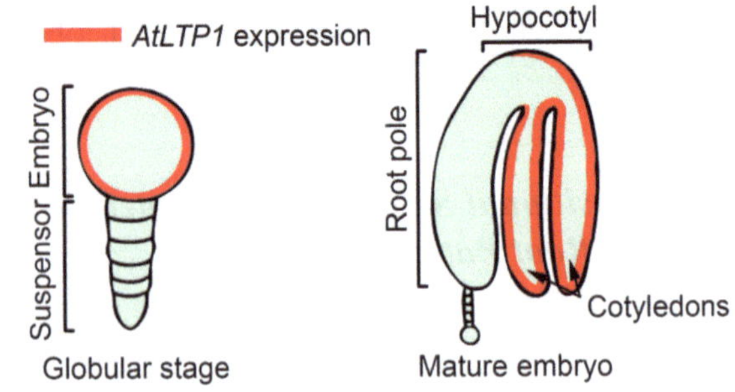

Fig. 8 Schematic of polarized accumulation of *AtLTP1* transcripts in the *Arabidopsis thaliana* embryo at globular and bent-cotyledon stages. Adapted from Altamura et al. (2007)

(Friml 2003; Ljung et al. 2005; Grieneisen et al. 2007; Benjamins and Scheres 2008; Stepanova et al. 2008; Ikeda et al. 2009; Petràsek and Friml 2009). Moreover, recent work has highlighted an antagonistic and, at times, transient interaction between auxin and cytokinins in determining positional information for specifying cell fate and organ pattern (Dello Ioio et al. 2007, 2008; Muller and Sheen 2008; Perilli et al. 2009).

4.3.1 RAM Establishment

Early during embryogenesis, the polarization of different auxin carriers determines the formation of auxin gradients and their directional flow (Benková et al. 2003; Friml et al. 2003; Petrášek and Friml 2009; Zazímalová et al. 2010). In particular, auxin transport is under the control of influx AUXIN RESISTANT1/LIKE AUX1 (AUX1/LAX) symporters, efflux PINFORMED (PIN) transporters, and ABCB/P-GLYPROTEIN (PGP) efflux/conditional transporters (Vieten et al. 2007; Petrášek and Friml 2009). Notably, the transcription of all these carriers is influenced by auxin itself, an action that is only a component of the fine-tuning mechanism by which auxin acts as one of the key regulators of its own transport (Vieten et al. 2007; Petrášek and Friml 2009). A coordinate action of different efflux carriers is essential in embryo patterning with a major role for PIN-dependent auxin transport. At the proembryo bicellular stage, *PIN* genes are already expressed and a non-polarized distribution of PIN1 protein is detected in the apical cell, while PIN7 localizes along the upper plasmalemma of the basal cell upper side, thus driving auxin flux toward the proembryo (Fig. 9a). At the eight cell stage, PIN7 is localized along the plasmalemma of the upper side of the hypophysis, the founder cell of the root stem-cell niche, and suspensor contacting cell (Fig. 9b). Thus, the auxin flux continues to be driven toward the developing embryo. At the globular stage, PIN protein localization changes dramatically. PIN1 relocalizes to the membrane of the lower side of the basal cells, while PIN7 shifts to the membrane of the lower side of the hypophysis and suspensor contacting cell (Fig. 9c). In such a way, auxin accumulates strongly at the base of the embryo (hypophysis) (Friml et al. 2003). As mentioned above, positional information is also related to a polarized repression of cytokinin signaling. In particular, early cytokinin activity is detectable in the hypophysis and maintained in its apical, lens-shaped derivative cell, which will form the QC. Conversely, cytokinin output signal is suppressed in the basal cell lineage through an auxin-mediated transcriptional activation of *A*-type *ARABIDOPSIS RESPONSE REGULATOR 7* and *15* (type-*A ARR7* and *ARR15*), which encode two negative regulators of cytokinin signaling (Fig. 9d) (Muller and Scheen 2008; To and Kieber 2007). Consistent with this, *arr7* and *arr5* mutants are unable to specify a stem-cell niche required for the establishment of a normal embryogenic root (Muller and Sheen 2008).

As embryogenesis proceeds, auxin preferentially accumulates in the developing root and at the torpedo stage, when the QC is already established; an auxin maximum is achieved in columella initial cells below it (Fig. 9e, f). *pin1* and

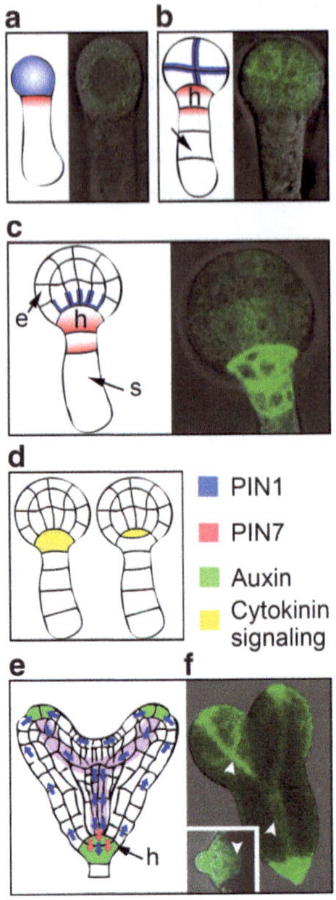

Fig. 9 (**a**, **b**, **c**, **e**) Schematic of the localization of PIN1 and PIN7 auxin carriers and whole-mount visualization of auxin accumulation through DR5rev::GFP expression during embryo development in *Arabidopsis thaliana*. (**d**) Schematic of cytokinin signaling in *Arabidopsis* globular embryo. (**a**) bicellular stage; (**b**) octant stage; (**c**, **d**) globular stage; (**e**) heart stage; (**f**) torpedo stage; *h* hypophysis; *e* embryo; *s* suspensor. Adapted from Altamura et al. (2007), Friml et al. (2003), and Petrášek and Friml (2009)

pin7 mutants fail to develop a polar longitudinal axis, thus supporting the key role for PIN-dependent auxin transport in the morphogenetic process (Friml et al. 2003). In this context, we may recall that subcellular trafficking and targeting of auxin transporters to direct auxin flow, as well as transport activity, relies on several cellular and molecular mechanisms, which involve posttranslational modifications of transporters themselves, membrane composition, endocytosis, and endosomal sorting/recycling (rewied by Petràsek and Friml 2009, Zazimalova et al. 2010). Consistent with this is the similarity between *pin* and *gn* mutants, which are unable to organize a longitudinal axis, a deleterious phenotype that begins with the first division of the zygote (Mayer et al.1993). *GNOME* (*GN*) encodes an ARF-GEF (ADP Rybosylation Factor-Guanilic Exchange Factor) protein involved in vesicular trafficking from the trans-Golgi (TGN Trans Golgi Network) to the plasmalemma. Moreover, the *gn* phenotype is linked to nonpolar accumulation of high auxin concentrations due to a wrong localization of PIN1 protein in embryo cells during development (Steinmann et al. 1999; Richter et al. 2010). Confirming that vesicular trafficking is integral to the polar distribution of auxin carriers, these observations indicate how *PIN* and *GN* genes operate on the same genetic pathway.

Notably, three relevant genes for embryo pattern such as *MONOPTEROS* (*MP*), *BODENLOS* (*BDL*), and *AUXINRESISTANCE6* (*AXR6*) are either under the control of, or are responsive to, high auxin levels (Hardtke and Berleth 1998; Hamann et al. 2002; Hellmann et al. 2003; Weijers et al. 2005). Indeed, *MP* encodes a transcription factor belonging to the AUXIN RESPONSIVE FACTORS (ARF) family, which interacts with specific DNA sequences, named Auxin Responsive Elements (AuxRE) (Fig. 10). *BDL* encodes an AUXIN/INDOLE-3-ACETIC ACID (Aux/IAA) protein, which is able to bind ARF transcription factors, such as MP, forming an inactive complex. AXR6 protein is a component of a ubiquitin-ligase complex involved in the degradation of BDL protein. Moreover, high auxin levels on the one hand promote the interaction of MP-AuxRE but on the other inhibits BDL-MP binding (Fig. 10) According to this network, in both loss-of-function *mp* mutants, gain-of-function *bdl* mutants and loss-of-function *axr6* mutants, target AuxRE sequences are not transcribed despite the presence of auxin. This is consistent with the similar phenotypes of single mutants defective in the above genes; they all fail to exhibit an embryonic root and all have a reduced vascular system (Hardtke and Berleth 1998; Hamman et al. 1999; Hobbie et al. 2000).

Possible gene targets of MP and BDL activity belong to the *WUSCHEL-RELATED HOMEOBOX* (*WOX*) gene family, which includes 14 members. Among these, four are differentially expressed during embryogenesis both spatially and temporally (Fig. 11) (Haecker et al. 2004). On the basis of their expression pattern, and as well as by the *wox* phenotypes, specific roles have been postulated for each member. In particular, at the early globular stage, a role in the specification of stem-cells, which will form the QC and embryo root, has been defined for WOX

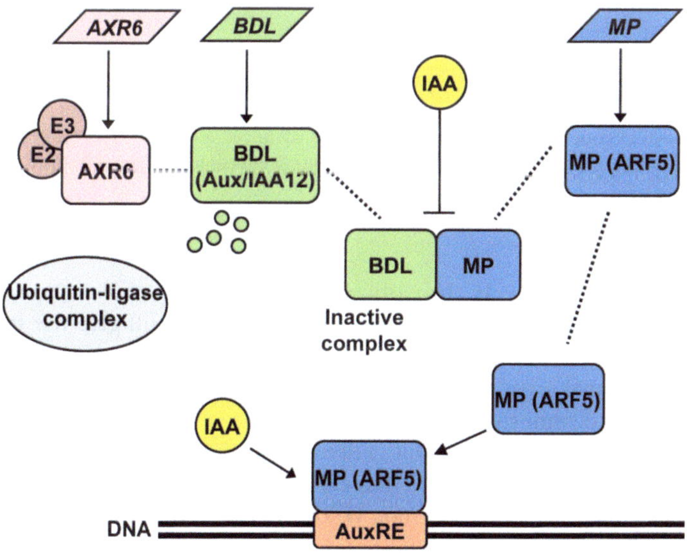

Fig. 10 Scheme depicting the interaction between *MP*, *BDL*, and *AXR6* gene activities in the context of auxin signaling (see text for further details). Adapted from Altamura et al. (2007)

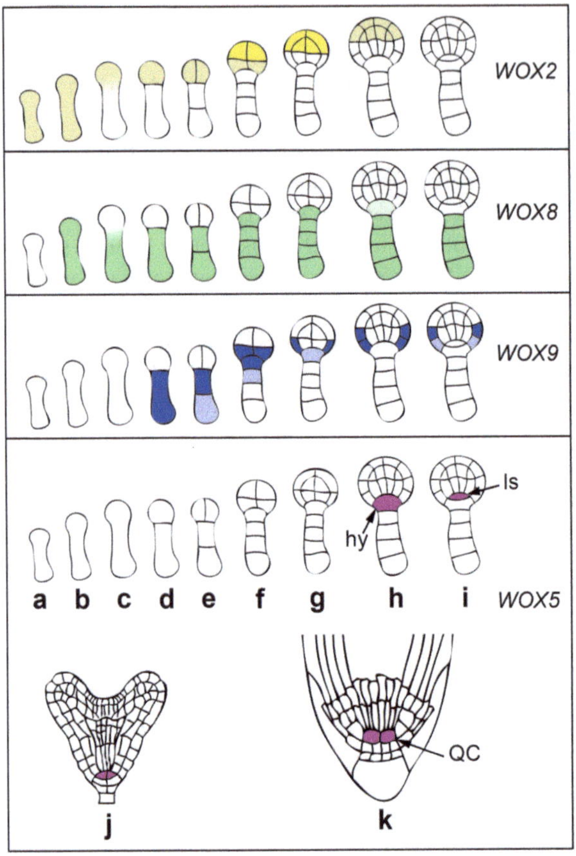

Fig. 11 Schematic of spatial and temporal expression pattern of (**a–j**) *WOX* genes during early embryogenesis and (**j–k**) of *WO5* gene at embryo heart stage and adult root of *Arabidopsis thaliana*. (**a–c**) zygotic cell; (**d–i**) from bicellular to advanced globular stage; (**j**) heart stage; (**k**) adult roots; *hy* hypophysis; *ls* lens-shaped cell; *QC* quiescent centre. Adapted from Altamura et al. (2007)

5 and WOX 9, respectively. In particular, at the early globular stage, WOX 5 expression first occurs in the embryonic cell lineage that will give rise to the QC. After the division of the hypophysis, *WOX5* transcripts are present in the upper lens-shaped cell that gives rise to the QC, while they are absent in the lower cell that gives rise to the central root cap (Fig. 11h, i). Subsequently, at the heart and bent cotyledon stages, *WOX5* expression is detectable in the direct descendants of the lens-shaped cell represented by the four cells of the QC (Fig. 11j, k). As discussed in further detail later, in the adult root, *WOX5* is involved in maintaining cell-stem state in a noncell-autonomous manner (Haecker et al. 2004; Sarkar et al. 2007; Breuninger et al. 2008). Evidence is available that the BDL/MP-mediated auxin signaling is required for *WOX5* and *WOX9* expression. For example, *WOX5*

expression was rarely detected in either *mp* or *bdl* mutants and *WOX9* expression was affected in *mp bdl* double mutants (Haecker et al. 2004; Sarkar et al. 2007).

4.3.2 RAM Maintenance

During postembryonic development, root distal pattern is related to auxin distribution and its antagonistic interplay with cytokinins. In growing roots, auxin polar transport and local biosynthesis act together to produce auxin gradients and maxima at the root tip, with its highest concentration in the QC and likely columella initials (Benkova et al. 2003; Blilou et al. 2005; Grieneisen 2007; Ikeda et al. 2009; Petersson et al. 2009). Several members of PIN gene family are differentially expressed in the root, and each transporter exhibits both a tissue-specific and cell-polarized localization, which, in effect, configures the root pole as a sink for auxin (Benková et al. 2003; Friml et al. 2003; Blilou et al. 2005; Petrášek and Friml 2009).

Just like during embryogenesis, the auxin gradient modulates the activity of the auxin-inducible *PLETHORA 1* and *2* (*PLT1* and *2*) genes (Fig. 12a, b), which encode AP2 class transcription factors that have an essential role in the specification of stem-cells. In particular, stem-cell fate is promoted by high *PLT* expression,

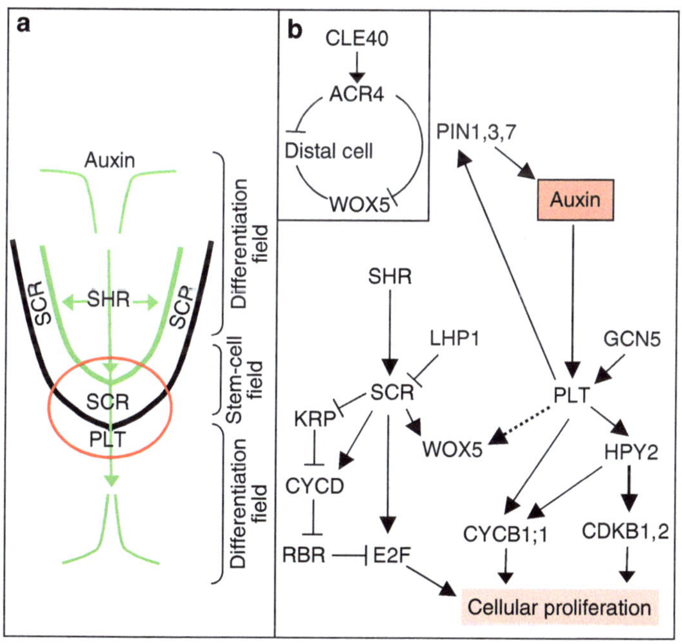

Fig. 12 Scheme depicting (**a**) auxin polar flux (adapted from Vernoux and Benfey 2005) and (**b**) its interaction with a transcriptional network in root patterning (see text for further details). *Arrows* indicate positive regulation; *barred lines* indicate negative regulation

whereas lower and lowest *PLT* expression promotes daughter cell proliferation and differentiation, respectively (Aida et al. 2004; Blilou et al. 2005; Galinha et al. 2007). Through a feedback mechanism, PLT proteins promote *PIN* expression, assuring a steady-state high level of auxin and therefore maintaining the stem-cell niche. A histone acetyltransferase activity, General Control Nonrepressed protein5 (GCN5), is also necessary to promote *PLT2* expression (Fig. 12b) (Kornet and Scheres 2009). Accordingly, the *gcn5* mutants exhibit a short root meristem and fail to maintain a QC, but their phenotype can be reversed by *PLT2* overexpression. Plant specific Cyclin-dependent kinases CDKB1 and 2 and CYCLIN B1;1 (CYCB1), which act at the G2-to-M phase transition of cell cycle, have been suggested as putative target genes of PLT2 (Fig. 12b) (Aida et al. 2004; Ishida et al. 2009; Stahl and Simon 2010). It is known that CYCs and CDKs are the major drivers of the mitotic cell cycle, and a reduction in CDK activity is a primary feature of cells that enter the endocycle, in which cells replicate their DNA without undergoing cell division (Larkins et al. 2001). The transition from the mitotic cell cycle into the endocycle is often coupled with the switch from cell proliferation to cell differentiation, and therefore, it is tightly integrated into plant developmental programs (Francis 2007). Interestingly, *PLT1* and *2* induce the expression and/or accumulation of HIGH PLOIDY2 (HPY2), a nuclear-localized SMALL UBIQUITIN-RELATED MODIFIER E3-ligase (SUMO E3-ligase), which is expressed in proliferating cells of the RAM and is a negative regulator of endoreduplication (Ishida et al. 2009). CDKB1 and 2 and CYCB1 levels were reduced in *hpy2-1* mutants suggesting that they are among the targets of HPY2 for sumoylation in order to modulate their activities (Ishida et al. 2009). Consequently, a picture emerges in which auxin maintains root meristem homeostasis through *PLT1/2* expression and high *HPY2* expression, which prevents endoreduplication and promotes cell proliferation (Ishida et al. 2009).

The activity of plant-specific GRAS family of transcription factors such as SCARECROW (SCR) and SHORTROOT (SHR) is also essential for root meristem patterning (Pysh et al. 1999; Nakajima et al. 2001, Sabatini et al. 2003). GRAS factors are required for correct function of the stem niche: *SHR* is expressed in the provascular cells of the stele and its encoded protein moves to adjacent QC cells and the endodermal initial cell layer (Fig. 12a) where it interacts with SCR through its central binding domain and is sequestered in the nucleus thereby preventing further SHR movement (Nakajima et al. 2001; Cui et al 2007). The interaction results in the activation of target genes, including *SCR* itself (Fig. 12b), which is required for the specification of QC identity and stem-cell homeostasis as well as for the promotion of asymmetric cell division, which underlies tissue specification in the E/C initials surrounding the QC (Di Laurenzio et al. 1996; Sabatini et al. 2003).

In this context, we must underline a trait that is distinctive of the *Arabidopsis* root compared to other plant species, that is, the absence of further asymmetric cell division after endodermis specification and the consequent maintenance of a two-layer ground tissue (i.e., endodermis and cortex) for a long period after germination (2 weeks) in spite of the presence of the SHR and SCR transcription factors

(Helariutta et al. 2000; Cui et al. 2007). However, more than 2 weeks after germination, additional divisions occur and a second or third layer of cortical cells form the middle cortex (Paquette and Benfey, 2005). Note that *scr* mutants are characterized by a premature middle cortex elevating the role of SCR in promoting the first but repressing additional asymmetric division in the endodermal cell layer. Posttranslational modification as well as interactions with other proteins could account for SCR's dual activities. Recently, a number of putative SCR-interacting proteins have been identified, which specifically interact with its N-terminal domain. Unlike the C-terminal domain, which interacts only with SHR thus activating asymmetric cell division, the N-terminal region was found to be a versatile interaction domain necessary to repress further cell divisions. The transcriptional repressor LIKE HETEROCHROMATIN PROTEIN 1 (LHP1), highly expressed in the root elongation zone, was included among the SCR interacting proteins (Cui and Benfey 2009). In such a way, SCR could either activate or inhibit gene expression depending on the interacting partner (Fig. 13). Notably, both LHP1 and SCR bind to the promoter of MAGPIE (MGP), a zinc finger protein previously confirmed as an SCR target, which has a role in ground tissue patterning by restricting SHR action (Welch et al. 2007). Moreover, *MGP* and *SCR* expression are enhanced in the *lhp1* mutant, which exhibits a premature middle cortex. This led to the conclusion that LHP1 plays a role in cortex formation by acting together with SCR in preventing further asymmetric cell divisions (Fig. 12b). (Cui and Benfey 2009; Welch et al. 2007). On the basis of the above-described genetic network, the stem-cell niche can be identified as the domain where the highest expression levels of *PLT1*, *SHR*, and *SCR* overlap (Fig. 12a) (Sabatini et al. 2003; Aida et al. 2004; Nawy et al. 2005; Galinha et al. 2007). Moreover, *SCR* acts cell autonomously to

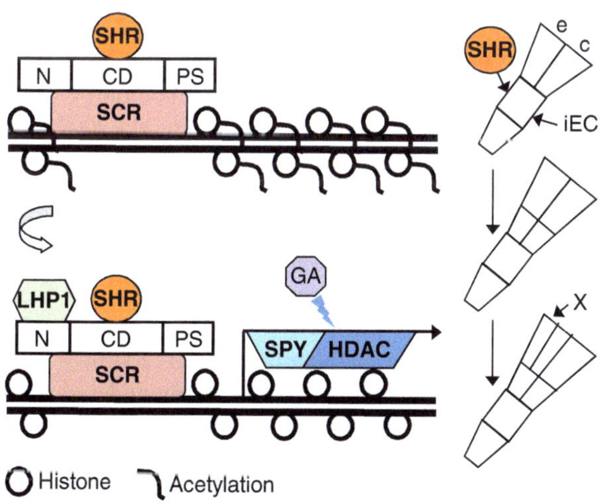

Fig. 13 Scheme depicting the interaction of LHP1, SPY, and GA signaling in a common epigenetic regulation of SCR activity in cortex formation (see text for further details). *c* cortex; *iEC* endodermis/cortex initial; *e* endodermis. Adapted from Cui and Benfey (2009)

maintain QC cell identity, whereas a nonautonomous signal from the QC maintains adjacent stem-cell division as demonstrated through cell ablasion experiments. Indeed, the ablation of one QC cell induces a precocious differentiation of columella initials in surface contact with the ablated cell (Van den Berg et al. 1997). Besides featuring the QC as the root OC, this genetic network is consistent with the reestablishment of a new functional QC from proximal initials after cell ablation: initials undergo division and auxin accumulates in these new cells. This in turn induces the expression of *PLT* genes, which activate *SCR*, thus respecifying QC identity.

Although *PLT* and *SHR–SCR* pathways are essential for the maintenance of the stem-cell population, loss-of-function of each these genes differentially affects it. Presumably, all these transcription factors do not control the same target genes with the possible exception of *WOX5*. Expression of *WOX* is confined to the QC (Fig. 11k) and plays a role in maintaining the stem-cell state and, most likely, controls QC-specific gene expression (Haecker et al. 2004; Sarkar et al. 2007.) Consistent with such a key role, *wox5-1* mutants exhibit precocious differentiation of columella cells, whereas the proximal meristem remains unaffected; ubiquitous *WOX5* activation maintains an undifferentiated state (Sarkar et al. 2007). However, *WOX5* expression is differentially modulated by the above-mentioned pathways. In particular, *WOX5* transcripts are undetectable or reduced in *shr* and *scr* mutants. Conversely, *PLT1* and *PLT2* play only a minor role since in *plt1*, *plt2* double mutant, while *WOX5* expression can be occasionally expanded, but *PLT1* expression is normal in *wox5-1*.

Note the unidirectionality of *WOX5* signaling, which is only required for columella stem-cell maintenance that resembles the unidirectional control played by *WUS* on the stem-cell niche in the SAM (Sarkar et al. 2007). Moreover, *WUS* and *WOX* genes exhibit comparable patterns of spatial expression being confined to the organizing centers of the SAM and RAM, respectively (Mayer et al. 1998; Sarkar et al. 2007). All these features, together with the high degree of sequence similarity between the two genes, have been interpreted as a homology in the homeostasis mechanisms operating in the SAM and RAM (Sablowski 2007). That both genes can functionally replace each other offers further support of such an interpretation (Gallois et al. 2004; Sarkar et al. 2007). This is consistent with the hypothesis that the RAM has evolved from the SAM as an adaptive strategy for water and nutrient uptake and soil anchorage, following water emersion (Jiang and Feldman 2005). Based on recent evidence showing that gymnosperms exhibit only one *WUS*/*WOX5* proortholog, separation of *WUS* and *WOX* genes has occurred relatively recently during the evolution of angiosperms (Nardmann et al. 2009).

As in the SAM, RAM totipotent cells progressively undergo a determinate fate. As a consequence, SAM/RAM homeostasis relies on a balance between the number of cells that are maintained in the stem-cell state and the number of TA cells, progressively displaced toward the differentiation zone. In the SAM, a feedback loop is established between *WUS* and members of the *CLAVATA* gene family that dynamically confines the organizing center into a defined number of cells (Brand et al. 2000). In particular, *WUS* is expressed in the organizing center keeping stem-cells undifferentiated and inducing the expression of *CLV3* that encodes a small ligand polypeptide for the *CLV1* encoded receptor-like kinase that, in turn, assists in maintaining the size of the *WUS*-expressing region (Mayer et al. 1998; Brand et al. 2000; Schoof et al. 2000). The closest homologs of *CLV3* that are expressed in root systems are members of *CLV3/ENDOSPERM SURROUNDING REGION* (*CLE*) family. Some of the *CLE* genes act to reduce root meristem size (e. g., *CLV3, CLE19, CLE40*) whilst others promote cell proliferation in the stele (*CLE41*) (Hobe et al. 2003; Fiers et al. 2006). More recently, a clearer role has been assigned to *CLE40* in controlling cell proliferation in the distal root meristem. It was proposed that CLE40 protein is secreted from columella cells into the QC and represses *WOX5* expression therein (Fig. 12b). Seemingly, this effect is brought about through CLE40's putative receptor, ARABIDOPSIS CRINKLY4 (ACR4), which is mainly expressed in the distal meristem and locally restricts cell division activity interfering with columella stem-cell maintenance (De Smet et al. 2008). Accordingly, *cle40* and *acr4* mutants exhibit opposite phenotypes (De Smet et al. 2008; Stahl and Simon 2009; Stahl et al. 2009). Hence, in the RAM, a CLE40/WOX5 pathway parallels the CLV3/WUS pathway in the SAM. However, a clear difference exists in that the feedback signal from differentiated descendant cells control stem-cell proliferation in the root, while in the shoot, signaling is derived within the stem-cell area (Brand et al. 2000; Schoof et al. 2000). Moreover, in the RAM but not in the SAM, this role is exerted through a cell-to-cell signal analogous, perhaps to cell to cell signaling in stem cell niches in animals (e.g., Spradling et al. 2001).

5 Hormonal Circuitry in Determining RAM Size and Pattern

A distinctive feature of plant hormones is to act synergically or antagonistically in several morphogenetic events during plant development. Recent advances in resolving the cross-talk between different hormone classes in controlling root patterning are assessed here.

5.1 Auxin/Cytokinin Interplay

Root meristem size is determined by the rate of cell division of stem and TA cells in the division zone and by the rate of cell elongation prior to differentiation of cells in the differentiation zone. Clearly, there is widespread evidence for auxin in promoting and controlling cell division in the RAM. Auxin action is mediated by auxin receptors, such as TRANSPORT INHIBITOR RESPONSE1 (TIR1). This protein is a component of the E3 ubiquitin ligase complex, which targets the transcriptional repressors of the Aux/IAA (AUXIN/INDOLE-3-ACETIC ACID) family for proteasome-mediated degradation (Fig. 14) (Kepinski and Leyser 2005). In such

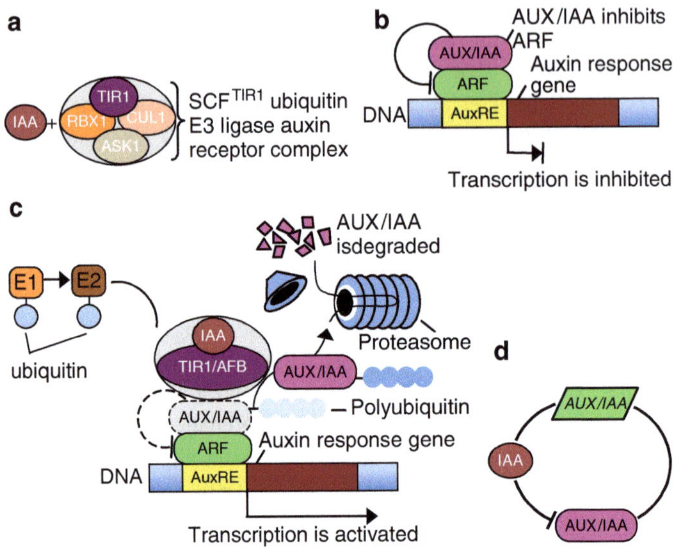

Fig. 14 Scheme depicting mechanisms of auxin on the coregulator-transcription factor duo AUX/IAA-ARF. Adapted from Taiz and Zeiger (2002)

a way, auxin response factors (ARFs), which are inhibited by Aux/IAA, are released to activate the transcription of auxin-responsive genes. A negative-feedback loop is active since auxin negatively regulates Aux/IAA abundance while their respective genes are themselves auxin-induced (Fig. 14) (Benjamins and Scheres 2008). Auxin action has also been linked to changes in cellular redox status (De Tullio et al. 2009; Eckardt 2010; Jiang et al. 2010), as well as alternative auxin-responsive pathways, and an even more complex signaling has been described (Benjamins and Scheres 2008; Lau et al. 2008; Strader et al. 2008). However, detailed reviewing of this mechanism is beyond the scope of the current chapter.

As recently demonstrated, from the earliest phases of embryogenesis, cytokinins act antagonistically to auxin in controlling root pattern through to postembryonic development (Muller and Sheen 2008; Moubayidin et al. 2009; Perilli et al 2009; Werner and Schmülling 2009). In particular, during embryogenesis, cytokinin/auxin interplay has an essential role for the specification of the first root stem-cell niche, which is marked by an auxin concentration maximum occurring in a single cell (i.e., hypophysis) (Fig. 9). As mentioned above (see Sect. 4.3.1), a high level of auxin at the level of the hypophysis activates the expression of two negative regulators of cytokinin signaling, thus suppressing cytokinin output (Muller and Sheen 2008).

As far as postembryonic root development is concerned, cytokinins promote cell differentiation at the boundaries between the division and elongation zones (i.e., the transition zone) by suppressing auxin signaling and transport, while auxin promotes cell division by suppressing cytokinin signaling (Blilou et al. 2005; Dello Ioio et al 2007, 2008; Perilli et al. 2009; Ruzicka et al. 2009). In particular, increased

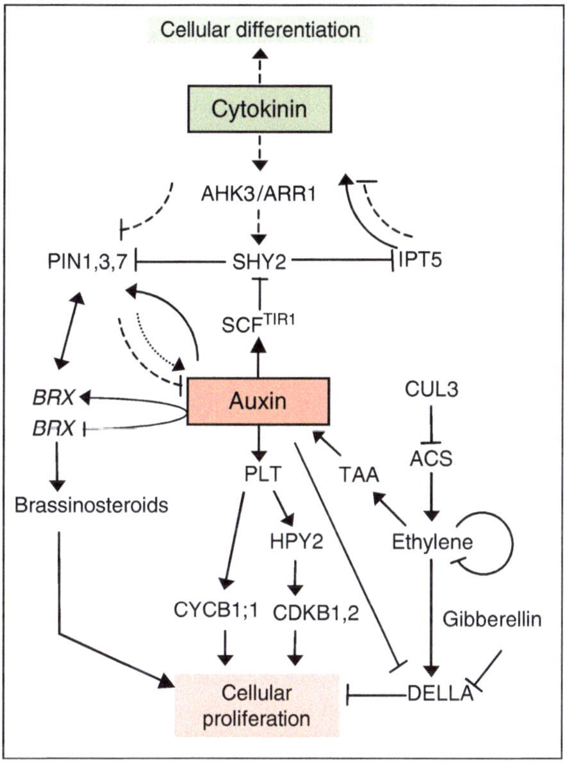

Fig. 15 Scheme depicting hormone interaction in transcriptional network underlying root patterning (see text for further details). *Arrows* indicate positive regulation; *barred lines* indicate negative regulation. Adapted from Stahl and Simon (2009)

cytokinin levels reduce root meristem size and inhibit root growth, by modulating PIN expression and therefore auxin distribution. This interplay relies on the convergence of both hormone classes on the same target gene, *SHORTHYPOCOTYL* (*SHY2*), which encodes an IAA-class repressor protein of the auxin signaling pathway (Fig. 15) (Benjamins and Scheres 2008). In particular, SHY2 prevents the activation of auxin-responsive genes by negatively regulating *PIN 1, 3,* and *7* and therefore repressing auxin export at the vascular transition zone (Dello Ioio et al. 2008). Note that auxin and cytokinins control SHY2 in opposite ways; auxin drives SHY2 protein degradation, through a ubiquitin-ligase complex (SCF^{TIR1}), thus stabilizing PIN expression level and auxin distribution. Conversely, cytokinins promote *SHY2* expression, via a two-component signaling pathway (AHK3/AA1), which activates ARR1, a primary type-B cytokinin response factor that binds to the *SHY2* promoter specifically at the vascular tissue transition zone (Fig. 15) (Benjamin and Scheres 2008; Dello Ioio et al. 2008; Moubayidin et al. 2009). Thus, in the presence of cytokinins, auxin transport is limited, auxin-dependent cell division is antagonized, and cell elongation occurs. Recently, it has been shown in maize that an ARR-mediated cytokinin signal is inactive in the QC and such repression could be related to the maintenance of the QC (Jiang et al. 2010). Finally, it must be underlined that a negative feedback control is active, since SHY2 protein downregulates *ISOPENTENYLTRANSFERASE* (*AtIPT*), which encodes an enzyme involved in a rate-limiting step of cytokinin biosynthesis (Fig. 15) (Sakakibara 2006; Dello Ioio et al. 2008). Note also that cytokinins antagonize auxin signaling and distribution during lateral root formation (Fukaki and Tasaka 2009). However, cytokinin-induced inhibition of root primordia development remains unaltered in auxin mutants, and auxin cannot reverse the cytokinin-induced inhibitory effect, suggesting that auxin and cytokinins likely control lateral root initiation through discrete pathways (Werner and Schmülling 2009)

5.2 Ethylene, Gibberellin, Abscisic Acid, and Brassinosteroids

In addition to auxin/cytokinins, an even more complex cross-talk between other hormone classes and transcription factors is active in the control of root pattern.

Ethylene and auxin interplay in maintaining RAM size. In particular, ethylene cooperates to inhibit cell elongation by stimulating localized auxin synthesis through TRYPTOPHAN AMINONOTRANFERASE (TAA1) activity; *TAA1* is expressed in the QC and is involved in the indol-3-pyruvic acid (IPA) branch of the auxin biosynthetic pathway (Stepanova et al. 2008). This relationship between local auxin synthesis and ethylene response is supported by phenotypic analyses of *taa1* and *tar2* mutants, which exhibit root-specific ethylene insensivity and a reduction in auxin concentration; *TRYPTOPHAN AMINONOTRANFERASE RELATED2* (*TAR2*) shows very close sequence homology to *TAA1*. Notably, both mutants exhibit an arrest of root growth due to precocious differentiation of meristematic cells leading to the complete loss of the stem-cell niche (Stepanova et al. 2008; Ortega-Martinez et al. 2007). On the other hand, ethylene production is modulated by the activity of CULLIN 3 (CUL3) proteins, members of a ubiquitin-ligase degrading complex, which targets 1-AMINOCYCLOPROPANE-1CARBOXYLATE SYNTHASE (ACS) for degradation. Normally, synthase (ACS) proteins are rate-limiting for ethylene biosynthesis. A negative feedback mechanism between ethylene signaling and response is also active (Fig. 15) (Chae and Kieber 2005; Thomann et al. 2009). In line with the role of CUL3 protein, the *cul3* mutant shows a reduced root meristem size and cell number, an effect due to a premature transition of cells to the elongation zone rather than to any alterations in cell division activity (Thomann et al. 2009). Notably, CUL3 proteins also control distal root patterning by modulating the division of the stem-cell niche and columella cap cells. Although this role should be consistent with the involvement of CUL3 proteins in ethylene synthesis and consequent increase of auxin level, an ethylene-independent role and a broader connection with auxin action has also been suggested for CUL3 proteins (Ortega-Martinez et al. 2007; Thomman et al. 2009).

The interplay between auxin and ethylene in root patterning is strengthened by their convergence on the activity of DELLA proteins, a subgroup of the GRAS protein family, which stimulate the production of cell cycle inhibitors thus limiting cell proliferation rate, although without affecting the stem-cell niche (Achard et al. 2003). In particular, ethylene delays the degradation of DELLA proteins, whereas auxin negatively modulates their stability (Achard et al. 2003; Fu and Harberd 2003). In the context of hormone circuitry, it is also worth noting that the degradation of DELLA proteins is controlled by another important hormone class, the Gibberellins (GAs) (Achard et al. 2009). This is consistent with a promotive effect of auxin and a repressive effect of ethylene on GA regulation of root elongation (Fig. 15).

The role of GA signaling in *Arabidopsis* root development has been largely documented in relation to cortex cell proliferation and therefore radial pattern. We described above (see Sect. 4.3.2) the role of SCR in promoting the first but repressing additional asymmetric division in the endodermal cell layer, thus assuring the maintenance of a two-layer ground tissue (Cui et al. 2007). This occurs through the activity of LHP1, a transcriptional repressor of *SCR* expression (see Sect. 4.3.2), which physically interacts with SCR itself (Fig. 13) (Welch et al. 2007; Cui and Benfey 2009). Note that the *scr* phenotype, which exhibits a premature middle cortex, is suppressed by GA treatment but enhanced through the inhibition of GA biosynthesis (Paquette and Benfey 2005). Notably, the *lhp1* phenotype is also rescued by GA treatment, suggesting that LHP1 and SCR share a common mechanism based on a major role of GA in repressing premature cortex proliferation (Sung et al. 2006; Cui and Benfey 2009). Moreover, a premature middle cortex root phenotype is exhibited by mutants deficient in key components of GA signaling (Cui and Benfey 2009). Surprisingly, a normal ground tissue phenotype has been detected in the root of both *sleepy* (*sly*) 1-10 and *sneezy* (*sne*) mutants in which DELLA proteins are not degraded ,and, therefore, a high steady-state level of GAs is believed to be present (McGinnis et al. 2003; Strader et al. 2004; Cui and Benfey 2009). Conversely, radial patterning defects were observed in *spindly* (*spy*) mutants in which GA signaling is not compromised (Swain et al. 2002; Cui and Benfey 2009). The possibility that *SPINDLY* (*SPY*), which encodes an O-GLNAc transferase, could act by postranscriptionally modulating SHR and SCR activities has been excluded based on experimental grounds (Cui and Benfey 2009). Note that in animal systems, a SPY homolog interacts with a histone deacetylases (HADC) (Yang et al. 2002). Since in wild-type roots a premature middle cortex phenotype has been also induced by inhibiting HADC activity, it was suggested that SPY could play a role in middle cortex formation through an epigenetic mechanism, perhaps negatively modulating cell division genes in the ground tissue (Cui and Benfey 2009). In the context of a more complex hormone circuitry, note also that a cross talk between GA and cytokinins has been suggested, which partially converge on SPY function (Greenboim-Wainberg et al. 2005). Indeed cytokinin signaling is affected in the *spy* mutant and GAs also repress the effects of cytokinins, by inhibiting induction of the cytokinin primary-response gene, type-A Arabidopsis response regulator 5 (ARF5). Thus, SPY acts as both a repressor of GA responses and a positive regulator of cytokinin signaling (Greenboim-Wainberg et al. 2005).

Cross-talk between auxin and brassinosteroids (BR) is also active in root patterning (Fig. 15) There is evidence that auxin strongly promotes the expression of *BREVIS RADIX* (*BRX*), which encodes a transcription factor that is rate-limiting for BR biosynthesis and auxin responsive gene expression. *brx* mutants are characterized by impaired root growth although they maintain a normal capacity to initiate lateral roots and respond normally to gravitropic stimulus. Moreover, in *brx* mutants, the expression of auxin-responsive genes is impaired, although they maintain an intact auxin signaling pathway (Mouchel et al. 2006; Scacchi et al. 2009). This is consistent with the BRX pathway lowering the level of constitutive repression of auxin-induced genes by impinging on the DNA-binding capacity of the repressive ARF2 (Vert et al. 2008). An autoregulatory feedback loop of *BRX* expression is also active in that BRX protein abundance is negatively regulated by auxin, likely being a target for auxin-induced proteasome degradation while BRX is itself an auxin-induced gene (Fig. 15). Recently, it has been also demonstrated that BRX protein colocalizes with PIN1 at the plasma membrane and moves to the nucleus in an auxin concentration- or flux-dependent manner (Mouchel et al. 2006; Scacchi et al. 2009). In spite of its reduced level in the nucleus, BRX plays a transcriptional modulatory role in cooperation with the B3 domain-type transcription factor NGATHA (NGA1), which in turn is related to ARFs. BRX protein and NGA 1 factors might represent a novel coregulator-transcription factor duo, analogous to AUX/IAA-ARF, which might act in conveying auxin efflux modulation into a different pattern of gene expression (Scacchi et al. 2009). The auxin response necessary for cytokinin-induced inhibiton of lateral root formation was hypothesized to be lost in the *brx* mutant, which if correct, would establish a link between BRX and cytokinin action (Li et al. 2009). Moreover, an interaction occurs between brassinosteroids, auxin, and abscisic acid (ABA) in root development. It is well known that ABA-related glucose signaling is involved in the control of RAM maintenance and development through pleiotropic effects (Brocard-Gifford et al. 2004; Shishkova et al. 2008). Recently, mutants affected in BR signaling also showed enhanced ABA sensitivity while transcriptomic analysis revealed that ABA negatively affects auxin signaling (Rodrigues et al. 2009).

6 Stem-Cell State and Chromatin Remodelers

Chromatin remodeling, accomplished through posttranslational histone modifications and changes in DNA methylation level, is an important mechanism for regulating gene activity in both animals and plants (Huck-Hui and Bird 1999; Riechman 2002; Suzuki and Bird 2008; Law and Jacobsen 2009, 2010). In animals, there is evidence that a stem-cell-specific state of chromatin is related to the repression of genes, which promote differentiation while simultaneously promoting stem-cells and proliferation factors (Boyer et al. 2006). Moreover, in murine embryonic stem-cells, repressive and promotive chromatin motifs have been identified within key development genes (Bernstein et al. 2006; Mikkelsen et al. 2007). This could be a mechanism for silencing differentiation genes while keeping them primed for activation. In mammalian stem-cells, a role has been demonstrated for the Polycomb group (PCg) proteins complexes PRC2 and PRC1 in gene silencing. Notably, PRC-complex-binding sites harbor trimethylated histone (H3K27met) (Azuara et al. 2006; Boyer et al. 2006; Schwaz et al. 2006; Schwarz and Pirrotta 2007).

Whether reversibility in chromatin accessibility is also related to stem-cell function in plants is of current debate. Despite the presence in plants of several PRC2 complexes with specific functions and the interaction of a PRC2 complex in repressing *SHOOTMERISTEMLESS* (*STM*) in the differentiated cells of the SAM, evidence for their putative role in a plant stem-cell niche is not yet available (Hsieh et al. 2003; Katz et al. 2004). However, there is evidence of a role for chromatin organization in plant stem-cell specification. For example, in the SAM, there is an involvement of plant–CHROMATIN-ASSEMBLY FACTOR-1 (CAF1), which is necessary for maintaining chromatin integrity, and SPLAYED (SYD), a chromatin remodeling ATPase belonging to a class of SFN2 transcriptional regulators (Kaya et al. 2001; Kwon et al. 2005). Further observations in RAMs deal with the activity of PICKLE (PKL), a CHD3 chromatin remodeling factor, and TOPLESS (TPL), a nuclear protein that bears some similarity to a transcriptional corepressor. PKL represses embryonic genes in germinating seedlings, while TPL seemingly participates in a chromatin-based stabilization of the embryo axis through the repression of root-promoting genes (Ogas et al. 1999; Long et al. 2006). Consistent with this role, *tpl* mutants were rescued by mutation of *HAG1*, which encodes a histone acetyltransferase (HAT), and this was enhanced by a mutation in *HAD19*, which encodes a histone deacetylase, acting in promoting and repressing target genes, respectively (Long et al. 2006). RNA-directed DNA methylation, which leads to chromatin modification and represents an additional mechanism of epigenetic regulation, has also been related to a correct development of the SAM (Kidner and Martienssen 2005).

Concerning the RAM, LHP1, which as mentioned above (see Sects. 4.3.2 and 4.2) has a role in ground tissue patterning, seemingly acts as a transcriptional repressor by modifying chromatin conformation (Sung et al. 2006; Turck et al. 2007; Welch et al. 2007). However, HLP1 seems not to change histone modification, and unlike the animal homolog, HP1, LHP1 targets both heterochromatin and euchromatin and binds to H3 histones that are trimethylated at lysine 27 (Turck et al. 2007). Moreover, the weakness of *lhp1* phenotypes compared to that of both *spy-3* and roots treated with HDAC inhibitor, suggested that LHP1 enhances the silencing effect of other mechanisms rather than directly causing gene silencing itself (Cui and Benfey 2009). So, it has been proposed that LHP1, SPY, and GA signaling act together in a common epigenetic mechanism involving histone deacetylation, which allows SCR to acts as a repressor of cell division gene activity (Fig. 13) (Cui and Benfey 2009). Notably and as described above (see Sect. 4.3.2), a histone acetyltransferase activity (GCN5) has also been found to act in the PLT pathway by modulating a PLT gradient (Kornet and Scheres 2009).

Additional evidence of epigenetic control of the stem-cell condition in plants is through the understanding of the RETINOBLASTOMA-related (RBR) protein, a master negative regulator of cell-cycle progression, which in animal systems interacts with differentiation promoting factors and chromatin remodeling proteins (Macaluso et al. 2006). In plants, reduced RBR activity induced additional division of columella daughter cells, which acquired stem-cell potential, thus expanding the stem-cell pool, while increased RBR expression promoted premature differentiation (Wildwater et al. 2005). Notably, RBR, which can be locally modulated by a root-specific RNAi, primarily affected cell differentiation rather than the cell cycle as seen by unaffected cell cycle progression in the expanding columella stem-cell population under reduced RBR activity (Wildwater et al. 2005). A model that provides robustness to the involvement of epigenetic regulation mechanisms has been proposed, in which KRYPTONYTE2 (KPR2), a histone H3 methyltransferase (Jackson et al. 2002), inhibits cyclinD/kinase, which inhibits RBR, which in turn inhibits cell cycle-promoting transcription factors (E2F), thus modulating the progression of a stem-cell to differentiation (Fig. 12b) (Wildwater et al. 2005). Recently, increased *RBR2;1* expression in the QC, together with *ARGONAUTE* (*AGO*)4-mediated siRNA events have been postulated to play a role in maize root repatterning (Jiang et al. 2010).

Finally, an overexpression of genes that play a role in chromosome organization and biogenesis, including genes involved in histones and DNA modifications, has been detected in the SAM stem-cell niche of *Arabidopsis*, through a high-resolution gene expression map. On this basis, it has been proposed that in plants, stem-cell chromatin is maintained in a flexible state in order to dynamically balance gene expression (Yadav et al. 2009).

7 Conclusions and Perspectives

Despite the huge quantity of data on transcriptional networks that underlie root developmental pattern and its interplay with hormone signaling, the spatiotemporal complexity and the dynamic framework of root transcriptional program are still far from being resolved. In the future, major understanding could be achieved by extending recent approaches, which have been refined to address and define expression profiles at the level of individual cell types (Birnbaum et al. 2003; Schmid et al. 2005; Levesque et al. 2006; Brady et al. 2007; Yadav et al. 2009), as well as by combining different tools. For example, genome-wide expression analysis has very high resolution and is efficient in identifying rare transcripts. Multiplex *in situ* hybridization allows one to simultaneously localize transcripts at the cellular level. Live-imaging has been developed to follow intracellular and spatial distribution of signaling molecules and regulatory factors, and therefore the dynamics of cell identity. ChIP/chip and ChIP/seq techniques enable identification of large numbers of transcription factors and target genes. Genome-wide insertional mutagenesis allows large gene functional characterization. Informatic elaboration enables the management of huge data sets and the elaboration of predictive computational models (Heisler et al. 2005; Brady et al. 2007; Kiefer 2009; Yadav et al. 2009).

With the development of such techniques, we appear to be on the cusp of major breakthroughs in understanding plant development. However, in the context of current knowledge, two major unanswered questions are, on the one hand, how transcriptional networks are temporally regulated during the progression of cells toward a particular developmental fate and, on the other, how such networks are modulated in relation to environmental stimuli and selection pressures. In particular, besides defining transcriptional changes, it will become increasingly important to identify how different genes interact in determining adaptive traits. Moreover, a related and poorly explored field deals with genetic networks and signaling that underlie stem-cell reprogramming within the context of differentiated tissues, which in plants frequently occurs both during intrinsic developmental programs (i.e., lateral root formation, activity of lateral meristems, axillary buds), as well as in relation to adventitious morphogenesis and in vitro regeneration processes (Chiappetta et al. 2006, 2009; Costa and Shaw 2006; Xu et al. 2006; Sena et al. 2009). Such knowledge should also be relevant for the development of refined plant biotechnology.

Finally, since root patterning is largely diversified within the plant kingdom, it will be interesting to extend our knowledge to different species and use natural variation to understand regulatory networks that control developmental processes. In particular, the transfer of this knowledge to crop plants should be of benefit in improving crop yield and quality and making them adapted to survive in our ever-changing global environment.

Acknowledgments We would like to thank Dennis Francis for fruitful discussions, helpful comments, and for critical reading of the manuscript. We also apologize to authors whose relevant work could not be cited owing to space constraints. Finally, M.B.B. is grateful to her Ph.D student Domenico Iaria for Photoshop drawings.

References

Achard P, Vriezen WH, Van Der Straeten D, Harberd NP (2003) Ethylene regulates *Arabidopsis* development via the modulation of DELLA protein growth repressor function. Plant Cell 15:2816–2825

Achard P, Gusti A, Cheminant S, Alioua M, Dhondt S, Coppens F, Beemster GT, Genschik P (2009) Gibberellin signaling controls cell proliferation rate in *Arabidopsis*. Curr Biol 19:1188–1193

Aida M, Beis D, Heidstra R, Willemsen V, Blilou I, Galinha C, Nussaume L, Noh Y, Amasino R, Scheres B (2004) The PLETHORA genes mediate patterning of the Arabidopsis root stem-cell niche. Cell 119:109–120

Altamura MM, Biondi S, Colombo L, Guzzo F (2007) Elementi di biologia dello sviluppo delle piante. EdiSES Publishers, Napoli, Italy

Armstrong JE, Heimsch C (1976) Ontogenetic reorganization of the root meristem in the *Compositae*. Am J Bot 63:212–219

Audus LJ (1960) Magnetotropism: a new plant growth response. Nature 185:132–134

Azuara V, Perry P, Sauer S, Spivakov M, Jørgensen HF, John RM, Gouti M, Casanova M, Warnes G, Merkenschlager M, Fisher AG (2006) Chromatin signatures of pluripotent cell lines. Nat Cell Biol 8:532–538

Barlow PW (1974) Regeneration of the cap of primary roots of *Zea mays*. New Phytol 73:937–954

Barlow PW (1978) The concept of the stem-cell in the context of plant growth and development. In: Lord BI, Potter CS, Cole RJ (eds) Stem-cells and tissue homeostasis. Cambridge University Press, Cambridge, pp 87–113

Barlow PW (1997) Stem-cells and founder zones in plants, particularly their roots. In: Potten CS (ed) Stem-cells. Academic, London, pp 29–57

Behrens HM, Weisenseel MH, Sievers A (1982) Rapid changes in the pattern of electric current around the root tip of *Lepidium Sativum* L. following gravistimulation. Plant Physiol 70:1079–1083

Benjamins R, Scheres B (2008) Auxin: the looping star in plant development. Ann Rev Plant Biol 59:443–465

Benková E, Michniewicz M, Sauer M, Teichmann T, Seifertová D, Jürgens G, Friml J (2003) Local, efflux-dependent auxin gradients as a common module for plant organ formation. Cell 115:591–596

Bernstein BE, Mikkelsen TS, Xie X, Kamal M, Huebert DJ, Cuff J, Fry B, Meissner A, Wernig M, Plath K, Jaenisch R, Wagschal A, Feil R, Schreiber SL, Lander ES (2006) A bivalent chromatin structure marks key developmental genes in embryonic stem-cells. Cell 125:315–326

Birnbaum K, Shasha DE, Wang JY, Jung JW, Lambert GM, Galbraith DW, Benfey PN (2003) A gene expression map of the *Arabidopsis* root. Science 302:1956–1960

Bitonti MB, Mazzuca S, Ting T, Innocenti AM (2006) Magnetic field affects meristem activity and cell differentiation in *Zea mays* roots. Plant Biosyst 140:87–93

Blilou I, Xu J, Wildwater M, Willemsen V, Paponov I, Friml J, Heidstra R, Aida M, Palme K, Scheres B (2005) The PIN auxin efflux facilitator network controls growth and patterning in *Arabidopsis* roots. Nature 433:39–44

Boyer LA, Plath K, Zeitlinger J, Brambrink T, Medeiros L, Lee TI, Levine SS, Wernig M, Tajonar A, Ray MK, Bell GW, Otte AP, Vidal M, Gifford DK, Young RA, Jaenisch R (2006) Polycomb complexes repress developmental regulators in murine embryonic stem-cells. Nature 441:349–353

Brady SM, Orlando DA, Lee JY, Wang JY, Koch J, Dinneny JR, Mace D, Ohler U, Benfey PN (2007) A high-resolution root spatiotemporal map reveals dominant expression patterns. Science 318:801–806

Brand U, Fletcher JC, Hobe M, Meyerowitz EM, Simon R (2000) Dependence of stem-cell fate in *Arabidopsis* on a feedback loop regulated by CLV3 activity. Science 289:617–619

Breuninger H, Rikirsch E, Hermann M, Ueda M, Laux T (2008) Differential expression of WOX genes mediates apical-basal axis formation in the *Arabidopsis* embryo. Dev Cell 14:867–876

Brocard-Gifford I, Linch TJ, Garcia ME, Malhotra B, Finkelstein RR (2004) The *Arabidopsis thaliana ABSCISIC-ACID-INSENSITIVE8* encodes a novel protein mediating abscisic acid and sugar response essential for growth. Plant Cell 16:406–421

Chae HS, Kieber J (2005) Eto Brute? The role of ACS turnover in regulating ethylene biosynthesis. Trends Plant Sci 10:291–296

Chiappetta A, Petrarulo M, Fambrini M, Pugliesi C, Dewitte W, Van Onckelen H, Innocenti AM, Bitonti MB (2006) Zeatin accumulation and misexpression of a class I knox gene are intimately linked in the epiphyllous response of the interspecific hybrid EMB-2 (*Helianthus annuus H. tuberosus*). Planta 223:917–931

Chiappetta A, Fambrini M, Petrarulo M, Rapparini F, Michelotti V, Bruno L, Greco M, Baraldi R, Salvini M, Pugliesi C, Bitonti MB (2009) Ectopic expression of LEAFY COTYLEDON1-LIKE gene and localized auxin accumulation mark embryogenic competence in epiphyllous plants of *Helianthus annuus* x *H. tuberosus*. Ann Bot 103:735–747

Clowes FLA (1954) The promeristem and the minimal constructional centre in grass root apices. New Phytol 53:108–116

Clowes FLA (1956) Nucleic acids in root apical meristems of *Zea*. New Phytol 55:29–34

Clowes FLA (1959) Reorganization of root apices after irradiation. Ann Bot 23:205–210

Clowes FLA (1961) Effect of β-radiation on meristems. Exp Cell Res 25:529–534

Clowes FLA (1975) The quiescent centre. In: Torrey JG, Clarkson DT (eds) The development and function of roots. Academie, London, pp 3–19

Clowes FLA (1976) Estimation of growth fractions in meristems of *Zea mays* L. Ann Bot 40:933–938

Clowes FLA (1981) The difference between open and closed meristems. Ann Bot 48:761–767

Clowes FAL (2000) Pattern in root meristem development in angiosperms. New Phytol 146:83–94

Clowes FAL, Stewart HE (1967) Recovery from dormancy in roots. New Phytol 66:115–123

Costa S, Shaw P (2006) Chromatin organization and cell fate switch respond to positional information in *Arabidopsis*. Nature 439:493–496

Cui H, Benfey PN (2009) Interplay between SCARECROW, GA and LIKE HETEROCHROMATIN PROTEIN 1 in ground tissue patterning in the *Arabidopsis* root. Plant J 58:1016–1027

Cui H, Levesque MP, Vernoux T, Jung JW, Paquette AJ, Gallagher KL, Wang JY, Blilou I, Scheres B, Benfey PN (2007) An evolutionarily conserved mechanism delimiting SHR movement defines a single layer of endodermis in plants. Science 316:421–425

Davies PJ (1995) Plant hormones physiology biochemistry and molecular biology. Kluwer, Dordrecht

De Smet I, Vassileva V, De Rybel B, Levesque MP, Grunewald W, Van Damme D, Van Noorden G, Naudts M, Van Isterdael G, De Clercq R (2008) Receptor-like kinase ACR4 restricts formative cell divisions in the *Arabidopsis* root. Science 322:594–597

De Tullio MC, Jiang K, Feldman LJ (2009) Redox regulation of root apical meristem organization: connecting root development to its environment. Plant Physiol Biochem. doi:10.1016/j.plaphy.2009.11.005

Dello Ioio R, Linhares FS, Scacchi E, Casamitjana-Martinez E, Heidstra R, Costantino P, Sabatini S (2007) Cytokinins determine *Arabidopsis* root-meristem size by controlling cell differentiation. Curr Biol 17:678–682

Dello Ioio R, Nakamura K, Moubayidin L, Perilli S, Taniguchi M, Morita MT, Aoyama T, Costantino P, Sabatini S (2008) A genetic framework for the control of cell division and differentiation in the root meristems. Science 322:1380–1384

Di Laurenzio L, Wysocka-Diller J, Malamy JE, Pysh L, Helariutta Y, Freshour G, Hahn MG, Feldmann KA, Benfey PN (1996) The SCARECROW gene regulates an asymmetric cell division that is essential for generating the radial organization of the *Arabidopsis* root. Cell 86:423–433

Dinneny JR, Benfey PN (2008) Plant stem-cell niches: standing the test of time. Cell 132:553–557

Dolan L, Janmaat K, Willemsen V, Linstead P, Poethig S, Roberts K, Scheres B (1993) Cellular organisation of the *Arabidopsis thaliana* root. Development 119:71–84

Doncheva S, Amenós M, Poschenrieder C, Barceló J (2005) Root cell patterning: a primary target for aluminium toxicity in maize. J Exp Bot 56:1213–1220

Eckardt NA (2010) Redox regulation of auxin signalling and plant development. Plant Cell 22:295. doi:10.1105/tpc.110.220212

Feldman LJ, Torrey GJ (1975) The quiescent center and primary vascular tissue patter formation in cultured roots *Zea*. Can J Bot 53:2796–2803

Fiers M, Golemiec E, van der Schors R, van der Geest L, Li KW, Stiekema WJ, Liu CM (2006) The CLAVATA3/ESR motif of CLAVATA3 is functionally independent from the nonconserved flanking sequences. Plant Physiol 141:1284–1292

Francis D (2007) The plant cell cycle – 15 years on. New Phytol 174:261–278

Francis D, Halford NG (2006) Nutrient sensing in plant meristems. Plant Mol Biol 60:981–993

Friml J (2003) Auxin transport shaping the plant. Curr Opin Plant Biol 6:7–12
Friml J, Vieten A, Sauer M, Weijers D, Schwarz H, Hamann T, Offringa R, Jürgens G (2003) Efflux-dependent auxin gradients establish the apical-basal axis of *Arabidopsis*. Nature 426:147–153
Fu XD, Harberd NP (2003) Auxin promotes Arabidopsis root growth by modulating gibberellin response. Nature 421:740–743
Fukaki H, Tasaka M (2009) Hormone interaction during lateral root formation. Plant Mol Biol 69:437–449
Fusconi A, Gnavi E, Trotta A, Berta G (1999) Apical meristems of tomato roots and their modifications induced by arbuscolar mychorrhizal and soliborne pathogenic fungi. New Phytol 142:505–516
Galinha C, Hofhuis H, Luijten M, Willemsen V, Blilou I, Heidstra R, Scheres B (2007) PLETHORA proteins as dose-dependent master regulatorsof *Arabidopsis* root development. Nature 449:1053–1057
Gallois JL, Nora FR, Mizukami Y, Sablowski R (2004) WUSCHEL induces shoot stem-cell activity and developmental plasticity in the root meristem. Genes Dev 18:375–380
Goodman AM, Ennos AR (1996) A comparative study of the response of the roots and shoots of sunflower and maize to mechanical stimulation. J Exp Bot 47:1499–1507
Greenboim-Wainberg Y, Maymon I, Borochov R, Alvarez J, Olszewski N, Ori N, Eshed Y, Weiss D (2005) Cross talk between gibberellin and cytokinin: the *Arabidopsis* GA response inhibitor SPINDLY plays a positive role in cytokinin signalling. Plant Cell 17:92–102
Grieneisen VA, Xu J, Maree AF, Hogeweg P, Scheres B (2007) Auxin transport is sufficient to generate a maximum and gradient guiding root growth. Nature 449:1008–1013
Guttenberg HV (1968) Der prïmare Bau der Angiospermenwurzel. Handbuch der Pflanzenanatomie. Band 8, Teil 5. Gebrüder Borntraeger, Berlin, Germany
Haecker A, Grob-Hardt R, Geiges B, Sarkar A, Breuninger H, Herrmann A, Laux T (2004) Expression dynamics of WOX genes mark cell fate decisions during early embryonic patterning in *Arabidopsis thaliana*. Development 131:657–668
Hall Q, Cannon MC (2002) The cell wall hydroxyproline-rich glycoprotein RSH is essential for normal embryo development in *Arabidopsis*. Plant Cell 14:1161–1172
Hamann T, Mayer U, Jürgens G (1999) The auxin-insensitive bodenlos mutation affects primary root formation and apical-basal patterning in the *Arabidopsis* embryo. Development 126:1387–1395
Hamann T, Benkova E, Bäurle I, Kientz M, Jürgens G (2002) The *Arabidopsis* BODENLOS gene encodes an auxin response protein inhibiting MONOPTEROS-mediated embryo patterning. Genes Dev 16:1610–1615
Hardtke CS, Berleth T (1998) The Arabidopsis gene MONOPTEROS encodes a transcription factor mediating embryo axis formation and vascular development. EMBO J 17:1405–1411
Heimsch C, Seago JL Jr (2008) Organization of the root apical meristem in angiosperms. Am J Bot 95:1–21
Heisler MG, Ohno C, Das P, Sieber P, Reddy GV, Long JA, Meyerowitz EM (2005) Patterns of auxin transport and gene expression during primordium development revealed by live imaging of the *Arabidopsis* inflorescence meristem. Curr Biol 15:1899–1911
Helariutta Y, Fukaki H, Wysocka-Diller J, Nakajima K, Jung J, Sena G, Hauser MT, Benfey PN (2000) The SHORT-ROOT gene controls radial patterning of the *Arabidopsis* root through radial signaling. Cell 101:555–567
Hellmann H, Hobbie L, Chapman A, Dharmasiri S, Dharmasiri N, Del Pozo C, Reinhardt D, Estelle M (2003) *Arabidopsis* AXR6 encodes CUL1 implicating SCF E3 ligases in auxin regulation of embryogenesis. EMBO J 22:3314–3325
Hobbie L, McGovern M, Hurwitz LR, Pierro A, Liu NY, Bandyopadhyay A, Estelle M (2000) The axr6 mutants of *Arabidopsis thaliana* define a gene involved in auxin responses and early development. Development 127:23–32
Hobe M, Muller R, Grunewald M, Brand U, Simon R (2003) Loss of CLE40, a protein functionally equivalent to the stem-cell restricting signal CLV3, enhances root waving in *Arabidopsis*. Dev Genes Evol 213:371–381

Hsieh TF, Hakim O, Ohad N, Fischer RL (2003) From flour to flower: how Polycomb group proteins influence multiple aspects of plant development. Trends Plant Sci 8:439–445

Huck-Hui N, Bird A (1999) DNA methylation and chromatin modification. Curr Opin Genet Dev 9:158–163

Ikeda Y, Men S, Fischer U, Stepanova AN, Alonso JM, Ljung K, Grebe M (2009) Local auxin biosynthesis modulates gradient-directed planar polarity in Arabidopsis. Nat Cell Biol 11:731–738

Innocenti AM, Bitonti MB, Arrigoni O, Liso R (1990) The size of quiescent centre in roots of *Allium cepa* L. grown with ascorbic acid. New Phytol 114:507–509

Ishida T, Fujiwara S, Miura K, Stacey N, Yoshimura M, Schneider K, Adachi S, Minamisawa K, Umeda M, Sugimoto K (2009) Sumo E3 ligase HIGH PLOIDY2 regulates endocycle onset and meristems maintenance in *Arabidopsis*. Plant Cell 2:2284–2297

Jackson D (2005) Transcription factor movement through plasmodesmata. In: Oparka K (ed), Plasmodesmata. Ann Plant Rev 18. Blackwell, Dundee, pp 113–130

Jackson JP, Lindroth AM, Cao X, Jacobsen SE (2002) Control of CpNpG DNA methylation by the KRYPTONITE histone H3 methyltransferase. Nature 416: 556–560

Jiang K, Feldman LJ (2005) Regulation of root apical meristem development. Annu Rev Cell Dev Biol 21:485–509

Jiang K, Zhu T, Diao Z, Huang H, Feldman LJ (2010) The maize root stem cell niche: a partnership between two sister cell populations. Planta 231:411–424

Katz A, Oliva M, Mosquna A, Hakim O, Ohad N (2004) FIE and CURLY LEAF polycomb proteins interact in the regulation of homeobox gene expression during sporophyte development. Plant J 37:707–719

Kaya H, Shibahara KI, Taoka KI, Iwabuchi M, Stillman B, Araki T (2001) FASCIATA genes for chromatin assembly factor-1 in *Arabidopsis* maintain the cellular organization of apical meristems. Cell 104:131–142

Kepinski S, Leyser O (2005) The *Arabidopsis* F-box protein TIR1 is an auxin receptor. Nature 435:446–451

Kidner CA, Martienssen R (2005) The developmental role of microRNA in plants. Curr Opin Plant Biol 8:38–44

Kiefer JC (2009) Primer and interviews: gene regulation in *Arabidopsis thaliana*. Dev Dyn 238:2249–2458

Kornet N, Scheres B (2009) Stem-cell factors in plants: chromatin connections. Cold Spring Harb Symp Quant Biol 73:235–242

Kropf DL, Bisgrove SR, Hable WE (1999) Establishing a growth axis in fucoid algae. Trends Plant Sci 4:490–494

Kwon CS, Chen C, Wagner D (2005) WUSCHEL is a primary target for transcriptional regulation by SPLAYED in dynamic control of stem-cell fate in *Arabidopsis*. Genes Dev 19:992–1003

Larkins BA, Dilkes BP, Dante RA, Coelho CM, Woo Y, Liu Y (2001) Investigating the hows and whys of DNA endoreduplication. J Exp Bot 52:183–192

Lau S, Jürgens G, De Smet I (2008) The evolving complexity of the auxin pathway. Plant Cell 20:1738–1746

Lauber MH, Waizenegger I, Steinmann T, Schwarz H, Mayer U, Hwang I, Lukowitz W, Jürgens G (1997) The *Arabidopsis* KNOLLE protein is a cytokinesis-specific syntaxin. J Cell Biol 139:1485–1493

Law JA, Jacobsen SE (2009) Dynamic DNA methylation. Science 323:1568–1569

Law JA, Jacobsen SE (2010) Establishing, maintaining and modifying DNA methylation patterns in plants and animals. Nat Rev Genet 11:204–220

Lee JY, Cui W (2009) Non-cell autonomous RNA trafficking and long-distance signaling. J Plant Biol 52:10–18

Levesque MP, Vernoux T, Busch W, Cui H, Wang JY, Blilou I, Hassan H, Nakajima K, Matsumoto N, Lohmann JU (2006) Whole-genome analysis of the SHORT-ROOT developmental pathway in *Arabidopsis*. Plant Biol 4:739–752

Li L, Xie T (2005) Stem-cell niche: structure and function. Annu Rev Cell Dev Biol 21:605–631
Li H, Cheng Y, Murphy A, Hagen G, Wilfoyle TJ (2009) Constitutive repression and activation of auxin signaling in Arabidopsis. Plant Physiol 149:1277–1288
Liso R, Innocenti AM, Bitonti MB, Arrigoni O (1988) Ascorbic acid-induced progression of quiescent centre cells from G1 to S phase. New Phytol 110:469–471
Ljung K, Hull AK, Celenza J, Yamada M, Estelle M, Normanly J, Sandberg G (2005) Sites and regulation of auxin biosynthesis in *Arabidopsis* roots. Plant Cell 17:1090–1104
Long JA, Ohno C, Smith ZR, Meyerowitz EM (2006) TOPLESS regulates apical embryonic fate in *Arabidopsis*. Science 312:1520–1523
Lucas WJ, Lee JY (2004) Plasmodesmata as a supracellular control network in plants. Nat Rev Mol Cell Biol 5:712–726
Macaluso M, Montanari M, Giordano A (2006) A: Rb family proteins as modulators of gene expression and new aspects regarding the interaction with chromatin remodeling enzymes. Oncogene 25:5263–5267
Mayer U, Büttner G, Jürgens G (1993) Apical–basal pattern formation in the *Arabidopsis* embryo: studies on the role of the gnom gene. Development 117:149–162
Mayer KFX, Schoof H, Haecker A, Lenhard M, Jurgens G, Laux T (1998) Role of WUSCHEL in regulating stem-cell fate in the *Arabidopsis* shoot meristem. Cell 95:805–815
McCulloch EA, Till JE (2005) Perspectives on the properties of stem-cells. Nat Med 11:1026–1028
McGinnis KM, Thomas SG, Soule JD, Strader LC, Zale JM, Sun TP, Steber CM (2003) The *Arabidopsis* SLEEPY1 gene encodes a putative F-box subunit of an SCF E3 ubiquitin ligase. Plant Cell 15:1120–1130
Meyerowitz EM (2002) Plants compared to animals: the broadest comparative study of development. Science 295:1482–1485
Mikkelsen TS, Ku M, Jaffe DB, Issac B, Lieberman E, Giannoukos G, Alvarez P, Brockman W, Kim TK, Koche RP, Lee W, Mendenhall E, O'Donovan A, Presser A, Russ C, Xie X, Meissner A, Wernig M, Jaenisch R, Nusbaum C, Lander ES, Bernstein BE (2007) Genome-wide maps of chromatin state in pluripotent and lineage-committed cells. Nature 448:553–560
Morrison SJ, Spradling AC (2008) Stem-cells and niches: mechanisms that promote stem-cell maintenance throughout life. Cell 132:598–611
Moubayidin L, Di Mambro R, Sabatini S (2009) Cytokinin-auxin crosstalk. Trends Plant Sci 14:557–562
Mouchel CF, Osmont KS, Hardtke CS (2006) BRX mediates feedback between brassinosteroid levels and auxin signaling in root growth. Nature 443:458–461
Muller B, Sheen J (2008) Cytokinin and auxin interaction in root stem-cell specification during early embryogenesis. Nature 453:1094–1097
Nakajima K, Sena G, Nawy T, Benfey P (2001) Intercellular movement of the putative transcription factor SHR in root patterning. Nature 413:307–311
Nardmann J, Reisewitz P, Werr W (2009) Discrete shoot and root stem-cell promoting WUS/WOX5 functions are an evolutionary innovation of angiosperms. Mol Biol Evol 26:1745–1755
Nawy T, Lee JY, Colinas J, Wang JY, Thongrod SC, Malamy JE, Birbaum K, Benfey PN (2005) Transcriptional profile of the Arabidopsis root quiescent center. Plant Cell 17:1908–1925
Ogas J, Kaufmann S, Henderson J, Somerville C (1999) PICKLE is a CHD3 chromatin-remodeling factor that regulates the transition from embryonic to vegetative development in *Arabidopsis*. Proc Natl Acad Sci USA 96:13839–13844
Ohlstein B, Kai T, Decotto E, Spradling A (2004) The stem-cell niche: theme and variations. Curr Opin Cell Biol 16:693–699
Ortega-Martinez O, Pernas M, Carol RJ, Dolan L (2007) Ethylene modulates stem-cell division in the *Arabidopsis thaliana* root. Science 317:507–510
Paquette AJ, Benfey PN (2005) Maturation of the ground tissue of the root is regulated by gibberellin and SCARECROW and requires SHORT-ROOT. Plant Physiol 138:636–640
Perilli S, Moubayidin L, Sabatini S (2009) The molecular basis of cytokinin function. Curr Opin Plant Biol 13:21–26

Petersson SV, Johansson AI, Kowalczyk M, Makoveychuk A, Wang JY, Moritz T, Grebe M, Benfey PN, Sandberg G, Ljung K (2009) An auxin gradient and maximum in the *Arabidopsis* root apex shown by high-resolution cell-specific analysis of IAA distribution and synthesis. Plant Cell 21:1659–1668

Petrášek J, Friml J (2009) Auxin transport routes in plant development. Development 136: 2675–2688

Potters G, Pasternak TP, Guisez Y, Jansen MA (2009) Different stresses, similar morphogenic responses: integrating a plethora of pathways. Plant cell Environ 32:158–169

Pysh LD, Wysocka-Diller JW, Camilleri C, Bouchez D, Benfey PN (1999) The GRAS gene family in *Arabidopsis*: sequence characterization and basic expression analysis of the SCARECROW-LIKE genes. Plant J 18:111–119

Qu LJ, Zhu YX (2006) Transcription factor families in *Arabidopsis*: major progress and outstanding issues for future research. Curr Opin Plant Biol 9:544–549

Richter S, Anders N, Wolters H, Beckmann H, Thomann A, Heinrich R, Schrader J, Singh MK, Geldner N, Mayer U, Jürgens G (2010) Role of the GNOM gene in *Arabidopsis* apical-basal patterning – from mutant phenotype to cellular mechanism of protein action. Eur J Cell Biol 89:138–144

Riechman JL (2002) Transcriptional regulation: a genomic overview. In: Somerville CR, Meyerowitz EM (eds) The Arabidopsis book. American Society of Plant Biologists, Rockville, MD

Robert HS, Friml J (2009) Auxin and other signals on the move in plants. Nat Chem Biol 5:325–332

Rodrigues A, Santiago J, Rubio S, Saez A, Osmont KS, Gadea J, Hardtke CS, Rodriguez PL (2009) The short-rooted fhenotype of the *brevis radix* mutant partly reflects root Abscisic Acid hypersensitivity[1[C][W][OA]]. Plant Physiol 149:1917–1928

Ruzicka K, Simaskova M, Duclercq J, Petrasek J, Zazimalova E, Simon S, Friml J, Van Montagu MCE, Benkova E (2009) Cytokinin regulates root meristem activity via modulation of the polar auxin transport. Proc Natl Acad Sci USA 106:4284–4289

Sabatini S, Heidstra R, Wildwater M, Scheres B (2003) SCARECROW is involved in positioning the stem-cell niche in the *Arabidopsis* root meristem. Genes Dev 17:354–358

Sablowski R (2007) The dynamic plant cell niches. Curr Opin Plant Biol 10:639–644

Sakakibara H (2006) CYTOKININS: activity biosynthesis, and translocation. Ann Rev Plant Biol 57:431–449

Sanchez-Calderon L, Lopez-bucio J, Chacon-Lopez A, Cruz-Ramirez A, Nieto-Jacobo F, Dubrrovsky JG (2005) Phosphate starvation induces a determinate developmental program in the roots of *Arabidopsis thaliana*. Plant Cell Physiol 46:174–184

Sarkar AK, Luijten M, Miyashima S, Lenhard M, Hashimoto T, Nakajima K, Scheres B, Heidstra R, Laux T (2007) Conserved factors regulate signaling in *Arabidopsis thaliana* shoot and root stem-cell organizers. Nature 446:811–814

Scacchi E, Osmont KS, Beuchat J, Salinas P, Navarrete-Gomez M, Trigueros M, Ferrandiz C, Hardtke CS (2009) Dynamic, auxin-responsive plasma membrane-to-nucleus movement of *Arabidopsis* BRX. Development 136:2059–2067

Scheres B (2001) Plant cell identity. The role of position and lineage. Plant Physiol 125:112–114

Scheres B (2007) Stem-cell niches: nursery rhymes across kingdoms. Nat Rev Mol Cell Biol 8:345–354

Schmid M, Davison TS, Henz SR, Pape UJ, Demar M, Vingron M, Schölkopf B, Weigel D, Lohmann JU (2005) A gene expression map of *Arabidopsis* thaliana development. Nat Genet 37:501–506

Schoof H, Lenhard M, Haecker A, Meyer KFX, Jürgens G, Laux T (2000) The stem-cell population of *Arabidopsis* shoot meristema is maintained by a regulatory loop between the CLAVATA and WUSCHEL genes. Cell 100:635–644

Schwartz YB, Pirrotta V (2007) Polycomb silencing mechanisms and the management of genomic programmes. Nat Rev Genet 8:9–22

Schwartz YB, Kahn TG, Nix DA, Li XY, Bourgon R, Biggin M, Pirrotta V (2006) Genome-wide analysis of Polycomb targets in *Drosophila melanogaster*. Nat Genet 38:700–705

Seago JL, Heimsch C (1969) Apical organization in roots of the *Convolvulaceae.* Am J Bot 56:131–138
Sena G, Wang X, Liu HY, Hofhuis H, Birnbaum KD (2009) Organ regeneration does not require a functional stem-cell niche in plants. Nature 457:1150–1153
Sharma P, Dubey RS (2007) involvement of oxidative stress and role of antioxidative defense system in growing rice seedlings exposed to toxic concentrations of aluminium. Plant Cell Rep 26:2027–2038
Shishkova S, Rost TL, Dubrovsky JG (2008) Determinate root growth and meristem maintenance in Angiosperms. Ann Bot 101:319–340
Spradling A, Drummond-Barbosa D, Kai T (2001) Stem-cells find their niche. Nature 414: 98–104
Stahl Y, Simon R (2009) Is the Arabidopsis root niche protected by sequestration of the CLE40 signal by its putative receptor ACR4. Plant Signal Behav 4:634–635
Stahl Y, Simon R (2010) Plant primary meristems: shared functions and regulatory mechanisms. Curr Opin Plant Biol 13:53–58
Stahl Y, Wink RH, Ingram GC, Simon R (2009) A signaling module controlling the stem-cell niche in *Arabidopsis* root meristems. Curr Biol 19:909–914
Stange BC, Rowland RE, Rapley BI, Podd JV (2002) ELF magnetic fields increase amino acid uptake into *Vicia faba* L. roots and alter ion movement across the plasma membrane. Bioelectromagnetics 23:347–354
Steinmann T, Geldner N, Grebe M, Mangold S, Jackson CL, Paris S, Gälweiler L, Palme K, Jürgens G (1999) Coordinated polar localization of auxin efflux carrier PIN1 by GNOM ARF GEF. Science 8:316–318
Stepanova AN, Robertson-Hoyt J, Yun J, Benavente LM, Xie DY, Dolezal K, Schlereth A, Jurgens G, Alonso JM (2008) TAA1-mediated auxin biosynthesis is essential for hormone crosstalk and plant development. Cell 133:177–191
Strader LC, Ritchie S, Soule JD, McGinnis KM, Steber CM (2004) Recessive-interfering mutations in the gibberellin signaling gene SLEEPY1 are rescued by overexpression of its homologue, SNEEZY. Proc Natl Acad Sci USA 101:12771–12776
Strader LC, Monroe-Augustus M, Bartel B (2008) The IBR5phosphatase promotes *Arabidopsis* auxin responses through a novel mechanism distinct from TIR1 mediated repressor degradation. BMC Plant Biol 8:41–56
Sung S, He Y, Eshoo TW, Tamada Y, Johnson L, Nakahigashi K, Goto K, Jacobsen SE, Amasino RM (2006) Epigenetic maintenance of the vernalized state in *Arabidopsis thaliana* requires LIKE HETEROCHROMATIN PROTEIN 1. Nat Genet 38:706–710
Suzuki MM, Bird A (2008) DNA methylation landscapes: provocative insights from epigenomics. Nat Rev Genet 9:465–476
Swain SM, Tseng TS, Thornton TM, Gopalraj M, Olszewski NE (2002) SPINDLY is a nuclearlocalized repressor of gibberellin signal transduction expressed throughout the plant. Plant Physiol 129:605–615
Swarup R, Kramer EM, Perry P, Knox K, Geyser MO, Haseloff J, Beemster GTS, Bhalerao R, Bennett MJ (2005) Root gravitropism requires lateral root cap and epidermal cells for transport and response to a mobile auxin signal. Nat Cell Biol 11:1057–1064
Taiz L, Zeiger E (1998) Plant physiology, 2nd edn. Sinauer Associates, Sunderland, MA
Taiz L, Zeiger E (2002) Fisiologia vegetale, 2nd ed, Piccin, Padova, Italy
Thomann A, Lechner E, Hansen M, Dumbliauskas E, Parmentier Y, Kieber J, Scheres B, Genschik P (2009) *Arabidopsis* CULLIN3 genes regulate primary root growth and patterning by ethylene-dependent and – independent mechanisms. Plos Genet 5:e1000328
To JPC, Kieber JJ (2007) Cytokinin signaling: two-components and more. Trends Plant Sci 13:85–92
Turck F, Roudier F, Farrona S, Martin-Magniette ML, Guillaume E, Buisine N, Gagnot S, Martienssen RA, Coupland G, Colot V (2007) *Arabidopsis* TFL2/LHP1 specifically associates with genes marked by trimethylation of histone H3 lysine 27. PLoS Genet 3(6):e86
Van den Berg C, Willemsen V, Hage W, Weisbeek P, Scheres B (1995) Cell fate in the *Arabidopsis* root meristem determined by directional signaling. Nature 378:62–65

Van den Berg C, Willemsen V, Hage W, Weisbeek P, Scheres B (1997) Short-range control of cell differentiation in the *Arabidopsis* root meristem. Nature 390:287–289

Veit B (2006) Stem-cell signaling networks in plants. Plant Mol Biol 60:793–810

Vernon DM, Meinke DW (1994) Embryogenic transformation of the suspensor in twin, a polyembryonic mutant of *Arabidopsis*. Dev Biol 165:566–573

Vernoux T, Benfey PN (2005) Signal that regulate stem-cell activity during plant development. Curr Opin Genet Dev 15:388–394

Vert G, Walcher CL, Chory J, Nemhauser JL (2008) Integration of auxin and brassinosteroid pathways by Auxin Response Factor 2. Proc Natl Acad Sci USA 105:9829–9834

Vieten A, Sauer M, Brewer PB, Friml J (2007) Molecular and cellular aspects of auxin-transport-mediated development. Trends Plant Sci 12:160–168

Vroemen CW, Langeveld S, Mayer U, Ripper G, Jürgens G, Van Kammen A, De Vries SC (1996) Pattern formation in the *Arabidopsis* embryo revealed by position-specific lipid transfer protein gene expression. Plant Cell 8:783–791

Waisel Y, Eshel A, Kafkafi U (1997) Plant root: the hidden half, 2nd edn. Marcel Dekker, New York, NY, pp 483–811

Waizenegger I, Lukowitz W, Assaad F, Schwarz H, Jürgens G, Mayer U (2000) The Arabidopsis KNOLLE and KEULE genes interact to promote vesicle fusion during cytokinesis. Curr Biol 10:1371–1374

Webster PL, Langenauer HD (1973) Experimental control of the activity of the quiescent centre in excised root tips of *Zea mays*. Planta 112:91–100

Webster PL, Mc Leod RD (1996) The root apical meristems and its margins. In: Waisel Y, Eshel A, Kafkafi U (eds) Plant root: the hidden half, 2nd edn. Marcel Dekker, New York, NY, pp 51–76

Weijers D, Benkova E, Jäger KE, Schlereth A, Hamann T, Kientz M, Wilmoth JC, Reed JW, Jürgens G (2005) Developmental specificity of auxin response by pairs of ARF and Aux/IAA transcriptional regulators. EMBO J 24:1874–1885

Welch D, Hassan H, Blilou I, Immink R, Heidstra R, Scheres B (2007) *Arabidopsis* JACKDAW and MAGPIE zinc finger proteins delimit asymmetric cell division and stabilize tissue boundaries by restricting SHORT-ROOT action. Genes Dev 21:2196–2204

Werner T, Schmülling T (2009) Cytokinin action in plant development. Curr Opin Plant Biol 12:527–538

Wildwater M, Campilho A, Perez-Perez JM, Heidstra R, Blilou I, Korthout H, Chatterjee J, Mariconti L, Gruissem W, Scheres B (2005) The RETINOBLASTOMA-RELATED gene regulates stem-cell maintenance in *Arabidopsis* roots. Cell 123:1337–1349

Xu J, Hofhuis H, Heidstra R, Sauer M, Friml J, Scheres B (2006) A molecular framework for plant regeneration. Science 311:385–388

Yadav RK, Girke T, Pasala S, Xie M, Reddy GV (2009) Gene expression map of the *Arabidopsis* shoot apical meristems stem-cell niche. Proc Natl Acad Sci USA 106:4941–4946

Yang X, Zhang F, Kudlow JE (2002) Recruitment of O-GlcNac transferase to promoters by corepressor mSin3A: coupling protein O-GlcNcylation to transcriptional repression. Cell 110:69–80

Zazímalová E, Murphy AS, Yang H, Hoyerová K, Hosek P (2010) Auxin transporters-why so many? Cold Spring Harb Perspect Biol 2(3):a001552

Evolution, Physiology and Phytochemistry of the Psychotoxic Arable Mimic Weed Darnel (*Lolium temulentum* L.)

Howard Thomas, Jayne Elisabeth Archer, and Richard Marggraf Turley

Contents

Abstract Darnel (*Lolium temulentum* L.), the subject of this review, is botanically and culturally significant because of its evolutionary origin as a mimic weed of cereals and its reputation as a source of potent psychoactive toxins. Evidence from molecular phylogeny, palaeontology and archaeology allows the source and spread

H. Thomas (✉)
IBERS, Edward Llwyd Building, Penglais Campus, Aberystwyth University, Ceredigion SY23 3DA, UK
hot@aber.ac.uk

J.E. Archer and R. Marggraf Turley
Department of English and Creative Writing, Penglais Campus, Aberystwyth University, Ceredigion SY23 3DY, UK

U. Lüttge et al. (eds.), *Progress in Botany 72*,
DOI 10.1007/978-3-642-13145-5_3, © Springer-Verlag Berlin Heidelberg 2011

of darnel in time and space to be reconstructed. Contemporaneously with the progenitors of wheat and barley, at the dawn of agriculture in the Fertile Crescent region of Mesopotamia and the Eastern Mediterranean, darnel was derived from a perennial ancestor that was subject to the same human-mediated selection pressures as the earliest cereal species, and shares with them the domestication traits of annuality, self-fertility, high harvest index and non-shattering grains. Because it combines the characteristics of cereals with those of forage species of the *Lolium–Festuca* complex, *L. temulentum* is a useful experimental subject for the study of the physiology of temperate grasses. In particular, it has been a model for research on the control of flowering by day length, as well as investigations of vegetative development, resource allocation and responses to abiotic stresses. Recent studies of the chemical basis for darnel's noxious reputation reveal a complex picture in which endophytic fungi, nematodes and pathogenic bacteria separately or in combination account for the toxicity of the darnel grain. The relationship, and frequent historical conflation, of darnel and ergot poisoning is considered in detail. Finally, some examples of literary allusions to *L. temulentum*, from Aristophanes to Shakespeare, are given.

1 Introduction

Darnel (*Lolium temulentum* L. Fig. 1) is an annual weed, formerly common in cereal fields across the world, now rare or extinct in intensive arable agriculture but still significant in developing countries and counted among 'The World's Worst Weeds' (Holm et al. 1977; Senda and Tominaga 2004). It has the reputation of being one of the very few poisonous grasses and, as a consequence of its toxic character, it occurs widely in history, folklore and literature (Ellacombe 1884; Camporesi 1989; Drury 1992; Musselman 2000). A search of publication databases returns more than 3000 hits at the time of writing and reveals that *L. temulentum* has been the subject of many studies across the range of plant biology, from molecular genetics and phylogeny through crop physiology and biochemistry to plant pathology, phytochemistry and pharmacology. The present paper reviews these aspects of the biology of *L. temulentum*, giving particular consideration to its origin and evolution as a cereal analogue, its value as a model species for the study of cereal and forage grass physiology, and the pathological and chemical origins of its toxicity.

2 Phylogeny and Evolution of *L. temulentum*

2.1 The L. temulentum *Genome*

Lolium species have been among the favoured subjects for investigating genome size and organisation in relation to interspecific and intergeneric hybridisation. The annual inbreeding *Lolium* spp. (*L. temulentum*, *L. remotum* and *L. persicum*) have nuclear DNA contents about 50% greater than that of the perennial outbreeder

Fig. 1 Morphological and physiological features of a range of *Lolium temulentum* genotypes. (**a**) Late heading accession. (**b**) Early heading accession. (**c**) Awned type. (**d**) Awnless type. (**e**, **f**) Genotypic variation for regrowth after cutting. (**g**) Seed morphology compared with other grass species. (**i**) maize, (**ii**) oat, (**iii**) rye, (**iv**) barley, (**v**) wheat, (**vi**) *Lolium perenne*, (**vii**) Miscanthus, (**viii**) different *L. temulentum* genotypes. Photographs courtesy of John Harper, IBERS, Aberystwyth University

L. perenne (Table 1; Hutchinson et al. 1979). *L. temulentum* can be made to hybridise with other *Lolium* and *Festuca* species, indicating close taxonomic relationships. Jenkin (1933) described limited success in early attempts at crossing *L. temulentum* (♀) and *L. perenne* (♂). Subsequently the same author gave a detailed account of reciprocal crosses between the two species (Jenkin 1935). This paper merits close re-reading in the light of modern understanding and research objectives.

L. temulentum is a strict annual. Jenkin found that, following flowering and harvest, practically all F1 hybrids with *L. perenne* were capable of vigorous regrowth in the following season, even after cutting back to ground level. This evidence points to a quantitative basis for the genetic determination of perenniality (Thomas et al. 2000), but in so far as it is realistic to speak of dominance relationships, Jenkin's results suggest that annuality is on the whole recessive to perenniality in the *Lolium* genus. This in turn implies that annuality in *Lolium* is a loss-of-function trait derived from, and therefore evolved more recently than, perenniality (Naylor 1960). This is discussed further below in connection with domestication of arable species and their weeds.

Table 1 Nuclear genome sizes of *Lolium* spp.

Species	Common name	pg DNA per 2C nucleus
Lolium perenne	Perennial ryegrass	4.16
L. multiflorum[a]	Italian ryegrass	4.31
L. rigidum[b]	Annual ryegrass, Wimmera ryegrass	4.33
L. canariense	Canary Islands ryegrass	4.23
L. loliaceum[b]		5.49
L. remotum	Hardy ryegrass	6.04
L. temulentum	Darnel, poisonous ryegrass	6.23
L. persicum	Persian darnel	6.35

After Hutchinson et al. (1979)

[a]Bennett and Leitch 2004, quoting Schifino and Winge (1983), list the 2C value for this species as 8.2pg

[b]*L. loliaceum* has been classified as *L. rigidum* var. *rottbollioides*, or as a hybrid between *L. rigidum* and *L. temulentum* – see Bennett (1997)

Jenkin (1935) also points out that natural hybrids between *L. temulentum* and *L. perenne* are not found, but that Italian ryegrass (now *L. multiflorum*, but at the time Jenkin preferred to consider it to be a 'variety' of *L. perenne*) has many of the characteristics of such a hybrid. *L. multiflorum*'s weak capacity for vegetative regeneration is of particular note in this regard. As discussed below, *L. multiflorum* has *temulentum*-like arable weed tendencies. The Kew database (Bennett and Leitch 2004) lists the C value for *L. multiflorum* as large – 8.2 pg per 2C compared with 4.16 for *L. perenne* – based on a publication by Schifino and Winge (1983). These information sources continue to be used in some studies of variation in C value (for example Caetano-Anollés 2005). If they are correct, it might suggest a relationship between genome expansion and the annuality tendencies of *Lolium* species that occur as arable weeds. We have been unable, however, to find another source that supports the data of Schifino and Winge, and would argue that the numbers in Hutchinson et al. (1979), which show *L. perenne* and *L. multiflorum* to be similar in C value (4.16 and 4.31 pg per 2C respectively – Table 1), are more reliable. The mismatch in size between the *L. perenne* and *L. temulentum* genomes undoubtedly causes physical problems at meiosis in the products of interspecific crosses (Jenkins and White 1990) and accounts for low fertility and the need for embryo rescue to recover hybrids. Nevertheless, C value is not a straightforward index of taxonomic relationships, nor is it a predictable outcome of combining two genomes of divergent size, nor is it consistently related to any particular phenotype (Murray 2005).

2.2 *Molecular Systematics of* Lolium

The evidence from a number of molecular phylogeny studies agrees that the genus *Lolium* is monophyletic, closely related to the broad-leaved fescues and probably evolved recently from a European *Schedonorus* ancestor (Catalan et al. 2004;

Table 2 Timeline for the evolution of *Lolium temulentum*

Years before present day	Palaeobiological event
77M	Grasses appear
50M	Maize and rice diverge
21M	Oats and barley diverge
13M	First record of Loliinae (temperate grasses)
2–3M	Split between *Festuca* and *Lolium*
1–2M	Diversification of *Lolium* spp.
1M	Speciation in the genus *Triticum*
10,000	Appearance of *Lolium temulentum* in the Fertile Crescent; beginning of temperate agriculture
9,500	Earliest emmer, einkorn remains
7,500	Earliest bread wheat remains
>5,000	Darnel, wheat preserved together

Inda et al. 2008). Most estimates date divergence of the genus *Lolium* at around 2–3 mya, and its differentiation into species about 1 mya (Charmet et al. 1997; Inda et al. 2008), which, significantly, is similar to the timing reported for speciation in the genus *Triticum* (Stebbins 1981). Inda et al. (2008) located the centre of origin of *L. perenne* and *L. rigidum* in the pan-Mediterranean SW Asia biogeographical area and suggest an interglacial Pleistocene origin for these most recently evolved Mediterranean forage ryegrasses.

Senda et al. (2005c) used DNA markers to determine the genetic relationships between 48 geographically diverse populations of *L. temulentum* collected from eight countries. They inferred from their phylogenetic analyses that the species originated between Southwestern Asia and the Mediterranean basin. Darnel evolved out of the centre of diversity of the genus *Lolium* (Balfourier et al. 2000), probably around 10,000 years ago, coincident with the beginnings of temperate cereal cultivation (Archatlas 2006). Examination of Egyptian and middle-eastern prehistoric sites dating from earlier than 2000 BC has revealed darnel grains amongst cereal residues (Täckholm et al. 1973; McCreery 1979). Table 2 presents a tentative timeline for the evolutionary origin of *L. temulentum*, based on evidence from molecular phylogeny, palaeontology and archaeology. The spread of *Lolium* spp. along three major routes from the Fertile Crescent region to Europe and North Africa is shown in Fig. 2.

2.3 *Selection of Domestication Traits in* L. temulentum

Compared with its near relatives amongst the outbreeding, perennial *Lolium* species, darnel is strikingly cereal-like in form and life-cycle (Fig. 1). Its distinctive features – self-fertility, annual habit, hypertrophied grains and foliage, non-shattering heads, rapid establishment and growth, high harvest-index and the relatively

Fig. 2 Colonisation routes for *Lolium* inferred from historical data on human migration during the spread of agriculture from the Fertile Crescent into Europe and North Africa. (**1**) Danubian movement. (**2**) Mediterranean movement. (**3**) North African continental route. After Balfourier et al. (2000)

uncomplicated environmental sensitivities of its floral induction mechanism – collectively represent the so-called domestication syndrome that also distinguishes cereal species such as wheat and barley from their wild progenitors (De Wet and Harlan 1975; Evans 1993; Allaby et al. 2008). It is clear that co-selection of crops and their weeds under cultivation has converted the ancestor of darnel into a cereal analogue, thereby fitting it so closely to its niche that its geographical range precisely followed the spread of temperate arable agriculture in prehistory (Harlan 1981). Cultivated oat and rye are thought to have originated, like darnel, as weeds of wheat but to have taken the extra step of becoming adopted as cereals in their own right (Ladzinsky 1998). Catalan et al. (2004) identified an evolutionary trend in reduction of life-cycle and habit from the perennial, robust, and paniculate *Schedonorus* taxa, through the perennial, slender, subracemose *Micropyropsis* taxon to the mostly annual, ephemeral, racemose and single glumed taxa of *Lolium*. It is reasonable to suppose that human intervention has been instrumental in driving this process to the extremes seen in the inbreeding cereal mimics *L. temulentum*, *L. remotum* and *L. persicum* (Balfourier et al. 2000).

Some genotypes of darnel have awns. Others are awnless (Fig. 1c, d). Senda and Tominaga (2003) demonstrated a relationship between the frequency of contamination of the two forms, the nature and agronomy of the associated crop species and threshing and seed cleaning methods employed. They found the awnless form to be a contaminant mainly of bread wheat in Greece and of bread wheat and barley in Ethiopia, whereas awned darnel was associated with barley and emmer wheat in Greece and Ethiopia, respectively. In a detailed study of cereal cultivation practices in Malo village, Ethiopia, Tominaga and Fujimoto (2004) concluded that awned darnel is a close morphological mimic of emmer wheat grains. Darnel that lacks an awn resembles free-threshing wheat, and the awnless trait functions as a dispersal mechanism in cultivation systems for this crop. Inheritance studies led Senda and Tominaga (2003) to conclude that the wild progenitor of darnel, before it was strongly associated with wheat and barley, was awned. Awnless arises by mutation in a single dominant major gene or a close linkage group, and the resulting contamination of fields and granaries by both awned and awnless forms ensures propagation through wheat and barley crops of the most highly adapted genotypes in subsequent seasons.

2.4 The Spread and Perpetuation of L. temulentum *in Cereal Grain Stocks*

As Tominaga and Fujimoto (2004) have shown, darnel is one of a select group of weeds that persists through maintenance of a seed-bank, not in the soil but in grain stocks. The use of herbicides, advanced crop varieties and seed cleaning technology in intensive cereal production has more or less completely eliminated darnel from the agriculture of developed countries, where it is now classified as a rare species (Tominaga and Yamasue 2004). But darnel continues to be an abundant weed of cereals in North Africa and South and West Asia. For example: Ghanem and Hershko (1981), in a study of liver disease in Israeli Arabs, reported the presence of *Lolium* (presumably darnel) in the cereal grain stores of two out of nine villages surveyed; Tominaga and Fujimoto (2000) recorded contamination of barley grain stocks by darnel in an Ethiopian village market to be as high as 9.4%; and Musselman (2000) found that darnel was plentiful in the wheat fields of Northern Syria, in the mountains east of Latakia and at a threshing site near Quneitra. The occurrence of the many words for darnel in the languages of lands to which it was carried by the spread of agriculture and the trade in cereals (Harlan 1981, Balfourier et al. 2000; Fig. 2) testifies to the agricultural and social significance of this species (Table 3).

2.5 Other Lolium *spp. Following a Similar Evolutionary Pathway*

There are seven or eight diploid species within the genus *Lolium* (the taxonomy of this complex is fluid and debate about the species/subspecies status of members is ongoing – Terrell 1968; Bennett 1997). The perennial, outbreeding species include *L. perenne*, *L. multiflorum* (syn *L. italicum*) and *L. rigidum* (syn *L. loliaceum*? – Terrell 1968). Along with *L. temulentum*, the complex of annual inbreeders comprises *L. remotum*, *L. persicum* and *L. canariensis*. *L. remotum* is close to *L. temulentum* in terms of form, lifecycle, ecophysiology, cytogenetics and

Table 3 Words for *Lolium temulentum* in world languages

Language	Words for *L. temulentum*
English	Bearded darnel, Poison darnel, Annual darnel, Red darnel, Poison ryegrass, Darnel ryegrass, Ray-grass, Tares, Drake, Drawke, Drunk, Dragge, Sturdy ryle, Cheat, Wenwort, Chess, Virginian oat
Old English	Cokil, Cockle, Evir
Arabic	Zirwan, Samma, Aqoullab, Zawan, Zuwan, Shaylam, Suwal, Sikra, Danaqa
Basque	Iraka
Bresciano	Fraina, Lergheta, Loi
Breton	Draog, Ivre, Pigal, Pilgerc'h
Calabrese	Giògghju
Chinese	毒麦 (Du mai)
Colombia	Ballico
Czech	Jílek
Dutch	Dolik, Hondsdravik
Esperanto	Lolo, Zizanio
Estonian	Uimastav raihein
Finnish	Myrkkylust, Myrkkyrraiheinä
French	Ivraie annuelle, Ivraie enivrante, Herb á couteau, Herb d'ivrogne, Zizanie
Furlan	Vrae
Guarani	Ñati'ũna, Kapiatĩ
German	Taumellolch,Taumel-Raygras
Greek	Αίρα η μεθνστική, Ήρα η μεθνστική
Hebrew	סוג עשב
Hungarian	Konkoly
India	Machni, Mochni, Mostaki
Italian	Loglio del Veleno, Loglio ubriacante, Zizzania
Latin	Lolium temulentum
Mapunzugun	Wezakachu
Morocco	Zwân, Zuwân, Gesmatâ, I-medhun, Sîkrân, Sîkrâ, Saylam, Shaylam, Laichour
Persian	لوليم سيخكى ; چچم
Peru	Ballico, Cerisuelo, Sirisuela
Polish	Kąkol, Życica roczna
Portuguese	Joio
Russian	Плевел опьяющий
Spanish	Borrachuela, Cizaňa commún, Cizaňa embriagante, Cominillo, Joyo, Trigollo, Mala hierba, Rabillo
South Africa	Drabok raaigras, Dronkgras, Drabok
Swedish	Dårrepe
Romagnolo	Zizagna, Zizania
Valencian	Brossa
Welsh	Efrau, Efryn, Yd meddw, Efrau coliog, Pabi'r gwenith, Drewg, Pabi gwenith, Ller, Graban yr hwylydd, Lleren
Zulu	Shesi

From Porcher (2007) and other sources

geographical distribution (Jenkin and Thomas 1938; Naylor 1960; Terell 1968; Bulinska-Radomska and Lester 1985; Bennett 1997; Gaut et al. 2000); it also shares with *L. temulentum* and *L. persicum* an inflated nuclear DNA C-value (Hutchinson et al. 1979; Table 1). *L. remotum* has all the features of a weed of cereals that has been subjected to anthropophytic selection for domestication traits. In so far as it is recognisable as a separate species, it does not have the psychotoxic associations of darnel, but given the capricious nature of traditional plant identification, it is difficult to believe that history has not thoroughly scrambled the accounts of these two arable weeds.

As its name suggests, *L. persicum* is a native of Western Asia, whence it has become dispersed to other continents to pestilential effect (Forcella and Harvey 1988; Holman et al. 2004; Shinomo and Konuma 2008). According to Loos (1993), when it first appeared in Canada, it was classified as *L. temulentum*. It tends to lack some of the domestication-type features of *L. temulentum* and *L. remotum*. For example, its habit is more spreading than erect and its comparatively small grains shatter at maturity. But morphologically (Mirjalili et al. 2008), cytologically (Hutchinson et al. 1979; Senda et al. 2005a) and phylogenetically (Senda et al. 2005b), it is close to the other annual inbreeding *Loliums* and probably shares their anthropophytic origin.

Nature never relents: *L. multiflorum* (Italian ryegrass) is emerging as a successful weed of winter wheat in some parts of the US and Europe (Appleby et al. 1976; Wilson and Wright 1990). We may be witnessing a contemporary recapitulation of the prehistoric process in which the cereal crop is experiencing invasion by a perennial, outbreeding-type *Lolium*.

3 Physiology and Biochemistry of *L. temulentum*

3.1 A Model for the Study of Photoperiodic Control of Flowering

The control of floral induction by daylength was described by Garner and Allard (1920). Photoperiod-perceptive plants are divided into long-day and short-day types, and in most cases, sensitivity to photocycle is quantitative, that is, flowering is accelerated by, but not critically dependent on, the relative durations of day and night. For experimental studies of flowering physiology, a qualitative response is desirable and a number of model species that do not flower at all unless exposed to one or more short days have been identified and analysed (for example Saji et al. 1983; Salisbury 1963). *Lolium* spp. are long-day plants, mostly of the quantitative kind (Cooper 1950, 1951, 1952, 1954). In 1958, Lloyd Evans described the characteristics of a Canadian accession of *L. temulentum* that has the very useful property of remaining vegetative indefinitely under short days but responding to a single long day by switching to the reproductive state (Evans 1958; Fig. 3). This ecotype (now referred to as forma Ceres, after the Canberra Phytotron where much of the early research on this model was carried out – Evans 2003) has been the subject of most of the subsequent work on the mechanism of flowering in *Lolium*, and since the physiology of this arable weed is synchronised with that of the cereal species it mimics, *L. temulentum* continues to be a useful model for the study of reproductive development in monocot crops.

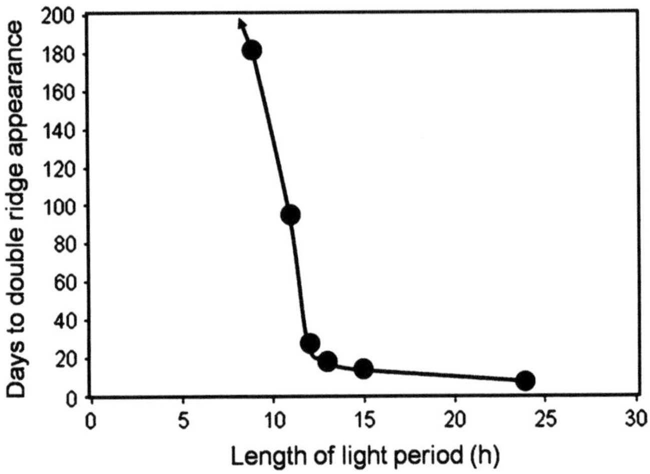

Fig. 3 *Lolium temulentum* forma Ceres is a qualitative long-day plant, requiring one inductive photocycle for floral induction. Plot of data from Evans (1958)

Flowering in *L. temulentum* Ceres has the classical features of a photoperiodically determined developmental response in being sensitive to darkness (Perilleux et al. 1997) and related to circadian rhythm (Perilleux et al. 1994). *L. temulentum* generally has a weak or no vernalisation requirement. A focus for research on flowering in *L. temulentum* Ceres has been the role of gibberellin (Mander et al. 1999; King et al. 2003), which is implicated in the transmission of the flowering stimulus to the shoot apex from the site of photoperception in the leaves. Since the arrival of Arabidopsis (a quantitative long-day species) as a genomic model, the use of mutants and molecular genetics has enabled the mechanism of developmental, photoperiodic and temperature regulation of floral induction to be minutely dissected and integrated with the processes of meristem and floral organ differentiation. Recently (Corbesier et al. 2007; Tamaki et al. 2007), this has culminated in research that establishes the 20 kDa protein FT as a major component of the signal formerly known as florigen, the stimulus for floral induction that moves from leaves to the shoot apex (Chailakhyan 1936). The *L. temulentum* homologue of FT has been identified (King et al. 2006), as have a number of other genes in the floral induction network, including: LEAFY (*LFY*), TERMINAL FLOWER 1 (*TFL1*), APETALA1 (*AP1*), CONSTANS (*CO*) and the TALE homeobox gene *ATH1* (van der Valk et al. 2004). In some cases, flowering genes and QTL for traits related to reproductive development, such as heading date, perenniality and so on, have been co-located on the *Lolium* linkage map. For example, Donnison et al. (2002) identified expression of a cDNA with homology to the *CO* gene of Arabidopsis in apices of *L. temulentum* Ceres exposed to long days, and mapped it in *L. perenne* close to a major QTL for heading date (Armstead et al. 2004; Fig. 4). This raises the prospect of using genetic manipulation to analyse and modify reproductive and

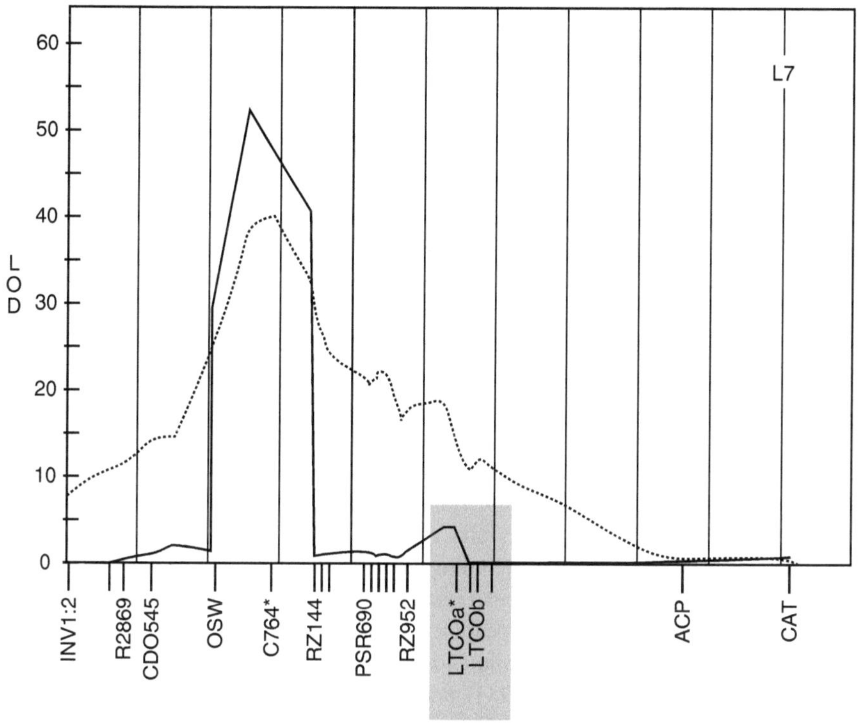

Fig. 4 Positions of heading-date QTL on *Lolium perenne* linkage group L7, identified by LOD (log of odds) profile using interval mapping (*dotted line*) or MQM (multiple QTL model) mapping (*solid line*). The map locations of the *CONSTANS* gene of *L. temulentum* (*LtCO*) are highlighted. From Armstead et al. (2004). doi: 10.1007/s00122-003-1495-6

vegetative development in the grasses. In such a study, transgenic *L. perenne* plants expressing *ATH1* from Arabidopsis were shown to be late-heading or completely non-flowering (van der Valk et al. 2004). The high degree of synteny across the genomes of grasses and cereals means that this information from studies of darnel can be directly exploited in the breeding of grain and forage monocot crops.

3.2 *Leaf Development and Senescence in* L. temulentum

Because *L. temulentum* combines the physiological and genetic characteristics of a cereal and a forage grass, it has been a favoured subject for studies of the development and senescence of foliage, the primary determinants of grain and fodder production. It is particularly useful as a model for the agronomic *Lolium* spp. (*perenne* and *multiflorum*), since its genetic uniformity and rapid rate of development makes it more experimentally convenient than its slower growing, outbreeding relatives. Borrill (1961) studied anatomical differences in successive leaves along the reproductive tiller of *L. temulentum*. Leaf size (blade plus sheath), which increased progressively up to the flag leaf, could be resolved into the width component (determined principally by the number of cell rows) and length (a function of cell volume). Inflorescence initiation was associated with a marked reduction in the ratio of blade to sheath length in the youngest and subsequent leaves. The total blade+sheath area of the flag, the largest leaf on the shoot, was virtually constant under different temperature treatments. Borrill (1961) concluded that the developmental anatomy of *L. temulentum* leaves closely resembles that of the cereals it infests, and has arisen by a common process of selection for domestication traits.

Because the extension of grass leaves is essentially one-dimensional, leaf growth is conveniently measured as length changes over time, either conventionally with a ruler or at high resolution with electronic position sensors. Fitted growth curves developed for foliar expansion measured as area or mass (for example, the various logistic-type functions described by Causton and Venus 1981) have been usefully adapted to model grass leaf extension. Thomas (1983a) and Thomas and Potter (1985) fitted generalised logistic functions to measurements of *L. temulentum* leaf lengths to produce growth curves and estimates of a number of physiologically meaningful growth parameters including curve inflexion point (time of maximal extension rate), final leaf length, mean relative and absolute extension rates, duration of growth, mean rate of leaf appearance and mean plastochron duration. High-resolution extension data for *L. temulentum* leaves were obtained by Thomas et al. (1984), Thomas and Stoddart (1984) and Stoddart et al. (1986) using electronic position transducers. These revealed subtleties of growth behaviour, including marked day–night variation in extension rate, which was shown to be a photoperiodically entrained endogenous circadian rhythm with a period of about 27 h.

The growing zone of the extending grass leaf is a region of a few millimetres just above the insertion point on the shoot apex. New leaves are enclosed in the sheaths at the bases of preceding leaves such that the growing zone remains protected until expansion is complete. As a consequence of the linear mode of tissue development, there is a direct relationship between cell age and position along the leaf, from youngest at the base to oldest at the tip. Several studies have exploited this gradient to analyse the biochemical and gene expression correlates of cell growth and differentiation in *L. temulentum* leaves. Ougham et al. (1987), Davies et al. (1989) and Ougham and Davies (1990) visualised the transition from cell division and growth at the base of the expanding *L. temulentum* leaf to assembly of the photosynthetic apparatus as young tissue emerges into the light by determining progressive changes in the abundances of diagnostic proteins and transcripts. Using an inhibitor of pigment metabolism (Davies et al. 1990) and a slow-to-green mutant (Ougham et al. 1992; Moses et al. 1997), chloroplast assembly during leaf emergence was shown to be co-regulated with chlorophyll biosynthesis.

In an early study of rates of protein synthesis in the fourth leaf of *L. temulentum* seedlings from birth to death, Hedley and Stoddart (1972) observed a distinct peak of amino acid incorporation during yellowing. They concluded that senescence is an active process, requiring new proteins to be made. Following this work, Thomas (1975) showed that inhibitors of protein synthesis applied to excised leaf tissue of

L. temulentum prevented yellowing and other senescence symptoms. Further studies on this species also established the pathway of proteolysis and amino acid metabolism associated with nitrogen mobilisation during leaf senescence and suggested that it is regulated at both the transcriptional and the post-transcriptional levels (Thomas and Stoddart 1974; Thomas 1975, 1978; Thomas and Feller 1993; Morris et al. 1996; Thomas et al. 2003). Much subsequent research on many species, gathering pace with the application of molecular biological approaches (Smart 1994), the genomics resources of Arabidopsis and other model plants (Buchanan-Wollaston et al. 2003) and the tools of systems biology (Wingler 2007) have confirmed these early indications of the role and regulation of gene expression in leaf senescence.

Mutants are essential subjects for research on senescence, and *Lolium* spp., including *L. temulentum*, have played an important part in the discovery of new mechanisms through the use of heritable genetic variation. Yellowing is diagnostic of senescence (Ougham et al. 2008), and isolation of staygreen (*SGR*), a key gene of chlorophyll-protein catabolism, was achieved in part by exploiting an introgression mutant of *L. temulentum*. Inheritance studies of a non-yellowing genotype of the outbreeding perennial *Festuca pratensis* showed the existence of a single Mendelian locus determining pigment-protein breakdown in this species (Thomas 1987). The mutant gene was transferred into a range of *Lolium* spp., including *L. temulentum*, via hybridisation between *F. pratensis* and *L. multiflorum*. This allowed biochemical and gene expression studies of the locus to be carried out in darnel's more uniform and experimentally convenient genetic background (Thomas et al. 1999, 2002; Roca et al. 2004; Gay et al. 2008 – Fig. 5). *SGR* (which was shown to be the homologue of the gene for green/yellow cotyledons of pea, originally described by Gregor Mendel) was finally isolated by a combination of comparative mapping between the *Lolium* and rice genomes and functional testing in Arabidopsis (Armstead et al. 2006, 2007).

3.3 Carbohydrate Metabolism and Carbon Partitioning

Leaf development and senescence have been related to physiological function in *L. temulentum* through a number of studies of photosynthesis. Some of the earlier investigations were attempts to identify the floral stimulus moving from leaves to apices, by administering radiolablelled CO_2 to photosynthesising plants and chasing the ^{14}C from source tissue through the translocation system to its destination (Evans and Wardlaw 1966; Ryle 1972; Ryle and Powell 1972). These studies concluded that the kinetics of assimilate distribution are incompatible with the observable properties of the flowering signal. *L. temulentum* has also been the subject of studies of the climate change-relevant issue of plant response to atmospheric CO_2 concentration. Gay and Hauck (1994) found that photosynthesis in this species under CO_2 enrichment acclimated better to a light flux of 150 than 350 μmol m^{-2} s^{-1}, and Lewis et al. (1999) observed that elevated CO_2 failed to

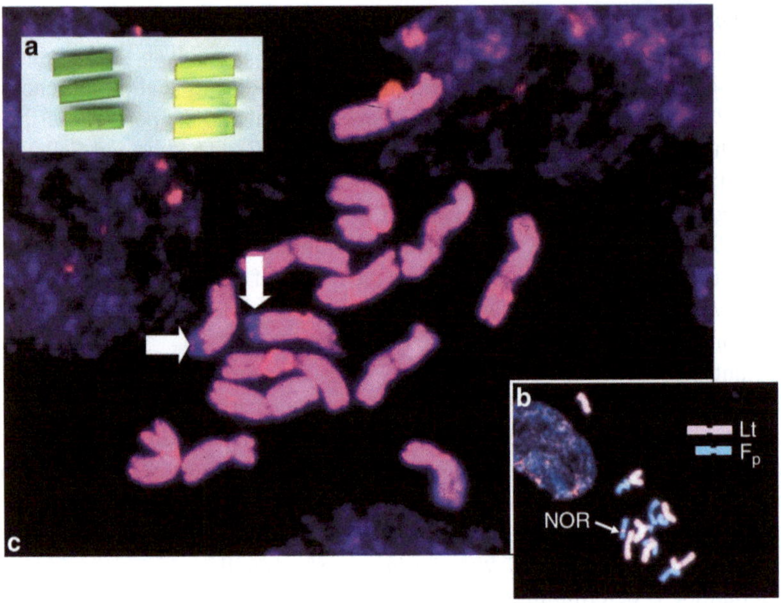

Fig. 5 Introgression of *stay-green*, a mutant leaf senescence gene, from *Festuca pratensis* into *Lolium temulentum*. (**a**) Senescing leaves of the introgression mutant line remain green while wild-type leaf tissue turns *yellow*. (**b**) The chromosomes of *F. pratensis* (Fp – *blue*) are readily distinguished from those of *L. temulentum* (Lt – *pink*) in an F1 hybrid by genomic in-situ hybridization. *NOR* nucleolar organizer region. (**c**) Repeated backcrossing reduces the size of the introgressed Fp segment in the Lt background to a terminal region on a single pair of chromosomes (corresponding to *Lolium* C5)

invoke a response in biomass, relative growth rate or biosynthesis of leaf storage carbohydrates. The latter authors concluded that for CO_2 enrichment to increase biomass in temperate grasses like *L. temulentum*, total light input has favourably to increase the rate of photosynthesis relative to that of respiration.

Gay and Thomas (1995) employed curve fitting together with measurements of gas exchange and chemical composition to describe the pattern of development from initiation to senescence in the fourth leaf of *L. temulentum* seedlings under non-floral inductive conditions. By combining photosynthesis and carbon content data, the carbon balance of the leaf could be calculated with respect both to the atmosphere and to the rest of the plant. Net gas exchange became positive about 7 days before full leaf expansion and leaf weight continued to increase so that the carbon balance with the rest of the plant did not become positive until just after final leaf size had been attained. Based on the model from this study, Thomas and Howarth (2000) showed that the net carbon contribution of each leaf of *L. temulentum* is sufficient to make about 3.7 more leaves (Fig. 6). Extrapolating to the point of absurdity, if each of these leaves in turn were able without limit to make a further 3.7 leaves and so on and on, it would require only about 33 leaves to achieve the estimated net primary productivity of the entire biosphere! The serious

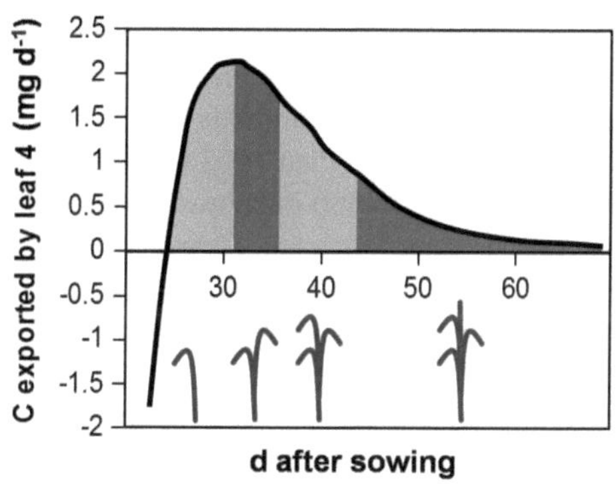

Fig. 6 Net carbon import and export from birth to death of the fourth leaf of *Lolium temulentum*. Making a leaf requires about 10 mg of carbon. The curve is segmented to show the periods over which the production of successive leaves is supported by the photosynthate contributed by leaf 4. Data from Gay and Thomas (1995), Thomas and Howarth (2000)

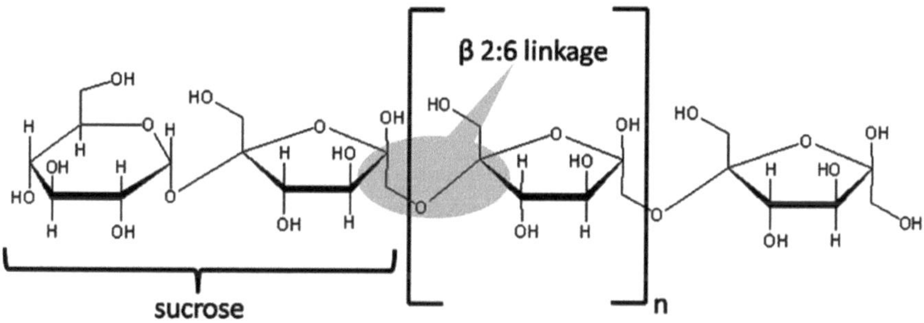

Fig. 7 Chemical structure of grass fructan. The molecule consists of a sucrose moiety extended by the serial addition of β 2:6-linked fructose residues. In *Lolium*, n has been determined to be around 75

point of this exercise is to emphasize the degree to which the repression of uncontrolled potential is the essence of biology.

L. temulentum is a typical temperate grass in that carbon in its vegetative parts is stored principally in the form of fructan rather than starch (Cairns et al. 2002). Because of the importance of the plant carbon economy in the productivity and nutritional quality of forage grasses, *L. temulentum* has been intensively studied as a model for the molecular and physiological control of fructan metabolism. Fructans are soluble carbohydrates whose chemical structure is based on sucrose, which in turn is a disaccharide comprising a glucose and a fructose moiety (Fig. 7). In fructans, the fructose 'end' of sucrose is extended by the addition of further fructose residues resulting in macromolecules, which, in *L. temulentum*, can extend to a degree of polymerisation in excess of 70. In *L. temulentum*, the linkages between adjacent sugar moieties are generally of one kind (β 2:6 – Fig. 7), though other bonds have been detected at low levels (Sims et al. 1992). Fructan biosynthesis is different from the process leading to the formation of other sugar polymers such as starch and cellulose, in that the monomeric carbohydrate precursors are not in the activated sugar nucleotide form. Instead, fructans are made by direct transfer of fructose residues from sucrose to a sucrose or sucrose-(fructose)$_n$ oligosaccharide acceptor (Vijn and Smeekens 1999). Thus fructan biosynthesis is driven by sucrose concentration, and treatments that cause buildup of sucrose in the tissue lead to accumulation of fructan (Pollock 1984). This has been shown for *L. temulentum* plants in which export of sucrose from leaves and/or the demand for sucrose in sink tissues has been restricted, for example, by cold treatment (Pollock and Lloyd 1998; Pollock et al. 1989) or excision (Housley et al. 1991; Cairns et al. 1997). Conversely, fructan is absent from unifoliated plants with high sink strength under floral induction conditions (Perilleux and Bernier 1997). Sucrose accumulation is directly or indirectly associated with expression of specific genes (Koch 1996), including those encoding enzymes of sucrose and fructan metabolism in *L. temulentum* (Winters et al. 1995; Gallagher et al. 2004). In temperate grasses like *L temulentum*, sucrose and fructan turnover is central to the integration of carbon partitioning in the whole plant (Farrar et al. 2000), and some of the genes for carbohydrate metabolism have been co-located with QTL for growth and productivity characters on the *Lolium* genetic map (Turner et al. 2006).

3.4 **L. temulentum** ***and Abiotic Stress***

Several studies have shown that suboptimal temperatures constrain vegetative growth in temperate grasses by inhibiting the extension of leaves rather than their initiation. Thus, over-wintering *Lolium* spp. develop a 'stored growth' potential that becomes expressed as a surge of canopy production when spring temperatures cross a threshold, usually about 5°C or an equivalent period of thermal time (Peacock 1975, 1976; Pollock and Eagles 1988, Fournier et al. 2005). *L. temulentum* has been a frequent subject for analysing the temperature relations of leaf extension. At 20°C under an 8 h (non-inductive) daylength in controlled-environment conditions, leaf 5 of *L. temulentum* line Ba3081 emerges 28 days after seed germination and reaches full expansion about 11 days later (Thomas 1983b; Thomas and Potter 1985). If plants are transferred from 20 to 5°C 21 days after germination, emergence of the fifth leaf is delayed by around 20 days and this leaf takes about 45 days to achieve a fully expanded size about 58% of that of 20°C-grown leaves. If plants are transferred from 20 to 2°C at 21 days, the fifth leaf fails to emerge at all. The response of *L. temulentum* leaf extension on exposure to 2°C is qualitative and almost immediate: high-resolution kinetic measurements with growth transducer instruments showed that growth fell precipitously to about 4% of the rate at 20°C and became confined to the dark phase of the photoperiod (Thomas and Stoddart 1984; Stoddart et al. 1986).

Chilling temperatures constrain leaf extension by throttling back demand for, rather than supply of, photosynthate. Pollock et al. (1983) showed that the Q_{10} for photosynthesis in *L. temulentum* between 20°C, 5°C and 2°C was in the range 2.1–4.6, whereas the corresponding range for relative growth rates was 1.7–44.6, and plants held at 5°C or 2°C accumulated high levels of soluble sugars, including fructans. Prolonged growth at 5°C reduced leaf cell number, chloroplast number and chlorophyll content but did not reduce photosynthetic capacity expressed on a unit area, fresh weight or chlorophyll basis (Pollock et al. 1984). Unlike exposure to supra-optimal temperatures, which induces the synthesis of heat shock proteins, there was no evidence of major qualitative changes in gene expression when *L. temulentum* was exposed to growth-limiting chilling temperatures (Ougham 1987).

Leaf extension is acutely sensitive to tissue water status, and the physiological characteristics of the limitation imposed on *L. temulentum* leaf extension by chilling temperatures are consistent with a direct response at the level of cell turgor and the constraints of cell wall rigidity. Thomas et al. (1989) measured turgor pressure parameters in the expanding zone of *L. temulentum* leaves over the temperature range 20°C–2°C and concluded that the site of thermal perception is the cell wall, the rheological properties of which are particularly sensitive to chilling temperatures. Bacon et al. (1997) came to a similar conclusion in their studies of the inhibition of leaf extension in *L. temulentum* by drought. These authors observed abrupt increases in the activity of cell wall-associated peroxidases in the extension zone and suggested that these enzymes have an important role in tissue rheology by making stiffening cross-links between cell wall polymers. Other plasticity-modifying factors associated with cell walls in growing leaf tissues of grasses include extensins, glycosyl transferases and gibberellins (Farell et al. 2006). Growth responses to salinity stress are also mediated in part by turgor. Baldwin and Dombrowski (2006) isolated a large number of salinity-responsive genes from leaves and crowns of *L. temulentum* exposed for 20 h to 500 mM NaCl administered to the roots. Of the 528 unique sequences identified, 167 were orthologues of previously identified plant stress response genes. Interestingly, genes with functions in cell wall cross-linking were absent from the collection. Subsequently, Dombrowski et al. (2008) focused on a gene from the library encoding smGTP, a small guanosine triphosphate-binding protein. On the basis of comparative specificity of its expression pattern with respect to salinity and other abiotic challenges, they concluded that smGTP is part of a dehydration stress signalling pathway.

It is normal for forage and amenity grasses to experience defoliation, through either grazing, or cutting for conservation, or harvest for seed production or mowing to maintain low sward height. Physiological changes in the cut herbage depend on post-harvest conditions. Herbivory by ruminants almost instantaneously changes the environment of the grazed tissue from mild temperatures in daylight and an aerobic atmosphere to darkness, anoxia and about 40°C. The immediate response of the still-viable ingesta is to trigger a characteristic cell death-like programme with far-reaching consequences for animal nutrition and its impact on the environment (Kingston-Smith and Theodoou 2000). Baldwin et al. (2007) made subtractive gene libraries from cut *L. temulentum* straw held under post-harvest conditions simulating those used in the field for seed production. They identified almost 600 unique cutting-specific sequences (Fig. 8). As might be expected, there was a fair amount of similarity with the profile of genes associated with leaf

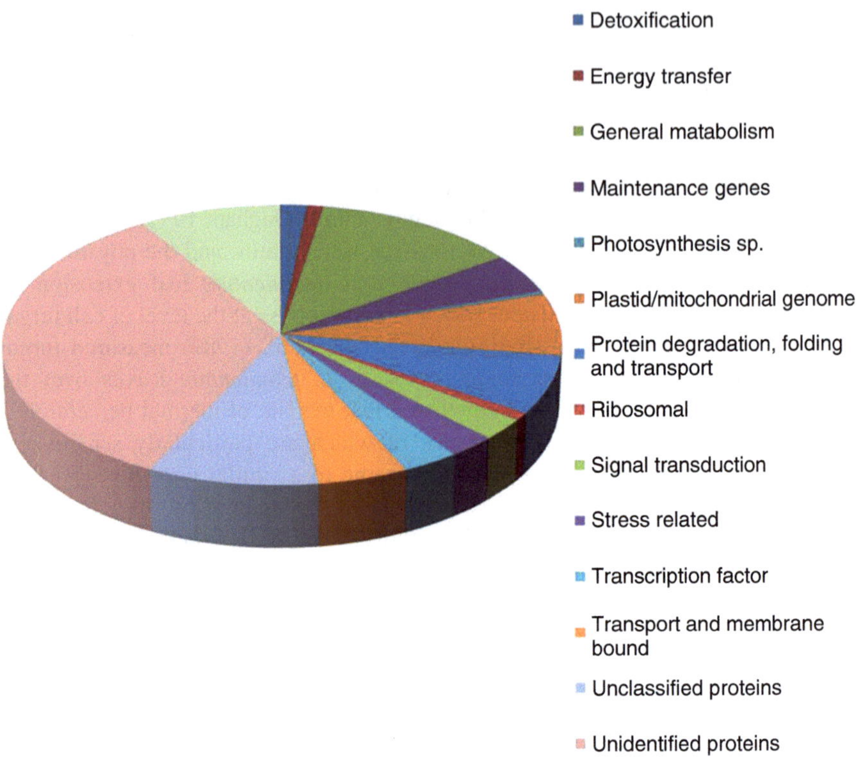

Fig. 8 Proportions of genes in different functional categories expressed post-harvest in seed-bearing shoots of *Lolium temulentum*. Data of Baldwin et al. (2007), doi:10.1016/j.plantsci.2007.04.001

senescence. For example, a number of genes encoding enzymes of protein degradation were identified, including several cysteine-, serine- and metallo-endopeptidases, Clp proteases and components of the ubiquitin/proteasome system. Responses of *L. temulentum* to defoliation and subsequent regrowth were studied by Ourry et al. (1996). The pool sizes and mobilization patterns of C and N compounds respond to defoliation through their control by modified source/sink relationships. Nitrogen uptake and mobilization by defoliated plants were determined by ^{15}N labelling. The difference in regrowth yield was accounted for by the initial availability of N reserves (Louahlia et al. (1999) reported similar observations for contrasting cultivars of *L. perenne* in a follow-up study). Changes in gene expression were visualised by comparing in vitro translation profiles of RNA extracted from meristem, sheath and lamina tissues after defoliation with those from intact *L. temulentum* plants. Modulations in expression pattern were observed to be dependent on the tissue and elapsed time since cutting. Some changes were detectable within 1h of defoliation, but the largest shifts in abundance of translatable transcripts occurred between 12 and 72 h after shoot removal. Amongst the activities that increase in leaves during regrowth are enzymes of amino acid and sucrose/fructan metabolism, whereas in the stubble, which acts as a remobilizing source to support the production of new foliage, peptidases and carbohydrate-recycling enzymes become elevated.

4 Biotic Interactions and Toxicity of *L. temulentum*

4.1 Symptoms of Darnel Poisoning

Throughout history, darnel has been known, and sometimes feared, for its psychotoxic properties. The chemical basis of darnel's potency has been the subject of intensive study, but pinning down the exact nature of the toxic principle and its mode of action in humans has been difficult. Rizk and Hussiney (1991) describe the symptoms of darnel poisoning as 'dizziness, headache, mental confusion with a sense of apprehension and difficulty in thinking, visual and speech difficulties (even loss of speech), decrease in salivary secretion, vomiting, inability to walk, griping, rarely diarrhea, tremor, general weakness and finally coma'. These authors summarise the results of tests of ethanolic extracts administered to rodents and conclude that acute lethal toxicity is caused by the cocktail of alkaloids present in the darnel grain.

Deliberate administration or consumption of herbal phytotoxins is a recurring feature of traditional medical treatments (according to the principle of pharmakon or hormesis[1]) and other cultural practices. Darnel was known in Classical Greece as αίρα (*aira*), the plant of frenzy, and associated with the Eleusinian rituals of the Demeter and Persephone cults (Wasson et al. 1978). John Gerard (1597) in the Herbal states: 'Red Darnell (as Dioscorides writeth) being drunke in sowre or harsh red Wine, stoppeth the laske, and the ouermuch flowing of the flowers or menses, and is a remedie for those that pisse in bed'. An account of further medicinal uses of darnel in the early modern period is given in the herbal of Rembert Dodoens (1595). The traditional use of darnel as an anaesthetic in Middle Eastern medicine is recorded (Haddad 2005). Camporesi (1989) wrote of the 'unsuspected artificial paradises (that) were opened up to the undernourished and starving' by the deliberate consumption of 'dazed bread' across whole strata of the population of Europe in the pre-industrial era. Curiously, the first century roman author Columella mentions darnel as a good food for domesticated fowl such as pigeons (Jashemski and Meyer 2002), and the Bolognese naturalist Aldrovandi (1600)

[1]Hormesis is the term for physiological responses to toxins delivered at low doses (Kaiser 2003). In contemporary usage, these responses are usually beneficial. However, in pre-modern medicine, the aim of the healer was to bring about a change – whether for good or ill – in the condition of the patient. In this respect, pharmakon, signifying both remedy and poison, is perhaps the more appropriate term – as demonstrated, for example, in Paracelsian iatrochemistry (Debus 1965).

describes how exposure to darnel may be used as a kind of fumigation treatment against disease in chickens.

4.2 *Fungal Endophytes and the Chemistry of* Lolium *Toxins*

The origin of the toxins appeared to become clear around the turn of the twentieth century when microscopists identified the presence of a mutualistic fungal mycelium located immediately outside the aleurone layer of the darnel grain (Freeman 1904). After various taxonomic revisions, the fungus is now named *Neotyphodium occultans* (Moon et al. 2000). *N. occultans* appears to be an asexually propagated congener of the *Epichloë* spp. that form endophytic associations with many grasses. Significantly, *Neotyphodium* and *Epichloë* are genera within the fungal family Clavicipitaceae, which includes the ergot organism *Claviceps purpurea* (Kuldau et al. 1997).

Hofmeister (1892) was the first to isolate alkaloids from darnel. He called a novel compound purified from grain extracts 'temuline' (it is now named norloline). Subsequently, temuline was shown to be just one representative of a family of structurally related alkaloids from *Lolium* spp. known collectively as the lolines and chemically classified as saturated pyrrolizidines (Fig. 9a). It is well established

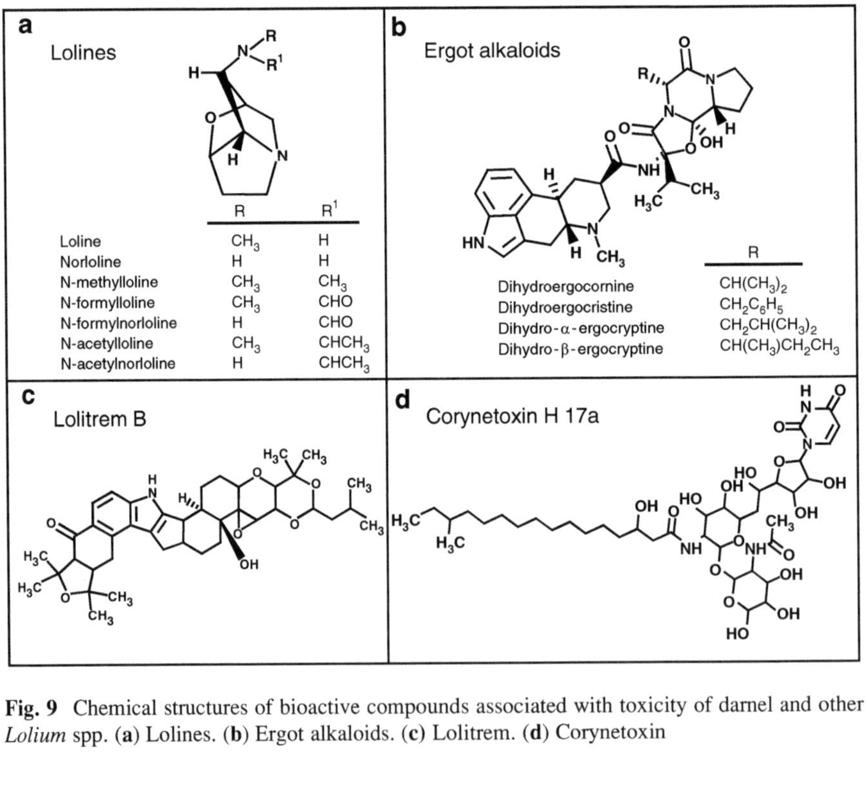

Fig. 9 Chemical structures of bioactive compounds associated with toxicity of darnel and other *Lolium* spp. (**a**) Lolines. (**b**) Ergot alkaloids. (**c**) Lolitrem. (**d**) Corynetoxin

that the lolines are biosynthesised by the fungal symbiont, and toxicity studies on insects and nematodes strongly suggest that the plant receives the benefit of protection from pests as its part of the symbiotic bargain with the fungus (Schardl et al. 2007). For much of the time since their discovery, it was assumed that the lolines must also be responsible for the symptoms of darnel poisoning in humans, and for similar diseases of livestock such as ryegrass staggers; but this has not stood up to testing by modern techniques and experimental design (Schardl et al. 2007). Apart from some nugatory vascular, hormonal and mitogenic effects at impractically high concentrations, human and animal physiologies are unresponsive to the loline alkaloids, and so we must look elsewhere for the source of darnel's toxicity.

4.3 Darnel and Ergot

The symptomology of darnel poisoning closely resembles that of ergot. It is likely that historical accounts of the scourge of contamination in the cereal food chain have entangled the two sources of corruption (see, for example, Barger 1931; Wasson et al. 1978). The etymology of the word 'darnel' offers an intriguing perspective on this relationship. Skeat (1899) persuasively argues that the word is a compound of 'dar' and 'nelle'. In many old and middle languages of Europe, 'dar' or 'dor' means stupefied, drunken, foolish. 'Nelle' is related to 'nigella', which in turn describes any of a number of weeds with black (Latin: niger) seeds. Definitions in the Oxford English Dictionary (2nd ed) tell us that the colloquial names of arable weeds were formerly applied interchangeably to a group of disparate species, including cockle, poppy, hemlock, nigella and darnel. Skeat, citing Henslow, Fuchsius and Gerard's Herball, considers the 'nel' element of the name to have originated in this free and easy approach to botanical nomenclature. The symptoms of fungal, bacterial and invertebrate infestations of crops and their weeds further confuse the picture. Camporesi (1989), quoting sixteenth century Italian sources, writes of 'grasses with stupefying seeds such as darnel, and a related variety ... a herb with black grains, used as forage or feed for chickens and capons'. Our colleagues Riccardo Battelli (University of Pisa) and Adriana Ravagnani (Aberystwyth University) have discovered that this 'variety' (called in dialect 'g(h)i(o)ttone' and related to the Italian for 'glutton') is in fact wheat infested with the nematode *Vibrio tritici* (now *Anguina tritici*– Curtis 1860; Riley 1992 – Fig. 10). To quote an early source: 'Prepare it (wheat) well and do not allow it to be contaminated by *Lolium* or ghiottone...If on the crust (of the bread) there are small black spots, there is ghiottone'. Evidently, darnel and ghiottone are closely connected in their agronomy and occurrence in the food chain. It is easy to imagine etymological confluence leading to association of *L. temulentum* with black grains.

But perhaps a more plausible explanation for the 'nel' element of darnel is simply that black seedheads is a symptom of ergot in the crop. We may speculate that such is *L. temulentum*'s propensity to become infested with ergot (Grieve 1931) that it was identified by, and came to be named after, its black seeds. The race of

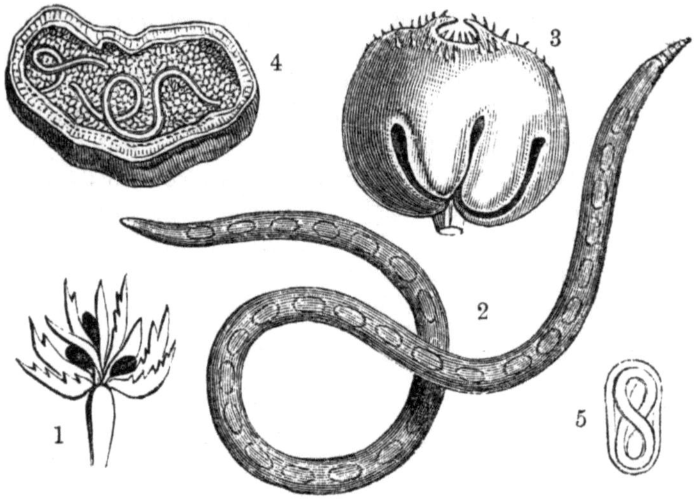

Fig. 10 Illustration from Curtis (1860) of *Vibrio* (now *Anguina*) *tritici*. (**1**) Drawing representing wheat suffering from 'ear cockle' or 'purples'. (**2**) 'The largest worms are 1/4 in. long at least, of a yellowish-white colour...'. (**3**) After releasing the worms '...the grains eventually assumed a dark brown colour, and were as hard as wood'. (**4**) '...section of a grain, exhibiting some worms and a multitude of eggs'. (**5**) Egg showing young worm visible through transparent skin

Claviceps purpurea specific to wheat, G1, is the same as that for *Lolium* spp. (Pazoutová 2003). It is conceivable that contaminating darnel is a reservoir of infestation for wheat under conditions of low crop hygiene (Carruthers 1875). Furthermore, as implied by the close taxonomic relationship between species within the family Clavicipitaceae (Kuldau et al. 1997), the genes for biosynthesis of ergot toxins (Fig. 9b) are represented in the genomes of *Lolium* endophytes (Wanga et al. 2004). The pathway leading to synthesis of *Claviceps*-type alkaloids may become active in the darnel endophyte under particular environmental conditions (Lyons et al. 1986), leading to ergot-like contamination of the cereal food chain.

4.4 Other Possible Sources of Toxicity in L. temulentum

Other kinds of bioactive chemical might be part of darnel's poisonous brew. The indolediterpene neurotoxins (lolitrems – Miles et al. 1992 – Fig. 9c) are a group of tremorigenic metabolites, produced by *Lolium* endophytes, with high potency in mammals. These chemicals are strongly implicated in the grass staggers disease of livestock and could well contribute to the toxicity of darnel towards humans. Another class of bioactive chemicals that might be significant is the corynetoxins (Fig. 9d). The seedheads of *L. rigidum* are prone to infection by a bacterium,

Clavibacter toxicus (Riley and McKay 1990), which produces corynetoxins, extremely poisonous tunicaminyluracil antibiotics. These are known to cause severe toxic reactions in livestock (Edgar et al. 1982; Culvenor and Jago 1985). Darnel is reported to be a source of corynetoxin (European Food Safety Authority 2007). Interestingly, the nematode *A. tritici* (see above) is a vector for the spread of *Clavibacter* in *Lolium* species and is implicated in the Annual Ryegrass Toxicity disease of livestock (Riley 1992). Contamination of the human food chain by corynetoxins has only recently begun to be researched but is already causing disquiet (Edgar 2004). Amongst other obscure but suspect organisms associated with *Lolium* is the fungus *Gloeotinia temulenta*, which causes blind seed disease (Alderman 1998). It is said that infected grass is not toxic to grazing animals, but no systematic study appears to have been made of the consequences of allowing the fungus into the human food chain.

5 *L. temulentum* in History and Literature

This concluding section is a brief and selective treatment drawn from a more comprehensive review by the authors to be published elsewhere (see Marggraf Turley et al. 2010). Darnel has been written about throughout history in many cultures. The medicinal and entheogenic uses of this species have been mentioned above. There are numerous references to darnel as a pernicious weed of cereals in Classical narratives (see Smith and Secoy 1975). The principal Classical sources on the subject of darnel in the Medieval and early modern periods are Pliny's *Natural History*, Dioscorides's *De Materia Medica* and Theophrastus's *Enquiry into Plants*. Amongst the earliest appearances of darnel in the literary context is reference in the play *Merchant Ships* by Aristophanes to 'αἱρόπινον', the 'darnel sieve', which is thought to be a comic allusion to the impression that the sieve, through which cereal grains are passed to clean them, has drunk the contaminating darnel seeds (Amigues 2003). Perhaps the best-known literary references to darnel occur in Shakespeare (*Complete Works*, 2008 edition), notably *Henry VI part 1*, *Henry V* and *King Lear*. It is significant that there are strong themes of (self) deception, treachery, political corruption and fomented unrest in these plays, implicit in the words of 'the English scourge' Joan la Pucelle (*Henry VI part 1* III ii):

> Good morrow, gallants! want ye corn for bread?
> I think the Duke of Burgundy will fast
> Before he'll buy again at such a rate:
> 'Twas full of darnel; do you like the taste?

Lear's madness also has more than a passing resemblance to the behaviour expected of someone who has, deliberately or inadvertently, consumed the 'furrow weeds' (including 'Darnel') with which he is crowned (*King Lear* IV iii). Figure 11 is a rare example of a representation of Lear that apparently depicts darnel in his crown.

Fig. 11 King Lear. Engraving from a pirated edition of Charles and Mary Lamb, *Tales From Shakespeare*, printed by Philadepia publisher Henry Altemus, in editions that appeared between 1895 and 1901. The engraver was Frederick Wentworth

Darnel was not only all too familiar to people in the early modern period as a dietary threat and a potential (although highly problematic) medicine or 'simple', it was also, as a malign mimic of cereals, a frequent literary trope for subversion and treasonous behaviour. This tradition in part derives from the long-standing use of darnel in religious writings and scriptural exegesis, where the wheat in the Biblical parable (Matthew 13: 24–30, 36–43; King James Version) represents the godly and faithful (those who read the Bible 'correctly'), even as tares (darnel – Table 3) stands for the heretic and schismatic (who read the Bible 'incorrectly'). The herbal of Levinus Lemnius (1587) is an example of the way in which early modern discourses of religion and politics come together in the shape of darnel. The coupling of religious and political sedition is reflected in the following example from William Gamage (1613):

> Joseph, with his Apostles twelve first plants,
> In Englands Soile, Religion pure to grow;
> But thou, and thy twise sixe infernall Wants,
> Didst this endeavour to supplant; and Sow
> Thy Popish Darnell; but the season fail'd,
> And thou with thine, to Tyburns post wast nail'd.
> (*Epig. 8. Garnet, with his Twelue Apostles*).

It refers to the Jesuit missionary Father Henry Garnett, convicted as one of the conspirators in the Gunpowder Plot of 1605, which sought to reinstate Roman Catholicism in the Protestant England of King James I (Caraman 1995).

L. temulentum's rich cultural and scientific history speaks eloquently of the enduring relationship between humans, no matter how technologically sophisticated and urbanised, and the agricultural and natural environments that sustain and sometimes threaten them.

Acknowledgements We are grateful to Ricardo Battelli and Adriana Ravagnani for information about ghiottone, Chris Pollock for answering queries about fructan and John Harper for providing illustrative material for Fig. 1. Constructive suggestions received during the review and editorial process were much appreciated. Howard Thomas acknowledges support from the Leverhulme Trust.

References

Alderman SC (1998) *Gloeotinia temulenta* and *G. granigena*, two distinct species. Mycologia 90:422–426

Aldrovandi U (1600) Ornithologiae tomus alter. Liber Decimusquartus qui est de Pulveratricibus Domesticis. [Digitized version at http://gdz.sub.uni-goettingen.de/no_cache/dms/load/toc/?IDDOC=234603, cited 4 May 2009]

Allaby RG, Fuller DQ, Brown TA (2008) The genetic expectations of a protracted model for the origins of domesticated crops. Proc Natl Acad Sci USA 105:13982–13986

Amigues S (2003) Le crible à ivraie d'Aristophane FR. 497 K. – A. Revue de philologie, de littérature et d'histoire anciennes 77:17–22

Appleby AP, Olson PO, Colbert DR (1976) Winter wheat yield reduction from interference by Italian ryegrass. Agron J 68:463–466

ArchAtlas (2006) 2nd edition, http://www.archatlas.org/, cited 12 May 2009

Armstead I, Turner LB, Farrell M, Skøt L, Gomez P, Montoya T, Donnison IS, King IP, Humphreys MO (2004) Synteny between a major heading-date QTL in perennial ryegrass (*Lolium perenne* L.) and the Hd3 heading-date locus in rice. Theor Appl Genet 108:822–828

Armstead I, Donnison I, Aubry S, Harper J, Hörtensteiner S, James C, Mani J, Moffet M, Ougham H, Roberts L, Thomas A, Weeden N, Thomas H, King I (2006) From crop to model to crop: identifying the genetic basis of the staygreen mutation in the forage grass *Festuca pratensis* (Huds.). New Phytol 172:592–597

Armstead I, Donnison I, Aubry S, Harper J, Hörtensteiner S, James C, Mani J, Moffet M, Ougham H, Roberts L, Thomas A, Weeden N, Thomas H, King I (2007) Cross-species identification of Mendel's I locus. Science 315:73

Bacon MA, Thompson DS, Davies WJ (1997) Can cell wall peroxidase activity explain the leaf growth response of *Lolium temulentum* L. during drought? J Exp Bot 48:2075–2085

Baldwin JC, Dombrowski JE (2006) Evaluation of *Lolium temulentum* as a model grass species for the study of salinity stress by PCR-based subtractive suppression hybridization analysis. Plant Sci 171:459–469

Baldwin JC, Dombrowski JE, Martin RC, Banowetz GM (2007) Differentially expressed genes associated with post-harvest processing in *Lolium temulentum* L. Plant Sci 173:73–83

Balfourier F, Imbert C, Charmet G (2000) Evidence for phylogeographic structure in *Lolium* species related to the spread of agriculture in Europe. A cpDNA study. Theor Appl Genet 101:131–138

Barger G (1931) Ergot and ergotism: a monograph. Gurney and Jackson, London

Bennett SJ (1997) A phenetic analysis and lateral key of the genus *Lolium* (Gramineae). Genet Resour Crop Evol 44:63–72

Bennett MD, Leitch IJ (2004) Angiosperm DNA C-values database (release 5.0, Dec. 2004) [http://data.kew.org/cvalues/homepage.html, cited 6 May 2009]

Borrill M (1961) The developmental anatomy of leaves in *Lolium temulentum*. Ann Bot 25:1–11

Buchanan-Wollaston V, Earl S, Harrison E, Mathas E, Navabpour S, Page T, Pink D (2003) The molecular analysis of leaf senescence - a genomics approach. Plant Biotechnol J 1:3–22

Bulinska-Radomska Z, Lester RN (1985) Relationships between five species of *Lolium* (Poaceae). Plant Syst Evol 148:169–175

Caetano-Anollés G (2005) Evolution of genome size in grasses. Crop Sci 45:1809–1816

Cairns AJ, Bonnett GD, Gallagher JA, Simpson RJ, Pollock CJ (1997) Fructan biosynthesis in excised leaves of *Lolium temulentum*. New Phytol 136:61–72

Cairns AJ, Cookson A, Thomas BJ, Turner LB (2002) Starch metabolism in the fructan-grasses: Patterns of starch accumulation in excised leaves of *Lolium temulentum* L. J Plant Physiol 159:293–305

Camporesi P (1989) Bread of dreams. Polity Press, Cambridge

Caraman P (1995) A study in friendship: Saint Robert Southwell and Henry Garnet. Institute of Jesuit Sources, Saint Louis, MO

Carruthers W (1875) On ergot. Am Nat 9:450–465

Catalan P, Torrecilla P, Rodrıguez JAL, Olmstead RG (2004) Phylogeny of the festucoid grasses of subtribe Loliinae and allies (Poeae, Pooideae) inferred from ITS and trnL–F sequences. Mol Phylogenet Evol 31:517–541

Causton DR, Venus JC (1981) The biometry of plant growth. Edward Arnold, London

Chailakhyan MK (1936) New facts in support of the hormonal theory of plant development. Comptes Rendus de l'Academie des Sciences, URSS 13:79–83

Charmet G, Ravel C, Balfourier F (1997) Phylogenetic analysis in the *Festuca-Lolium* complex using molecular markers and ITS rDNA. Theor Appl Genet 94:1038–1046

Cooper JP (1950) Day-length and head formation in the ryegrasses. J British Grassland Soc 5:105–112

Cooper JP (1951) Studies on growth and development in *Lolium*. II. Pattern of bud development on the shoot apex and its ecological significance. J Ecol 39:228–270

Cooper JP (1952) Studies on growth and development in *Lolium*. III. Influence of season and latitude on ear emergence. J Ecol 40:352–379

Cooper JP (1954) Studies on growth and development in *Lolium*. IV. Genetic control of heading responses in local populations. J Ecol 42:522–556

Corbesier L, Vincent C, Jang S, Fornara F, Fan Q, Searle I, Giakountis A, Farrona S, Gissot L, Turnbull C, Coupland G (2007) FT protein movement contributes to long-distance signaling in floral induction of Arabidopsis. Science 316:1030–1033

Culvenor CCJ, Jago MV (1985) Annual ryegrass toxicity. In: Lacey J (ed) Tricothenes and other mycotoxins. Wiley, London, pp 159–168

Curtis J (1860) Farm insects: being the natural history and economy of the insects injurious to the field crops of Great Britain and Ireland, and also those which infest barns and granaries, with suggestions for their destruction. Blackie and Son, London

Davies TGE, Ougham HJ, Thomas H, Rogers LJ (1989) Leaf development in *Lolium temulentum* - plastid membrane polypeptides in relation to assembly of the photosynthetic apparatus and leaf growth. Physiol Plant 75:47–54

Davies TGE, Rogers LJ, Thomas BJ, Thomas H (1990) Leaf development in *Lolium temulentum*: formation of the photosynthetic apparatus in the presence of the porphyrin synthesis inhibitor gabaculine. J Exp Bot 41:905–917

De Wet JMJ, Harlan JR (1975) Weeds and domesticates: evolution in the man-made habitat. Econ Bot 29:99–107

Debus AG (1965) The English paracelsians. Oldbourne, London

Dioscorides (2005) De Materia Medica (translated by LY Beck), 2005 edition. Olms-Weidmann, Hildesheim, NY

Dodoens R (1595) A new herball, or historie of plants (translated H Lyte). Edmund Bollifant, London, pp 541–42

Dombrowski JE, Baldwin JC, Martin RC (2008) Cloning and characterization of a salt stress-inducible small GTPase gene from the model grass species *Lolium temulentum*. J Plant Physiol 165:651–661

Donnison I, Cisneros P, Montoya T, Armstead I, Thomas B, Thomas A, Jones N, Morris P (2002) The floral transition in model and forage grasses. Flowering Newsl 33:42–48

Drury S (1992) Plants and pest control in England circa 1400–1700: a preliminary study. Folklore 103:103–106

Edgar J (2004) Future impact of food safety issues on animal production and trade: implications for research. Aust J Exp Agric 44:1073–1078

Edgar JA, Frahn JL, Cockrum PA, Anderton N, Jago MV, Culvenor CCJ, Jones AJ, Murray K, Shaw KJ (1982) Corynetoxins, causative agents of annual ryegrass toxicity; their identification as tunicamycin group antibiotics. J Chem Soc Chem Commun 222–224

Ellacombe HN (1884) Plant-lore and garden-craft of Shakespeare, 2nd edn. Simpkin, Marshall and Company, London

European Food Safety Authority (2007) Opinion of the Scientific Panel on contaminants in the food chain on a request from the European Commission related to pyrrolizidine alkaloids as undesirable substances in animal feed (Question N° EFSA-Q-2003-065). EFSA J 447:1–51

Evans LT (1958) *Lolium temulentum* L., a long-day plant requiring only one inductive photocycle. Nature 182:197–198

Evans LT (1993) Crop evolution adaptation and yield. Cambridge University Press, Cambridge

Evans LT (2003) Conjectures, refutations, and extrapolations. Annu Rev Plant Biol 54:1–21

Evans LT, Wardlaw IF (1966) Independent translocation of ^{14}C-labelled assimilates and of the floral stimulus in *Lolium temulentum*. Planta 68:310–326

Farrar J, Pollock C, Gallagher J (2000) Sucrose and the integration of metabolism in vascular plants. Plant Sci 154:1–11

Farrell AD, Ougham HJ, Tomos AD (2006) The effect of gibberellic acid on the response of leaf extension to low temperature. Plant Cell Environ 29:1329–1337

Forcella F, Harvey SJ (1988) Patterns of weed migration in Northwestern USA. Weed Sci 36:194–201

Fournier C, Durand JL, Ljutovac S, Schäufele R, Gastal F, Andrieu B (2005) A functional-structural model of elongation of the grass leaf and its relationships with the phyllochron. New Phytol 166:881–894

Freeman EM (1904) The seed-fungus of *Lolium temulentum* L., the darnel. Philos Trans R Soc Lond B 196:1–27, Containing Papers of a Biological Character

Gallagher JA, Cairns AJ, Pollock CJ (2004) Cloning and characterization of a putative fructosyltransferase and two putative invertase genes from the temperate grass *Lolium temulentum* L. J Exp Bot 55:557–569

Gamage W (1613) Linsi-woolsie. Or Two centuries of epigrammes. Oxford: Joseph Barnes, sigs A7r-A7v

Garner WW, Allard HA (1920) Effect of the relative length of day and night and other factors of the environment on growth and reproduction in plants. J Agric Res 18:553–606

Gaut BS, Tredway LP, Kubik C, Gaut RL, Meyer W (2000) Phylogenetic relationships and genetic diversity among members of the *Festuca-Lolium* complex (Poaceae) based on ITS sequence data. Plant Syst Evol 224:33–53

Gay AP, Hauck B (1994) Acclimation of *Lolium temulentum* to enhanced carbon dioxide concentration. J Exp Bot 45:1133–1141

Gay AP, Thomas H (1995) Leaf development in *Lolium temulentum*: photosynthesis in relation to growth and senescence. New Phytol 130:159–168

Gay A, Thomas H, Roca M, James C, Taylor J, Rowland J, Ougham H (2008) Non-destructive analysis of senescence in mesophyll cells by spectral resolution of protein synthesis-dependent pigment metabolism. New Phytol 179:663–674

Gerard J (1597) The Herball. London, John Norton, p 72

Ghanem J, Hershko C (1981) Veno-occlusive disease and primary hepatic vein thrombosis in Israeli Arabs. Israel J Med Sci 17:339–347

Grieve M (1931). A modern herbal. 1971 edition. NY: Dover. [Online version at http://www.botanical.com/botanical/mgmh/mgmh.html, cited 16 January 2009]

Haddad FS (2005) A middle age Arab/Islamic anesthetic herb. Middle East J Anesthesiol 18:249–264

Harlan JR (1981) The early history of wheat: earliest traces to the sack of Rome. In: Evans LT, Peacock WJ (eds) Wheat science – today and tomorrow. Cambridge University Press, Cambridge, pp 1–19

Hedley CL, Stoddart JL (1972) Patterns of protein synthesis in *Lolium temulentum* L. I. Changes occurring during leaf development. J Exp Bot 23:490–501

Hofmeister F (1892) The active constituents of *Lolium temulentum*. Arch Exp Path Pharmakol 30:203–230

Holm LG, Plucknett DL, Pancho JV, Herberger JP (1977) The World's worst weeds. University Press of Hawaii, Honolulu

Holman JD, Menalled FD, Bussan AJ (2004) Persian darnel: identification, biology and ecology. Montana Sate University Extension Service. [http://www.montana.edu/wwwpb/pubs/mt200411.html, cited 21 April 2009]

Housley TL, Gibeaut DM, Carpita NC, Sims I, Pollock CJ (1991) Fructosyl transfer from sucrose and oligosaccharides during fructan synthesis in excised leaves of *Lolium temulentum* L. New Phytol 119:491–497

Hutchinson J, Rees H, Seal AG (1979) Assay of the activity of supplementary DNA in *Lolium*. Heredity 43:411–421

Inda LA, Segarra-Moragues JG, Müller J, Peterson PM, Catalán P (2008) Dated historical biogeography of the temperate Loliinae (Poaceae, Pooideae) grasses in the northern and southern hemispheres. Mol Phylogenet Evol 46:932–957

Jashemski WMF, Meyer FG (2002) The natural history of Pompeii. Cambridge University Press, Cambridge

Jenkin TJ (1933) Interspecific and intergeneric hybrids in herbage grasses. Initial crosses. J Genet 28:205–264

Jenkin TJ (1935) Interspecific and intergeneric hybrids in herbage grasses. II Lolium perenne x Lolium temulentum. J Genet 31:379–411

Jenkin TJ, Thomas PT (1938) The breeding affinities and cytology of *Lolium* species. J Bot 76:10–12

Jenkins G, White J (1990) Elimination of synaptonemal complex irregularities in a *Lolium* hybrid. Heredity 64:45–53

Kaiser J (2003) Hormesis: sipping from a poisoned chalice. Science 302:376–379

King RW, Evans LT, Mander LN, Moritz T, Pharis RP, Twitchin B (2003) Synthesis of gibberellin GA_6 and its role in flowering of *Lolium temulentum*. Phytochemistry 62:77–82

King RW, Moritz T, Evans LT, Martin J, Andersen CH, Blundell C, Kardailsky I, Chandler PM (2006) Regulation of flowering in the long-day grass *Lolium temulentum* by gibberellins and the FLOWERING LOCUS T gene. Plant Physiol 141:498–507

Kingston-Smith AH, Theodoou MK (2000) Post-ingestion metabolism of fresh forage. New Phytol 148:37–55

Koch KE (1996) Carbohydrate-modulated gene expression in plants. Ann Rev Plant Physiol Plant Mol Biol 47:509–540

Kuldau GA, Liu J-S, White JF, Siegel MR, Schardl CL (1997) Molecular systematics of Clavicipitaceae supporting monophyly of genus *Epichloë* and form genus *Ephelis*. Mycologia 89:431–444

Ladizinsky G (1998) Plant evolution under domestication. Springer, Berlin

Lemnius L (1587) An herbal for the Bible. London, Edmund Bollifant, p 230

Lewis CE, Peratoner G, Cairns AJ, Causton DR, Foyer CH (1999) Acclimation of the summer annual species, *Lolium temulentum* to CO_2 enrichment. Planta 210:104–114

Loos BP (1993) Morphological variation in *Lolium* (Poaceae) as a measurement of species relationships. Plant Syst Evol 188:87–99

Louahlia S, Macduff JH, Ourry A, Humphreys M, Boucaud J (1999) Nitrogen reserve status affects the dynamics of nitrogen remobilization and mineral nitrogen uptake during recovery of contrasting cultivars of *Lolium perenne* from defoliation. New Phytol 142:451–462

Lyons PC, Plattner RD, Bacon CW (1986) Occurrence of peptide and clavine ergot alkaloids in tall fescue grass. Science 232:487–489

Mander LN, Sherburn M, Camp D, King RW, Evans LT, Pharis RP (1999) Effects of D-ring modified gibberellins on flowering and growth in *Lolium temulentum*. Phytochemistry 49:2195–2206

Marggraf Turley R, Thomas H, Archer, JE (2010) A tragedy of idle weeds. Times Literary Supplement No. 5577 (19 February 2010) pp 14–15

McCreery DW (1979) Flotation of the Bab edh-Dhra and Numeria plant remains. Ann Am Sch Orient Res 46:165–169

Miles CO, Wilkins AL, Gallagher RT, Hawkes AD, Munday SC, Towers NR (1992) Synthesis and tremorgenicity of paxitriols and lolitriol: possible biosynthetic precursors of lolitrem B. J Agric Food Chem 40:234–238

Mirjalili SA, Bennett SJ, Poorazizi E (2008) A phenetic analysis on the genus *Lolium* (Poaceae) in Iran. Plant Syst Evol 274:203–208

Moon CD, Scott B, Schardl CL, Christensen MJ (2000) The evolutionary origins of *Epichloë* endophytes from annual ryegrasses. Mycologia 92:1103–1118

Morris K, Thomas H, Rogers LJ (1996) Endopeptidases during the development and senescence of *Lolium temulentum* leaves. Phytochemistry 41:377–384

Moses L, Ougham HJ, Francis D (1997) The effect of the slow-to-green mutation on cell division during leaf initiation and early leaf growth in *Lolium temulentum*. New Phytol 135:51–57

Murray BG (2005) When does intraspecific C-value variation become taxonomically significant? Ann Bot 95:119–125

Musselman LJ (2000) Zawan and tares in the Bible. Econ Bot 54:537–542

Naylor B (1960) Species differentiation in the genus *Lolium*. Heredity 15:219–233

Ougham HJ (1987) Gene-expression during leaf development in *Lolium temulentum* - patterns of protein-synthesis in response to heat-shock and cold-shock. Physiol Plant 70:479–484

Ougham HJ, Davies TGE (1990) Leaf development in *Lolium temulentum* - gradients of RNA complement and plastid and nonplastid transcripts. Physiol Plant 79:331–338

Ougham HJ, Thomas H, Hilditch P (1987) Leaf development in *Lolium temulentum* L.: gradients of expression of growth and photosynthesis related polypeptides revealed by immunoblotting. J Plant Physiol 129:181–186

Ougham HJ, Thomas AM, Thomas BJ, Roberts PC, Mutinda C, Hayward MD, Dalton SJ (1992) Leaf development in *Lolium temulentum* L - characterization of a slow-to-green mutant. New Phytol 122:261–272

Ougham H, Hörtensteiner S, Armstead I, Donnison I, King I, Thomas H, Mur L (2008) The control of chlorophyll catabolism and the status of yellowing as a biomarker of leaf senescence. Plant Biol 10(suppl 1):4–14

Ourry A, Macduff JH, Ougham HJ (1996) The relationship between mobilization of N reserves and changes in translatable messages following defoliation in *Lolium temulentum* L and *Lolium perenne* L. J Exp Bot 47:739–747

Pazoutová S (2003) Evolutionary strategy of *Claviceps*. In: White JF, Bacon CW, Hywel-Jones NL, Spatafora JW (eds) Clavicipitalean Fungi: evolutionary biology, chemistry, biocontrol and cultural impacts. Marcel Dekker, Basel, NY, pp 329–354

Peacock JM (1975) Temperature and leaf growth in *Lolium perenne*. II. The site of temperature perception. J Appl Ecol 12:115–123
Peacock JM (1976) Temperature and leaf growth in four grass species. J Appl Ecol 13:225–232
Perilleux C, Bernier G (1997) Leaf carbohydrate status in *Lolium temulentum* during the induction of flowering. New Phytol 135:59–66
Perilleux C, Bernier G, Kinet J-M (1994) Circadian-rhythms and the induction of flowering in the long-day grass *Lolium temulentum* L. Plant Cell Environ 17:755–761
Perilleux C, Bernier G, Kinet J-M (1997) Darkness promotes flowering in the absolute long-day requiring plant. Lolium temulentum L. Ceres. J Exp Bot 48:349–351
Pliny the Elder (1938) Natural History (10 volumes, translated by H Rackham, WHS Jones, AC Andrews), 1938–63 edition. Harvard University Press for LOEB Classical Library, Cambridge, MA
Pollock CJ (1984) Sucrose accumulation and the initiation of fructan biosynthesis in *Lolium temulentum* L. New Phytol 96:527–534
Pollock CJ, Eagles CF (1988) Low temperature and the growth of plants. Symp Soc Exp Biol 42:157–180
Pollock CJ, Lloyd EJ (1998) The effect of low temperature upon starch, sucrose and fructan synthesis in leaves. Ann Bot 60:231–235
Pollock CJ, Lloyd EJ, Stoddart JL, Thomas H (1983) Growth, photosynthesis and assimilate partitioning in *Lolium temulentum* exposed to chilling temperatures. Physiol Plant 59: 257–262
Pollock CJ, Lloyd EJ, Thomas H, Stoddart JL (1984) Changes in photosynthetic capacity during prolonged growth of *Lolium temulentum* at low temperature. Photosynthetica 18:478–481
Pollock CJ, Cairns AJ, Collis BE, Walker RP (1989) Direct effects of low temperature upon components of fructan metabolism in leaves of *Lolium temulentum* L. J Plant Physiol 134:203–208
Porcher MH (2007) Multilingual multiscript plant name database: *Lolium*. [http://www.ars-grin.gov/misc/mmpnd/Lolium.html, cited 12 May 2009]
Riley IT (1992) *Anguina tritici* is a potential vector of *Clavibacter toxicus*. Australas Plant Pathol 2:147–149
Riley IT, McKay AC (1990) Specificity of the adhesion of some plant pathogenic micro-organisms to the cuticle of nematodes in the genus *Anguina* (Nematoda: Anguinidae). Nematologia 36:90–103
Rizk A-FM, Hussiney HA (1991) Chemistry and toxicity of *Lolium* species. In: Rizk A-FM (ed) Poisonous plant contamination of edible plants. CRC Press, Boca Raton, FL, pp 95–106
Roca M, James C, Pružinská A, Hörtensteiner S, Thomas H, Ougham H (2004) Analysis of the chlorophyll catabolism pathway in leaves of an introgression senescence mutant of *Lolium temulentum*. Phytochemistry 65:1231–1238
Ryle GJA (1972) A quantitative analysis of the uptake of carbon and of the supply of ^{14}C-labelled assimilates to areas of meristematic growth in *Lolium temulentum*. Ann Bot 36:497–512
Ryle GJA, Powell CE (1972) The export and distribution of ^{14}C-labelled assimilates from each leaf on the shoot of *Lolium temulentum* during reproductive and vegetative growth. Ann Bot 36:363–375
Saji H, Vince-Prue D, Furuya M (1983) Studies on the photoreceptors for the promotion and inhibition of flowering in dark-grown seedlings of *Pharbitis nil* Choisy. Plant Cell Physiol 67:1183–1189
Salisbury FB (1963) Biological timing and hormone synthesis in flowering of *Xanthium*. Planta 49:518–524
Schardl CL, Grossman RB, Nagabhyru P, Faulkner JR, Mallik UP (2007) Loline alkaloids: currencies of mutualism. Phytochemistry 68:980–996
Schifino MT, Winge H (1983) Karyoptypes and nuclear DNA content of species of the *Briza* complex and some other genera of Poeae (Gramineae). Rev Bras Genet 6:245–259
Senda T, Tominaga T (2003) Inheritance mode of the awnlessness of darnel (*Lolium temulentum* L.). Weed Biol Manag 3:46–48

Senda T, Tominaga T (2004) Genetic diversity of darnel (*Lolium temulentum* L.) in Malo, Ethiopia depends on traditional farming systems. Econ Bot 58:568–577

Senda T, Hiraoka Y, Tominaga T (2005a) Cytological affinities and interfertilities between *Lolium temulentum* and *L. persicum* (Poaceae) accessions. Hereditas 142:45–50

Senda T, Ohsako T, Tominaga T (2005b) Interspecific differentiation and phylogenetic relationships of poison ryegrass (*Lolium temulentum* L.) and Persian darnel (*L. persicum* Boiss. & Hohen. ex Boiss.). Can J Plant Sci 85:963–970

Senda T, Saito M, Ohsako T, Tominaga T (2005c) Analysis of *Lolium temulentum* geographical differentiation by microsatellite and AFLP markers. Weed Res 45:18–26

Shakespeare W (2008) *Complete works*, 2008 edition. Bate J, Rasmussen E (eds). Basingstoke: Macmillan

Shinomo Y, Konuma A (2008) Effects of human-mediated processes on weed species composition in internationally traded grain commodities. Weed Res 48:10–18

Sims IM, Pollock CJ, Horgan R (1992) Structural analysis of oligomeric fructans from excised leaves of *Lolium temulentum*. Phytochemistry 31:2989–2992

Skeat WW (1899) VI Notes on English etymology. Trans Phil Soc 24:261–291

Smart CM (1994) Gene expression during leaf senescence. New Phytol 126:419–448

Smith AE, Secoy DM (1975) Forerunners of pesticides in classical Greece and Rome. Agric Food Chem 23:1050–1055

Stebbins GL (1981) Co-evolution of grasses and herbivores. Ann Mo Bot Garden 68:75–86

Stoddart JL, Thomas H, Lloyd EJ, Pollock CJ (1986) The use of a temperature profiled position transducer for the study of low temperature growth in Gramineae: equipment design and output interpretation. Planta 167:359–363

Täckholm V, Drar M, Täckholm G (1973) Flora of Egypt. Cairo University Press, Cairo

Tamaki S, Matsuo S, Wong HL, Yokoi S, Shimamoto K (2007) Hd3a protein is a mobile flowering signal in rice. Science 316:1033–1036

Terrell EE (1968) A taxonomic revision of the genus *Lolium*. Tech Bull US Dep Agric 1392

Theophrastus (1916). Enquiry into plants. Cambridge, MA: Harvard University Press for LOEB Classical Library. (2 volumes, translated by AF Hort)

Thomas H (1975) Regulation of alanine aminotransferase in leaves of *Lolium temulentum* during senescence. Z Pflanzenphysiol 74:208–212

Thomas H (1978) Enzymes of nitrogen mobilization in detached leaves of *Lolium temulentum* during senescence. Ann Bot 51:363–371

Thomas H (1983a) Analysis of the nitrogen response of leaf extension in *Lolium temulentum* seedlings. Physiol Plant 57:509–513

Thomas H (1983b) Analysis of the response of leaf extension to chilling temperatures in *Lolium temulentum* seedlings. Physiol Plant 57:509–513

Thomas H (1987) Sid: a Mendelian locus controlling thylakoid membrane disassembly in senescing leaves of *Festuca pratensis*. Theor Appl Genet 73:551–555

Thomas H, Feller U (1993) Leaf development in *Lolium temulentum*: differential susceptibility of transaminase isoenzymes to proteolysis. J Plant Physiol 142:37–42

Thomas H, Howarth CJ (2000) Five ways to stay green. J Exp Bot 51:329–337

Thomas H, Potter JF (1985) Fitting logistic type curves to extension growth data for leaves of grass species by means of the Maximum Likelihood Program: analysis of leaf extension in *Lolium temulentum* at optimal and chilling temperatures. Environ Exp Bot 25:157–163

Thomas H, Stoddart JL (1974) Subcellular distribution of alanine aminotransferase in leaves of *Lolium temulentum*. Phytochemistry 13:1053–1058

Thomas H, Stoddart JL (1984) Kinetics of leaf growth in *Lolium temulentum* at optimal and chilling temperatures. Ann Bot 53:341–347

Thomas H, Rowland JJ, Stoddart JL (1984) A microprocessor based instrument for measuring plant growth. J Microcomput Appl 7:217–223

Thomas A, Tomos AD, Stoddart JL, Thomas H, Pollock CJ (1989) Cell expansion rate, temperature and turgor pressure in growing *Lolium temulentum* leaves. New Phytol 112:1–5
Thomas H, Morgan WG, Thomas AM, Ougham HJ (1999) Expression of the stay-green character introgressed into *Lolium temulentum* Ceres from a senescence mutant of *Festuca pratensis*. Theor Appl Genet 99:92–99
Thomas H, Thomas HM, Ougham H (2000) Annuality, perenniality and cell death. J Exp Bot 51:1–8
Thomas H, Ougham H, Canter P, Donnison I (2002) What stay-green mutants tell us about nitrogen remobilisation in leaf senescence. J Exp Bot 53:801–808
Thomas H, Ougham HJ, Wagstaff C, Stead AJ (2003) Defining senescence and death. J Exp Bot 54:1127–1132
Tominaga T, Fujimoto T (2000) Evolution of the diversity in darnel (*Lolium temulentum* L.) in Malo, Ethiopia. J Weed Sci Technol 45(suppl):176–177
Tominaga T, Fujimoto T (2004) Awn of darnel (*Lolium temulentum* L.) as an anthropogenic dispersal organ: a case study in Malo, south-western Ethiopia. Weed Biol Manag 4:218–221
Tominaga T, Yamasue Y (2004) Crop-associated weeds. The strategy for adaptation. In: Inderjit (ed) Weed biology and management. Kluwer, Amsterdam, pp 47–63
Turner LB, Cairns AJ, Armstead IP, Ashton J, Skøt K, Whittaker D, Humphreys MO (2006) Dissecting the regulation of fructan metabolism in perennial ryegrass (*Lolium perenne*) with quantitative trait locus mapping. New Phytol 169:45–57
van der Valk P, Proveniers MCG, Pertijs JH, Lamers JTWH, van Dun CMP, Smeekens JCM (2004) Late heading of perennial ryegrass caused by introducing an Arabidopsis homeobox gene. Plant Breed 123: 531–535
Vijn I, Smeekens S (1999) Fructan: more than a reserve carbohydrate? Plant Physiol 120:351–359
Wanga J, Machadoa C, Panaccioneb DG, Tsaia H-F, Schardl CL (2004) The determinant step in ergot alkaloid biosynthesis by an endophyte of perennial ryegrass. Fungal Gen Biol 41:189–198
Wasson RG, Ruck CAP, Hofmann A (1978) The road to Eleusis: unveiling the secret of the mysteries. Harcourt Brace Jovanovich, New York
Wilson BJ, Wright KJ (1990) Predicting the growth and competitive effects of annual weeds in wheat. Weed Res 30:201–211
Wingler A (2007) Transcriptional or posttranscriptional regulation – how does a plant know when to senesce? New Phytol 175:1–4
Winters AL, Gallagher J, Pollock CJ, Farrar JF (1995) Isolation of a gene expressed during sucrose accumulation in leaves of *Lolium temulentum* L. J Exp Bot 46:1345–1350

"Omics" Technologies and Their Input for the Comprehension of Metabolic Systems Particularly Pertaining to Yeast Organisms

L. Strack and U. Stahl

Contents

Abstract "Omics" technologies comprise genomics, transcriptomics, proteomics, metabolomics and phenomics. In this review, these techniques that concentrate on aspects of the "course from genotype to phenotype" are surveyed.

With the aid of these global methods, it is possible to combine a collective knowledge of the investigated organism, which is necessary to understand the details of its metabolic system. Hence, the challenge is to introduce the above-mentioned studies for the determination of targets and approaches for the improvement of several organisms.

In particular, for yeasts, "omics" technologies can be applied well because research is advanced. For this eukaryotic model organism, an in-depth knowledge is indispensable in order to understand the metabolic fluxes better. Herein, the yeast *Saccharomyces cerevisiae* as well as brewing yeasts are reviewed with concern to the determination of their "ome" levels.

U. Stahl (✉) and L. Strack
Institute for Biotechnology, Berlin University of Technology, 13353 Berlin, Germany
e-mail: Ulf.Stahl@lb.tu-berlin.de, L.Strack@tu-berlin.de

U. Lüttge et al. (eds.), *Progress in Botany 72*,
DOI 10.1007/978-3-642-13145-5_4,

1 Introduction

The improvement of organisms towards certain traits has been enabled by the progress in DNA recombinant technology. But before one can start with genetic work, the objective target for the improvement of organisms has to be considered. Today, there are different approaches that concentrate on aspects of the "course from gene to metabolites" for which the strategy of *metabolic engineering* (*ME*) is used. By means of *ME*, the cellular activities can be improved by manipulation of enzymatic, transport, and regulatory functions of the cell (Bailey et al. 1990, 2002). Indeed, it is necessary to have a specific knowledge about the genes, the encoded enzymes, and their regulation for the use of a genotypic trait as a starting point or further improvements. But in many cases, there is insufficient information concerning cellular functions. For this reason, an alternative approach for the improvement of an organism towards desired traits is required for which there is no need of the aforementioned knowledge. The so-called *inverse metabolic engineering* (*IME*) starts with the phenotypic trait in contrast to rational *ME*, which uses the genotype. The *IME* strategy comprises (1) the construction of a desired phenotype and (2) the identification of the genetic basis for the different peculiarities of the traits that is required to (3) allow genetic manipulations. With the aid of these technological advances, it should be possible to understand how a phenotype is determined by a genotype.

But these approaches do not combine a collective knowledge of the organism, which is necessary to understand the details of its metabolic system. Hence, the challenge is to introduce global studies for the comprehension of an organism's metabolism in order to determine targets and approaches for the improvement of several organisms.

2 "Omics" Technologies

The availability of complete genomes for an increasing number of organisms brought about a strong need for comprehensive methods of analysis to take advantage of these complete inventories of genes (Wu et al. 2004). Because only sequencing an organism's genome does not give any details of its enzymes and their encoding genes, the field of "omics" technologies was created to understand the sequence information better (Oliver 2002). These techniques have become important tools in many *ME* strategies and furthermore have facilitated the understanding of organisms' metabolism (Bro and Nielsen 2004). These are mainly genomics, transcriptomics, proteomics, and metabolomics. Genomics concentrate on the study of genomes of an organism. Within functional genomics, the levels of the transcriptome, the proteome, and the metabolome reveal the complete set of the mRNA molecules, the proteins, or the metabolites of a cell. These techniques have allowed the identification of genetic differences (transcripts) and gave insight into their cellular effects (proteins, metabolites). Furthermore, they have provided an insight into the cellular response to genetic alterations or environmental changes. The accumulation of such in-depth knowledge is highly advantageous for elucidating functions of novel genes and gives cause for thought about their genetic manipulation with regard to organisms' improvement (Oliver 2002).

An improvement is still needed for many organisms such as yeast, which has been used over centuries for the production of wines, bakery products, and beer. And although today yeasts are used precisely, various yeast species do not combine every desired trait in one strain. Hence, the application of "omics" technologies to obtain sufficient information on the cellular level especially for imperfect organisms, which are not matable, is indispensable.

For the brewing industry, it is crucial that yeast improvement is the main focus of research purposes. Because over 90% of the beers produced worldwide are lager brews, the research concentrates on lager brewing yeasts (Kodama et al. 2006).

The challenge for the application of "omics" technologies to strains of lager brewer's yeast is that these yeasts are hybrids. The genome constitution differs for various brewer's yeasts (Rainieri et al. 2006; Dunn and Sherlock 2008). Therefore, the application of "omics" technologies in brewer's yeast studies has mostly been performed by employing the current knowledge of the *Saccharomyces cerevisiae* genome sequence. Nevertheless, recent approaches regarding the "ome" levels of brewer's yeast studies had been carried out (Bro and Nielsen 2004).

Recently, one of the known lager yeasts (W34/70) has been suggested to be a hybrid between *S. cerevisiae* and *Saccharomyces bayanus* with two subgenomes and a *S. bayanus*-based mitochondrial genome (Nakao et al. 2009). As the genome sequence of brewer's yeast W34/70 has now been publicly available, the possibility for the application of new technologies and basic research is given (Nakao et al. 2009).

2.1 Genomics

Genomics promote the understanding of the structure, function, and evolution of genomes and the application of genome technologies to challenging problems in biology and medicine. Intensive effort is now given to decode the entire DNA sequence of organisms and to fine-scale genetic mapping. Furthermore, studies of intragenomic phenomena such as heterosis, epistasis, pleiotropy, and other interactions between loci and alleles within the genome are included. In contrast, the investigation of single genes and their function is not the object of "genomics."

Useful tools and fingerprinting methods are available for global-scale genome analysis. Microarray-based comparative genomic hybridization (CGH) has been used successfully in detecting gene deletions, quantification of gene copy numbers, and in searching for information on chromosomal aneuploidies as well as translocations of gene parts (Winzeler et al. 1999a; Daran-Lapujade et al. 2003). The identification of genomic differences between species showing different degrees of a certain phenotype can directly reveal target genes for an organism's improvement.

Another firmly established method among these genomic fingerprinting methods is the amplified fragment length polymorphism (AFLP), which is based on the selective PCR amplification of restriction fragments from a total digest of genomic DNA (Vos et al. 1995; Meudt and Clarke 2007). AFLP has the capability to detect various polymorphisms in different genomic regions simultaneously.

These technologies have become widely used for the identification of genetic variation in strains or closely related species of plants, fungi, animals, and bacteria.

Especially for yeast, one can find many research studies dealing with species determination and the relatedness among them. A diverse collection of up to 70 *S. cerevisiae* strains and its closest relative sampled from different ecological niches were examined using microarrays. Herein, their variations in gene content, single nucleotide polymorphisms, nucleotide insertions and deletions, copy numbers, and transposable elements had been revealed (Liti et al. 2009; Schacherer et al. 2009).

For the diversity and complexity studies of brewer's yeast genomes, CGH using microarrays was carried out. The results showed the diversity of genome composition and possible occurrence of hybridization events between different lager brewer's yeast strains suggesting their hybrid genomes (Dunn and Sherlock 2008). In addition, large segments of the *S. cerevisiae* DNA were absent in lager brewer's yeast (Bond et al. 2004). Pope et al. (2007) combined both microarray and AFLP techniques to discriminate between various lager strains, which resulted in their relatively good characterization.

It can be concluded that DNA microarray hybridization has a high resolution to discriminate between species, even when the strains are closely related. Furthermore, it is very useful in the analysis of evolutionarily genetic distance among species (Watanabe et al. 2004) and facilitates finding correlations of phenotypic variation with global genome-wide phylogenetic relationships. However, to obtain meaningful information at the individual gene level, this method should be confirmed by additional techniques such as AFLP (Daran-Lapujade et al. 2003).

2.2 Functional Genomics

After sequencing the genome, the next logical step is to elucidate the functions and interactions of the enzymes, which the genes are coding for. Functional genomics (FG) is a field of molecular biology that describes gene and protein functions and interactions. The area focuses on the dynamic aspects such as gene transcription, translation, and protein–protein interactions, as opposed to the genomic information such as DNA sequences or structures (see Sect. 2.1). FG includes function-related aspects of the genome itself such as mutation and polymorphism (same as Single Nucleotide Polymorphism, SNP) analysis as well as measurement of molecular activities. Furthermore, it comprises "omics" technologies as transcriptomics (gene expression), proteomics (protein expression), and metabolomics (metabolite profiles) (Oliver 2002). By means of FG, it is possible to quantitate the steps from transcriptome to metabolome and to improve our understanding of gene and protein functions and interactions.

FG approaches involve the use of large-scale and/or high-throughput methods to understand genome-scale functions and regulations of transcriptomes, proteomes, and resulting metabolomes. These methods are (1) the Gene Chip technology (DNA microarrays) and Serial analysis of gene expression (SAGE), which provide various images of the transcriptome (mRNA) (Jansen and Gerstein 2000), (2) the two-dimensional gel electrophoresis (2D gel electrophoresis) and mass spectrometry (MS or coupled MS/MS) for the proteome level (Gygi et al. 2000), and (3) MS-based techniques such as GC/MS, LC/MS, or NMR for measurements of the metabolites leading to pathway analysis (Ostergaard et al. 2000).

2.2.1 Transcriptomics

Among the different levels of FG, the transcriptome has attracted the most attention (Wu et al. 2004). Transcriptomics deal with the transcripts of a given organism and reveal the whole set of mRNA molecules in the cell. With the aid of transcriptome analysis, it is possible to study the expression of all genes in a single experiment (Pugh et al. 2002). Moreover, transcriptome analyses can disclose external environmental conditions and reflect the genes that are being actively expressed at any point in time.

For the global-scale transcriptome analysis, the DNA microarray is a powerful tool that has revolutionized the field of genetics (Pugh et al. 2002). Microarray-based transcriptome analysis enables the examination of abundance of all transcripts in the cell at a given state or condition and thus allows both the identification of genes, which are coregulated, and the analysis of global responses to genomic mutations.

DNA microarrays can be used to detect DNA or RNA. For the detection of copy number, changes or losses in comparison to another related organism DNA is first stained and then hybridized to specific DNA probes. The dye ratio along the chromosomes can observe regions of DNA gain or loss in the analyzed sample. The method has been referred to as comparative genomic hybridization.

The measurement of RNA is most commonly done after reverse transcription into cDNA. The so-called expression profiling allows simultaneous monitoring of the expression levels of thousands of genes in order to study the effects of certain conditions and to get a deep insight about the regulation of genes (Khatri and Draghici 2005).

For standard microarrays, first of all, the probes for the hybridization of complementary DNA have to be designed and synthesized, which are then attached to a solid surface by a covalent bond to a chemical matrix. This solid surface can be glass or a silicon chip (gene chip). Other microarray platforms use microscopic beads. The small chips contain series of thousands of microscopic spots of DNA, each containing picomoles of the specific DNA sequence. Therefore, sample-specific oligonucleotides (25–70 mers long), which can be short sections of a gene or other DNA elements such as open reading frames (ORF), are spotted on the surface to hybridize to the sample-cDNA (target). Probe-target hybridization is usually detected and quantified by fluorescence-based detection of fluorophore-labeled targets to determine relative abundance of nucleic acid sequences of the target.

For the detection of hybridization events, one can choose between a one-color or a two-color system. The advantage of single channel experiments is the easy comparison of arrays from different experiments even months or years after the procedure. When samples shall be directly compared to each other, hybridization by two-channeling is carried out.

For the single channel hybridization, the sample cDNA is biotin-labeled, and then hybridization can be induced. During the staining step, the chip is dyed with a fluorescent molecule, which binds to biotin ("antibody staining"). Many series of washes and stainings provide an amplified fluorophore that emits light, which can be scanned with a control laser. The result is a specific signal pattern of the tested organism at a certain point in time, which indicates the status of cellular processes (Eisen et al. 1998).

When using the two-channel hybridization, the cDNA is labeled with two different fluorophores, generally Cy3 and Cy5 (Cyanine dyes), and then mixed (Shalon et al. 1996). Relative intensities of each fluorophore may then be used in ratio-based analysis to identify upregulated and downregulated genes (Tang et al. 2007). With the two-color system, relative differences in expression among different spots within a sample and between samples can be measured.

The use of miniaturized microarrays for gene expression profiling was first reported in 1995 (Schena et al. 1995). The result was a huge increase in interest shown by researchers regarding this new method. The obtained expression levels using microarrays provide the researchers with the opportunity to broaden their knowledge about the correlation of genotypes and their functional roles in the cell. But before one can correlate genotypic settings and resulting effects, the great quantities of data have to be evaluated. A big challenge is the development of efficient software for organizing and interpreting such large amounts of data (DeRisi et al. 1997).

Hence, initially, it is necessary to normalize and summarize the values. The expression signals are set to a certain value and calculated from the background-adjusted value to get a trimmed mean. Then, the statistical differential analysis can follow, which tests whether there are differentially expressed probes. In general, the significance is defined by statistical analysis via *t*-testing (for two-sampling). Most commonly, One-way Analysis of Variance (ANOVA, for at least three independent groups) is used, which filters out genes that have the same expression level across groups (samples, datasets) followed by a pairwise comparison for the identification of genes which differ from each other.

The probability value is traditionally set to $p\text{-value} = 0.05$ and states that random sampling from identical groups would lead to a difference smaller than one observed in 95% of experiments and larger than one observed in 5% of experiments. In some cases, genes are classified as significantly different when actually they are not ("false positives") (Reiner et al. 2003). To avoid such multiple testing errors, follow-up tests have to be done. The False Discovery Rate (FDR) can resolve this problem and enabling construction of statistically reliable gene lists. FDR has become an essential tool in any study that has a very large multiplicity problem (Benjamini and Yekutieli 2005).

To exploit a list of differentially expressed genes, it is essential to translate it into a functional profile that can offer insights into the cellular mechanisms activated in a given condition (Khatri and Draghici 2005). These detected genes have to be annotated because otherwise there is no use for the data. To get functional genomic information, manifold tools and databases have been established (see Review Khatri and Draghici (2005)).

One important ontological analysis tool is Gene Ontology (GO), which features a large amount of annotations for genes of many species (Ashburner et al. 2000; Doniger et al. 2003). GO as well as the Kyoto Encyclopedia of Genes and Genomes (KEGG) facilitate functional interpretations (Al-Shahrour et al. 2006). KEGG for instance maintains the gene catalogs for all organisms with completely sequenced genomes and selected organisms with partial genomes (Ogata et al. 1999). These databases help to resolve function of the detected genes, functional roles between them, and offer pathway maps, which are hierarchically classed (Kanehisa et al. 2004). A further bioinformatics resource is MAPPFinder, which enables the creation of gene expression profiles from microarray data with the aid of both GO terms and GenMAPP and therefore, eases data analysis (Doniger et al. 2003). Other web tools that use GO terms for the functional analysis of groups of genes are Babelomics (Al-Shahrour et al. 2005b, 2006) and FatiGO (Al-Shahrour et al. 2004, 2005a, 2007). The tools are integrated in the analysis package Gene Expression Profile Analysis Suite (GEPAS), which analyzes gene expression patterns, and is currentlythe most complete web-based resource (Vaquerizas et al. 2005).

Despite the growing number of programs available for microarray data analysis, there are still many aspects with incomplete coverage (Herrero et al. 2004). One reason is knowledge-limitation based on only a subset of genes of sequenced organisms (King et al. 2003). Another problem is the manual curation process of annotations and its time lag or even incorrectness (Khatri and Draghici 2005) whose weak point is poorly coordinated use of different tools and linking them together. This can lead to problems with input/output formats caused by different task-performances (Herrero et al. 2004). Until now, there is no unified analysis for data evaluation available (Khatri and Draghici 2005).

Another important step in the analysis of gene expression data is the identification of gene groups and clusters (Ray et al. 2007). Array and sample qualities as well as the degree of potential relationships between organisms or within the same species can be approximated using cluster analysis. Therefore, clusters show the diversity of characteristic values, and ideally, common samples should cluster into similar classes. In contrast, the more the distant organisms are in relation to each other, the less the accordances that will be found.

The complete genome of *S. cerevisiae* on a microarray was published in 1997 (Lashkari et al. 1997), which made this yeast the outrider for all organisms concerning transcriptome analyses. Furthermore, transcriptomics became the method of choice especially for the comparison of yeast's transcriptional profiles. Various yeasts showing different phenotypes were analyzed to find the genetic basis relevant to these differences. Using microarrays, target genes for strain improvement could be identified (Bro et al. 2005; Jin et al. 2005; Gorsich et al. 2006; Petersson et al. 2006; Cordier et al. 2007).

Moreover, yeasts were exposed to different media compositions to measure how the latter have an influence on the transcripts of the yeast. Special conditions were zinc, iron, or nutrient deprivation where it was demonstrated that the transcriptional activities of particular involved *S. cerevisiae*-genes differ from those at stable conditions (ter Linde et al. 1999; Gasch et al. 2000; de Nobel et al. 2001; Yale and Bohnert 2001; Yoshimoto et al. 2002; Higgins et al. 2003b; Shakoury-Elizeh et al. 2004; Wu et al. 2004).

Transcriptional profiles of brewer's yeast during the fermentation process highlighting different points in time were also studied by the application of DNA microarrays (Olesen et al. 2002; Pugh et al. 2002; Higgins et al. 2003a; James et al. 2003). So far, these studies have been performed with *S. cerevisiae* gene chips. Brewer's yeast consists of a hybrid genome, and the information (gene expression) for the lager genome part is not available. However, using *S. cerevisiae* chip technology, two-species microarrays for the detection of the brewer's yeast transriptome level have been developed (Dunn and Sherlock 2008). Microarray approaches have provided valuable insights according to the involvement of genes in various cellular processes. Even for the detection of genes that were expressed under certain conditions, or which were involved in particular pathways, microarray technology could help to understand the yeast physiology during brewing processes (Higgins et al. 2003b; Blieck et al. 2007).

It is also possible to measure genes expressed in an organism using SAGE, which was developed by Velculescu in 1995 (Velculescu et al. 1995, 1997). The general goal of the technique is similar to the DNA microarray. The benefit of this method is the revelation of new, unknown transcripts while measuring expression levels without reference to a standard so that data are definitively acquired and cumulative. However, a major limitation of SAGE is (1) that it can hardly be performed on multiple samples and (2) its high costs and labor input compared with microarrays (Marti et al. 2002). Nevertheless, the application of DNA microarray experiments in combination with SAGE analysis can reveal gene expression profiles and allow networking of gene regulations (Skoneczna 2006).

Follow up experiments play key roles in successful expression profiling experiments. Therefore, another method that is often used for confirmation of the microarray data is real time PCR (RT-PCR) where messenger RNA in cells is sensitively quantified. The technology is used to determine gene expression over time, for example in response to changes in environmental conditions. Therefore, RT-PCR is a routinely used tool in molecular biology for determination of mRNA concentrations of specific genes (Pfaffl 2004).

Transcriptome analyses are able to reflect changes on a molecular level and check whether requirements and conditions in industrial processes are well adjusted. Recording the expression levels enables the implication of changes in phenotype with associated changes in genotype of an organism (Gill 2003).

2.2.2 Proteomics

Although microarray studies can reveal the relative amounts of different mRNAs in the cell, these levels are not directly proportional to the expression level of the proteins they code for. The number of protein molecules synthesized using a given mRNA molecule as a template is highly dependent on translation-initiation features of the mRNA sequence and will vary with time and environmental conditions to which a cell or organism is exposed. The whole set of proteins of a cell is known as the proteome, which brings us one level closer to the phenotype (Wilkins et al. 1996; Bro and Nielsen 2004). The study of the protein-entirety is described by proteomics (Pandey and Mann 2000). Hence, the proteomics technology comprises highly sensitive and high throughput methods, which are able to quantitatively display and analyze all the proteins present in a sample (Haynes and Yates 2000). Proteome analysis is mostly accomplished by two-dimensional gel electrophoresis (2DE) and MS (Gygi et al. 2000).

The 2DE is a powerful technique whereby proteins are separated initially on their charge by isoelectric focusing (IEF) and subsequently according to their molecular weight by sodium dodecyl sulfate polyacrylamide gel electrophoresis (SDS-PAGE) (Beranova-Giorgianni 2003).

2DE is considered more laborious, less sensitive, and less reproducible than DNA microarray despite its potential in the study of gene functions. However, the optimization of the 2DE and hence the development of new methods for global quantification of proteins based on MS will bring about more perspectives for global-scale proteome studies (Aebersold and Mann 2003). Within MS, it is possible to identify unknown molecules while determining the structure of a compound by observing its fragmentation and quantifying the amount of a compound in a sample. For the determination of phosphorylation sites in proteins (which implies the latter being an intermediate structure in a pathway), MALDI-TOF/MS can also be used (Graves and Haystead 2002). Its high sensitivity and applicability makes it the method of choice for investigating complex molecules in biology and chemistry.

Recently, a gel-free method for the specific isolation of phosphorylated peptides using diagonal, reversed-phase chromatography with further peptide-characterization by MS has been described (van Damme et al. 2008; Gevaert and Vandekerckhove 2009). This technology points to changes in protein expression levels and has been widely used as an important tool for exploring complex biological systems (Nie et al. 2008).

Since the introduction of 2DE and MS, these technologies have been successfully used in a number of studies in many biological fields (Beranova-Giorgianni 2003).

One of the most exciting applications of proteomics involves studies of yeasts. Indeed, the number of identified protein spots could be largely extended to create protein maps for reference purposes (Boucherie et al. 1995; Perrot et al. 1999, 2007).

For the detection of the translated proteins (proteomics) in brewer's yeast strains, 2D gel electrophoresis as well as MS techniques have been used. The first proteome maps of lager and ale brewer's yeast were presented by Joubert et al. (2000) and Kobi et al. (2004). Notably, the relatedness between different brewer's yeasts and *S. cerevisiae* and the comigration of the proteins could be observed. Furthermore, different fermentation stages such as the lag and early exponential phase were investigated by proteomics to determine early-induced proteins (Brejning et al. 2005). These proteome analyses allowed the identification of many novel non-*S. cerevisiae* proteins of lager brewer's yeast attributable to their hybrid nature (Joubert et al. 2001). Although the results obtained via transcriptome analyses of brewing yeast strains could be confirmed by proteome analyses, the expression levels of the genes and their encoded proteins differed (Brejning et al. 2005). Once again, these findings indicate that the amount of the gene product (protein) cannot be predicted simply on the basis of the expression signals of the activated genes.

These approaches demonstrate that proteomics have the ability to give further information for the understanding of cellular functions of an organism (Kolkman et al. 2005; Pham and Wright 2007). Even mitochondrial proteins as well as phenotypic differences can be identified with these methods as a result of deletion events (Rogowska-Wrzesinska et al. 2001; Sickmann et al. 2003) (the topic "phenomics" will be treated in Sect. 2.3).

For the research on the so-called interactome, the "interactomics" couples MS techniques with additional methods. As proteins of a cell interact with each other and form multiprotein complexes, which link biological molecules, it is necessary to investigate new methods for detecting protein–protein interactions and, hence, gain more information concerning gene functions and protein-regulated mechanisms (Kiemer and Cesareni 2007). Therefore, methods such as the yeast two-hybrid (Y2H) screening and the tandem affinity purification (TAP) have been developed (Puig et al. 2001; Maple and Møller 2007; Lu et al. 2008). Y2H is a powerful and precise method and is used primarily for initial identification of interacting proteins (Young 1998). In contrast to Y2H, TAP is a high throughput method and seems to be the best adapted strategy for protein characterization. Firstly, the proteins are purified efficiently in order to subsequently identify them by MS (Dziembowski and Séraphin 2004). Besides the above mentioned methods, there is another technical straightforward technique named Bimolecular Fluorescence Complementation (BiFC). The advantage of BiFC compared to Y2H and TAP is the direct visualization of protein interactions in living cells (Kerppola 2008).

In this method, one protein is tagged with one half of Yellow Fluorescent protein (YFP), and the putative interactor is tagged with the other half of the YFP. Interaction is resolved microscopically by the detection of yellow fluorescence as a result of fusion of the two halves of the YFP.

With the aid of these methods, it is possible to reveal the protein–protein interactions and their consequences for the cell. Meanwhile, a huge amount of data exists (for review see (Kiemer and Cesareni 2007)).

Especially for the yeast, *S. cerevisiae,* over 90% of transcribed proteins under native conditions have been screened and characterized so that this organism is nearly fully specified (Krogan et al. 2006). In different approaches, the above-mentioned techniques could provide high-quality binary interaction information for yeast (Uetz et al. 2000; Ito et al. 2001; Gavin et al. 2002; Sung and Huh 2007; Yu et al. 2008). In addition, a large part of the identified *S. cerevisiae* interactions were linked together in a yeast two-hybrid map (Schwikowski et al. 2000). Furthermore, datasets of protein interactions in yeast have been curated and ordered for better system-level inferences (Reguly et al. 2006).

The identified protein networks enable studies and improve functional prediction on individual gene products (Krogan et al. 2006). Furthermore, they give reason for finding parallels not only to yeasts but also to other organisms and can help to clarify regulatory mechanisms.

2.2.3 Metabolomics

The metabolome level refers to the complete set of small-molecule metabolites, which can be found within a biological sample (Oliver et al. 1998). It includes metabolites such as metabolic intermediates, hormones, and other signaling molecules. A range of analytical techniques are required in order to maximize the number of metabolites that can be identified in a matrix (Griffin 2006). For this purpose, separation and detection methods are applied. Popular metabolic profiling tools are gas chromatography (GC) and high performance liquid chromatography (HPLC) coupled with MS. Using HPLC, a wider range of molecules can potentially be measured, but compared to GC, the chromatographic resolution is lower. The disadvantage for GC-analyses is that a chemical derivatization is required for most compounds in order to measure them, which, in some cases, is unfeasible. Hence, not every molecule can be detected. Moreover, the sample evaporates and cannot be used for further analysis. Although the nuclear magnetic resonance (NMR) spectroscopy allows for additional analyses, it is relatively insensitive compared to MS-based techniques. However, a full metabolic picture requires the combination of several analytical techniques (Smedsgaard and Nielsen 2005).

The above-mentioned technologies opened up new vistas and were used to characterize specific enzymes and genes in the cell, to map effects of specific mutations in the genome and to measure metabolite profiles.

Due to their complex metabolism, yeast organisms produce a broad range of metabolites (Smedsgaard and Nielsen 2005). These metabolites have to be profiled in order to understand the metabolic fluxes in the cell. NMR spectroscopy and GC/MS methods were applied to *S. cerevisiae* especially for achieving new insights into central carbon metabolism (Gombert et al. 2001; Maaheimo et al. 2001; dos Santos et al. 2003). Moreover, there are a lot of *ME* strategies elucidated in the literature for improving yeasts concerning product formation or cellular properties (Nevoigt 2008). Within these approaches, the metabolites have been measured with different techniques such as HPLC and LC/MS (Olsson and Nielsen 2000; Dietvorst et al. 2005; Lafaye et al. 2005).

Besides transcriptomics and proteomics, metabolic footprinting methods such as GC/MS have also been applied for the discrimination between brewer's yeasts (Pope et al. 2007). The obtained information helped to design genetic manipulations for improving brewer's yeasts and accordingly monitoring the adjustment of industrial fermentation conditions (Penttilä 2001). Numerous attempts are described for *ME* strategies for brewer's yeasts (Dequin 2001; Nielsen 2001; Donalies et al. 2008; Nevoigt 2008).

While analyzing the metabolite spectrum, it is possible to determine specific changes as a result of *ME* in order to clarify cellular functions and metabolic fluxes. The technologies can generally be applied to any organism.

2.3 Phenomics

In order to determine the function of genes that have not been classified, it is possible to generate mutations such as deletions of specific genes. For giving answers about how these gene disruptions influence cellular capabilities and spatial location of the gene product, there is a clear need for phenotypic analysis (Shoemaker et al. 1996). These approaches are pooled within "phenomics," which study the gene dispensability by quantitative analysis of phenotypes (e.g., growth curve display) (Fernandez-Ricaud et al. 2005).

The construction of the yeast *S. cerevisiae*-deletion collection and the reception of functional data helped to find out what happens after gene loss at the phenotypic level. Since the disruption of nearly all of the genes of the yeast, *S. cerevisiae* genome had been completed. The phenotypic consequences of more than 500 genes have been established (Winzeler et al. 1999b; Giaever et al. 2002). With the application of phenotypic analysis, unknown genes and their functions could be identified after gene deletion, e.g., under different growth conditions (saline stress, growth inhibitors, etc.) or UV irradiation (Birrell et al. 2001; Warringer and Blomberg 2003; Warringer et al. 2003).

The integration of phenotypic data on a genome-wide scale helped to form functional networking and predict cellular roles for previously uncharacterized genes (Warringer et al. 2003). In addition, it is possible to find parallels to other organisms as it was the case for two genes in *S. cerevisiae* whose human homologs are also involved in cancer formation (Birrell et al. 2001).

So far, there are no phenotypic studies on a multifunctional level since only particular observations such as investigations on the growth behavior were considered. This means that measurements of the metabolites are essential and have to be taken into account for future research to understand how the phenotype is determined by the genotype.

3 Conclusions

Applying the global "omics" technologies has already enabled the accumulation of in-depth knowledge about cellular activities of microorganisms. Within these methods, it is not only possible to resolve the gene transcripts but also to categorize the entirety of metabolites. The combination of the "omics" systems engineering has an enormous potential to provide new and profound insights into cell biology and illustrates the power of integrating different types of data obtained from the same sample (Griffin et al. 2002). The advancements of the "omics" technologies will help to obtain precise qualitative and quantitative results and offers new perspectives to broaden the understanding of diverse organisms. Such analyses have therefore facilitated the development of new products, e.g., for medical purposes.

Regarding yeast improvement programs, "omics" technologies have enabled their characterization at different "ome" levels.

Implementing "omics" for further improvements of brewer's yeasts is much more challenging because of the diversity of *S. cerevisiae* and brewer's yeast's genomes. Since these analyses are based on the current knowledge of the *S. cerevisiae* genome sequence, they cannot be simply applied to lager brewer's yeast studies. Nevertheless, the availability of the sequenced brewer's yeast genome will afford scientists to carry out comprehensive expression analyses and genome structural analyses (Kodama et al. 2006). Furthermore, the construction of the bottom-fermenting yeast DNA microarray will strongly facilitate the enlargement of the basic knowledge. Both the progress of *genetic engineering* and the application of "ome" analyses have led to the creation of numerous novel brewer's yeast strains with high benefit for the brewing industry and now raise fresh hope to control beer processes and their sustainability.

But as long as the connections of genes with their accompanying proteins and metabolites are not entirely clarified, the knowledge about the cell as a whole system will remain incomplete.

References

Aebersold R, Mann M (2003) Mass spectrometry-based proteomics. Nature 422:198–207

Al-Shahrour F et al (2004) FatiGO: a web tool for finding significant associations of gene ontology terms with groups of genes. Bioinformatics 20:578–580

Al-Shahrour F et al (2005a) Discovering molecular functions significantly related to phenotypes by combining gene expression data and biological information. Bioinformatics 21:2988–2993

Al-Shahrour F et al (2005b) BABELOMICS: a suite of web tools for functional annotation and analysis of groups of genes in high-throughput experiments. Nucleic Acids Res 33:W460–W464

Al-Shahrour F et al (2006) BABELOMICS: a systems biology perspective in the functional annotation of genome-scale experiments. Nucleic Acids Res 34:W472–W476

Al-Shahrour F et al (2007) FatiGO+: a functional profiling tool for genomic data. Integration of functional annotation, regulatory motifs and interaction data with microarray experiments. Nucleic Acids Res 35:W91–W96

Ashburner M et al (2000) Gene ontology: tool for the unification of biology. Nat Genet 25:25–29

Bailey JE et al (1990) Strategies and challenges in metabolic engineering a. Ann N Y Acad Sci 589:1–15

Bailey JE et al (2002) Inverse metabolic engineering: a strategy for directed genetic engineering of useful phenotypes. Biotechnol Bioeng 79:568–579

Benjamini Y, Yekutieli D (2005) Quantitative trait loci analysis using the false discovery rate. Genetics 171:783–790

Beranova-Giorgianni S (2003) Proteome analysis by two-dimensional gel electrophoresis and mass spectrometry: strengths and limitations. Trends Anal Chem 22:273–281

Birrell GW et al (2001) A genome-wide screen in *S. cerevisiae* for genes affecting UV radiation sensitivity. Proc Natl Acad Sci USA 98:12608–12613

Blieck L et al (2007) Isolation and characterization of Brewer's yeast variants with improved fermentation performance under high-gravity conditions. Appl Environ Microbiol 73:815–824

Bond U et al (2004) Aneuploidy and copy number breakpoints in the genome of lager yeasts mapped by microarray hybridisation. Curr Genet 45:360–370

Boucherie H et al (1995) Two-dimensional protein map of *S. cerevisiae*: construction of a gene-protein index. Yeast 11:601–613

Brejning J et al (2005) Identification of genes and proteins induced during the lag and early exponential phase of lager brewing yeasts. J Appl Microbiol 98:261–271

Bro C, Nielsen J (2004) Impact of 'ome' analyses on inverse metabolic engineering. Metab Eng 6:204–211

Bro C et al (2005) Improvement of galactose uptake in *S. cerevisiae* through overexpression of phosphoglucomutase: example of transcript analysis as a tool in *inverse metabolic engineering*. Appl Environ Microbiol 71:6465–6472

Cordier H et al (2007) A metabolic and genomic study of engineered *S. cerevisiae* strains for high glycerol production. Metab Eng 9:364–378

Daran-Lapujade P et al (2003) Comparative genotyping of the *S. cerevisiae* laboratory strains S288C and CEN.PK113-7D using oligonucleotide microarrays. FEMS Yeast Res 4:259–269

de Nobel H et al (2001) Parallel and comparative analysis of the proteome and transcriptome of sorbic acid-stressed *S. cerevisiae*. Yeast 18:1413–1428

Dequin S (2001) The potential of genetic engineering for improving brewing, wine-making and baking yeasts. Appl Microbiol Biotechnol 56:577–588

DeRisi JL et al (1997) Exploring the metabolic and genetic control of gene expression on a genomic scale. Science 278:680–686

Dietvorst J et al (2005) Maltotriose utilization in lager yeast strains: MTT1 encodes a maltotriose transporter. Yeast 22:775–788

Donalies U et al (2008) Improvement of Saccharomyces Yeast strains used in brewing, wine making and baking. In Adv Biochem Engin/Food Biotechnology 111:67–98

Doniger S et al (2003) MAPPFinder: using gene ontology and GenMAPP to create a global gene-expression profile from microarray data. Genome Biol 4:R7

dos Santos MM et al (2003) Identification of in vivo enzyme activities in the cometabolism of glucose and acetate by *S. cerevisiae* by using 13C-labeled substrates. Eukaryot Cell 2:599–608

Dunn B, Sherlock G (2008) Reconstruction of the genome origins and evolution of the hybrid lager yeast *S. pastorianus*. Genome Res 18:1610–1623

Dziembowski A, Séraphin B (2004) Recent developments in the analysis of protein complexes. FEBS Lett 556:1–6

Eisen MB et al (1998) Cluster analysis and display of genome-wide expression patterns. Proc Natl Acad Sci USA 95:14863–14868

Fernandez-Ricaud L et al (2005) PROPHECY-a database for high-resolution phenomics. Nucleic Acids Res 33:D369–D373

Gasch AP et al (2000) Genomic expression programs in the response of yeast cells to environmental changes. Mol Biol Cell 11:4241–4257

Gavin A-C et al (2002) Functional organization of the yeast proteome by systematic analysis of protein complexes. Nature 415:141–147

Gevaert K, Vandekerckhove J (2009) Reverse-phase diagonal chromatography for phosphoproteome research. In Methods in Molecular Biology/Phospho-Proteomics 527:219–227

Giaever G et al (2002) Functional profiling of the *S. cerevisiae* genome. Nature 418:387–391

Gill RT (2003) Enabling inverse metabolic engineering through genomics. Curr Opin Biotechnol 14:484–490

Gombert AK et al (2001) Network identification and flux quantification in the central metabolism of *S. cerevisiae* under different conditions of glucose repression. J Bacteriol 183:1441–1451

Gorsich S et al (2006) Tolerance to furfural-induced stress is associated with pentose phosphate pathway genes ZWF1, GND1, RPE1, and TKL1 in *S. cerevisiae*. Appl Microbiol Biotechnol 71:339–349

Graves PR, Haystead TAJ (2002) Molecular biologist's guide to proteomics. Microbiol Mol Biol Rev 66:39–63

Griffin JL (2006) Review. The Cinderella story of metabolic profiling: does metabolomics get to go to the functional genomics ball? Philos Trans R Soc B: Biol Sci 361:147–161

Griffin TJ et al (2002) Complementary profiling of gene expression at the transcriptome and proteome levels in *S. cerevisiae*. Mol Cell Proteomics 1:323–333

Gygi SP et al (2000) Evaluation of two-dimensional gel electrophoresis-based proteome analysis technology. Proc Natl Acad Sci USA 97:9390–9395

Haynes PA, Yates JRI (2000) Proteome profiling – pitfalls and progress. Yeast 17:81–87

Herrero J et al (2004) New challenges in gene expression data analysis and the extended GEPAS. Nucleic Acids Res 32:W485–W491

Higgins VJ et al (2003a) Yeast genome-wide expression analysis identifies a strong ergosterol and oxidative stress response during the initial stages of an industrial lager fermentation. Appl Environ Microbiol 69:4777–4787

Higgins VJ et al (2003b) Application of genome-wide expression analysis to identify molecular markers useful in monitoring industrial fermentations. Appl Environ Microbiol 69:7535–7540

Ito T et al (2001) A comprehensive two-hybrid analysis to explore the yeast protein interactome. Proc Natl Acad Sci USA 98:4569–4574

James TC et al (2003) Transcription profile of brewery yeast under fermentation conditions. J Appl Microbiol 94:432–448

Jansen R, Gerstein M (2000) Analysis of the yeast transcriptome with structural and functional categories: characterizing highly expressed proteins. Nucleic Acids Res 28:1481–1488

Jin Y-S et al (2005) Improvement of xylose uptake and ethanol production in recombinant *S. cerevisiae* through an inverse metabolic engineering approach. Appl Environ Microbiol 71:8249–8256

Joubert R et al (2000) Two-dimensional gel analysis of the proteome of lager brewing yeasts. Yeast 16:511–522

Joubert R et al (2001) Identification by mass spectrometry of two-dimensional gel electrophoresis-separated proteins extracted from lager brewing yeast. Electrophoresis 22:2969–2982

Kanehisa M et al (2004) The KEGG resource for deciphering the genome. Nucleic Acids Res 32: D277–D280

Kerppola TK (2008) Bimolecular fluorescence complementation (BiFC) analysis as a probe of protein interactions in living cells. Annu Rev Biophys 37:465–487

Khatri P, Draghici S (2005) Ontological analysis of gene expression data: current tools, limitations, and open problems. Bioinformatics 21:3587–3595

Kiemer L, Cesareni G (2007) Comparative interactomics: comparing apples and pears? Trends Biotechnol 25:448–454

King OD et al (2003) Predicting gene function from patterns of annotation. Genome Res 13:896–904

Kobi D et al (2004) Two-dimensional protein map of an "ale"-brewing yeast strain: proteome dynamics during fermentation. FEMS Yeast Res 5:213–230

Kodama Y et al (2006) Lager brewing yeast. In Topics in Current Genetics/Comparative Genomics 15:145–164

Kolkman A et al (2005) Development and application of proteomics technologies in *S. cerevisiae*. Trends Biotechnol 23:598–604

Krogan NJ et al (2006) Global landscape of protein complexes in the yeast *S. cerevisiae*. Nature 440:637–643

Lafaye A et al (2005) Combined proteome and metabolite-profiling analyses reveal surprising insights into yeast sulfur metabolism. J Biol Chem 280:24723–24730

Lashkari DA et al (1997) Yeast microarrays for genome wide parallel genetic and gene expression analysis. Proc Natl Acad Sci USA 94:13057–13062

Liti G et al (2009) Population genomics of domestic and wild yeasts. Nature 458(7236): 337–341

Lu JP et al (2008). Dual expression recombinase based (DERB) single vector system for high throughput screening and verification of protein interactions in living cells. Available from nature proceedings <http://hdl.handle.net/10101/npre.2008.1550.1>

Maaheimo H et al (2001) Central carbon metabolism of *S. cerevisiae* explored by biosynthetic fractional (13)C labeling of common amino acids. Eur J Biochem 268:2464–2479

Maple J, Møller SG (2007) Yeast two-hybrid screening. Methods Mol Biol 362:207–223

Marti J et al (2002) Transcriptomes for serial analysis of gene expression. J Soc Biol 196:303–307

Meudt HM, Clarke AC (2007) Almost forgotten or latest practice? AFLP applications, analyses and advances. Trends Plant Sci 12:106–117

Nakao Y et al (2009) Genome sequence of the lager brewing yeast, an interspecies hybrid. DNA Res 16:115–129

Nevoigt E (2008) Progress in metabolic engineering of *S. cerevisiae*. Microbiol Mol Biol Rev 72:379–412

Nie L et al (2008) Statistical application and challenges in global gel-free proteomic analysis by mass spectrometry. Crit Rev Biotechnol 28:297–307

Nielsen J (2001) Metabolic engineering. Appl Microbiol Biotechnol 55:263–283

Ogata H et al (1999) KEGG: Kyoto encyclopedia of genes and genomes. Nucleic Acids Res 27:29–34

Olesen K et al (2002) The dynamics of the *S. carlsbergensis* brewing yeast transcriptome during a production-scale lager beer fermentation. FEMS Yeast Res 2:563–573

Oliver SG et al (1998) Systematic functional analysis of the yeast genome. Trends Biotechnol 16:373–378

Oliver SG (2002) Functional genomics: lessons from yeast. Philos Trans R Soc Lond B Biol Sci 357:17–23

Olsson L, Nielsen J (2000) The role of metabolic engineering in the improvement of *S. cerevisiae*: utilization of industrial media. Enzyme Microb Technol 26:785–792

Ostergaard S et al (2000) Metabolic engineering of *S. cerevisiae*. Microbiol Mol Biol Rev 64:34–50

Pandey A, Mann M (2000) Proteomics to study genes and genomes. Nature 405:837–846

Penttilä M (2001) Metabolic engineering approaches-opportunities for brewing Proc Congr Eur Conv Budapest 28:505–513

Perrot M et al (1999) Two-dimensional gel protein database of *S. cerevisiae* (update 1999). Electrophoresis 20:2280–2298

Perrot M et al (2007) Yeast proteome map (update 2006). Proteomics 7:1117–1120

Petersson A et al (2006) A 5-hydroxymethyl furfural reducing enzyme encoded by the *S. cerevisiae* ADH6 gene conveys HMF tolerance. Yeast 23:455–464

Pfaffl MW (2004) Real-time RT-PCR: Neue Ansätze zur exakten mRNA Quantifizierung. BIOspektrum 1:92–95

Pham TK, Wright PC (2007) Proteomic analysis of *S. cerevisiae*. Expert Rev Proteomics 4:793–813

Pope GA et al (2007) Metabolic footprinting as a tool for discriminating between brewing yeasts. Yeast 24:667–679

Pugh T et al (2002) Global analysis of yeast gene expression during a brewery fermentation. ASBC 2002 Annual Meeting, Savannah, Georgia

Puig O et al (2001) The tandem affinity purification (TAP) method: a general procedure of protein complex purification. Methods 24:218–229

Rainieri S et al (2006) Pure and mixed genetic lines of *S. bayanus* and *S. pastorianus* and their contribution to the lager brewing strain genome. Appl Environ Microbiol 72:3968–3974

Ray SS et al (2007) Gene ordering in partitive clustering using microarray expressions. J Biosci 32:1019–1025

Reguly T et al (2006) Comprehensive curation and analysis of global interaction networks in *S. cerevisiae*. J Biol 5:11

Reiner A et al (2003) Identifying differentially expressed genes using false discovery rate controlling procedures. Bioinformatics 19:368–375

Rogowska-Wrzesinska A et al (2001) Comparison of the proteomes of three yeast wild type Strains: CEN.PK2, FY1679 and W303. Comp Funct Genomics 2:207–225

Schacherer J et al (2009) Comprehensive polymorphism survey elucidates population structure of *S. cerevisiae*. Nature 458(7236):342–345

Schena M et al (1995) Quantitative monitoring of gene expression patterns with a complementary DNA microarray. Science 270:467–470

Schwikowski B et al (2000) A network of protein–protein interactions in yeast. Nat Biotechnol 18:1257–1261

Shakoury-Elizeh M et al (2004) Transcriptional remodeling in response to iron deprivation in *S. cerevisiae*. Mol Biol Cell 15:1233–1243

Shalon D et al (1996) A DNA microarray system for analyzing complex DNA samples using two-color fluorescent probe hybridization. Genome Res 6:639–645

Shoemaker DD et al (1996) Quantitative phenotypic analysis of yeast deletion mutants using a highly parallel molecular bar-coding strategy. Nat Genet 14:450–456

Sickmann A et al (2003) The proteome of *S. cerevisiae* mitochondria. Proc Natl Acad Sci USA 100:13207–13212

Skoneczna A (2006) Decade of genomics-methods for genome investigation in yeast *S. cerevisiae*. Postepy Biochem 52:435–447

Smedsgaard J, Nielsen J (2005) Metabolite profiling of fungi and yeast: from phenotype to metabolome by MS and informatics. J Exp Bot 56:273–286

Sung M-K, Huh W-K (2007) Bimolecular fluorescence complementation analysis system for in vivo detection of protein–protein interaction in *S. cerevisiae*. Yeast 24:767–775

Tang T et al (2007) Expression ratio evaluation in two-colour microarray experiments is significantly improved by correcting image misalignment. Bioinformatics 23:2686–2691

ter Linde JJM et al (1999) Genome-wide transcriptional analysis of aerobic and anaerobic chemostat cultures of *S. cerevisiae*. J Bacteriol 181:7409–7413

Uetz P et al (2000) A comprehensive analysis of protein–protein interactions in *S. cerevisiae*. Nature 403:623–627

van Damme P et al (2008) Protein processing characterized by a gel-free proteomics approach. In Methods in Molecular Biology/ Functional Proteomics 484:245–262

Vaquerizas JM et al (2005) GEPAS, an experiment-oriented pipeline for the analysis of microarray gene expression data. Nucleic Acids Res 33:W616–W620

Velculescu VE et al (1995) Serial analysis of gene expression. Science 270:484–487

Velculescu VE et al (1997) Characterization of the yeast transcriptome. Cell 88:243–251

Vos P et al (1995) AFLP: a new technique for DNA fingerprinting. Nucleic Acids Res 23:4407–4414

Warringer J, Blomberg A (2003) Automated screening in environmental arrays allows analysis of quantitative phenotypic profiles in *S. cerevisiae*. Yeast 20:53–67

Warringer J et al (2003) High-resolution yeast phenomics resolves different physiological features in the saline response. Proc Natl Acad Sci USA 100:15724–15729

Watanabe T et al (2004) A new approach to species determination for yeast strains: DNA microarray-based comparative genomic hybridization using a yeast DNA microarray with 6000 genes. Yeast 21:351–365

Wilkins MR et al (1996) From proteins to proteomes: large scale protein identification by two-dimensional electrophoresis and amino acid analysis. Biotechnol Bioeng 14:61–65

Winzeler EA et al (1999a) Whole genome genetic-typing in yeast using high-density oligonucleotide arrays. Parasitology 118:73–80

Winzeler EA et al (1999b) Functional characterization of the *S. cerevisiae* genome by gene deletion and parallel analysis. Science 285:901–906

Wu J et al (2004) Global analysis of nutrient control of gene expression in *S. cerevisiae* during growth and starvation. Proc Natl Acad Sci USA 101:3148–3153

Yale J, Bohnert HJ (2001) Transcript expression in *S. cerevisiae* at high salinity. J Biol Chem 276:15996–16007

Yoshimoto H et al (2002) Genome-wide analysis of gene expression regulated by the calcineurin/Crz1p signaling pathway in *S. cerevisiae*. J Biol Chem 277:31079–31088

Young KH (1998) Yeast two-hybrid: so many interactions, (in) so little time. Biol Reprod 58:302–311

Yu H et al (2008) High-quality binary protein interaction map of the yeast interactome network. Science 322:104–110

Part III
Physiology

Rhizosphere Signals for Plant–Microbe Interactions: Implications for Field-Grown Plants

Ulrike Mathesius and Michelle Watt

Contents

Abstract This review presents an analysis of rhizosphere signals important in plant–microbial interactions that have been studied in controlled conditions and how they may function on field-grown roots. We define rhizosphere signals to be molecules on or emitted from microorganism or root cells that are recognized by other cells and trigger a response. A well-known example are the flavonoids from legume roots, which bind to transcriptional activators in rhizobia bacteria triggering the release of Nod factors (lipochitin oligosaccharides) that bind to root hairs or cortical cells initiating nodule development. Many other signals are reported, e.g.,

U. Mathesius (✉)
ARC Centre of Excellence for Integrative Legume Research, Division of Plant Science, Research School of Biology, Australian National University, Canberra, ACT 0200, Australia
e-mail: mathesius@anu.edu.au

M. Watt
CSIRO Plant Industry, Black Mountain Laboratories, GPO Box 1600, Canberra, ACT 2601, Australia

U. Lüttge et al. (eds.), *Progress in Botany 72*,
DOI 10.1007/978-3-642-13145-5_5, © Springer-Verlag Berlin Heidelberg 2011

phytohormones, quorum sensing signals (QSS) and their mimics, strigolactones, and exopolysaccharides. Some are involved in infection of roots by symbionts or pathogens; others in growth, physiological, and immune responses caused by commensal organisms that do not invade the root. The signals have so far been largely studied in the classical host–microbe framework in controlled conditions. Field rhizospheres, however, are host–microbe communities that change in space and time. Root surfaces, and thus binding sites for signals, change as the root ages and differentiates, as do chemicals released, which may be energy substrates, signals, or toxins to soil microorganisms in a local patch of soil. The nutritional status and disease susceptibility of the plant and the soil properties are also changing with space and time in the field. These dynamics explain in part why it has been so difficult to translate laboratory effects of signals to the field. Our analysis suggests that signal function in field rhizospheres would depend on (1) compatibility between microorganism populations and root tissue age and cell surfaces for signal–receptor binding, (2) proximity to the root, (3) moisture, and (4) plant nutrition and predisposition to response to microorganisms due to biotic or abiotic factors. These ideas need to be tested. Two major gaps in knowledge are the microorganisms that colonize the rhizosphere and their signals – only a small percentage have been sequenced, and how the plant genetics is regulating the perception and response to the diverse signals that may bind successfully to the root surfaces in the field.

1 Introduction

Soil grown plants are surrounded by a myriad of soil microorganisms, some of which have evolved to form symbiotic or pathogenic relationships. Other organisms are commensals living on root substrates such as exudates, mucilages, and dead plant material. While the soil environment is relatively nutrient-poor, the rhizosphere can contain high concentrations of nutrients from the root. It has been estimated that up to 20% of the carbon fixed by the shoot could be transferred into the rhizosphere (Knee et al. 2001), and thus microbial numbers, as well as their signals, can be high around the root surfaces. In addition to exudation of nutrients, vitamins, and minerals, the plant also exudes various secondary metabolites, many of which act as signals (Walker et al. 2003a) or as compounds inhibitory to the growth and functioning of microorganisms or roots called allelochemicals (Bertin et al. 2003). Before plants make physical contact with microbial partners, chemical signal exchange has often preceded their interaction and prepared the partner for a successful interaction. This rhizosphere signaling is necessary to coordinate bacterial behaviors, to signal the presence of microbial partners to plant hosts, and to elicit specific responses in hosts and microbial partners that contribute to symbiotic and pathogenic interactions and to commensal interactions (Hirsch et al. 2003).

The rhizosphere has a very wide range of signal molecules from the root or microorganisms that have diverse chemical structures and functions (Table 1;

Fig. 1). We consider in this review signals to be molecules that occur on the surface of, or are released from the cell of one organism, and to bind to another cell, triggering a response. They can be highly specific to cell partners and processes. The identity, function, and response to a number of rhizosphere signals have been well described, e.g., Costacurta and Vanderleyden (1995); Hirsch et al. (2003); Somers et al. (2004), and this review will concentrate on some of the best characterized rhizosphere signals to highlight recent progress as well as gaps in our knowledge about the function and potential application of rhizosphere signals to improve crop plant performance. We will consider the role of rhizosphere signals in specific signaling processes like recognition, attraction of partners, elicitation of host responses, and coordination of bacterial behaviors. We will highlight that signals are not just used for the purpose they might have originally been intended for. There are examples of eavesdropping, signal mimicry, signal interference, and nonsignaling function of "signaling" molecules. In addition, the functioning of rhizosphere signals depends on soil and root properties (Watt et al. 2006b). Signal breakdown can be a physicochemical process, but it can also actively be influenced by hosts and soil microorganisms. An important aspect of rhizosphere signaling is the dependence of signaling on spatial properties of the root, including surface properties and developmental zones, as well as the rhizosphere composition. This paper will review the modes of action of a few key signals in controlled conditions and consider field rhizospheres and the factors that would favor signal exchange to improve crop performance.

2 Examples for the Roles of Rhizosphere Signals Under Controlled Conditions

2.1 Plant Signals Regulating Interactions with Soil Microbes

2.1.1 Plant Signals in the Root Nodule Symbiosis

One of the best studied plant–bacterial interactions is that between most members of the legume family with nitrogen-fixing soil bacteria generically called rhizobia (Long 1989; Sprent 2008). Rhizobia are α-proteobacteria that can live feely in the soil as saprophytes, but form a facultative symbiosis with legumes and the nonlegume genus *Parasponia*. However, other bacteria can also nodulate legumes, including species of the β-proteobacteria *Burkholderia*, which were traditionally known as plant pathogens (Moulin et al. 2001). Rhizobia invade the roots of host plants and induce formation of root nodules (RNs), in which they fix atmospheric nitrogen that is transferred to the plant in exchange for carbohydrates. This symbiosis provides the majority of biologically fixed nitrogen and is especially important for low input agricultural systems in Africa and South America (Peoples et al. 2009). Legume–rhizobia symbioses are not the only nitrogen-fixing symbioses, as

Table 1 Examples of well-studied signals found in the rhizosphere that are important for plant–microbial interactions

Chemical name	Source	Functions	Properties of molecule
Acyl homoserine lactones (AHLs)	Gram-negative bacteria	Quorum sensing signal (QSS) among Gram-negative bacteria; sensed by plants, triggering multiple responses	Diffuses in biofilms, stable under acid conditions only, small AHLs with short acyl chains water soluble, enzyme degradable
Exopolysaccharides	Surface molecules of bacteria	Regulate recognition in and infection of hosts and elicitation of defense responses	Usually bound on bacterial surface and not released into rhizosphere
Flavonoids	Made and exuded by plants	Depending on structure, induction of *nod* gene expression in rhizobia can be toxic to bacteria and fungi; can be utilized as C source by bacteria; can interfere with quorum sensing in bacteria	Water soluble as glycosides, less water soluble as aglycones
Lumichrome	Breakdown product of riboflavin	Can be used as a signal to interfere with quorum sensing in bacteria; has effects on plant physiology	Poorly water soluble
Nod factors (lipochitin oligosaccharides)	Made by most rhizobia	Necessary for nitrogen-fixing symbioses in legumes, triggering nodule initiation and infection thread development	Large lipophilic molecule, can be modified by bacteria in soil
Phytohormones, e.g., auxins (mainly indole-3-acetic acid, indole-3-butyric acid, 4-chloroindole-3-acetic acid, 2-phenylacetic acid) and cytokinins (adenine derivatives).	Plant or bacterial	Modify plant growth; can be used as C source by bacteria	Auxins and cytokinins poorly water soluble; can be broken down enzymatically in soil
Riboflavin	Made and exuded by plants and bacteria	Vitamin, can be used as a signal to interfere with quorum sensing in bacteria	Poorly water soluble, unstable under alkaline conditions
Strigolactones	Made and exuded by plants	Cause hyphal branching of mycorrhizal fungi and germination of plant parasite seeds; plant branching hormone	Lipophilic; can be hydrolyzed in the rhizosphere

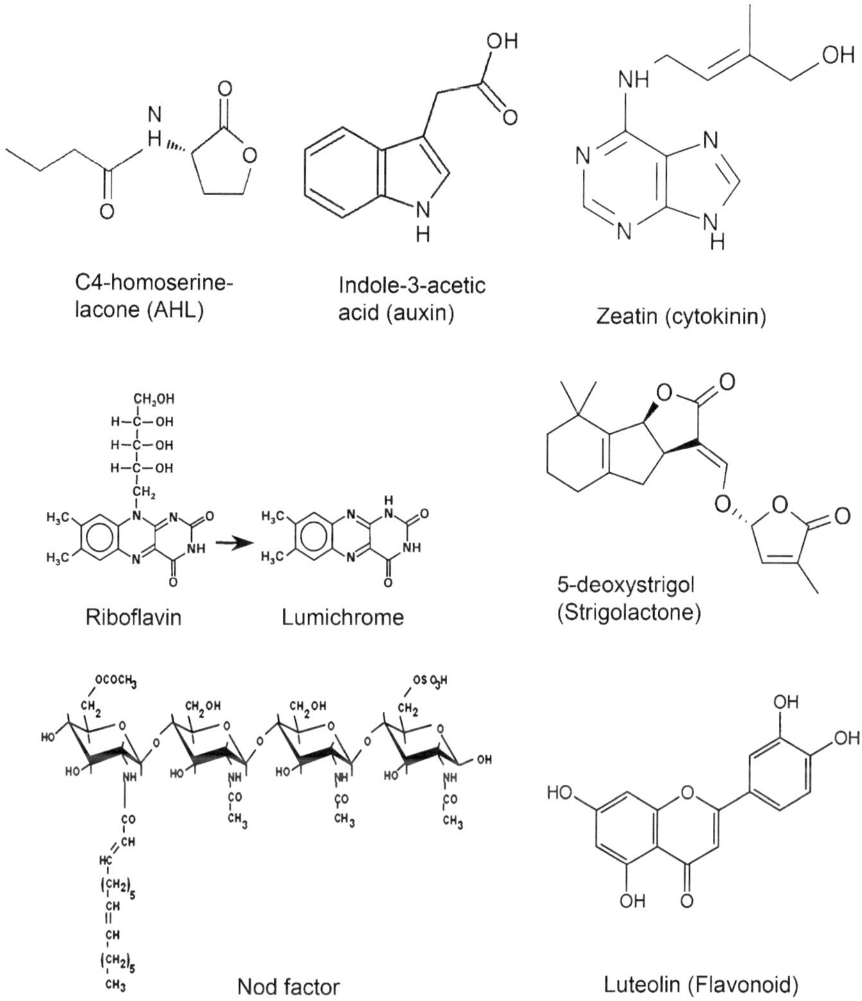

Fig. 1 Chemical structures of examples of rhizosphere signal molecules mentioned in the text

members of several plant families of the plant orders Fagales, Cucurbitales, and Rosales form the so-called actinorhizal symbioses with nitrogen-fixing Actinobacteria of the genus *Frankia* (Swensen 1996). In addition, a wider range of nitrogen-fixing bacteria can form nitrogen-fixing symbioses with grasses that do not involve formation of nodules but intercellular colonization (Reinhold-Hurek and Hurek 1998).

The RN symbiosis starts by attraction of the rhizobia to the host surface, followed by biofilm formation of bacteria on the root surface, and infection of root hairs by rhizobia. Only young, emerging root hairs are susceptible to *Rhizobium* infection (Bhuvaneswari et al. 1981) and the growing root hairs curl around the bacteria. Rhizobia then invaginate the plasma membrane of the root hair and form an infection thread that grows toward the cortex of the root. However, some legumes are infected by rhizobia through crack entry at lateral root bases where the bacteria invade intercellularly, and in the case of the legume *Sesbania rostrata*, the mode of infection depends on whether plants grow under flooded or dry conditions (Goormachtig et al. 2004). Different types of nodulation occur in different legumes. Many tropical legumes like soybean and bean form determinate nodules, in which cortical cells of the outer cortex first divide to form a nodule primordium. Nodule primordia grow into determinate nodules that lose the activity of the terminal meristem. In many temperate legumes, nodules originate from inner cortical cells and are characterized by a meristem that remains active in the growing nodule (Rolfe and Gresshoff 1988; Hirsch 1992).

This symbiosis shows a high specificity between the plant and bacterial partners. To establish a successful symbiosis, the legume host plant first needs to attract its specific symbiont, which in turn needs to recognize the plant and produce signal molecules that are perceived by the plant (Fig. 2a). Flavonoids (Fig. 1) and betaines were identified as the signal molecules exuded by legume hosts and recognized by rhizobia (Peters et al. 1986; Redmond et al. 1986). Different legume species exude specific flavonoids, and only their symbionts recognize their host flavonoids (Aoki et al. 2000). Flavonoids are also released from actinorhizal plants and activate the *Frankia* symbiont (Benoit and Berry 1997). Flavonoids bind to and activate the bacterial transcriptional activators NodD and SyrM, which then bind to promoter elements containing a so-called *nod* box (Mulligan and Long 1985). Several NodD proteins exist in different species of rhizobia and these have different specificities for flavonoids, betaines, and sometimes simpler phenolics (Schell 1993). While some flavonoids activate NodD, others are inhibitory. Genes activated by NodD include *NodA*, *B*, and *C*, which encode enzymes necessary for Nod factors synthesis (see Sect. 2.2.1), exopolysaccharide synthesis, and expression of a range of proteins in rhizobia (Chen et al. 2000). While flavonoids are continuously released from legume roots and seeds, their exudation and induction inside the plant increase in response to rhizobia (Zuanazzi et al. 1998). Flavonoid exudation patterns change along the root and *NodC–LacZ* fusions can be used in reporter strains to evaluate spatial aspects and identity of root exudates that can activate *Nod* gene expression in the rhizosphere (Innes et al. 1985; Mulligan and Long 1985). Flavonoid-deficient legumes have been used to support separate roles for flavonoids as *nod* gene inducers and as regulators of auxin transport during nodule initiation (Subramanian et al. 2006; Wasson et al. 2006; Zhang et al. 2009), and it would be interesting to use these plants to examine the role of flavonoid exudates on rhizosphere microorganisms in situ. In addition to their roles as *nod* gene inducers, flavonoids act as defense compounds in the rhizosphere and could affect viability of bacteria and fungi (Dakora and Phillips 1996).

Another class of plant molecules playing a role in the RN symbiosis is lectins, the carbohydrate-binding proteins on the surface of root hairs. Lectins are thought to facilitate the symbiosis between the specific plant and bacterial partners by binding specific surface polysaccharides of the bacteria and therefore ensuring

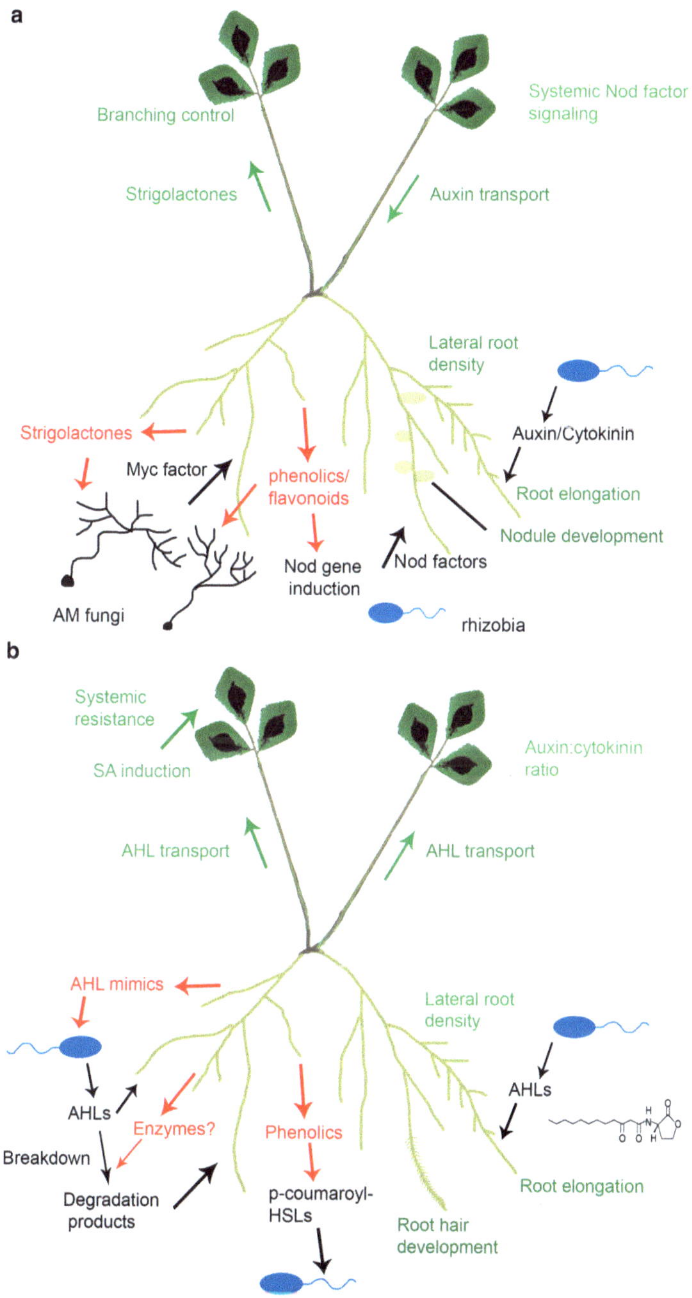

Fig. 2 Signal interactions between roots and microorganisms. (**a**) Signal interactions in symbiotic associations of roots with rhizobia and AM. (**b**) Signal interactions between roots and acyl homoserine lactone (AHL)-producing rhizosphere bacteria. *Red arrows and letters*: Root exudates and signals. *Black arrows and letters*: Signals from microorganisms. *Green arrows and letters*: Local and systemic effects of signals in the plant

good attachment of the bacteria to the root hair (Bohlool and Schmidt 1974). Lectins determine part of the specificity of the RN symbiosis, as demonstrated by an experiment in which transfer of a pea lectin gene to white clover conferred the ability of the transgenic white clover plants to be nodulating by pea-specific rhizobia (Díaz et al. 1989). However, the developing nodules were not fully invaded and functional, even though the rhizobial symbionts of pea and white clover are closely related. Similarly, when a soybean lectin gene was introduced to *Lotus japonicus*, nodule-like outgrowth were observed after infection of the transgenic *L. japonicus* plants with the soybean-nodulating strain *Bradyrhizobium japonicum*, which is only distantly related to the *Mesorhizobium loti* symbiont of *L. japonicus* (van Rhijn et al. 1998). This required a functional sugar-binding moiety of the lectin and the presence of exopolysaccharides on the bacterial surface; however, Nod factor effects on the elicitation of nodules were independent of the presence of lectins (van Rhijn et al. 1998). While lectins might partially control the host range of rhizobia, Nod factors are the critical determinants of symbiosis. This was demonstrated in *Medicago sativa* plants, which were transformed with either soybean or pea lectin genes and then infected with either soybean or pea rhizobia, respectively (van Rhijn et al. 2001). The authors of this study showed that infection threads or infected nodules were only formed when the soybean or pea nodulating rhizobia carried the functional *nod* genes from the *M. sativa* symbiont, *Sinorhizobium meliloti*. In addition, the presence of exopolysaccharides on the rhizobial surface was necessary for successful invasion (van Rhijn et al. 2001). In the rhizosphere, lectins are likely to be important to facilitate successful contact between symbiotic partners and to increase concentrations of Nod factors at site of bacterial attachments to the root hairs.

2.1.2 Plant Signals in the Arbuscular Mycorrhizal Symbiosis

Another ecologically significant and widespread mutualistic symbiosis is that between the majority of higher plant species with mycorrhizal fungi. This ancient symbiosis is beneficial but not obligatory for the plant partner, whereas the fungal symbiont cannot complete its life cycle in the absence of a plant host. In contrast to the *Rhizobium*-legume symbioses, mycorrhizal symbioses are much less specific, i.e., most higher plant species can be mycorrhized and fungal partners show a wide host range. One of the best-studied symbioses is the arbuscular mycorrhizal (AM) association of many land plants with fungal partners of the Glomeromycota. The symbiosis involves plant signals that stimulate fungal spore germination, hyphal branching, and growth toward the plant. Upon physical contact, fungal hyphae form appressoria on the root surface and then penetrate the root and form intercellular hyphal networks as well as intracellular arbuscules inside host cortical cells. The arbuscules are highly branched or coiled structures that aid in the exchange of mineral nutrients between plant and fungal partners, particularly, phosphorus mined by the fungal partner and transferred to the host in exchange for carbohydrates (Harrison 1997, 1998).

Plant signals that act as host recognition signals in the AM symbiosis are strigolactones (Fig. 1), carotenoid-derived sesquiterpene signals that stimulate branching of fungal partners (Akiyama and Hayashi 2006). While fungal spores can germinate in the absence of plant signals, their presymbiotic growth toward the plant is stimulated by strigolactones, which have been shown to activate mitochondrial metabolism, respiration, and mitosis in the fungal partner (Besserer et al. 2006; Besserer et al. 2008). Strigolactones can be synthesized by a large range of land plants (Bouwmeester et al. 2007) and the same strigolactone molecule was shown to activate a range of mycorrhizal fungi, suggesting that the wide spread occurrence of AM symbioses could be due to the lack of specificity of the strigolactone signals (Besserer et al. 2006). Phosphate starvation, which stimulates AM symbioses, also increases strigolactone exudation by roots (Lopez-Raez et al. 2008); however, this response was not observed in nonmycorrhizal host species (Yoneyama et al. 2008). Further support for the importance of strigolactones was presented with the recent identification of strigolactone mutants, which were shown to be deficient in stimulating hyphal branching (Gomez-Roldan et al. 2008). A major future task remains to identify a fungal strigolactone receptor and signaling pathways that mediate the observed stimulation of spore germination and hyphal branching.

Flavonoids are not necessary for the formation of AM symbioses (Becard et al. 1995), but there are multiple reports that they stimulate hyphal branching and fungal spore germination (Tsai and Phillips 1991; Scervino et al. 2005; Antunes et al. 2006; Kikuchi et al. 2007). Infection of plants by mycorrhizal fungi stimulates flavonoid synthesis inside the root, possibly providing a feedback loop, and this depends on the developmental stage of the plant (Larose et al. 2002). Apart from AM, signals necessary for other types of mycorrhizal associations, for example, ectomycorrhiza, are not well understood.

2.2 *Perception and Response to Microbial Signals by Plants*

2.2.1 Nod Factors

In response to the transcriptional activation of *Nod* genes by flavonoids or betains, rhizobia synthesize molecules that in most cases are necessary for a successful symbiosis with legumes, the so-called Nod factors (Fig. 2a). Nod factors are lipochitin oligosaccharides (Fig. 1) and consist of a β-1-4 linked *N*-acetylglucosamine backbone of typically four or five subunits, a variable acyl side chain of 16–18 carbon molecules linked to the nonreducing end, and various "decorations" of the chitosan backbone, which can include sulfation, glycosylation, and acetylation. Nod factor structures are species-specific, but most rhizobia synthesize a mixture of several different Nod factor molecules. The structure of the Nod factor is a critical determinant of their host range, and transfer of *Nod* genes from a specific rhizobial symbiont to a nonsymbiont enables the nonsymbiont to nodulate the host (Spaink et al. 1991; Denarié and Débellé 1996; Spaink 1996). Purified Nod factors stimulate many of the early responses of legume hosts to their rhizobial symbionts, for example, root hair deformation, calcium spiking, early nodulin gene expression, and induction of cortical cell division, although in most legumes Nod factors are not capable of inducing full nodules. Nod factors act at very low concentrations (10^{-9}–10^{-12} M) and only have effects on specific hosts, suggesting that they bind a receptor. These receptors have been identified in model legumes (Limpens et al. 2003; Madsen et al. 2003; Radutoiu et al. 2003, 2007; Indrasumunar et al. 2010). The receptors are characterized by a transmembrane domain, an intracellular serine/threonin receptor-like kinase, and extracellular LysM motifs typical for the binding of peptidoglycans, which, like Nod factors, contain *N*-acetylglucosamine. Legumes most likely require two separate receptors or receptor dimers, one pair with high stringency for the correct Nod factor structure, mediating entry of rhizobia into the root hair via an infection thread, the other, lower stringency pair, mediating early signaling events (Ardourel et al. 1994; Arrighi et al. 2006; Smit et al. 2007; Indrasumunar et al. 2010). Nod factor perception via the signaling receptor(s) leads to the activation of a leucine-rich repeat receptor-like kinase, and subsequent calcium spiking and transcriptional responses that coordinate the early infection and nodule organogenesis (Oldroyd and Downie 2006, 2008). This "symbiosis signaling" cascade is thought to have ancient origins, as several of the signal transduction genes, for example, the leucine-rich repeat receptor-like kinase, is also required by AM and actinorhizal symbioses (Gherbi et al. 2008; Markmann et al. 2008; Markmann and Parniske 2009). The predisposition for RN symbioses is likely to be twofold: First, a longer structural version of SYMRK (SYMbiosis Receptor Kinase) with several leucine-rich repeats appears only in Rosids (which include all nodulating plant species) and is present in all examined nodulating legumes (Markmann et al. 2008). Second, the Nod factor receptor is likely to be a new 'invention' that specifically allowed RN symbioses. Nod factor receptors could have evolved from closely related receptors found in the nonlegumes *Arabidopsis thaliana* and rice, which are required for chitin perception during defense responses (Kaku et al. 2006; Wan et al. 2008).

While it has traditionally been assumed that legume nodulation requires the action of Nod factors, the recent identification of a photosynthetic *Bradyrhizobium* species that does not encode the canonical genes for Nod factor synthesis suggests that Nod factors are not necessary. It is likely that these bacteria use purines or their derivatives to activate nodulation, and it has been suggested that cytokinins could be the active purine derivatives used by these bacteria (Giraud et al. 2007; Masson-Boivin et al. 2009). Cytokinin signaling was found to be necessary for the formation of nodules in legumes and likely to be a downstream response to Nod factor signaling (Gonzalez-Rizzo et al. 2006; Murray et al. 2007). Plant mutants with a constitutively active cytokinin receptor nodulate spontaneously (Tirichine et al. 2007). Rhizobia and *Frankia* can synthesize cytokinins (Fig. 1) (Stevens and Berry 1988; Sturtevant and Taller 1989) and these might be important for successful symbiosis in addition to Nod factors. The *Frankia* genome also does not appear to encode any *Nod* genes, suggesting that actinorhizal symbioses rely on other types of signaling, even though signaling downstream of a putative Nod factor receptor are at least partially conserved between actinorhizal and RN symbioses (Normand et al. 2007).

Other signals from rhizobia that have important functions in successful nodulation are surface exopolysaccharides and capsular polysaccharides, which play a role in the (down)regulation of defense responses during symbiosis (Finan et al. 1985; Le Quere et al. 2006; Jones et al. 2008), in addition to binding to lectins on the host surface, as discussed above. In addition, effectors are directly secreted into the legume host, some of which resemble effectors from plant or human pathogens (Deakin and Broughton 2009; Kambara et al. 2009). Several of the symbiotic effectors are likely to interfere with signaling pathways and phosphorylation cascades within the host (Deakin and Broughton 2009). How the exact difference between effectors secreted into host cells by symbionts compared to pathogens shifts the balance between suppressing versus inducing host defense responses is still not well understood.

2.2.2 The Enigmatic "Myc" Factor

During the formation of AM symbioses, extensive reprogramming of host gene expression and protein and membrane synthesis is necessary to accommodate the invading fungus and coordinate nutrient exchange. For many years, the existence of a "Myc factor," analogous to the *Rhizobium* Nod factor, has been hypothesized, based on studies that showed that a diffusible factor from mycorrhizal fungi can induce host responses (Fig.2a). By separating roots of *Medicago truncatula* from growing mycorrhizal hyphae with a cellophane membrane, Kosuta and colleagues (Kosuta et al. 2003) showed that a reporter gene, *Enod11-gusA*, was activated in the host roots in the absence of physical contact with the fungus. *Enod11* is an early response gene activated during both the AM and RN symbioses. While different species of mycorrhizal fungi could induce the *Enod11* response in the host, pathogenic fungi did not have any effect, supporting the notion of a specific AM signal produced by the fungal partner (Kosuta et al. 2003). A diffusible signal from the fungal partner was subsequently shown to also activate lateral root formation (Olah et al. 2005) and elicit calcium elevation and calcium oscillations in host roots, which are part of the early plant response cascade necessary for the establishment of the AM symbiosis (Navazio et al. 2007; Kosuta et al. 2008). In addition, the membrane steroid-binding protein, *Mt*MSBP1, was transiently induced in epidermal and subepidermal host cells by a diffusible mycorrhizal signal (Kuhn et al. 2010). This study suggested that sterol homeostasis might be important for fungal infection and could play a role in plasma membrane invagination during the infection process. Transcriptional studies also revealed that some early AM response genes require action of the receptor-like kinase *MtDMI2*, while others do not, suggesting the existence of at least two perception pathways for Myc factors (Kuhn et al. 2010). So far, the Myc factor(s) remain(s) to be structurally identified. Similarly, receptor(s) for Myc factor(s) have not yet been identified, whereas some of the early signaling cascade triggered by Myc factor perception has been characterized and is partially identical to the one mediating Nod factor perception. This common "symbiotic pathway" contains a receptor kinase, a nuclear-localized potassium channel required for the induction of calcium spiking, a calcium/calmodulin-dependent protein kinase, and a protein of unknown function (Cyclops) (Oldroyd et al. 2009). Interestingly, the plant appears to decode signals from the two different symbionts through the detection of specific calcium oscillations resulting from either Nod factor or Myc factor perception (Kosuta et al. 2008). As the AM symbiosis has existed up to 400 million years longer than the RN symbiosis, it is very likely that Nod factor perception has coopted the preexisting perception system for Myc factors (Markmann and Parniske 2009).

2.2.3 Quorum Sensing Signals: Implications for Plants

While many previous studies have regarded the bacterial partner as a single-celled organism, it is becoming increasingly apparent that bacteria of a single species act in a coordinated fashion not unlike multicellular organisms. This coordination of cellular behavior has been named "quorum sensing" (QS) (Fuqua et al. 2001; Miller and Bassler 2001; Fuqua and Greenberg 2002). QS describes the coordination of bacterial behaviors by regulation of gene expression in response to population densities of bacteria. Many bacteria synthesize and release the so-called "autoinducer" molecules that can act as transcriptional regulators once they reach a threshold concentration, for example, when bacterial cell densities become high. This allows bacteria to monitor the presence and density of their own as well as other species of bacteria in the environment and respond appropriately. Some bacteria can detect QS signals only from their own species while others can detect signals from other species even without synthesizing QS signals themselves (Taga and Bassler 2003). Most Gram-negative bacteria use acyl-homoserine lactone (AHL) QS signals (Fig. 1), which contain a homoserine lactone moiety linked to a variable acyl side chain, depending on the bacterial species synthesizing the AHL. This is not unlike the specificity of Nod factors (see Sect. 2.2.1). Most bacteria synthesize a mixture of AHLs. Many Gram-positive bacteria use oligopeptide autoinducers (Dunny and Leonard 1997), while a range of Gram-negative and Gram-positive bacteria use a furanone termed AI2 (Autoinducer 2) (Chen et al. 2002). Many bacteria are known to use more than one QS system. Here, we will focus on AHL QS signals as these are known to be important for plant–bacteria interactions.

AHLs are produced constitutively and short AHLs diffuse into and out of cells, while longer chain AHLs are transported back into the cell via transporters. Once the AHL concentration inside a bacterial cell reaches a critical threshold ("quorum"), AHLs bind to their cognate receptors, which then act as transcription factors to activate hundreds of downstream genes, including those for enzymes that synthesize more AHLs (Whitehead et al. 2001). In plant–bacterial interactions, QS is important for regulating bacterial behaviors important for host infection, including motility, biofilm formation, plasmid transfer, nitrogen fixation, and synthesis of virulence factors, exopolysaccharides, and degradative enzymes

(Gonzalez and Marketon 2003; Marketon et al. 2003; von Bodman et al. 2003). A survey of soil and rhizosphere bacteria found that QS is more common in plant-associated than nonplant-associated bacteria, underlining the importance of QS for the colonization and infection of plants (Cha et al. 1998). It is thought that QS is important for coordinating bacterial behaviors of plant-associated bacteria in order to (a) coordinate behaviors that only make sense once a critical number of bacterial cells have accumulated, e.g., biofilm formation, and (b) to prevent the premature elicitation of plant defense responses. For example, if only a few bacteria were producing cell wall degrading enzymes on a host surface, the plant could easily mount defenses to overcome the bacteria, whereas if degradative enzymes were only produced once the bacterial population had reached a "quorum," it is more likely that the bacteria could overwhelm the host (von Bodman et al. 2003). In order to test hypothesis (b), Mäe et al. (2001) expressed the AHL synthesis gene *expI* from the soft rot bacterium *Erwinia carotovora* in transgenic tobacco plants. AHLs regulate the synthesis of cell wall-degrading enzymes in *E. carotovora*, which then triggers plant defense responses (Jones et al. 1993). The synthesis of 3-oxoacyl homoserine lactones (HSLs) in tobacco led to enhanced resistance to *E. carotovora*, and this could be mimicked through external application of AHLs to wild-type tobacco (Mäe et al. 2001). In addition, enhanced resistance of the plant could be overcome by using high inoculum numbers of *E. carotovora*. This suggested that premature triggering of AHL responses in a pathogen can cause plant defense responses that would protect the plant from a pathogen. In contrast, Toth et al. (2004) expressed the *yenI* gene from *Yersinia enterocolitica* in transgenic potato, leading to synthesis of 3-*oxo*-hexanoyl-HSL and hexanoyl-HSL in the transgenic plants, which are AHLs regulating QS-regulated genes in *E. carotovora*. This study showed that AHL synthesis by the host plant significantly promoted disease progression by even low inoculum numbers of *E. carotovora*. Therefore, the authors of this study suggested that the production of cell wall-degrading enzymes under control of AHLs may be a response to nutrient limitation at high bacterial densities during later stages of plant infection, not a mechanism to avoid plant defenses (Toth et al. 2004).

As AHLs are important signals regulating the behaviors of symbiotic and pathogenic bacteria that inhabit the rhizosphere and infect host plants, it is not surprising that plants can "eavesdrop" on and interfere with bacterial QS signals (Fig. 2b). AHLs are present in the rhizosphere, especially around bacteria biofilms on the root surface (Steidle et al. 2001; Gantner et al. 2006). Treatment of roots of the legume *M. truncatula* with nanomolar concentrations of synthetic and purified AHLs from its symbiont *S. meliloti* and from *Pseudomonas aeruginosa* led to specific and extensive alterations in root protein expression (Mathesius et al. 2003). AHL-regulated proteins in roots included potential defense-related proteins, metabolic enzymes, and hormone-response proteins. In addition, the expression of reporter constructs for chalcone synthase (CHS), the first enzyme of the flavonoid pathway, and an auxin response gene was induced in specific cell types in white clover (*Trifolium repens* L.) roots locally treated with AHLs (Mathesius et al. 2003). For example, the *CHS3:GUS* reporter was induced in inner root cortical cells underlying the site of treatment but not in epidermal or outer cortex cells.

This suggests that responses to AHLs are cell type-specific and that either the AHLs or secondary signals triggered by their perception can travel inside the root. AHLs also have systemic effects on plants, as it was shown that treatment of roots with AHL-producing bacteria (*Serratia liquefaciens*), but not their AHL-deficient mutants, increased systemic resistance against the fungal leaf pathogen *Alternaria alternata* in tomato shoots (Schuhegger et al. 2006). A transcriptomic approach in *Arabidopsis* showed gene expression changes for several hundred genes in shoots and roots in response to 10 μM concentrations of C_6-HSL (von Rad et al. 2008). This study showed evidence for the induction of auxin response genes and increased auxin levels in roots and shoots of C_6-HSL-treated plants, whereas cytokinin responses and concentrations were reduced. Treatment with C_6-HSL and C_4-HSL increased root growth in *Arabidopsis*, whereas longer chain AHLs reduced root growth, and shoot growth appeared not to be affected (von Rad et al. 2008). In contrast to the induction of systemic resistance by AHLs in tomato, AHL-treated *Arabidopsis* showed no enhanced resistance toward *Pseudomonas syringae* and only weak induction of typical defense related genes.

A different study with *Arabidopsis* demonstrated that growth of plants on AHL-containing medium inhibited root growth, depending on the length of the acyl side chain (Ortiz-Castro et al. 2008). In addition, some AHLs, particularly C_{10}-HSL, increased lateral root density and accelerated their emergence in *Arabidopsis* and caused changes in root hair development, although at fairly high concentrations (48–192 μM). The lateral root and root hair induction close to the root tip could indicate that the AHLs have caused a termination of the apical root meristem, and this was supported by a reduction in the expression of cell cycle markers (Ortiz-Castro et al. 2008).

Overall, the studies show that plant responses to AHLs depend on AHL structures and concentrations, on the plant species, that responses vary between root and shoot, that responses can be systemic as well as local and cell type-specific, and that responses involve defense-related as well as developmental changes. The response of plants to AHLs also depends on movement and possible active transport in the plant, AHL distribution, and concentration on plant surfaces. AHLs appear to enter and move systemically in plant tissues, although the mechanism of transport is unknown. Mass spectrometric detection of C_6- and C_{10}-HSL in *Arabidopsis* showed that C_6-HSL accumulated in shoots within 4 days after treatment of roots with AHLs, whereas the more hydrophobic C_{10}-HSL remained mainly in the root (von Rad et al. 2008). Likewise, C_6-HSL accumulated in shoots of root-treated barley (*Hordeum vulgare* L.) and yam bean (*Pachyrhizus erosus* (L.) Urban), while longer chain AHLs were less abundant in shoots (Gotz et al. 2007). Roots of barley and yam bean also discriminated different chiral forms of AHL, preferring the L-isomers over D-isomers (Gotz et al. 2007).

The specific responses of plants to low concentrations of AHLs and the ability to distinguish AHLs of different structures argue for the existence of an AHL receptor in plants. There is no evidence that plants encode receptor proteins similar to those of bacterial AHL receptors. Likewise, no other type of plant receptor has been identified in plants thus far. The ability of plants to detect the presence of AHLs from the rhizosphere could be advantageous to plants in order to intercept pathogens. It is thought that QS regulated expression of elicitors or virulence factors presents, for example, a strategy for bacteria to avoid early detection by the plant host because elicitors would only be produced after high densities of bacteria are present. However, if plants can detect low concentrations of AHLs, they might be able to respond before pathogen numbers build up. This concept would have to be tested in the future with plant mutants unable to respond to AHLs.

2.3 Plant Interference with Rhizosphere Signals

While plants produce and receive specific rhizosphere signals, they can also interfere with rhizosphere signals. A prominent example is the interference of plants with QS signals from bacteria (Bauer and Teplitski 2001; Bauer and Robinson 2002). As plants can detect and respond to AHLs, it makes sense that they have also evolved strategies to interfere with bacterial QS and associated behaviors. It has been known for some years that red algae produce QS mimic compounds that block QS in bacteria that form biofilms on the algal surfaces. These QS mimic compounds were identified as halogenated furanones, with similar chemical structures to AHLs (Givskov et al. 1996). It was shown that halogenated furanones block QS in bacteria by accelerating the breakdown of the AHL receptor protein in bacteria (Manefield et al. 1999). Subsequently, Teplitski et al. (2000) showed that higher plants also produce compounds that either block or stimulate AHL-regulated QS in bacteria. Multiple compounds were isolated from seedling exudates of pea (*Pisum sativum* L.), rice (*Oryza sativa*), soybean (*Glycine max*), crown vetch (*Coronilla varia*), tomato (*Lycopersicum esculentum*), and barrel medic (*M. truncatula* Gaertn.). While their structures have not been identified yet, some compounds could be degraded by bacterial lactonases, suggesting similar structures to bacterial AHLs (Gao et al. 2003). *M. truncatula* was shown to produce at least a dozen separable AHL mimic compounds, and their production changed with seedling age (Gao et al. 2003). The exudation of AHL mimic compounds by *M. truncatula* can moreover be increased upon exposure of roots with AHLs, suggesting that mimic exudation is a response of the plant to the presence of AHLs by bacteria in the rhizosphere (Gao et al. 2003; Mathesius et al. 2003). Alternatively, some of the induced "mimic" compounds could be degradation products of bacterial AHLs as the result of the induction of plant enzymes that can degrade AHLs.

Some studies have identified plant signals that can interfere directly or indirectly with bacterial QS. For example, L-canavanine, an arginine analog exuded by alfalfa (*M. sativa* L.) seeds, inhibited AHL signaling in a reporter strain, with no effect on bacterial growth. L-Cananavine also inhibited the production of exopolysaccharides by the alfalfa symbiont *S. meliloti*. Exopolysaccharides are important for host infection and nitrogen fixation, and their synthesis is regulated by AHL-mediated QS (Keshavan et al. 2005). The synthesis of AHLs by *S. meliloti* was not affected by L-canavanine. It is likely that L-canavanine does not directly compete with AHLs for their receptors, but is incorporated into proteins as an L-arginine analog, hindering protein folding. Thus, the effect of L-canavanine on AHL regulated behaviors might be general rather than specific. Another explanation for the effect of L-canavanine could be that some bacteria can degrade it to homoserine, which could then be converted to an AHL mimic (Keshavan et al. 2005).

Other plant secondary metabolites, which interfere with AHL signaling, include salicylic acid (SA), *p*-coumaric acid, catechin, riboflavin, and lumichrome. SA, a regulator of plant defense responses, can cause breakdown of AHLs by stimulating expression of a lactonase in *Agrobacterium tumefaciencs* (Yuan et al. 2007). *p*-Coumaric acid, a phenolic metabolite produced by plants as a lignin precursor and often exuded into the soil, inhibited activity of several QS biosensors in a concentration-dependent manner (Bodini et al. 2009). The vitamin riboflavin and its breakdown product lumichrome (Fig. 1) have traditionally been seen as bacterial metabolites. However, it has been found that riboflavin can also be produced by roots and exuded into the rhizosphere. Lumichrome and riboflavin can interfere with bacterial QS by binding to and activating the AHL receptor LasR (Rajamani et al. 2008). By altering the amino acid sequence of LasR, the authors of this report showed that even though lumichrome and riboflavin are structurally different to the LasR ligand 3-*oxo*-C_{12}-HSL, it appears to bind in the same binding pocket. The roles of lumichrome and riboflavin in the rhizosphere could be complex, as they can also be synthesized by bacteria, and have multiple effects on plants, including shoot and root physiology and plant growth (Phillips et al. 1999; Matiru and Dakora 2005; Khan et al. 2008; Gouws and Kossmann 2009).

While the identification of potential mimic compounds from plants is progressing, the concentration of these mimics in the rhizosphere and their biological relevance in regulating QS-related behaviors in rhizosphere settings await investigation.

In addition to the production of mimic compounds, plants can also interfere with QS by breaking down AHLs enzymatically. The *Arabidopsis* fatty acid amide hydrolase could be degrading AHLs, because overexpressing mutants were more resistant to AHLs whereas a fatty acid amide hydrolase mutant was more susceptible to developmental changes by AHLs (Ortiz-Castro et al. 2008). A study in the model legume *L. japonicus* showed that AHLs are less stable around roots than in bulk soil (Delalande et al. 2005). AHL breakdown products are also likely to act as signals in plants, as treatment of roots with homoserine and homoserine lactone altered transpiration and stomatal conductance in shoots of bean (*Phaseolus vulgaris* L.) (Joseph and Phillips 2003).

2.4 Bacterial Interference with Rhizosphere Signals

Signals produced by either plants or microorganisms can be modified or destroyed by bacteria, fungi, and other rhizosphere organisms, and this can affect the activity and solubility of rhizosphere signals. For example, plant exuded flavonoids can be metabolized by bacteria (Shaw et al. 2006). Flavonoids can be exuded as inactive glycosides, and certain bacteria can deglycosyate these, resulting in release of the active aglycone (Hartwig and Phillips 1991). This changes the solubility and mobility of the flavonoids in the rhizosphere, as glycosides are much more water soluble than the aglycones. Flavonoids are also destructively metabolized and used as a carbon source (Pillai and Swarup 2002) and this can be demonstrated to affect soil concentrations of flavonoids (Ozan et al. 1997). This could affect signaling between rhizobia and legumes in the soil and also reduce the potential of flavonoids acting as plant defense compounds (Rao and Cooper 1995; Shaw et al. 2006). In addition, bacteria may be able to convert flavonoids from *Nod* gene inducers to *Nod* gene repressors, or vice versa, which could affect the success of RN symbioses (Rao and Cooper 1995). As mentioned previously (Sects. 2.1.1 and 2.1.2), both rhizobia and mycorrhizal fungi can also alter rhizosphere concentration of flavonoids by stimulating their production in the host.

QS signals are also actively modified by bacteria, and this could be a strategy to interfere with QS signals of competing species (Taga 2007). Bacteria encode both lactonases and acylases that can inactivate AHL signals (Dong et al. 2001, 2002; Lin et al. 2003; Dong and Zhang 2005). In addition to possibly actively interfering with QS signaling, some bacteria can also use homoserine lactone as a carbon source (Yang et al. 2006). The destruction of AHLs by bacteria has been demonstrated in the rhizosphere of potato (Jafra et al. 2006).

2.5 *Adaptation of Signals for Multiple Purposes and Cross-Signaling*

There are many examples of signals that may have initially been "intended" for a certain purpose but then modified or coopted for secondary purposes. Often it is not easy to ascertain which was the original signaling mechanism and which the modification.

For example, similar to the cooption of symbiotic signaling pathways by the RN symbiosis (as discussed in Sect. 2.2.1), the function of flavonoids as stimulators of AM hyphal branching may have later been adapted to activate rhizobial symbionts. Flavonoids also have activity in stimulating haustoria formation in root parasitic plants (Albrecht et al. 1999), and it could be speculated that these parasites have exploited the presence of flavonoids to detect their hosts. Similarly, bacteria appear to have exploited the presence of flavonoids from host roots to coordinate bacterial–plant interactions regulated by QS: The flavonoid *p*-coumaric acid, which can be exuded by roots can be integrated into a newly discovered QS signal, *p*-coumaroyl-homoserine lactone (Schaefer et al. 2008). This study suggests that *p*-coumaroyl-HSL could have dual functions, to control density-dependent behaviors in bacteria as well as sense the proximity of a host plant. Complicating this view is that *p*-coumaric acid can also be produced by bacteria as a breakdown product of flavonoids, and thus it might not always be a reliable indicator for the presence of a host (Rao and Cooper 1995). Flavonoids could also more directly interfere with QS in bacteria. The flavonoid catechin has recently been identified as a potential QS mimic molecule isolated from bark of *Combretum albiflorum*, a medicinal plant with antimicrobial properties (Vandeputte et al. 2010). The induction of the flavonoid pathway in roots in response to external application of AHLs may indicate that AHLs elicit flavonoid responses aimed at interference with QS (Mathesius et al. 2003). However, whether flavonoids have more general roles as QS mimic compounds will need to be investigated more broadly, and its relevance in the rhizosphere tested with reporter strains.

Strigolactones are examples of another group of signals with multiple effects on a range of organisms. Similar to flavonoids, in addition to stimulating germination and branching in AM fungi, strigolactones also trigger the germination of seeds of the parasitic weeds *Striga* sp. (witchweed) and *Orobanche* sp. (Cook et al. 1966; Cook et al. 1972). However, some of the strigolactones active on parasitic plant germination were not active as stimulants of mycorrhizal hyphal branching (Besserer et al. 2006). Perhaps this is a consequence of the reaction of plants to requiring strigolactone exudation for stimulating the beneficial symbiosis with AM fungi, while at the same time responding to the deleterious effect of strigolactones on the germination of these parasitic weeds. In addition, certain strigolactones have recently also been identified as a new plant hormone regulating shoot branching (Gomez-Roldan et al. 2008; Umehara et al. 2008). In addition, strigolactones also have stimulating effects on nodulation in alfalfa (Soto et al. 2010) and on root growth in tomato (Koltai et al. 2009), although these studies need to be repeated making use of strigolactone mutants and in consideration of the natural concentration range of strigolactones. Since strigolactones are thought to be ubiquitously present in land plants, although again, this remains to be tested, it is difficult to determine whether they originated as a plant branching hormone and then got adapted for plant–plant and plant–microbe signaling or vice versa. Another crucial challenge is the identification of a strigolactone receptor in plants and microorganisms and to determine whether different organisms use a related receptor.

A third example for coopted signals is the plant hormones auxin and cytokinin (Fig. 1), which are also synthesized by many plant growth-promoting rhizobacteria and can alter plant growth and development (Fig. 2a) (Persello-Cartieaux et al. 2003). Only a few examples are described here because of an extensive review of this area (Costacurta and Vanderleyden 1995). Both symbiotic and pathogenic bacteria synthesize auxins and cytokinins, in many cases, to form infection structures in host roots. For example, rhizobia commonly synthesize auxin (Kefford et al. 1960) and cytokinin (Sturtevant and Taller 1989). While the early steps of nodule initiation can be induced by Nod factors alone, synthesis of IAA by rhizobia could be important at later stages of nodulation (Kefford et al. 1960). *Rhizobium* mutants deficient in IAA synthesis can have impaired nitrogen fixation, whereas increased nodulation efficiency has been observed in IAA overproducing strains (Pii et al. 2007). As mentioned previously, cytokinins or their derivatives could be important in rhizobia not synthesizing Nod factors (Giraud et al. 2007). Plant hormone synthesis is also important for plant growth-promoting bacteria that form loose associations with plants. Here, synthesis of plant hormones has been shown to stimulate root growth, which increases the host surface these bacteria can colonize (Dobbelaere et al. 1999).

3 Signaling in Field Rhizospheres

A goal of rhizosphere research is to identify robust processes between plants and soil organisms that can be repeatedly manipulated in the field to increase plant productivity (Watt et al. 2006b). This could be the exchange of signal molecules among root and soil organism cells within the rhizosphere triggering specific processes beneficial to the plant (Fig. 2). Symbiotic and pathogenic signaling clearly can be successful in the field. Symbiotic nitrogen-fixation by rhizobia is managed to increase success of infection by (a) identifying the best host-rhizobial species and strain combinations and (b) improving the viability of inoculum applied with the seed (Thrall et al. 2000). These two management strategies are done to increase the chance of a successful infection – that is, more living bacteria close to emerging root hairs, and the best signal molecule partnership between the root and the specific strain.

Signaling among commensal microorganisms and roots in the field for plant growth promoting rhizobacterial (PGPR) processes such as pathogen inhibition, nutrient acquisition, and plant growth promotion with hormones, however, has proven much more difficult to manipulate for increased crop productivity (Stewart 2001). These organisms do not form a symbiotic, biotrophic association within root tissues, and must elicit responses from outside the root in a field rhizosphere. Field rhizospheres are much more complex than those in controlled laboratory conditions (McCully 1999). They are surrounded by soil solids, spaces, and other roots and many diverse soil micro and macroorganisms (Buee et al. 2009). In a field, over seasons, roots are in close contact with the live and dead roots of current and previous season roots such that rhizospheres "overlap" (Watt et al. 2005; Watt et al. 2006c). Field rhizospheres have a much longer age and size – on a single root system, the root surfaces, exuded chemicals, and microorganism populations change over time and in space through the course of a plant's life and beyond (Fig. 4). There are very little data tracking rhizosphere signal fate on field growth roots (Fig. 3). In this section, we progress through the stages of signal fate in a field rhizosphere (Fig. 3) and focus on what is known of field-grown plants and their rhizospheres to identify the conditions that would favor successful signal exchange.

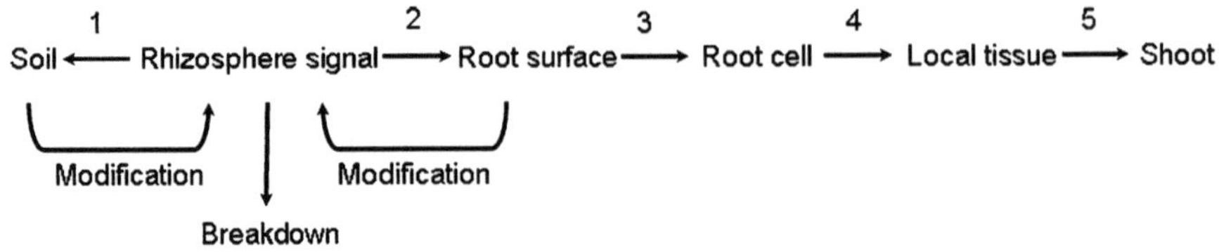

Fig. 3 The potential fates of a signal at the root–soil interface. (1) The signal reaches the soil, where it could get perceived by or modified by microorganisms, or be broken down through chemical processes. (2) The signal is perceived at the root surface. Root exudates and enzymes could further modify the signal at or near the surface. (3) The signal diffuses into or is transported into a root cell. (4) The signal or downstream signals generated by signal perception travels into the local tissue where it can have an effect. (5) The signal travels further into the shoot or ha systemic secondary effects

3.1 *Time, Distances, and Diffusivities Determine Chance of Signal Exchange*

Let's first consider step (1) in Fig. 3, where the signal may bind quickly to the soil and become inactivated, chemically through adsorption, or biologically through consumption by soil organisms. A helpful relationship for estimating the chances of a signal reaching a root in the field is shown below:

$$t^* = a^2/D.$$

The characteristic time t^* it would take a signal to be bioactively bound or adsorbed to a soil surface relative to the root surface depends on the distance a the signal is from the soil surface for a solute in a given medium and its diffusivity D (Crank 1975; Watt et al. 2006c). This relationship shows that distance is very important for determining if a signal molecule will reach a root surface, because time depends on distance squared. If a molecule is 30 μm from a root surface, trapped within a biofilm, it will arrive 1,000 times sooner on the root than if it is 1,000 μm from the root. Closer signals have much more chance of eliciting a response on a root or microorganism surface, and exchange of signals would be much quicker close to the root rather than far from the root. Further, microorganisms close to the root in a biofilm have much more chance of influencing the plant than those far from the root. Interactions right at the root surface in biofilms should be studied on field roots.

A study using fluorescent in situ hybridization (FISH) of total cellular and filamentous bacteria and *Pseudomonas* species on field grown roots confirmed that bacteria are indeed embedded in a biofilm matrix (ranging from 2 to 30 μm thick) that is resistant to washing (Watt et al. 2006a); they are distributed along the root depending on root developmental stage and contact with dead roots of previous crops. They were found to be clustered in microcolonies within or on top of biofilms on 40% of root regions observed, but only half of the clusters were associated with recognizable physical features on the root such as grooves between cells. The other half of the clusters were in regions where no physical feature was visible, suggesting chemical events such as signaling or nutrient gradients in the biofilm were affecting distribution (Kim et al. 2008). Signaling around root surfaces might depend on physical and chemical properties of biofilms of plant-associated bacteria forming on root surfaces (Ramey et al. 2004). Distribution of bacterial biofilms on root surfaces vary with bacterial species and developmental stage of the root and often occur in crevices between epidermal cells and in the root hair zone (Fujishige et al. 2006a, b). Biofilms can contain very dense accumulations of microbes as well as carbohydrates bound in a matrix (Branda et al. 2005). The bacterial composition of biofilms can greatly alter the matrix composition, and this could in turn alter diffusion, breakdown, and longevity of signals in a biofilm (Branda et al. 2005). Local conditions like pH and oxygen levels within biofilms can vary over very short distances (DeBeer et al. 1994; Horswill et al. 2007) and this may explain in part the observation on field grown roots that bacteria are not evenly distributed. Biofilm formation is dependent on QS signals (Davies et al. 1998), on the ability of bacteria to produce exopolysaccharides, flagella (Fujishige et al. 2006a, b), Nod factors (Fujishige et al. 2008), and on environmental parameters (Rinaudi et al. 2006). Biofilm formation could alter the diffusion and longevity of rhizosphere signals around a biofilm due to local changes in pH and enzymatic degradation of signals. Therefore, field biofilms are worthy of intense research.

The above equation also depends on the diffusivity *D*, which depends on the chemical structure of the molecule, the soil chemical and biological characteristics, and the water content. For example, if the molecule is positively charged and the soil type is clay, the molecule will have a high chance of binding to negatively charged sites on the high surface area of the clay particles. The biological characteristics include the density and diversity of microbial populations, and if there are organisms in the path of the signal molecule that can consume the signal. For example, AHL concentration can be affected by adsorption to soil particles, opening of the lactone ring via photo-catalyzed oxidation or alkaline pH of the surrounding soil, as well as breakdown by plant and bacterial enzymes. Calculations of the half-life of AHLs at different temperatures and pH values ranged from hours at alkaline pH and warm (>20°C) temperatures to many days under cold (<4°C) and acidic conditions (Delalande et al. 2005). Wang and Leadbetter (2005) showed that AHLs disappeared within hours in most soils. In a turf soil, they disappeared at a rate of 13.4 nmol per hour per gram fresh weight soil, and this was almost entirely attributed to microorganism consumption and very little to chemical adsorption to surfaces (Wang and Leadbetter 2005). Under very alkaline conditions, stability of AHLs could be in the range of minutes due to lactolysis (Horswill et al. 2007), but lactolysis is reversible under acidic conditions. In addition, the AHL 3-*oxo*-C_{12}-HSL can undergo Claisen-like condensations in aqueous environments, leading to the formation of the antimicrobial compound 3-hydroxydecylidene 5-(2-hydroxyethyl)pyrrolidine 2,4-dione, which is effective against Gram-positive bacteria (Kaufmann et al. 2005), suggesting that AHLs could be converted from a signal to a "weapon" against other microbes. This compound also has high affinity for iron and could be used to sequester iron from the rhizosphere (Kaufmann et al. 2005).

The water content of the soil influences the diffusivity of a molecule. If high, the molecule will diffuse much more rapidly (Watt et al. 2006c). For a glucose molecule, a doubling in soil water content in soil from 10 to 20% caused a tenfold increase in diffusion rate (Olesen et al. 2000). Protons diffused 1,000 times faster in agar compared to sandy soil with a moisture content of 7% (Nichol and Silk 2001). The effect of water on AHL signaling was elegantly demonstrated directly on leaf surfaces (Dulla and Lindow 2008). On wet leaves, the quorum induction was slower – it took longer for the signal concentration to reach a critical level, presumably because the signal was diluted as it traveled continuously away from the cells in the water, recalling also that with time, distance is squared (equation above). On dry leaves, QS induction was quicker because the signal diffusion was restricted around the cells and reached a critical concentration faster. This presents a paradoxical situation – water facilitates diffusion along longer distances faster, increasing the chances that the signal will reach a location on a biological surface before getting inactivated, yet the signal molecules are fewer and more dilute than if constrained to a discrete patch of moisture. The study of Dulla and Lindow (2008) could be extended to forming hypotheses about the role of moisture in signaling in field rhizospheres.

Examining the spatial aspects of the rhizosphere has led to further insights into the concept of QS. While most experiments on QS in bacteria have been carried out under in vitro conditions, often with bacteria grown in liquid cultures, the use of reporter strains of rhizosphere bacteria in soil grown roots has allowed the detection of QS responses in single cells colonizing roots (Steidle et al. 2001). Studies on QS under real rhizosphere conditions (Gantner et al. 2006) and on leaf surfaces (Dulla and Lindow 2008) have shown that regarding QS as a purely population density-dependent behavior, as it has been traditionally thought of, is not sufficient to describe QS in the field. Using QS reporter strains to monitor "calling distances" for QS in the rhizosphere of tomato and wheat, Gantner and colleagues showed that even single bacteria can be a sufficient source of AHLs to trigger QS in other cells. The authors found that bacteria signal over distances of usually 4–5 μm increasing to 20–30 μm at the root tip and up to 70–80 μm in the root hair zone (Gantner et al. 2006), although this will depend on water availability on the root surface as demonstrated by Dulla and Lindow (2008).

This led (Hense et al. (2007)) to propose the concept of "efficiency sensing" to explain the complexities of QS in structured microbial communities like micro-colonies growing in the rhizosphere, an open and mixed environment. Efficiency sensing combines the ideas of QS (i.e., regulating gene expression and coordinating group behavior in response to population density, with benefit for the group (Fuqua et al. 1994)) and diffusion sensing (i.e., regulating gene expression in response to mass flow in the surrounding environment and for benefit of the single cell (Redfield 2002)) to suggest that bacteria may sense the spatial distribution of other bacteria in their environment, for both group and single cell benefit. Efficiency sensing takes into account that the concentration of autoinducers that bacteria sense is a combination of autoinducer synthesis, chemical and biological destruction, and diffusion characteristics of the environment. Mathematical modeling showed that autoinducer concentration can be more strongly determined by spatial distribution of cells (e.g., microcolony formation on root surfaces) than by bacterial cell density (Hense et al. 2007). The fact that many bacteria occur in the rhizosphere as discrete microcolonies is likely to reduce the cross talk theoretically possible due to perception of QS signals from other species. Discrete spatial structures of bacterial microcolonies were found to be necessary for stable interactions between species, and this needs to be taken into account when aiming to "engineer" the rhizosphere (Kim et al. 2008).

3.2 *Surfaces and Epitopes for Signal Binding Are Constantly Changing*

Referring to Fig. 3, let us assume that the signal reaches a root surface and does not bind and become inactivated by soil biotic, physical, or chemical factors. To elicit a response within the plant, the signal must interact with the correct epitope on a cell surface. We consider here two factors that could disrupt binding: exudates from the roots and microorganisms that interfere with or mimic the signals chemically, and the presence of the epitope in the context of root aging and differentiation.

Mimic and interference molecules are highlighted above in the examples of signal exchange in symbiotic interactions (Sects. 2.3 and 2.4). In the field, we can assume that there are many more interfering and mimic signals because there are many more types of microorganisms, exudates, and ages and types of roots. It is extremely difficult to predict the activity of interfering and mimic activities in the field, but we may be able to say that the net effect would be to dampen the activities of individual signals compared to their effect in the laboratory, and that they contribute to lack of effects of individual signals in the field, especially those that regulate commensal organism activities.

Root aging and differentiation is constantly taking place in a given position along a root (Fig. 4) (Vermeer and McCully 1982; Jones et al. 2004; Watt et al. 2006c). Once the root tip arrives in a volume of soil, root cells start to go through a series of differentiation events. The two important implications for root-microorganism signaling are that first the root surfaces in contact with the rhizosphere progress through outer wall and cell membrane modifications, and thus the binding epitopes for signals change. Secondly, the local root cell physiology changes so that the exudates released to the rhizosphere and the biochemical responsiveness within the cells changes. This would suggest that the source cell of the signal and the receptor cells, either plant or microbial, need to be well-matched in space and time for an event to be triggered by a signal. This is true in the rhizobia interactions, where root hairs at the stage of emergence and underlying cortical cells must be at the right

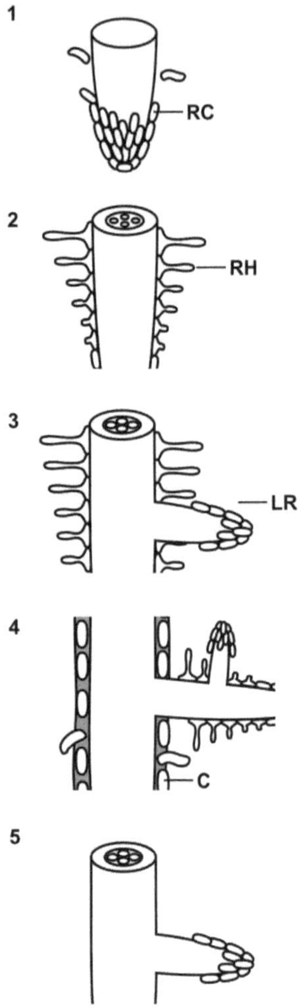

Fig. 4 Developmental stages of roots in a patch of soil. (1) Minutes to hours: The root tip reaches a patch of soil. The growing root tip is covered by root cap cells (RC), mucilage and is characterized by high rates of exudation and rhizodeposition. (2) Hours to days: Root hairs (RH) start to differentiate behind the root tip. This is a zone where roots have active interactions with soil microbes, for example during the RN and AM symbioses. (3) Days to weeks: Lateral roots (LR) start to emerge from the main root, which can locally create a new area of active exudation. (4) Weeks to months or longer: The root starts to senesce, often characterized by sloughing of cortical cells (C), so that only the original central vascular tissue remains of the tap root. (5) Months to years: Senescence and possible regeneration of root if plant is a perennial

stage to be induced to form a nodule by rhizobial signals (Bhuvaneswari et al. 1981). Because almost nothing is known of the receptors for most rhizosphere signals, especially those from a population of commensals, we cannot say if their presence on the outside of the root changes with time because of cell differentiation, or that signaling between roots and commensals resulting in a response depends on root aging.

However, root cell and rhizosphere differentiation has received some study on soil-grown roots (see McCully 1999). In grasses, root surfaces in a given volume of soil progress through a state of mucilage and cap cells when the root tip first arrives, to root hairs with bound rhizosheath and emerging branch roots, shed hairs and cortex, and then endodermal cells as the outer living surface (Vermeer and McCully 1982); see also Fig. 4. These changes are accompanied by changes in exudates and microbiological populations (Gochnauer et al. 1989; Jaeger et al. 1999; DeAngelis et al. 2009). Roots of dicotyledonous plants progress through root mucilage and cap cells when the root tip arrives in a volume of soil, root hairs with rarely a tightly bound rhizosheath, branch roots, and shed cortex, and a dividing pericycle as the outer living layer of the parent root (Sprague et al. 2007). These changes are also accompanied by shifts in associated bacteria (Marschner et al. 2002). It may be safe to say, that on field roots, signals must be at the right place at the right time to bind to root cells to trigger a response, and that the continual differentiation of outer root cell walls and membranes may offer another explanation as to why it has been difficult to manage rhizosphere signaling in the field without knowledge of the receptors and where and when they are present on the root surfaces. The continual differentiation of surfaces and exudates from roots may be a mechanism that plants have to tolerate pathogens and commensals that are inhibitory to plant growth.

Furthermore, the rate of differentiation of cells depends on the soil abiotic conditions of the plant and this alters interactions with rhizosphere microorganisms. For example, when the soil is harder, wheat roots extend more slowly than in loose soil, and within a patch of soil, differentiate root hairs and branch roots quicker, release more exudates per unit time, and have higher numbers of *Pseudomonas* compared to other slower-growing bacteria (Watt et al. 2003; Watt et al. 2006a). When soil is cooler, wheat roots also extend more slowly than when warm, but the time of root hair and branch root emergence is not earlier. *Rhizoctonia* hyphal length increases at cool temperatures on the root tips and the more mature regions of the root, but penetration only increases in the more mature region of the root, and this appears to depend more on the biochemical state of the cells and perhaps their ability to produce inhibitory exudates (Watt unpublished results). These two studies were conducted in controlled conditions in soil and focused on single groups of microorganisms and they highlight the complexity in predicting and managing rhizosphere signaling and events in field conditions.

3.3 How Do Root Cells Perceive and Respond to the Rhizosphere Microorganism Community?

In the field, roots are surrounded by many diverse microorganisms and many signals from this community may attach to root cells. Therefore at step (3) in Fig. 3, root cells are most definitely faced with continually perceiving and potentially responding to many rhizosphere signals at the same time. This area of community perception and response is not well understood partly because the total populations of organisms have not been characterized nor isolated for testing, and the spectrum of signal molecules is not known. However, research in animal guts has led the way to start understanding how root cells could perceive and respond to the commensals in the rhizosphere community – through the sensing of cell wall components and other surface molecules, common to many bacteria by pattern recognition receptors (PRRs) (Hooper and Gordon 2001; Hooper et al. 2001; Artis 2008). This model can be extended to the rhizosphere (Zipfel 2008; Boller and Felix 2009; Deakin and Broughton 2009; Faure et al. 2009). Signals such as lipopolysaccharides (LPSs) from diverse rhizosphere microorganisms may be perceived by the plant through PRRs – these molecule "signals" are called microbe-associated molecular patterns (MAMPs) or pathogen-associated molecular patterns (PAMPs). The PRR model may explain how the host cell decides which organisms to respond to in the field. Hooper et al. (2001) suggest that in the gut, the impact of the microbiota largely depends on the state of the host immunity, and interactions may result in an inflammatory response or an antiinflammatory response depending on the equilibrium of the organism's immune system at the time of perception of signals. A role for PRR in field grown roots has yet to be demonstrated, but the discovery that both symbionts and pathogens of roots use type III and IV secretion systems, which result in injection of effectors into host cells that can modulate PRR-mediated immune responses, argues for a similar situation as in animal guts (Deakin and Broughton 2009).

If root cells respond to signals, this response may extend to tissues and the shoot (steps four and five in Fig. 3). This is demonstrated in nodule development, where nodulation is regulated both locally and systemically (Kinkema et al. 2006), leading to shoot-to-root auxin transport changes (Kinkema et al. 2006; van Noorden et al. 2006), and in systemic effects of AHLs and homoserine lactone on transpiration and systemic resistance (Fig. 2) (Joseph and Phillips 2003; Schuhegger et al. 2006). Either the signal molecule must be able to travel between cells within the symplast or the apoplast, or the rhizosphere signal turns on plant signaling pathways that act downstream in root and shoot tissues. Local and long distance signaling in the plant in field conditions has been demonstrated for symbiotic and pathogenic rhizosphere interactions. For commensal interactions, signaling within the plant as a result from rhizosphere signals has not been demonstrated.

4 Future Approaches

The potential to engineer the rhizosphere to improve plant growth in the field has gained much recent attention, e.g., Ryan et al. (2009), but is limited by knowledge of the complexity of signals, organisms, and physicochemical heterogeneity. A range of new techniques will be necessary to unravel this complexity (Sorensen et al. 2009). The first challenge is to identify new signaling molecules in root exudates and secretions of microorganisms. This could be done with metabolomics approaches (Walker et al. 2003b; Mark et al. 2005), especially if experiments take into account that spatial and temporal changes that occur in the rhizospheres of field-grown plants that are highlighted in Sect. 3. Secondly, as explained above, most signals are likely to have multiple effects on different organisms, and even for known signals, not all functions have been discovered (Walker et al. 2003a, b). These studies will also need to involve identification of signal receptors in different organisms, which has been possible in culturable, sequenced, and easily transformable organisms, but lags behind, for example, in mycorrhizal and other fungi, many of which have complicated genetics and cannot be cultured in the absence of their host. To unravel new functions for rhizosphere signals, reporter constructs operating in bacteria and plants have been useful. These are limited in organisms that cannot be transformed, like many fungi and oomycetes. Another challenge for the future will be to identify microorganisms that are not currently culturable, and these "unknowns" have been estimated to comprise the majority of soil inhabitants (Sait et al. 2002). Molecular microbiology and sequence phylogeny is increasingly used to identify and characterize microorganisms from rhizosphere soil, e.g., DeAngelis et al. (2009); Osborne et al. (2010), but studying their biology will remain difficult if they cannot be cultured. We also need a more detailed understanding of the physical and chemical characteristics of the rhizosphere that determine the availability, effective concentration, and distribution of signals in the rhizosphere, which is not even known for well-studied signals such as Nod factors or flavonoids. Finally, functional tests of metabolites on target organisms need to be done eventually in real rhizospheres, e.g., Gantner et al. (2006). It will be important to correlate the molecular action of rhizosphere signals with the detailed anatomical and structural features of the root and rhizosphere, for example, with advanced electron microscopy and fluorescence techniques (McCully et al. 2009). To achieve the goal of managing the rhizosphere in the field to increase plant productivity, the new high throughput sequencing, metabolomic and in situ data will need to be applied in a new, predictive microbial ecology framework because of the complexity of the system (McMahon et al. 2007).

Acknowledgements UM was funded from the Australian Research Council through the Centre of Excellence for Integrative Legume Research (CE0348212) and an Australian Research Fellowship (DP0557692).

References

Akiyama K, Hayashi H (2006) Strigolactones: chemical signals for fungal symbionts and parasitic weeds in plant roots. Ann Bot 97:925–931

Albrecht H, Yoder JI, Phillips DA (1999) Flavonoids promote haustoria formation in the root parasite *Triphysaria versicolor*. Plant Physiol 119:585–591

Antunes PM, Rajcan I, Goss MJ (2006) Specific flavonoids as interconnecting signals in the tripartite symbiosis formed by arbuscular mycorrhizal fungi, *Bradyrhizobium japonicum* (Kirchner) Jordan and soybean (*Glycine max* (L.) Merr.). Soil Biol Biochem 38:533–543

Aoki T, Akashi T, Ayabe S (2000) Flavonoids of leguminous plants: structure, biological activity, and biosynthesis. J Plant Res 113:475–488

Ardourel M, Demot N, Debelle F, Maillet F, de Billy F, Prome J-C, Denaric J, Truchet G (1994) *Rhizobium meliloti* lipooligosaccharide nodulation factors: different structural requirements for bacterial entry into target root hair cells and induction of plant symbiotic developmental responses. Plant Cell 6:1357–1374

Arrighi JF, Barre A, Ben Amor B, Bersoult A, Soriano LC, Mirabella R, de Carvalho-Niebel F, Journet EP, Gherardi M, Huguet T, Geurts R, Denarie J, Rouge P, Gough C (2006)

The *Medicago truncatula* lysine motif-receptor-like kinase gene family includes NFP and new nodule-expressed genes. Plant Physiol 142:265–279

Artis D (2008) Epithelial-cell recognition of commensal bacteria and maintenance of immune homeostasis in the gut. Nat Rev Immunol 8:411–420

Bauer WD, Robinson JB (2002) Disruption of bacterial quorum sensing by other organisms. Curr Opin Biotechnol 13:234–237

Bauer WD, Teplitski M (2001) Can plants manipulate bacterial quorum sensing? Aust J Plant Physiol 28:913–921

Becard G, Taylor LP, Douds DD, Pfeffer PE, Doner LW (1995) Flavonoids are not necessary plant signal compounds in arbuscular mycorrhizal symbioses. Mol Plant Microbe Interact 8:252–258

Benoit LF, Berry AM (1997) Flavonoid-like compounds from seeds of red alder (*Alnus rubra*) influence host nodulation by *Frankia* (Actinomycetales). Physiol Plant 99:588–593

Bertin C, Yang XH, Weston LA (2003) The role of root exudates and allelochemicals in the rhizosphere. Plant Soil 256:67–83

Besserer A, Becard G, Jauneau A, Roux C, Sejalon-Delmas N (2008) GR24, a synthetic analog of strigolactones, stimulates the mitosis and growth of the arbuscular mycorrhizal fungus *Gigaspora rosea* by boosting its energy metabolism. Plant Physiol 148:402–413

Besserer A, Puech-Pages V, Kiefer P, Gomez-Roldan V, Jauneau A, Roy S, Portais JC, Roux C, Becard G, Sejalon-Delmas N (2006) Strigolactones stimulate arbuscular mycorrhizal fungi by activating mitochondria. PLoS Biol 4:1239–1247

Bhuvaneswari TV, Bhagwat AA, Bauer WD (1981) Transient susceptibility of root cells in four common legumes to nodulation by rhizobia. Plant Physiol 68:1144–1149

Bodini SF, Manfredini S, Epp M, Valentini S, Santori F (2009) Quorum sensing inhibition activity of garlic extract and *p*-coumaric acid. Lett Appl Microbiol 49:551–555

Bohlool BB, Schmidt EL (1974) Lectins: a possible basis for specificity in the *Rhizobium*-legume symbiosis. Science 185:269–271

Boller T, Felix G (2009) A Renaissance of elicitors: perception of microbe-associated molecular patterns and danger signals by pattern-recognition receptors. Annu Rev Plant Biol 60:379–406

Bouwmeester HJ, Roux C, Lopez-Raez JA, Becard G (2007) Rhizosphere communication of plants, parasitic plants and AM fungi. Trends Plant Sci 12:224–230

Branda SS, Vik A, Friedman L, Kolter R (2005) Biofilms: the matrix revisited. Trends Microbiol 13:20–26

Buee M, De Boer W, Martin F, van Overbeek L, Jurkevitch E (2009) The rhizosphere zoo: an overview of plant-associated communities of microorganisms, including phages, bacteria, archaea, and fungi, and of some of their structuring factors. Plant Soil 321:189–212

Cha C, Gao P, Chen YC, Shaw PD, Farrand SK (1998) Production of acyl-homoserine lactone quorum-sensing signals by gram-negative plant-associated bacteria. Mol Plant Microbe Interact 11:1119–1129

Chen HC, Higgins J, Oresnik IJ, Hynes MF, Natera S, Djordjevic MA, Weinman JJ, Rolfe BG (2000) Proteome analysis demonstrates complex replicon and luteolin interactions in *pSymA*-cured derivatives of *Sinorhizobium meliloti* strain 2011. Electrophoresis 21:3833–3842

Chen X, Schauder S, Potier N, Van Dorsselaer A, Pelczer I, Bassler BL, Hughson FM (2002) Structural identification of a bacterial quorum-sensing signal containing boron. Nature 415:545–549

Cook CE, Coggon P, McPhail AT, Wall ME, Whichard LP, Egley GH, Luhan PA (1972) Germination stimulants. 2. Structure of strigol – potent seed germination stimulant for witchweed (*Striga lutea* Lour). J Am Chem Soc 94:6198

Cook CE, Whichard LP, Turner B, Wall ME (1966) Germination of witchweed (*Striga lutea* Lour) – isolation and properties of a potent stimulant. Science 154:1189

Costacurta A, Vanderleyden J (1995) Synthesis of phytohormones by plant-associated bacteria. Crit Rev Microbiol 21:1–18

Crank J (1975) The mathematics of diffusion. Clarendon, Oxford

Dakora FD, Phillips DA (1996) Diverse functions of isoflavonoids in legumes transcend antimicrobial definitions of phytoalexins. Physiol Mol Plant Pathol 49:1–20
Davies DG, Parsek MR, Pearson JP, Iglewski BH, Costerton JW, Greenberg EP (1998) The involvement of cell-to-cell signals in the development of a bacterial biofilm. Science 280: 295–298
Deakin WJ, Broughton WJ (2009) Symbiotic use of pathogenic strategies: rhizobial protein secretion systems. Nat Rev Microbiol 7:312–320
DeAngelis KM, Brodie EL, DeSantis TZ, Andersen GL, Lindow SE, Firestone MK (2009) Selective progressive response of soil microbial community to wild oat roots. ISME J 3:168–178
DeBeer D, Stoodley P, Roe F, Lewandowski Z (1994) Effects of biofilm structures on oxygen distribution and mass transport. Biotechnol Bioeng 43:1131–1138
Delalande L, Faure D, Raffoux A, Uroz S, D'Angelo-Picard C, Elasri M, Carlier A, Berruyer R, Petit A, Williams P, Dessaux Y (2005) *N*-Hexanoyl-L-homoserine lactone, a mediator of bacterial quorum-sensing regulation, exhibits plant-dependent stability and may be inactivated by germinating *Lotus corniculatus* seedlings. FEMS Microbiol Ecol 52:13–20
Denarié J, Débellé F (1996) *Rhizobium* lipo-chitooligosaccharide nodulation factors: signaling molecules mediating recognition and morphogenesis. Annu Rev Biochem 65:503–535
Díaz CL, Melchers LS, Hooykaas PJJ, Lugtenberg EJJ, Kijne JW (1989) Root lectin as a determinant of host-plant specificity in the Rhizobium-legume symbiosis. Nature 338: 579–581
Dobbelaere S, Croonenborghs A, Thys A, Vande Broek A, Vanderleyden J (1999) Phytostimulatory effect of *Azospirillum brasilense* wild type and mutant strains altered in IAA production on wheat. Plant Soil 212:155–164
Dong YH, Gusti AR, Zhang Q, Xu JL, Zhang LH (2002) Identification of quorum-quenching *N*-acyl homoserine lactonases from *Bacillus* species. Appl Environ Microbiol 68:1754–1759
Dong YH, Wang LH, Xu JL, Zhang HB, Zhang XF, Zhang LH (2001) Quenching quorum-sensing-dependent bacterial infection by an *N*-acyl homoserine lactonase. Nature 411:813–817
Dong YH, Zhang LH (2005) Quorum sensing and quorum-quenching enzymes. J Microbiol 43:101–109
Dulla G, Lindow SE (2008) Quorum size of *Pseudomonas syringae* is small and dictated by water availability on the leaf surface. Proc Natl Acad Sci USA 105:3082–3087
Dunny GM, Leonard BAB (1997) Cell–cell communication in gram-positive bacteria. Annu Rev Microbiol 51:527–564
Faure D, Vereecke D, Leveau JHJ (2009) Molecular communication in the rhizosphere. Plant Soil 321:279–303
Finan TM, Hirsch AM, Leigh JA, Johansen E, Kuldau GA, Deegan S, Walker GC, Signer ER (1985) Symbiotic mutants of *Rhizobium meliloti* that uncouple plant from bacterial differentiation. Cell 40:869–877
Fujishige NA, Kapadia NN, De Hoff PL, Hirsch AM (2006a) Investigations of *Rhizobium* biofilm formation. FEMS Microbiol Ecol 56:195–206
Fujishige NA, Kapadia NN, Hirsch AM (2006b) A feeling for the micro-organism: structure on a small scale. Biofilms on plant roots. Bot J Linn Soc 150:79–88
Fujishige NA, Lum MR, De Hoff PL, Whitelegge JP, Faull KF, Hirsch AM (2008) *Rhizobium* common *nod* genes are required for biofilm formation. Mol Microbiol 67:504–515
Fuqua C, Greenberg EP (2002) Listening in on bacteria: acyl-homoserine lactone signalling. Nat Rev Mol Cell Biol 3:685–695
Fuqua C, Parsek MR, Greenberg EP (2001) Regulation of gene expression by cell-to-cell communication: acyl-homoserine lactone quorum sensing. Annu Rev Genet 35:439–468
Fuqua WC, Winans SC, Greenberg EP (1994) Quorum sensing in bacteria – the luxR-LuxI family of cell density-responsive transcriptional regulators. J Bacteriol 176:269–275
Gantner S, Schmid M, Durr C, Schuhegger R, Steidle A, Hutzler P, Langebartels C, Eberl L, Hartmann A, Dazzo FB (2006) *In situ* quantitation of the spatial scale of calling distances and

population density-independent *N*-acylhomoserine lactone-mediated communication by rhizobacteria colonized on plant roots. FEMS Microbiol Ecol 56:188–194

Gherbi H, Markmann K, Svistoonoff S, Estevan J, Autran D, Giczey G, Auguy F, Peret B, Laplaze L, Franche C, Parniske M, Bogusz D (2008) SymRK defines a common genetic basis for plant root endosymbioses with arbuscular mycorrhiza fungi, rhizobia, and *Frankia* bacteria. Proc Natl Acad Sci USA 105:4928–4932

Giraud E, Moulin L, Vallenet D, Barbe V, Cytryn E, Avarre JC, Jaubert M, Simon D, Cartieaux F, Prin Y, Bena G, Hannibal L, Fardoux J, Kojadinovic M, Vuillet L, Lajus A, Cruveiller S, Rouy Z, Mangenot S, Segurens B, Dossat C, Franck WL, Chang WS, Saunders E, Bruce D, Richardson P, Normand P, Dreyfus B, Pignol D, Stacey G, Emerich D, Vermeglio A, Medigue C, Sadowsky M (2007) Legumes symbioses: absence of nod genes in photosynthetic bradyrhizobia. Science 316:1307–1312

Givskov M, DeNys R, Manefield M, Gram L, Maximilien R, Eberl L, Molin S, Steinberg PD, Kjelleberg S (1996) Eukaryotic interference with homoserine lactone-mediated prokaryotic signaling. J Bacteriol 178:6618–6622

Gochnauer MB, McCully ME, Labbe H (1989) Different populations of bacteria associated with sheathed and bare regions of roots of field-grown maize. Plant Soil 114:107–120

Gomez-Roldan V, Fermas S, Brewer PB, Puech-Pages V, Dun EA, Pillot JP, Letisse F, Matusova R, Danoun S, Portais JC, Bouwmeester H, Becard G, Beveridge CA, Rameau C, Rochange SF (2008) Strigolactone inhibition of shoot branching. Nature 455:189–194

Gonzalez-Rizzo S, Crespi M, Frugier F (2006) The *Medicago truncatula* CRE1 cytokinin receptor regulates lateral root development and early symbiotic interaction with *Sinorhizobium meliloti*. Plant Cell 18:2680–2693

Gonzalez JE, Marketon MM (2003) Quorum sensing in nitrogen-fixing rhizobia. Microbiol Mol Biol Rev 67:574

Goormachtig S, Capoen W, Holsters M (2004) Rhizobium infection: lessons from the versatile nodulation behaviour of water-tolerant legumes. Trends Plant Sci 9:518–522

Gotz C, Fekete A, Gebefuegi I, Forczek ST, Fuksova K, Li X, Englmann M, Gryndler M, Hartmann A, Matucha M, Schmitt-Kopplin P, Schroder P (2007) Uptake, degradation and chiral discrimination of N-acyl-D/L-homoserine lactones by barley (*Hordeum vulgare*) and yam bean (*Pachyrhizus erosus*) plants. Anal Bioanal Chem 389:1447–1457

Gouws LM, Kossmann J (2009) The molecular analysis of the effects of the rhizosphere signal molecule lumichrome as a plant growth promoter. S Afr J Bot 75:402–403

Harrison MJ (1997) The arbuscular mycorrhizal symbiosis: an underground association. Trends Plant Sci 2:54–60

Harrison MJ (1998) Development of the arbuscular mycorrhizal symbiosis. Curr Opin Plant Biol 1:360–365

Hartwig UA, Phillips DA (1991) Release and modification of nod gene-inducing flavonoids from alfalfa seeds. Plant Physiol 95:804–807

Hense BA, Kuttler C, Mueller J, Rothballer M, Hartmann A, Kreft JU (2007) Opinion – does efficiency sensing unify diffusion and quorum sensing? Nat Rev Microbiol 5:230–239

Hirsch AM (1992) Developmental biology of legume nodulation. New Phytol 122:211–237

Hirsch AM, Bauer WD, Bird DM, Cullimore J, Tyler B, Yoder JI (2003) Molecular signals and receptors: controlling rhizosphere interactions between plants and other organisms. Ecology 84:858–868

Hooper LV, Gordon JI (2001) Commensal host–bacterial relationships in the gut. Science 292:1115–1118

Hooper LV, Wong MH, Thelin A, Hansson L, Falk PC, Gordon JI (2001) Molecular analysis of commensal host–microbial relationships in the intestine. Science 291:881–884

Horswill AR, Stoodley P, Stewart PS, Parsek MR (2007) The effect of the chemical, biological, and physical environment on quorum sensing in structured microbial communities. Anal Bioanal Chem 387:371–380

Indrasumunar A, Kereszt A, Searle I, Miyagi M, Li D, Nguyen CDT, Men A, Carroll BJ, Gresshoff PM (2010) Inactivation of duplicated Nod-Factor Receptor 5 (*NFR5*) genes in recessive loss-of-function non-nodulation mutants of allotetraploid soybean (*Glycine max* L. Merr.). Plant Cell Physiol 51:201–214

Innes RW, Kuempel PL, Plazinski J, Cantercremers H, Rolfe BG, Djordjevic MA (1985) Plant factors induce expression of nodulation and host-range genes in *Rhizobium trifolii*. Mol Gen Genet 201:426–432

Jaeger CH, Lindow SE, Miller S, Clark E, Firestone MK (1999) Mapping of sugar and amino acid availability in soil around roots with bacterial sensors of sucrose and tryptophan. Appl Environ Microbiol 65:2685–2690

Jafra S, Przysowa J, Czajkowski R, Michta A, Garbeva P, Van der Wolf JM (2006) Detection and characterization of bacteria from the potato rhizosphere degrading *N*-acyl-homoserine lactone. Can J Microbiol 52:1006–1015

Jones DL, Hodge A, Kuzyakov Y (2004) Plant and mycorrhizal regulation of rhizodeposition. New Phytol 163:459–480

Jones KM, Sharopova N, Lohar DP, Zhang JQ, VandenBosch KA, Walker GC (2008) Differential response of the plant *Medicago truncatula* to its symbiont *Sinorhizobium meliloti* or an exopolysaccharide-deficient mutant. Proc Natl Acad Sci USA 105:704–709

Jones S, Yu B, Bainton NJ, Birdsall M, Bycroft BW, Chhabra SR, Cox AJR, Golby P, Reeves PJ, Stephens S, Winson MK, Salmond GPC, Stewart GSAB, Williams P (1993) The *Lux* autoinducer regulates the production of exoenzyme virulence determinants in *Erwinia carotovora* and *Pseudomonas aeruginosa*. EMBO J 12:2477–2482

Joseph CA, Phillips DA (2003) Metabolites from soil bacteria affect plant water relations. Plant Physiol Biochem 41:189–192

Kaku H, Nishizawa Y, Ishii-Minami N, Akimoto-Tomiyama C, Dohmae N, Takio K, Minami E, Shibuya N (2006) Plant cells recognize chitin fragments for defense signaling through a plasma membrane receptor. Proc Natl Acad Sci USA 103:11086–11091

Kambara K, Ardissone S, Kobayashi H, Saad MM, Schumpp O, Broughton WJ, Deakin WJ (2009) Rhizobia utilize pathogen-like effector proteins during symbiosis. Mol Microbiol 71:92–106

Kaufmann GF, Sartorio R, Lee SH, Rogers CJ, Meijler MM, Moss JA, Clapham B, Brogan AP, Dickerson TJ, Janda KD (2005) Revisiting quorum sensing: discovery of additional chemical and biological functions for 3-oxo-*N*-acylhomoserine lactones. Proc Natl Acad Sci USA 102:309–314

Kefford NP, Brockwell J, Zwar JA (1960) The symbiotic synthesis of auxin by legumes and nodule bacteria and its role in nodule development. Aust J Biol Sci 13:456–467

Keshavan ND, Chowdhary PK, Haines DC, Gonzalez JE (2005) L-canavanine made by *Medicago sativa* interferes with quorum sensing in *Sinorhizobium meliloti*. J Bacteriol 187: 8427–8436

Khan W, Prithiviraj B, Smith DL (2008) Nod factor [Nod Bj V (C-18: 1, MeFuc)] and lumichrome enhance photosynthesis and growth of corn and soybean. J Plant Physiol 165:1342–1351

Kikuchi K, Matsushita N, Suzuki K, Hogetsu T (2007) Flavonoids induce germination of basidiospores of the ectomycorrhizal fungus *Suillus bovinus*. Mycorrhiza 17:563–570

Kim HJ, Boedicker JQ, Choi JW, Ismagilov RF (2008) Defined spatial structure stabilizes a synthetic multispecies bacterial community. Proc Natl Acad Sci USA 105:18188–18193

Kinkema M, Scott PT, Gresshoff PM (2006) Legume nodulation: successful symbiosis through short- and long-distance signalling. Funct Plant Biol 33:707–721

Knee EM, Gong FC, Gao MS, Teplitski M, Jones AR, Foxworthy A, Mort AJ, Bauer WD (2001) Root mucilage from pea and its utilization by rhizosphere bacteria as a sole carbon source. Mol Plant Microbe Interact 14:775–784

Koltai H, Dor E, Hershenhorn J, Joel DM, Weiniger S, Lekalla S, Shealtiel H, Bhattacharya C, Eliahu E, Resnick N, Berg R, Kapulnik Y (2009) Strigolactones' effect on root growth and root hair elongation may be mediated by auxin efflux carriers. J Plant Growth Regul. doi:10.1007/s00344-009-9122-7

Kosuta S, Chabaud M, Lougnon G, Gough C, Denarie J, Barker DG, Becard G (2003) A diffusible factor from arbuscular mycorrhizal fungi induces symbiosis-specific *MtENOD11* expression in roots of *Medicago truncatula*. Plant Physiol 131:952–962

Kosuta S, Hazledine S, Sun J, Miwa H, Morris RJ, Downie JA, Oldroyd GED (2008) Differential and chaotic calcium signatures in the symbiosis signaling pathway of legumes. Proc Natl Acad Sci USA 105:9823–9828

Kuhn H, Küster H, Requena N (2010) Membrane steroid-binding protein 1 induced by a diffusible fungal signal is critical for mycorrhization in Medicago truncatula. New Phytol 185: 716–733

Larose G, Chenevert R, Moutoglis P, Gagne S, Piche Y, Vierheilig H (2002) Flavonoid levels in roots of *Medicago sativa* are modulated by the developmental stage of the symbiosis and the root colonizing arbuscular mycorrhizal fungus. J Plant Physiol 159:1329–1339

Le Quere AJL, Deakin WJ, Schmeisser C, Carlson RW, Streit WR, Broughton WJ, Forsberg LS (2006) Structural characterization of a K-antigen capsular polysaccharide essential for normal symbiotic infection in Rhizobium sp NGR234 – deletion of the rkpMNO locus prevents synthesis of 5,7-diacetamido-3,5,7,9-tetradeoxy-non-2-ulosonic acid. J Biol Chem 281:28981–28992

Limpens E, Franken C, Smit P, Willemse J, Bisseling T, Geurts R (2003) LysM domain receptor kinases regulating rhizobial nod factor-induced infection. Science 302:630–633

Lin YH, Xu JL, Hu JY, Wang LH, Ong SL, Leadbetter JR, Zhang LH (2003) Acyl-homoserine lactone acylase from *Ralstonia* strain XJ12B represents a novel and potent class of quorum-quenching enzymes. Mol Microbiol 47:849–860

Long SR (1989) Rhizobium-legume nodulation – life together in the underground. Cell 56:203–214

Lopez-Raez JA, Charnikhova T, Gomez-Roldan V, Matusova R, Kohlen W, De Vos R, Verstappen F, Puech-Pages V, Becard G, Mulder P, Bouwmeester H (2008) Tomato strigolactones are derived from carotenoids and their biosynthesis is promoted by phosphate starvation. New Phytol 178:863–874

Madsen EB, Madsen LH, Radutoiu S, Olbryt M, Rakwalska M, Szczyglowski K, Sato S, Kaneko T, Tabata S, Sandal N, Stougaard J (2003) A receptor kinase gene of the LysM type is involved in legume perception of rhizobial signals. Nature 425:637–640

Mäe A, Montesano M, Koiv V, Palva ET (2001) Transgenic plants producing the bacterial pheromone N-acyl-homoserine lactone exhibit enhanced resistance to the bacterial phytopathogen *Erwinia carotovora*. Mol Plant Microbe Interact 14:1035–1042

Manefield M, de Nys R, Kumar N, Read R, Givskov M, Steinberg P, Kjelleberg SA (1999) Evidence that halogenated furanones from *Delisea pulchra* inhibit acylated homoserine lactone (AHL)-mediated gene expression by displacing the AHL signal from its receptor protein. Microbiology 145:283–291

Mark GL, Dow JM, Kiely PD, Higgins H, Haynes J, Baysse C, Abbas A, Foley T, Franks A, Morrissey J, O'Gara F (2005) Transcriptome profiling of bacterial responses to root exudates identifies genes involved in microbe–plant interactions. Proc Natl Acad Sci USA 102:17454–17459

Marketon MM, Glenn SA, Eberhard A, Gonzalez JE (2003) Quorum sensing controls exopolysaccharide production in *Sinorhizobium meliloti*. J Bacteriol 185:325–331

Markmann K, Giczey G, Parniske M (2008) Functional adaptation of a plant receptor-kinase paved the way for the evolution of intracellular root symbioses with bacteria. PLoS Biol 6:497–506

Markmann K, Parniske M (2009) Evolution of root endosymbiosis with bacteria: how novel are nodules? Trends Plant Sci 14:77–86

Marschner P, Neumann G, Kania A, Weiskopf L, Lieberei R (2002) Spatial and temporal dynamics of the microbial community structure in the rhizosphere of cluster roots of white lupin (*Lupinus albus* L.). Plant Soil 246:167–174

Masson-Boivin C, Giraud E, Perret X, Batut J (2009) Establishing nitrogen-fixing symbiosis with legumes: how many rhizobium recipes? Trends Microbiol 17:458–466

Mathesius U, Mulders S, Gao MS, Teplitski M, Caetano-Anolles G, Rolfe BG, Bauer WD (2003) Extensive and specific responses of a eukaryote to bacterial quorum-sensing signals. Proc Natl Acad Sci USA 100:1444–1449

Matiru VN, Dakora FD (2005) The rhizosphere signal molecule lumichrome alters seedling development in both legumes and cereals. New Phytol 166:439–444

McCully ME (1999) Roots in soil: unearthing the complexities of roots and their rhizospheres. Annu Rev Plant Physiol Plant Mol Biol 50:695–718

McCully ME, Canny MJ, Huang CX (2009) Cryo-scanning electron microscopy (CSEM) in the advancement of functional plant biology. Morphological and anatomical applications. Funct Plant Biol 36:97–124

McMahon KD, Martin HG, Hugenholtz P (2007) Integrating ecology into biotechnology. Curr Opin Biotechnol 18:287–292

Miller MB, Bassler BL (2001) Quorum sensing in bacteria. Annu Rev Microbiol 55:165–199

Moulin L, Munive A, Dreyfus B, Boivin-Masson C (2001) Nodulation of legumes by members of the beta-subclass of proteobacteria. Nature 411:948–950

Mulligan JT, Long SR (1985) Induction of *Rhizobium meliloti NodC* expression by plant exudates requires NodD. Proc Natl Acad Sci USA 82:6609–6613

Murray JD, Karas BJ, Sato S, Tabata S, Amyot L, Szczyglowski K (2007) A cytokinin perception mutant colonized by *Rhizobium* in the absence of nodule organogenesis. Science 315:101–104

Navazio L, Moscatiello R, Genre A, Novero M, Baldan B, Bonfante P, Mariani P (2007) A diffusible signal from arbuscular mycorrhizal fungi elicits a transient cytosolic calcium elevation in host plant cells. Plant Physiol 144:673–681

Nichol SA, Silk WK (2001) Empirical evidence of a convection-diffusion model for pH patterns in the rhizospheres of root tips. Plant Cell Environ 24:967–974

Normand P, Lapierre P, Tisa LS, Gogarten JP, Alloisio N, Bagnarol E, Bassi CA, Berry AM, Bickhart DM, Choisne N, Couloux A, Cournoyer B, Cruveiller S, Daubin V, Demange N, Francino MP, Goltsman E, Huang Y, Kopp OR, Labarre L, Lapidus A, Lavire C, Marechal J, Martinez M, Mastronunzio JE, Mullin BC, Niemann J, Pujic P, Rawnsley T, Rouy Z, Schenowitz C, Sellstedt A, Tavares F, Tomkins JP, Vallenet D, Valverde C, Wall LG, Wang Y, Medigue C, Benson DR (2007) Genome characteristics of facultatively symbiotic *Frankia* sp strains reflect host range and host plant biogeography. Genome Res 17:7–15

Olah B, Briere C, Becard G, Denarie J, Gough C (2005) Nod factors and a diffusible factor from arbuscular mycorrhizal fungi stimulate lateral root formation in *Medicago truncatula* via the *DMI1/DMI2* signalling pathway. Plant J 44:195–207

Oldroyd GED, Downie JA (2006) Nuclear calcium changes at the core of symbiosis signalling. Curr Opin Plant Biol 9:351–357

Oldroyd GED, Downie JM (2008) Coordinating nodule morphogenesis with rhizobial infection in legumes. Annu Rev Plant Biol 59:519–546

Oldroyd GED, Harrison MJ, Paszkowski U (2009) Reprogramming plant cells for endosymbiosis. Science 324:753–754

Olesen T, Moldrup P, Yamaguchi T, Nissen HH, Rolston DE (2000) Modified half-cell method for measuring the solute diffusion coefficient in undisturbed, unsaturated soil. Soil Sci 165:835–840

Ortiz-Castro R, Martinez-Trujillo M, Lopez-Bucio J (2008) N-acyl-L-homoserine lactones: a class of bacterial quorum-sensing signals alter post-embryonic root development in *Arabidopsis thaliana*. Plant Cell Enviro 31: 1497–1509

Osborne CA, Peoples MB, Janssen PH (2010) Detection of a reproducible, single-member shift in soil bacterial communities exposed to low levels of hydrogen. Appl Environ Microbiol 76:1471–1479

Ozan A, Safir GR, Nair MG (1997) Persistence of isoflavones formononetin and biochanin A in soil and their effects on soil microbe populations. J Chem Ecol 23:247–258

Peoples MB, Brockwell J, Herridge DF, Rochester IJ, Alves BJR, Urquiaga S, Boddey RM, Dakora FD, Bhattarai S, Maskey SL, Sampet C, Rerkasem B, Khan DF, Hauggaard-Nielsen

H, Jensen ES (2009) The contributions of nitrogen-fixing crop legumes to the productivity of agricultural systems. Symbiosis 48:1–17

Persello-Cartieaux F, Nussaume L, Robaglia C (2003) Tales from the underground: molecular plant–rhizobacteria interactions. Plant Cell Environ 26:189–199

Peters NK, Frost JW, Long SR (1986) The flavone, luteolin, induces expression of *Rhizobium meliloti* nodulation genes. Science 233:977–990

Phillips DA, Joseph CM, Yang GP, Martinez-Romero E, Sanborn JR, Volpin H (1999) Identification of lumichrome as a *Sinorhizobium* enhancer of alfalfa root respiration and shoot growth. Proc Natl Acad Sci USA 96:12275–12280

Pii Y, Crimi M, Cremonese G, Spena A, Pandolfini T (2007) Auxin and nitric oxide control indeterminate nodule formation. BMC Plant Biol 7:21

Pillai BVS, Swarup S (2002) Elucidation of the flavonoid catabolism pathway in *Pseudomonas putida* PML2 by comparative metabolic profiling. Appl Environ Microbiol 68:143–151

Radutoiu S, Madsen LH, Madsen EB, Felle HH, Umehara Y, Gronlund M, Sato S, Nakamura Y, Tabata S, Sandal N, Stougaard J (2003) Plant recognition of symbiotic bacteria requires two LysM receptor-like kinases. Nature 425:585–592

Radutoiu S, Madsen LH, Madsen EB, Jurkiewicz A, Fukai E, Quistgaard EMH, Albrektsen AS, James EK, Thirup S, Stougaard J (2007) LysM domains mediate lipochitin-oligosaccharide recognition and Nfr genes extend the symbiotic host range. EMBO J 26:3923–3935

Rajamani S, Bauer WD, Robinson JB, Farrow JM, Pesci EC, Teplitski M, Gao MS, Sayre RT, Phillips DA (2008) The vitamin riboflavin and its derivative lumichrome activate the LasR bacterial quorum-sensing receptor. Mol Plant Microbe Interact 21:1184–1192

Ramey BE, Koutsoudis M, von Bodman SB, Fuqua C (2004) Biofilm formation in plant–microbe associations. Curr Opin Microbiol 7:602–609

Rao JR, Cooper JE (1995) Soybean nodulating rhizobia modify nod gene inducers daidzein and genistein to yield aromatic products that can influence gene-inducing activity. Mol Plant Microbe Interact 8:855–862

Redfield RJ (2002) Is quorum sensing a side effect of diffusion sensing? Trends Microbiol 10:365–370

Redmond JW, Batley M, Djordjevic MA, Innes RW, Kuempel PL, Rolfe BG (1986) Flavones induce expression of nodulation genes in *Rhizobium*. Nature 323:632–635

Reinhold-Hurek B, Hurek T (1998) Life in grasses: diazotrophic endophytes. Trends Microbiol 6:139–144

Rinaudi L, Fujishige NA, Hirsch AM, Banchio E, Zorreguieta A, Giordano W (2006) Effects of nutritional and environmental conditions on *Sinorhizobium meliloti* biofilm formation. Res Microbiol 157:867–875

Rolfe BG, Gresshoff PM (1988) Genetic analysis of legume nodule initiation. Annu Rev Plant Physiol Plant Mol Biol 39:297–319

Ryan PR, Dessaux Y, Thomashow LS, Weller DM (2009) Rhizosphere engineering and management for sustainable agriculture. Plant Soil 321:363–383

Sait M, Hugenholtz P, Janssen PH (2002) Cultivation of globally distributed soil bacteria from phylogenetic lineages previously only detected in cultivation-independent surveys. Environ Microbiol 4:654–666

Scervino JM, Ponce MA, Erra-Bassells R, Vierheilig H, Ocampo JA, Godeas A (2005) Flavonoids exhibit fungal species and genus specific effects on the presymbiotic growth of *Gigaspora* and *Glomus*. Mycol Res 109:789–794

Schaefer AL, Greenberg EP, Oliver CM, Oda Y, Huang JJ, Bittan-Banin G, Peres CM, Schmidt S, Juhaszova K, Sufrin JR, Harwood CS (2008) A new class of homoserine lactone quorum-sensing signals. Nature 454:595–U596

Schell MA (1993) Molecular biology of the LysR family of transcriptional regulators. Annu Rev Microbiol 47:597–626

Schuhegger R, Ihring A, Gantner S, Bahnweg G, Knappe C, Vogg G, Hutzler P, Schmid M, Van Breusegem F, Eberl L, Hartmann A, Langebartels C (2006) Induction of systemic resistance in tomato by *N*-acyl-L-homoserine lactone-producing rhizosphere bacteria. Plant Cell Environ 29:909–918

Shaw LJ, Morris P, Hooker JE (2006) Perception and modification of plant flavonoid signals by rhizosphere microorganisms. Environ Microbiol 8:1867–1880

Smit P, Limpens E, Geurts R, Fedorova E, Dolgikh E, Gough C, Bisseling T (2007) Medicago LYK3, an entry receptor in rhizobial nodulation factor signaling. Plant Physiol 145:183–191

Somers E, Vanderleyden J, Srinivasan M (2004) Rhizosphere bacterial signalling: a love parade beneath our feet. Crit Rev Microbiol 30:205–240

Sorensen J, Nicolaisen MH, Ron E, Simonet P (2009) Molecular tools in rhizosphere microbiology-from single-cell to whole-community analysis. Plant Soil 321:483–512

Soto MJ, Fernandez-Aparicio M, Castellanos-Morales V, Garcia-Garrido JM, Ocampo JA, Delgado MJ, Vierheilig H (2010) First indications for the involvement of strigolactones on nodule formation in alfalfa (*Medicago sativa*). Soil Biol Biochem 42:383–385

Spaink HP (1996) Regulation of plant morphogenesis by lipo-chitin oligosaccharides. Crit Rev Plant Sci 15:559–582

Spaink HP, Sheeley DM, Vanbrussel AAN, Glushka J, York WS, Tak T, Geiger O, Kennedy EP, Reinhold VN, Lugtenberg BJJ (1991) A novel highly unsaturated fatty-acid moiety of lipo-oligosaccharide signals determines host specificity of *Rhizobium*. Nature 354:125–130

Sprague SJ, Watt M, Kirkegaard JA, Howlett BJ (2007) Pathways of infection of *Brassica napus* roots by *Leptosphaeria maculans*. New Phytol 176:211–222

Sprent JI (2008) Evolution and diversity of legume symbiosis. In: Dilworth MJ, James EK, Sprent JI, Newton WE (eds) Nitrogen-fixing leguminous symbioses, vol 7. Springer, Dordrecht, pp 1–21

Steidle A, Sigl K, Schuhegger R, Ihring A, Schmid M, Gantner S, Stoffels M, Riedel K, Givskov M, Hartmann A, Langebartels C, Eberl L (2001) Visualization of *N*-acylhomoserine lactone-mediated cell–cell communication between bacteria colonizing the tomato rhizosphere. Appl Environ Microbiol 67:5761–5770

Stevens GA, Berry AM (1988) Cytokinin secretion by *Frankia* sp. HFP ARI3 in defined medium. Plant Physiol 87:15–16

Stewart A (2001) Commercial biocontrol – reality or fantasy? Australas Plant Pathol 30:127–131

Sturtevant DB, Taller BJ (1989) Cytokinin production by *Bradyrhizobium japonicum*. Plant Physiol 89:1247–1252

Subramanian S, Stacey G, Yu O (2006) Endogenous isoflavones are essential for the establishment of symbiosis between soybean and *Bradyrhizobium japonicum*. Plant J 48:261–273

Swensen SM (1996) The evolution of actinorhizal symbioses: evidence for multiple origins of the symbiotic association. Am J Bot 83:1503–1512

Taga ME (2007) Bacterial signal destruction. ACS Chem Biol 2:89–92

Taga ME, Bassler BL (2003) Chemical communication among bacteria. Proc Natl Acad Sci USA 100:14549–14554

Teplitski M, Robinson JB, Bauer WD (2000) Plants secrete substances that mimic bacterial N-acyl homoserine lactone signal activities and affect population density-dependent behaviors in associated bacteria. Mol Plant Microbe Interact 13: 637–648

Thrall PH, Burdon JJ, Woods MJ (2000) Variation in the effectiveness of symbiotic associations between native rhizobia and temperate Australian legumes: interactions within and between genera. J Appl Ecol 37:52–65

Tirichine L, Sandal N, Madsen LH, Radutoiu S, Albrektsen AS, Sato S, Asamizu E, Tabata S, Stougaard J (2007) A gain-of-function mutation in a cytokinin receptor triggers spontaneous root nodule organogenesis. Science 315:104–107

Toth IK, Newton JA, Hyman LJ, Lees AK, Daykin M, Ortori C, Williams P, Fray RG (2004) Potato plants genetically modified to produce N-acylhomoserine lactones increase susceptibility to soft rot Erwiniae. Mol Plant Microbe Interact 17:880–887

Tsai SM, Phillips DA (1991) Flavonoids released naturally from alfalfa promote development of symbiotic *Glomus* spores *in vitro*. Appl Environ Microbiol 57:1485–1488

Umehara M, Hanada A, Yoshida S, Akiyama K, Arite T, Takeda-Kamiya N, Magome H, Kamiya Y, Shirasu K, Yoneyama K, Kyozuka J, Yamaguchi S (2008) Inhibition of shoot branching by new terpenoid plant hormones. Nature 455:195–U129

van Noorden GE, Ross JJ, Reid JB, Rolfe BG, Mathesius U (2006) Defective long distance auxin transport regulation in the *Medicago truncatula super numerary nodules* mutant. Plant Physiol 140:1494–1506

van Rhijn P, Fujishige NA, Lim PO, Hirsch AM (2001) Sugar-binding activity of pea lectin enhances heterologous infection of transgenic alfalfa plants by *Rhizobium leguminosarum* biovar *viciae*. Plant Physiol 126:133–144

van Rhijn P, Goldberg RB, Hirsch AM (1998) *Lotus corniculatus* nodulation specificity is changed by the presence of a soybean lectin gene. Plant Cell 10:1233–1249

Vandeputte OM, Kiendrebeogo M, Rajaonson S, Diallo B, Mol A, El Jaziri M, Baucher M (2010) Identification of catechin as one of the flavonoids from *Combretum albiflorum* bark extract that reduces the production of quorum-sensing-controlled virulence factors in *Pseudomonas aeruginosa* PAO1. Appl Environ Microbiol 76:243–253

Vermeer J, McCully ME (1982) The rhizosphere in Zea – new insight into its structure and development. Planta 156:45–61

von Bodman SB, Bauer WD, Coplin DL (2003) Quorum sensing in plant-pathogenic bacteria. Annu Rev Phytopathol 41:455–482

von Rad U, Klein I, Dobrev PI, Kottova J, Zazimalova E, Fekete A, Hartmann A, Schmitt-Kopplin P, Durner J (2008) Response of *Arabidopsis thaliana* to N-hexanoyl-DL-homoserine-lactone, a bacterial quorum sensing molecule produced in the rhizosphere. Planta 229: 73–85

Walker TS, Bais HP, Grotewold E, Vivanco JM (2003a) Root exudation and rhizosphere biology. Plant Physiol 132:44–51

Walker TS, Bais HP, Halligan KM, Stermitz FR, Vivanco JM (2003b) Metabolic profiling of root exudates of *Arabidopsis thaliana*. J Agric Food Chem 51:2548–2554

Wan JR, Zhang XC, Neece D, Ramonell KM, Clough S, Kim SY, Stacey MG, Stacey G (2008) A LysM receptor-like kinase plays a critical role in chitin signaling and fungal resistance in *Arabidopsis*. Plant Cell 20:471–481

Wang YJ, Leadbetter JR (2005) Rapid acyl-homoserine lactone quorum signal biodegradation in diverse soils. Appl Environ Microbiol 71:1291–1299

Wasson AP, Pellerone FI, Mathesius U (2006) Silencing the flavonoid pathway in *Medicago truncatula* inhibits root nodule formation and prevents auxin transport regulation by rhizobia. Plant Cell 18:1617–1629

Watt M, Hugenholtz P, White R, Vinall K (2006a) Numbers and locations of native bacteria on field-grown wheat roots quantified by fluorescence in situ hybridization (FISH). Environ Microbiol 8:871–884

Watt M, Kirkegaard JA, Passioura JB (2006b) Rhizosphere biology and crop productivity – a review. Aust J Soil Res 44:299–317

Watt M, Kirkegaard JA, Rebetzke GJ (2005) A wheat genotype developed for rapid leaf growth copes well with the physical and biological constraints of unploughed soil. Funct Plant Biol 32:695–706

Watt M, McCully ME, Kirkegaard JA (2003) Soil strength and rate of root elongation alter the accumulation of *Pseudomonas spp.* and other bacteria in the rhizosphere of wheat. Funct Plant Biol 30:483–491

Watt M, Silk WK, Passioura JB (2006c) Rates of root and organism growth, soil conditions, and temporal and spatial development of the rhizosphere. Ann Bot 97:839–855

Whitehead NA, Barnard AML, Slater H, Simpson NJL, Salmond GPC (2001) Quorum-sensing in gram-negative bacteria. FEMS Microbiol Rev 25:365–404

Yang WW, Han JI, Leadbetter JR (2006) Utilization of homoserine lactone as a sole source of carbon and energy by soil *Arthrobacter* and *Burkholderia* species. Arch Microbiol 185:47–54

Yoneyama K, Xie XN, Sekimoto H, Takeuchi Y, Ogasawara S, Akiyama K, Hayashi H (2008) Strigolactones, host recognition signals for root parasitic plants and arbuscular mycorrhizal fungi, from Fabaceae plants. New Phytol 179:484–494

Yuan ZC, Edlind MP, Liu P, Saenkham P, Banta LM, Wise AA, Ronzone E, Binns AN, Kerr K, Nester EW (2007) The plant signal salicylic acid shuts down expression of the vir regulon and activates quormone-quenching genes in *Agrobacterium*. Proc Natl Acad Sci USA 104: 11790–11795

Zhang JH, Subramanian S, Stacey G, Yu O (2009) Flavones and flavonols play distinct critical roles during nodulation of *Medicago truncatula* by *Sinorhizobium meliloti*. Plant J 57:171–183

Zipfel C (2008) Pattern-recognition receptors in plant innate immunity. Curr Opin Immunol 20:10–16

Zuanazzi JAS, Clergeot PH, Quirion JC, Husson HP, Kondorosi A, Ratet P (1998) Production of *Sinorhizobium meliloti nod* gene activator and repressor flavonoids from *Medicago sativa* roots. Mol Plant Microbe Interact 11:784–794

Impacts of Elevated CO_2 on the Growth and Physiology of Plants with Crassulacean Acid Metabolism

Johan Ceusters and Anne M. Borland

Contents

Abstract The photosynthetic specialization of crassulacean acid metabolism (CAM) employs both Rubisco and phosphoenolpyruvate carboxylase (PEPC) for uptake of CO_2 over the day and night. Temporal separation of the C_3 and C_4 carboxylases optimizes photosynthetic performance and carbon gain under water-limited environments. The water-conserving attributes of CAM has highlighted the potential of plants with this photosynthetic pathway as a means of carbon sequestration and biomass production on marginal lands. Sustainable agronomic and horticultural production of CAM species requires an understanding of how exposure to elevated atmospheric concentrations of CO_2 will affect growth and

J. Ceusters
Afdeling Plantenbiotechniek, Departement Biosystemen, Faculteit Bio-Ingenieurswetenschappen, KULeuven, Willem Decroylaan 42, 3001 Heverlee, Belgium

A.M. Borland (✉)
Molecular Plant Physiology, Newcastle Institute for Research on Sustainability, University of Newcastle Upon Tyne, Devonshire Building, NE1 7RU Newcastle, UK
e-mail: a.m.borland@newcastle.ac.uk

U. Lüttge et al. (eds.), *Progress in Botany 72*,
DOI 10.1007/978-3-642-13145-5_6, © Springer-Verlag Berlin Heidelberg 2011

productivity. In this review, the physiological responses of CAM plants to [CO_2] elevation will be assessed in terms of net carbon gain, growth, anatomy, morphology, and water use efficiency. Photosynthetic responses to elevated [CO_2] will be specifically discussed on a background of carbohydrate metabolism and partitioning toward the potentially competing sinks of nocturnal acid synthesis, respiration, and export for growth in CAM species.

1 Introduction

Crassulacean acid metabolism (CAM) is a specialized mode of photosynthetic carbon assimilation that is estimated to be present in approximately 7% of vascular plant species (Winter and Smith 1996), many of which dominate the plant biomass of arid, marginal regions of the world. CAM facilitates the uptake of CO_2 at night thereby improving water-use efficiency of carbon assimilation, particularly in water-limited habitats. Agronomically important CAM species include pineapple (*Ananas comosus*), *Opuntia ficus-indica*, an important fodder and forage crop, and several of the *Agave* species that are cultivated for fiber and for alcohol. Other CAM species with economic importance include tropical epiphytic and terrestrial bromeliads and orchids, which are cultivated as ornamentals. The potential of high yielding CAM species as feed stocks for bioenergy production on marginal and degraded lands has recently been highlighted (Borland et al. 2009) and requires an understanding of how the rising atmospheric [CO_2] associated with global climate change is likely to affect growth and productivity of CAM plants. Likewise, sustainable production of ornamental CAM crops needs information on how different species and cultivars respond to CO_2 enrichment, a method that is widely used for reducing the energy costs of greenhouse production. The aim of this review is to outline the biochemical and physiological properties of CAM and to consider how CAM species might be expected to respond to life in a higher [CO_2] world, in terms of net carbon gain, growth, anatomy, morphology, and water use efficiency (WUE).

2 Rising [CO_2] and Photosynthesis

Elevation of the atmospheric CO_2 concentration is a key issue underpinning global changes in the earth's climate. The current [CO_2] of 380 ppm is predicted to rise toward 560 ppm during the second half of this century due to anthropogenic influences that include fossil fuel combustion and forest destruction (IPCC Assessment Report 2007). As most modern plant genera evolved in the Tertiary under lower atmospheric [CO_2], i.e., 180–240 ppm, the current anthropogenic CO_2 enrichment represents a novel challenge to plant life, i.e., facing the sudden abundance of a traditionally scarce resource (Crowley and Berner 2001; Pagani et al. 2005; Körner 2006). In this respect, the growth and physiological performance of plants are affected by [CO_2] in a diverse and complex manner, depending on the pathway of carbon fixation and accompanying environmental conditions (Körner 2001; Bazzaz and Catovsky 2002).

Experimental data almost univocally show a stimulation of C_3 leaf photosynthesis under elevated CO_2 exposure, which only saturates at approximately 1,000 ppm CO_2 (Poorter and Navas 2003; Körner 2006; Körner et al. 2007). The C_3 response to elevated [CO_2] is attributed to the fact that ribulose-1,5-bisphosphate carboxylase/oxygenase (Rubisco) is not saturated at current atmospheric [CO_2], and photorespiration can result in significant reductions in plant yield (Isdo and Kimball 1992; Bowes 1993). It has also been suggested that elevated [CO_2] might draw down the light compensation point for C_3 photosynthesis and permit C_3 plants to grow in deeper shade (Würth et al. 1998; Granados and Körner 2002). The situation for C_4 plants under elevated [CO_2] is less clear since Rubisco is localized in the chloroplasts of the bundle sheath cells where the internal [CO_2] is 3–6 times higher than in the atmosphere. The C_4 carbon concentrating mechanism is a consequence of the prior uptake of CO_2 mediated via phosphoenolpyruvate carboxylase (PEPC), which results in an effective drawdown of CO_2 in the mesophyll cells and produces C_4 acids that are subsequently decarboxylated in the bundle sheath cells (Bowes 1993; Sage 2001; von Caemmerer and Furbank 2003). C_4 plants avoid photorespiration due to the saturation of Rubisco with CO_2, and on this biochemical background, C_4 plants would not be expected to display greater rates of photosynthesis in a higher [CO_2] environment. However, because of the typical reduction of stomatal conductance in plants exposed to elevated atmospheric [CO_2] (Ainsworth and Long 2005), yield improvements to C_4 plants could occur in a higher [CO_2] world through an amelioration of short-term drought stress via the conservation of soil moisture (Ghannoum et al. 2000; Leakey et al. 2004).

For CAM plants that use both Rubisco and PEPC to take up CO_2, and which are very conservative in their use of water, the response to elevated [CO_2] is less easy to predict. The carboxylation processes and water-conserving consequences of CAM may be described in four phases that occur over a 24 h light/dark cycle (Osmond 1978). Phase I represents the dark period when stomata are open and atmospheric and/or respiratory CO_2 is fixed in the cytosol by PEPC resulting in malate formation. The CO_2 acceptor PEP is formed via the glycolytic breakdown of storage polysaccharides (e.g., starch) or soluble sugars (e.g., glucose, fructose, sucrose) assembled during the previous day. Phase II at dawn indicates the switch between C_4 and C_3 carboxylation and atmospheric CO_2 uptake still largely appears to be dominated by PEPC (Griffiths et al. 1990). Rubisco remains at a low activation state until PEPC is inactivated by dephosphorylation (Maxwell et al. 1999). During the middle of the day (phase III), gas exchange is curtailed by stomatal closure as a result of high intercellular partial pressure of CO_2 generated by malate decarboxylation. The released CO_2 is fixed through Rubisco via the Calvin cycle into triose phosphates that are partitioned between the provision of carbohydrates for growth or as substrates for the subsequent production of PEP for phase I carboxylation. Toward the end of the light period when the decarboxylation of malate is complete, the internal partial pressure of CO_2 drops and stomata reopen. During this phase IV, CO_2 is mainly fixed and assimilated via

Rubisco with a major proportion of carbon fixed partitioned toward the synthesis of sugars for export and growth (Borland et al. 1994). Over the latter part of phase IV, C_4 carboxylation by PEPC can be detected (Griffiths et al. 1990; Borland and Griffiths 1996). The relative contributions from C_3 and C_4 carboxylation to 24 h net carbon gain is determined by interplay of circadian and metabolite control, assuring flexible modulation of the supply and demand for carbon under different environmental conditions (Borland et al. 1999; Borland and Taybi 2004). As a consequence of this metabolic flexibility, a wide and seemingly contradictory range of photosynthetic responses to elevated [CO_2] has been reported in CAM species.

3 Photosynthesis in CAM Plants Under Elevated [CO_2]

In comparison to C_3 and C_4 plants, relatively few studies have addressed the effects of elevated [CO_2] on gas exchange characteristics in CAM plants (Poorter and Navas 2003). However, in the different CAM morphotypes that have been studied (i.e., epiphytes, stem succulents, and leaf succulents), an average stimulation in CO_2 uptake over the diel cycle of about 35–40% due to CO_2 elevation has been reported (taken from references given in Table 1). However, beyond this average value, large variation exists between species, and different conclusions have been reached on how CAM plants might be expected to adjust their photosynthesis in a higher [CO_2] world, ranging from decreased night-time CO_2 uptake to increases of CO_2 uptake during both day and night (Table 1).

Increases in the duration and the rate of daytime net CO_2 uptake are consistent with the effects of higher internal [CO_2] concentrations on Rubisco, which dominates carboxylation during phase IV in the afternoon. Less expected for CAM plants are the reports of an increase in nocturnal (phase I) CO_2 uptake under elevated [CO_2], given that PEPC is thought to be saturated at current atmospheric [CO_2] (Ting 1994; Winter and Engelbrecht 1994). The majority of investigated CAM species show nocturnal increases in malate accumulation under elevated [CO_2], indicating increased nocturnal CO_2 uptake (Griffihs 1988). In *O. ficus-indica*, the maximal rate of nocturnal CO_2 uptake in daughter cladodes was found to be 96% higher under a doubled atmospheric CO_2 concentration (Cui et al. 1993). These results indicate that nocturnal uptake of CO_2 seems to be far from being saturated at the current atmospheric CO_2 concentration for a number of CAM species.

Diffusional constraints on CO_2 supply to the cellular sites of carbon assimilation are a likely consequence of the typically succulent photosynthetic tissues in CAM species and this could be a key factor underpinning the various reports of stimulation of diurnal and nocturnal CO_2 uptake by elevated [CO_2] in CAM plants. Tight cell packing and low amounts of intercellular air spaces are typical of photosynthetic CAM leaves, cladodes and stems, and this imposes significant constraints on the internal diffusive supply of CO_2 (out with the "regenerative" phase III, when internal CO_2 concentrations are likely to saturate Rubisco (Maxwell et al. 1997,

Table 1 General gas exchange responses of CAM plants to elevated atmospheric CO_2 (680–750 ppm) concentrations in comparison with ambient grown plants (340–370 ppm)

CO_2 uptake	Species	References
Increase of daytime uptake	*Aechmea* Maya	Ceusters et al. (2008a)
	Ferocactus acanthodes	Nobel and Hartsock (1986)
	Kalanchoë pinnata	Winter et al. (1997)
Increase of night-time uptake	*Agave vilmoriana*	Szarek et al. (1987)
	Mokara Yellow	Gouk et al. (1997), Li et al. (2002)
Increase of both day- and night-time uptake	*Agave deserti*	Graham and Nobel (1996)
	Agave salmiana	Nobel (1996)
	Ananas comosus	Cote et al. (1993), Zhu et al. (1999)
	Hylocereus undatus	Raveh et al. (1995)
	Opuntia ficus-indica	Nobel and Israel (1994), Wang and Nobel (1996)
	Stenocereus queretaroensis	Nobel (1996)
Unchanged	*Kalanchoë daigremontiana*	Osmond and Björkman (1975), Holtum et al. (1983)
Decrease of night-time uptake	*Portulacaria afra*	Huerta and Ting (1988)
	Clusia uvitana	Winter et al. (1992)

1998; Nelson and Sage 2008). At night, the carbon isotope signals associated with PEPC-mediated uptake of CO_2 suggest diffusion limitation in *Kalanchoë daigremontiana* (Griffiths et al. 2007). Additionally, mesophyll conductances derived during phase IV of gas exchange in the light in several CAM species are inversely related to the degree of leaf succulence, with internal CO_2 supply at Rubisco potentially as low as 110 μmol mol^{-1} (Griffiths et al. 1999).

Another potentially limiting factor for dark CO_2 uptake in CAM plants could be the activity of carbonic anhydrase (CA), which mediates the conversion of CO_2 into HCO_3^- which is the substrate for PEPC. In the C_4 plant *Flaveria bidentis*, it has been reported that leaf CA activity is in excess and does not limit CO_2 assimilation under ambient conditions (Cousins et al. 2006), but in several other C_4 species it would appear that CA activity is only just sufficient to support photosynthetic rates, especially in monocots (Hatch and Burnell 1990; Gillon and Yakir 2000). The situation in CAM plants with regards to CA activity and the partitioning of CA activity between cytosol (to support PEPC-mediated CO_2 uptake) and chloroplast (to support Rubisco-mediated CO_2 uptake) remains to be resolved.

4 Water Use Efficiency of CAM Plants Under Elevated [CO_2]

In all three photosynthetic groups, i.e., C_3, C_4 and CAM, exposure to elevated atmospheric [CO_2] has been associated with an increase in WUE (Jarvis et al. 1999; Ward et al. 1999; Ghannoum 2009). CAM plants are considered to be very conservative in their use of water as a consequence of restricting most transpirational water loss to the cooler nocturnal period. Thus, the relatively high WUE of CAM is usually attributed to PEPC-mediated CO_2 uptake during phase I

(Black 1973; Lüttge 2004). At first sight, the enhancement of WUE in CAM species under elevated [CO_2] might be somewhat controversial because of the generally greater relative contribution of daytime CO_2 uptake under elevated [CO_2], which is usually considered less water use efficient than nocturnal carboxylation. However, in the constitutive CAM plant *Aechmea* "Maya" (Ceusters et al. 2008b, 2009c), daytime CO_2 uptake doubled from 25 to 50% of total diel uptake, while diel WUE increased from 7 to 15 mmol CO_2 mol^{-1} H_2O under CO_2 enrichment (Ceusters et al. 2008a). Compared to plants under ambient CO_2, the gain in WUE during phase I in *A.* "Maya" under elevated CO_2 was achieved by a reduction of stomatal conductance without a concomitant reduction in CO_2 uptake. During the day-time phases, both increased Rubisco and PEPC-mediated CO_2 uptake under high [CO_2] have been found to result in equally high WUE values to those recorded at night, confirming the statement of Eller and Ferrari (1997) that high WUE is not exclusively associated with performing C_4 carboxylation at night (Ceusters et al. 2008a).

5 Metabolite Dynamics and Carbohydrate Partitioning in CAM Plants Under Elevated [CO_2]

The reciprocating pools of organic acids and carbohydrates are central to the metabolic control of CAM. The day/night turnover and availability of carbohydrates in particular play a key role in setting the phases of CAM by determining the magnitude of dark CO_2 uptake (Borland and Dodd 2002; Dodd et al. 2003; Borland and Taybi 2004). In addition to providing carbon skeletons for the synthesis of PEP, the substrate for nocturnal carboxylation, carbohydrates must also be partitioned to other sinks in order to sustain plant growth and maintain photosynthetic integrity. How CAM plants deal with this potential conflict of interests is still unclear but is likely to involve both coarse and fine control of the activity of a range of enzymes involved in carbohydrate synthesis, degradation, and transport (Häusler et al. 2000; Antony et al. 2008). The magnitude and dynamics of day/night turnover of organic acids also have repercussions in terms of the length of time that stomata remain closed during the day.

5.1 *Organic Acids*

In accordance with the plethora of gas exchange responses reported in Table 1, CAM plants have been found to show considerable variation in the nocturnal accumulation of organic acids in response to [CO_2] enhancement. Drennan and Nobel (2000) reported decreases from about 10 to 25% in nocturnal acidity in some CAM species (Holtum et al. 1983; Winter et al. 1997) yet up to 50% increases in nocturnal acidity in others under nearly doubled CO_2 concentrations (Szarek et al. 1987; Nowak and Martin 1995; Wang and Nobel 1996; Zhu et al. 1999). It is possible that in certain CAM species dark CO_2 uptake might be limited by restricted vacuolar storage capacity for malic acid. Thus, the increase in leaf succulence that has been reported for some CAM species under elevated [CO_2] may represent an increased capacity for nocturnal accumulation of organic acids (Nobel 2000).

The dynamics or timing of malate accumulation/degradation is also influenced by [CO_2] elevation as noted for CAM plants such as *K. daigremntiana* and *A.* "Maya" (Holtum et al. 1983: Ceusters et al. 2008a) while others seemed unresponsive in that respect (Wang and Nobel 1996; Li et al. 2002). Results obtained with *A.* "Maya" showed that malic acid degradation was completed earlier in the day under high CO_2 thereby allowing the afternoon phase IV to start 3 h earlier than under ambient [CO_2] (Ceusters et al. 2008a). In this case, stomata may reopen sooner under high [CO_2], thereby taking profit from the higher atmospheric CO_2 concentrations and maximizing direct carbon gain (Maxwell et al. 1997; Ceusters et al. 2008a). Whilst there is clearly a degree of linkage between the biochemistry of internal carbon supply and stomatal movements in CAM plants, recent investigations on *Kalanchoë* species showed that stomatal response to [CO_2] is not the primary determinant of the length of time that stomata remain closed during phase III (Von Caemmerer and Griffiths 2009). The same authors suggest that a single CO_2 sensor, which interacts with other signaling pathways, underpins the diel changes in stomatal conductance in CAM species. Additional studies are required to unravel the environmental factors that regulate the timing and rate of malate decarboxylation and to establish how the biochemistry of malate turnover and other signaling pathways interact to regulate stomatal conductance in CAM plants.

Most CAM species accumulate predominantly malic acid overnight, and thus the influence of [CO_2] elevation on citric acid oscillations has largely been overlooked or found to be of minor importance. Winter et al. (1997) reported no changes in the typical small diel citrate fluctuations in *Kalanchoë pinnata* (Milburn et al. 1968; Lüttge 1988), upon CO_2 enhancement and in *A.* "Maya" citric acid levels remained variable, yet similar to those measured under ambient [CO_2] (Ceusters et al. 2008a). However, these findings may not be generalized, and in a world of rising [CO_2], it may be interesting to contemplate the carbon reactions of CAM-performing *Clusia* species under CO_2 enrichment. The substantial day/night changes in citric acid levels reported for various *Clusia* species may represent an important hallmark of the photosynthetic flexibility in this genus (Lüttge 1996, 2006), and long-term responses to growth under elevated [CO_2] remain to be reported for these tropical trees.

5.2 *Storage Carbohydrates*

Nocturnal CO_2 uptake catalyzed via PEPC relies on the provision of the 3-carbon acceptor PEP that is generated by the glycolytic turnover of storage carbohydrates. The recovery of carbohydrates during the day proceeds via gluconeogenesis and the products from this pathway are subsequently transported either into the chloroplast and stored as starch or transported into the vacuole and stored as sucrose and/or hexose (Antony and Borland 2008). CAM species may be divided into two main groups in terms of the carbohydrate source used to provide PEP, i.e., starch versus sugar-using species (Black et al. 1996; Christopher and Holtum 1996, 1998). However, recent observations indicate that CAM plasticity is not only manifest at the level of diel gas exchange patterns but also encompasses flexibility in the use of different form of carbohydrate reserves as sources of PEP. Cushman et al. (2008) demonstrated that nocturnal malate accumulation could be restored in a starch and CAM-deficient mutant of *Mesembryanthemum crystallinum* by feeding leaves with either glucose or sucrose. Moreover, flexible use of different storage carbohydrate pools has been reported to mediate in drought adaptation in *A.* "Maya" with a transition from starch to sucrose as the primary source of PEP under progressing water stress (Ceusters et al. 2009a, b). However, in general and particularly under high [CO_2], the impact of environmental factors on carbohydrate metabolism in CAM plants has received relatively little coverage in the literature.

In the CAM species investigated to date, elevated [CO_2] does not seem to induce a shift in the use of different carbohydrates as sources of PEP (Wang and Nobel 1996; Ceusters et al. 2008a). In *A.* "Maya" under elevated [CO_2] (Ceusters et al. 2008a), sucrose content and daily turnover remained unaltered over ambient controls, but the day/night turnover and amounts of glucose and fructose were substantially (three and fivefold respectively) enhanced by long-term (5 months) exposure to elevated [CO_2], although these monosaccharides did not appear to be used as sources of PEP. The significant increase in monosaccharide content in *A.* "Maya" under elevated [CO_2] did not suppress net CO_2 uptake by either PEPC or Rubisco and may be attributable to vacuolar sequestration of glucose and fructose. Soluble sugars have been reported to accumulate in the photosynthetic tissues of other CAM species without a downregulation of photosynthesis (Drennan and Nobel 2000). Possible reasons for this include substantial vacuolar storage capacity, and the typically dilute cellular contents of CAM photosynthetic tissues represent a significant buffering capacity against rising concentrations of solutes (Nobel 1988). Moreover, several CAM species have been reported to possess a relatively low invertase activity (Goldschmidt and Huber 1992; Wang and Nobel 1996) and a reduced hexokinase activity under doubled [CO_2] concentrations (Wang and Nobel 1996; Moore et al. 1999); both invertase and hexokinase are implicated in sugar-mediated regulation of gene expression. Enhanced sugar content under elevated [CO_2] without a downward acclimation of photosynthesis would seem to be a desirable trait for CAM species and cultivars that might be considered as potential feedstocks for bioenergy production on marginal lands.

5.3 Carbohydrate Partitioning

Carbohydrate availability is considered to be a major limiting factor for dark CO_2 uptake and hence the expression of CAM (Borland and Dodd 2002; Dodd et al. 2003). Some 8–20% of leaf dry matter is committed each day and night to the cycle of carbohydrate turnover (Winter and Smith 1996). This investment of resources into storage carbohydrates implies restriction of carbohydrate partitioning to other metabolic activities, including dark respiration, growth, and acclimation to abiotic stress.

Plasticity in exploiting the different day–night phases of gas exchange (*sensu* Osmond 1978), in terms of magnitude and duration in response to external environmental conditions or endogenous developmental requirements, has been found to be ubiquitous among CAM species (Borland et al. 2000; Cushman and Borland 2002; Dodd et al. 2002). Moreover, several CAM crops that include species of *Agave*, *Ananas*, and *Opuntia* have been shown to achieve comparable and even higher productivities under well-watered conditions than a range of C_3 crops (Nobel 1991; Nobel 1996; Lüttge 2004). These data indicate that modulation of carbohydrate partitioning to the competing sinks of nocturnal carboxylation, respiration, and export under changing environmental conditions and/or shifts in development is a key determinant for the growth, productivity, and ecological success of CAM species (Borland and Dodd 2002).

To decipher the carbon partitioning between the different competing sinks, net 24 h carbon budgets can be constructed that measure (1) the source of carbon in the leaf, i.e., C_3 (Rubisco-mediated CO_2 uptake) or C_4 (from breakdown of malate where 1 mol malate = 4 mol C), (2) the partitioning of this carbon between starch or soluble sugars (sucrose, glucose, and fructose) with the excess going to daytime export where 1 mol Glc eq = 6 mol C, and (3) the partitioning of carbohydrates between the generation of substrates for CAM, respiratory CO_2 or dark export (Borland 1996).

This modeling approach was used by Ceusters et al. (2008a) to describe diel carbon budgets of the constitutive CAM plant *A.* "Maya" under ambient and doubled atmospheric [CO_2] (Fig. 1). Doubling of the CO_2 concentration resulted in an increase of daily carbon input by 35%, mainly due to enhanced atmospheric CO_2 uptake during the transitional phases II and IV that was mediated by both Rubisco and PEPC. The major sink for the acquired carbon was the daytime accumulation of starch. Under ambient CO_2, a further 23% of fixed carbon was allocated to the accumulation of soluble sugars, allowing ~3% of total acquired carbon to be exported during the day and a further ~6% exported at night. Under elevated CO_2, the net flux of carbon into soluble sugars (mainly glucose and fructose) doubled over that in ambient conditions. Net flux to starch accumulation remained the same under elevated and ambient CO_2, but daytime net export of sugars was abolished under elevated CO_2. Limited net nocturnal depletion of either glucose or fructose occurred and thus 25% of accumulated sugar carbon was withheld from the competing fates of PEPC mediated CO_2 uptake and export for growth. The results suggest that for *A.* "Maya" under elevated CO_2, PEPC was not carbohydrate limited or else the glucose and fructose reserves, presumably stored in the vacuole, were unavailable for the generation of PEP (Ceusters et al. 2008a). Previous considerations of ^{13}C signatures of different carbohydrate species have implied the existence of discrete pools of storage carbohydrates destined for CAM or export/respiration (Borland and Dodd 2002). The data obtained here for *A.* "Maya" are consistent with this suggestion and could provide a means by

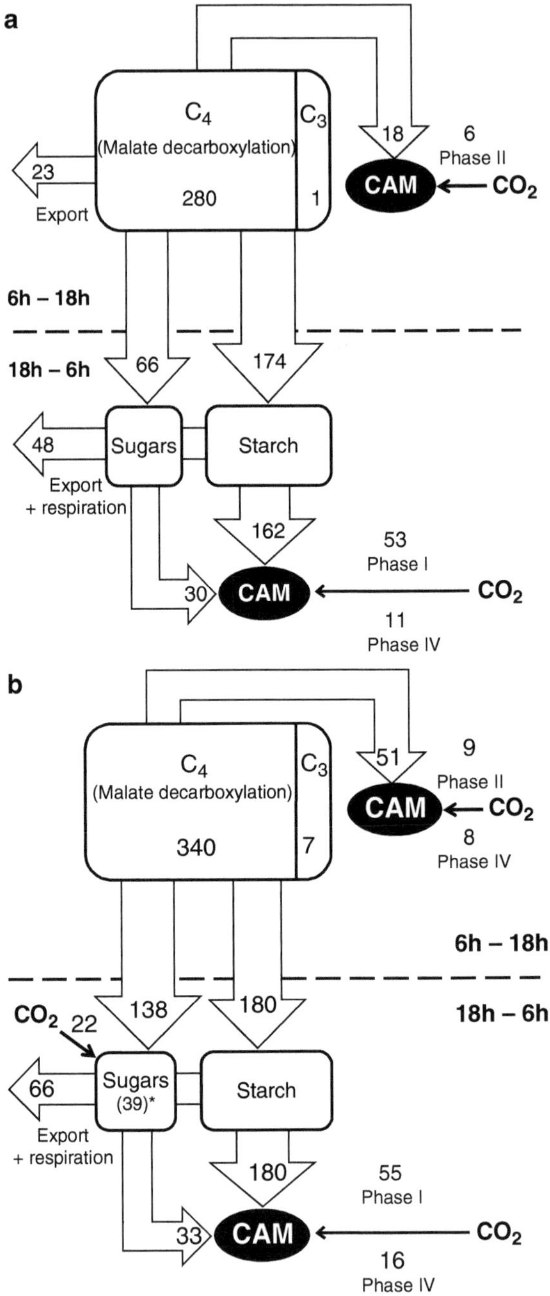

Fig. 1 Diel net carbon budgets for young, fully expanded leaves of Aechmea "Maya" at ambient (380 μmol mol^{-1}) (**a**) and elevated (700 μmol mol^{-1}) (**b**) CO_2. Fluxes shown above the horizontal dashed line mainly coincide with degradation of malic acid and concomitant accumulation of carbohydrate reserves, while those under the horizontal dashed line are characterized by the reverse process. Export includes growth and maintenance and all units are mmol C m^{-2} d^{-1}. *Asterisk* accumulated hexose carbon, which is not degraded or exported overnight (Ceusters et al. 2008a)

which the potentially competing fates of CAM and export/respiration for growth could be regulated independently, thereby facilitating photosynthetic plasticity in a changing environment.

As a consequence of the accumulation of hexoses under doubled CO_2, which remains enigmatic, the total net export of carbon for growth and maintenance during the diel cycle in *A.* "Maya" was similar in both regimes of CO_2 supply (Ceusters et al. 2008a). The data indicate that while *A.* "Maya" did not show enhanced biomass production under elevated [CO_2], higher sugar content was achieved with reduced inputs of water, since WUE increased twofold in this species under elevated [CO_2] (Ceusters et al. 2008a). This physiological response could be viewed as a desirable trait in considering the potential of CAM species as feedstocks for bioethanol production on marginal lands (Borland et al. 2009). Further investigations that apply the carbon balance model described here alongside a consideration of plant water use would be revealing for selecting other CAM species/cultivars as bioenergy feedstocks.

6 Morphology and Anatomy of CAM Plants Under Elevated [CO_2]

Morphological and anatomical characteristics of photosynthetic tissues are important determinants for gas exchange and light harvesting in all plants. Any changes in structural characteristics that are elicited by elevation of [CO_2] could significantly influence physiological performance and indeed alter the esthetic appeal of ornamentals. For the few CAM species where morphology has been studied under elevated [CO_2], an almost consistent effect is an increase in thickness of the photosynthetic organs. After exposure to a doubled atmospheric [CO_2], cladodes of *O. ficus-indica* were found to increase in thickness by ca. 15% (Cui et al. 1993; Nobel et al. 1994; North et al. 1995). Similarly, leaves of *Agave deserti* and *A. comosus* were ca. 11% thicker under a doubled atmospheric CO_2 concentration (Graham and Nobel 1996; Zhu et al. 1997). These increases were mainly accounted for by a substantial increase in the thickness of the chlorenchyma (Graham and Nobel 1996; Nobel et al. 1994), which might be related to higher CO_2 concentrations deeper within the leaves (Powles et al. 1980). Concomitantly, a decrease in stomatal density under elevated CO_2 in CAM plants (North et al. 1995) would potentially result in higher WUE. However, care is needed with generalizing too much in a group of taxa where plasticity in form and function may be considered of paramount importance for ecological success. Recently, Croonenborghs et al. (2009) showed an unaltered or even diminished leaf and chlorenchyma thickness for the CAM bromeliads *A.* "Maya" and *A. fasciata* "Primera," respectively, under [CO_2] enrichment. These results challenge some of the ideas discussed above and deserve further attention with respect to diffusional constraints and vacuolar storage capacity of organic acids in CAM plants under long-term [CO_2] enhancement. Moreover, Croonenborghs et al. (2009) controversially noticed a more compact plant shape in CAM bromeliads as a result of broadened and shortened leaves in a high [CO_2] environment. According to Farrar (1990), carbohydrate availability, i.e., sucrose, might be the critical factor in determining leaf shape by influencing signal transduction pathways ultimately resulting in altered patterns of cell proliferation and expansion. The results obtained with *A.* "Maya" and *A. fasciata* "Primera" seem to confirm this statement with the suggestion that rather than sucrose, the significant accumulation of glucose and fructose noted in these species may mediate in the observed morphological changes of bromeliads grown under elevated [CO_2] (Ceusters et al. 2008a; Ceusters J Unpublished results).

7 Growth and Biomass Enhancement of CAM Plants Under Elevated [CO_2]

In comparison with the C_3 and C_4 pathways, CAM still lacks sufficient representatives studied under elevated [CO_2] to allow robust comparisons of biomass enhancement potentials among different photosynthetic pathways in a higher CO_2 world. Based on the limited data available (i.e. ~ten species), Drennan and Nobel (2000) calculated an average increase in dry matter of 35% for CAM plants within 3 months at elevated [CO_2] of about 650–750 μmol mol^{-1} (Table 2). Elsewhere, analyses of biomass enhancement ratios (BER; ratio between total plant mass of high CO_2 grown plants and that of plants grown at control levels) showed a response of +23% for CAM, which is intermediate between that recorded for C_3 and C_4 plants (Poorter and Navas 2003). However, in the same paper, the community is warned about premature generalizations for CAM species. Indeed, if one should compare the BER ratios and more specifically their variability among the three photosynthetic pathways (Poorter and Navas 2003), it is surprising that a group of plants (i.e., CAM) that are characterized by such a medley of physiotypes and morphotypes (*sensu* Lüttge 2004) possesses a variability in response to [CO_2] that is several magnitudes smaller than herbaceous C_3, woody C_3 and C_4 species respectively. All CAM species tested at that time, of which most are terrestrial, showed significant stimulation of biomass accumulation under increasing CO_2. The only negative results (lack of biomass response or even less biomass accumulation) under CO_2 elevation that were reported for *Agave vilmoriana* (Isdo et al. 1986) and *A. comosus* (Ziska et al. 1991) were attributed to wet soil conditions and thus were not taken into account in reviewing CAM biomass responses under elevated [CO_2] (Zhu et al. 1999; Drennan and Nobel 2000).

However, more recent observations have shown nil and even negative biomass enhancement responses for CAM bromeliads upon [CO_2] elevation without excess wetting of the soil (Croonenborghs et al. 2009; Table 2), thereby expanding the variability in biomass responses of CAM species under an enriched CO_2 atmosphere. Other than an unaltered gas exchange pattern for well watered high CO_2 and ambient grown *A. vilmoriana* (Szarek et al. 1987), recent observations (Ceusters J Unpublished results) indicated that net CO_2 uptake was significantly

Table 2 Response of biomass of CAM plants to long-term (>1 month) exposure to nearly doubled atmospheric CO_2 concentrations as a percentage increase over controls maintained under ambient atmospheric CO_2 concentrations for the same period

Species	Biomass (% increase)	References
Agave deserti	30–31	Nobel and Hartsock (1986), Graham and Nobel (1996)
Agave salmiana	17	Nobel (1996)
Agave vilmoriniana	28	Isdo et al. (1986)
Aechmea fasciata "Primera"	−25	Croonenborghs et al. (2009)
Aechmea magdalenae	36	Ziska et al. (1991)
Aechmea "Maya"	0	Croonenborghs et al. (2009)
Ananas comosus	−10	Ziska et al. (1991)
	23	Zhu et al. (1997)
Ferrocactus acanthodes	30	Nobel and Hartsock (1986)
Opuntia ficus-indica	21–55	Cui et al. (1993), Cui and Nobel (1994), Nobel and Israel (1994)
Kalanchoë blossfeldiana	37	Mortensen and Moe (1992)
Kalanchoë pinnata	42–51	Winter et al. (1997)

enhanced on the long term for both *A.* "Maya" and *A. fasciata* "Primera" under elevated CO_2, mainly during day-time phases II and IV. However, these CAM plants seemed unable to successfully process the extra gained carbon into biomass. The curtailed biomass production for *A. fasciata* "Primera" under high [CO_2] in comparison with ambient grown plants, indicates an adverse affect of rising CO_2 levels for this CAM species, starting up the quest in delineating how the system processes the increased supply of carbon. The role of root respiration as a sink for carbon under elevated CO_2 is one area for further study. However, the data from *A.* "Maya" suggest that while some CAM species may not show enhanced biomass production in a higher CO_2 world, productivity could be maintained with reduced inputs of water.

8 Conclusions and Future Perspectives

Whilst data describing the physiological impacts of [CO_2] on CAM plants are limited in comparison to C_3 and C_4 species, it would appear that under elevated atmospheric [CO_2], CAM productivity may be expected either to increase or be maintained at levels comparable to those under ambient [CO_2] but with reduced inputs of water. Such predictions indicate a major ecophysiological advantage for CAM plants under a changing climate and lend further support to proposals that CAM species should be considered as a means of carbon sequestration and as feedstocks for bioenergy production from arid, marginal lands. In order to consolidate predictions regarding CAM biomass productivity under a changing climate, it will be necessary to investigate how exposure to elevated [CO_2] will impact on physiological responses to other elements of climate change, namely increased temperature, decreased precipitation, and exposure to other greenhouse gasses such as ozone. The physiological plasticity of CAM and the existence of diverse morphotypes represent a range of novel biological systems for investigating how plant form and function interact to determine ecological success in a changing environment. A key goal will be to achieve a systems-level understanding of how CAM species reconcile the conflicting demands for partitioning of resources between nocturnal carboxylation and biomass productivity in potentially limiting environments.

References

Ainsworth EA, Long SP (2005) What have we learned from 15 years of free-air CO_2 enrichment (FACE)? A meta-analytic review of the responses of photosynthesis, canopy properties and plant production to rising CO_2. New Phytol 165:351–371

Antony E, Borland AM (2008) The role and regulation of sugar transporters in plants with crassulacean acid metabolism. Prog Bot 70:127–143

Antony E, Taybi T, Courbot M, Mugford S, Smith JAC, Borland AM (2008) Cloning, localisation and expression analysis of vacuolar sugar transporters in the CAM plant *Ananas comosus* (pineapple). J Exp Bot 59:1895–1908

Bazzaz FA, Catovsky S (2002) Impact of global environmental change on plants: from cells to ecosystems. In: Mooney HA, Canadell JG (eds) Encyclopedia of global environmental change, vol 2. Wiley, Chichester, pp 94–111

Black CC (1973) Photosynthetic carbon fixation in relation to net CO_2 uptake. Annu Rev Plant Physiol 24:253–286

Black CC, Chen JQ, Doong RL, Angelov MN, Sung SJS (1996) Alternative carbohydrate reserves used in the daily cycle of crassulacean acid metabolism. In: Winter K, Smith JAC (eds) Crassulacean acid metabolism: biochemistry. Ecophysiology and evolution. Springer-Verlag, Berlin, pp 31–43

Borland AM (1996) A model for the partitioning of photosynthetically fixed carbon during the C_3–CAM transition in *Sedum telephium*. New Phytol 134:433–444

Borland AM, Griffiths H (1996) Variations in the phases of CAM and regulation of carboxylation patterns determined by carbon isotope discrimination techniques. In: Winter K, Smith JAC (eds) Crassulacean acid metabolism: biochemistry. Ecophysiology and evolution. Springer-Verlag, Berlin, pp 230–246

Borland AM, Dodd AN (2002) Carbohydrate partitioning in CAM plants: reconciling potential conflicts of interest. Funct Plant Biol 29:707–716

Borland AM, Taybi T (2004) Synchronization of metabolic processes in plants with crassulacean acid metabolism. J Exp Bot 55:1255–1265

Borland AM, Griffiths H, Broadmeadow MSJ, Fordham MC, Maxwell C (1994) Carbon-isotope composition of biochemical fractions and the regulation of carbon balance in leaves of the C_3-crassulacean acid metabolism plant *Clusia minor* L. growing in Trinidad. Plant Physiol 106:493–501

Borland AM, Griffiths H, Hartwell J, Smith JAC (2009) Exploiting the potential of plants with crassulacean acid metabolism for bioenergy production on marginal lands. J Exp Bot 60:2879–2896

Borland AM, Hartwell J, Jenkins GI, Wilkins MB, Nimmo HG (1999) Metabolite control overrides circadian regulation of PEPc kinase and CO_2 fixation in crassulacean acid metabolism (CAM). Plant Physiol 121:889–896

Borland AM, Maxwell K, Griffiths H (2000) Ecophysiology of plants with crassuacean acid metabolism. In: Leegood RC, Sharkey TD, von Caemmerer S (eds) Photosynthesis: physiology and metabolism. Kluwer Academic Publishers, Dordrecht, pp 583–605

Bowes G (1993) Facing the inevitable plant and increasing atmospheric CO_2. Annu Rev Plant Physiol Plant Mol Biol 44:309–332

Ceusters J, Borland AM, Londers E, Verdoodt V, Godts C, De Proft MP (2008a) Diel shifts in carboxylation pathway and metabolite dynamics in the CAM bromeliad *Aechmea* 'Maya' in response to elevated CO_2. Ann Bot 102:389–397

Ceusters J, Londers E, Verdoodt V, Ceusters N, De Proft MP (2008b) Seasonal impact on physiological leaf damage risk of *Aechmea* hybrid under greenhouse conditions. Sci Hortic 118:242–245

Ceusters J, Borland AM, De Proft MP (2009a) Drought adaptation in plants with crassulacean acid metabolism involves the flexible use of different storage carbohydrate pools. Plant Signal Behav 4:212–214

Ceusters J, Borland AM, Londers E, Verdoodt V, Godts C, De Proft MP (2009b) Differential usage of storage carbohydrates in the CAM bromeliad *Aechmea* 'Maya' during acclimation to drought and recovery from dehydration. Physiol Plant 135:174–184

Ceusters J, Londers E, Verdoodt V, Ceusters N, Godts C, De Proft MP (2009c) Impact of developmental stage on CAM expression and growth in an *Aechmea* hybrid under greenhouse conditions. J Hortic Sci Biotechnol 84:399–402

Cote FX, Folliot M, Andre M (1993) Photosynthetic crassulacean acid metabolism in pineapple: diel rhythm of CO_2 fixation, water use, and effect of water stress. Acta Hortic 344:113–129

Cousins AB, Badger MR, von Caemmerer (2006) Carbonic anhydrase and its influence on carbon isotope discrimination during C_4 photosynthesis. Insights from anti-sense RNA in *Flaveria bidentis*. Plant Physiol 141:232–242

Christopher JT, Holtum JAM (1996) Patterns of carbon partitioning in leaves of crassulacean acid metabolism species during deacidification. Plant Physiol 112:393–399

Christopher JT, Holtum JAM (1998) Carbohydrate partitioning in the leaves of Bromeliaceae performing C_3 photosynthesis or crassulacean acid metabolism. Aust J Plant Physiol 25:371–376

Croonenborghs S, Ceusters J, Londers E, De Proft MP (2009) Effects of elevated CO_2 on growth and morphological characteristics of ornamental bromeliads. Sci Hortic 121:192–198

Crowley TJ, Berner RA (2001) CO_2 and climate change. Science 292:870–872

Cui M, Nobel PS (1994) Gas exchange and growth responses to elevated CO_2 and light levels in the CAM species *Opuntia ficus-indica*. Plant Cell Environ 17:935–944

Cui M, Miller PM, Nobel PS (1993) CO_2 exchange and growth of the crassulacean acid metabolism plant *Opuntia ficus-indica* under elevated CO_2 in open-top chambers. Plant Physiol 103:519–524

Cushman JC, Agarie S, Albion S, Elliot SM, Taybi T, Borland AM (2008) Isolation and characterization of mutants of ice plant, *Mesembryanthemum crystallinum*, defective in crassulacean acid metabolism. Plant Physiol 147:228–238

Cushman JC, Borland AM (2002) Induction of crassulacean acid metabolism by water limitation. Plant Cell Environ 25:295–310

Dodd AN, Borland AM, Haslam RP, Griffiths H, Maxwell K (2002) Crassulacean acid metabolism: plastic fantastic. J Exp Bot 53:569–580

Dodd AN, Griffiths H, Taybi T, Cushman JC, Borland AM (2003) Integrating diel starch metabolism with the circadian and environmental regulation of crassulacean acid metabolism in *Mesembryanthemum crystallinum*. Planta 216:789–797

Drennan PM, Nobel PS (2000) Responses of CAM species to increasing atmospheric CO_2 concentrations. Plant Cell Environ 23:767–781
Eller BM, Ferrari S (1997) Water use efficiency of two succulents with contrasting CO_2 fixation pathways. Plant Cell Environ 20:93–100
Farrar JF (1990) Starch turnover: its role in source-sink relations and a comparison to sucrose. In: Bonnemain JL, Delrot S, Lucas WT, Dainty J (eds) Recent advances in phloem transport and assimilate compartmentation. Quest Edition, Nantes, pp 213–223
Ghannoum O (2009) C_4 photosynthesis and water stress. Ann Bot 103:635–644
Ghannoum O, von Caemmerer S, Ziska LH, Conroy JP (2000) The growth response of C_4 plants to rising atmospheric CO_2 partial pressure: a reassessment. Plant Cell Environ 23:931–942
Gillon JS, Yakir D (2000) Naturally low carbonic anhydrase activity in C_4 and C_3 plants limits discrimination against (COO)-O_18 during photosynthesis. Plant Cell Environ 20:1217–1230
Goldschmidt EE, Huber SC (1992) Regulation of photosynthesis by end-product accumulation in leaves of plants storing starch, sucrose, and hexose sugars. Plant Physiol 99:1443–1448
Gouk SS, Yong JWH, Hew CS (1997) Effects of superelevated CO_2 on the growth and carboxylating enzymes in an epiphytic CAM orchid plantlet. J Plant Physiol 151:129–136
Graham EA, Nobel PS (1996) Long-term effects of a doubled atmospheric CO_2 concentration on the CAM species *Agave deserti*. J Exp Bot 47:61–69
Granados J, Körner C (2002) In deep shade, elevated CO_2 increases the vigor of tropical climbing plants. Glob Change Biol 8:1109–1117
Griffihs H (1988) Crassulacean acid metabolism: a re-appraisal of physiological plasticity in form and function. Adv Bot Res 15:43–92
Griffiths H, Broadmeadow MSJ, Borland AM, Hetherington CS (1990) Short-term changes in carbon isotope discrimination identify transitions between C_3 and C_4 carboxylation during crassulacean acid metabolism. Planta 181:604–610
Griffiths H, Borland AM, Gillon JS, Harwood KG, Maxwell K, Wilson JM (1999) Stable isotopes reveal exchanges between soil, plants and the atmosphere. In: Press MC, Scholes JD, Barker MG (eds) Advances in physiological plant ecology. Blackwell Science, Oxford, pp 415–441
Griffiths H, Cousins AB, Badger MR, von Caemmerer S (2007) Discrimination in the dark. Resolving the interplay between metabolic and physical constraints to phosphoenolpyruvate carboxylase activity during the crassulacean acid metabolism cycle. Plant Physiol 143:1055–1067
Hatch MD, Burnell JN (1990) Carbonic anhydrase activity in leaves and its role in the first step of C4 photosynthesis. Plant Physiol 93:825–828
Häusler RE, Baur B, Scharte J, Teichmann T, Eicks M, Fischer KL, Flügge U-I, Schubert S, Weber A, Fischer K (2000) Plastidic metabolite transporters and their physiological functions in the inducible crassulacean acid metabolism plant *Mesembryanthemum crystallinum*. Plant J 24:285–296
Holtum JAM, O'Leary MH, Osmond CB (1983) Effect of varying CO_2 partial pressure on photosynthesis and on carbon isotope composition of carbon-4 of malate from the crassulacean acid metabolism plant *Kalanchoë daigremontiana* Hamet et Perr. Plant Physiol 71:602–609
Huerta AJ, Ting IP (1988) Effects of various levels of CO_2 on the induction of crassulacean acid metabolism in *Portulacaria afra* (L.) Jacq. Plant Physiol 88:183–188
IPCC (Intergovernmental Panel for Climate Change) (2007) Climate change. The second assessment report of IPCC working group I. Cambridge University Press, Cambridge
Isdo SB, Kimball BA, Anderson MG, Szarek SR (1986) Growth response of a succulent plant, *Agave vilmoriniana* to elevated CO_2. Plant Physiol 80:796–797
Isdo SB, Kimball BA (1992) Above-ground inventory of sour orange trees exposed to different atmospheric CO_2 concentrations for three full years. Agr Forest Meteorol 60:145–151
Jarvis AJ, Mansfield TA, Davies WJ (1999) Stomatal behaviour, photosynthesis and transpiration under rising CO_2. Plant Cell Environ 22:639–648
Körner C (2001) Biosphere responses to CO_2 enrichment. Ecol Appl 10:1590–1619

Körner C (2006) Plant CO_2 responses: an issue of definition, time and resource supply. New Phytol 172:393–411

Körner C, Morgan JA, Norby RJ (2007) CO_2 fertilization: when, where, how much? In: Canadell JP, Pataki DE, Pitelka LF (eds) Terrestrial ecosystems in a changing world, The IGBP series. Springer, Berlin, pp 9–21

Leakey ADB, Bernacchi CJ, Dohleman FG, Ort DR, Long SP (2004) Will photosynthesis of maize (*Zea mays*) in the US Corn Belt increase in future [CO_2] rich atmospheres? An analysis of diurnal courses of CO_2 uptake under free-air concentration enrichment (FACE). Glob Change Biol 10:951–962

Li CR, Gan LJ, Xia K, Zhou X, Hew CS (2002) Responses of carboxylating enzymes, sucrose metabolising enzymes and plant hormones in a tropical epiphytic CAM orchid to CO_2 enrichment. Plant Cell Environ 25:369–377

Lüttge U (1988) Day–night changes of citric-acid levels in crassulacean acid metabolism: phenomenon and ecophysiological significance. Plant Cell Environ 11:445–451

Lüttge U (1996) *Clusia*: plasticity and diversity in a genus of C_3/CAM intermediate tropical trees. In: Winter K, Smith JAC (eds) Crassulacean acid metabolism: biochemistry. Ecophysiology and evolution. Springer-Verlag, Berlin, pp 296–311

Lüttge U (2004) Ecophysiology of crassulacean acid metabolism (CAM). Ann Bot 93:629–652

Lüttge U (2006) Photosynthetic flexibility and ecological plasticity: questions and lessons from *Clusia*, the only CAM tree, in the neotropics. New Phytol 171:7–25

Maxwell K, von Caemmerer S, Evans JR (1997) Is a low internal conductance to CO_2 diffusion a consequence of succulence in plants with crassulacean acid metabolism? Austr J Plant Physiol 24:777–786

Maxwell K, Badger MR, Osmond CB (1998) A comparison of CO_2 and O_2 exchange patterns and the relationship with chlorophyll fluorescence during photosynthesis in C_3 and CAM plants. Austr J Plant Physiol 25:45–52

Maxwell K, Borland AM, Haslam RP, Helliker BR, Roberts A, Griffiths H (1999) Modulation of Rubisco activity during the diurnal Phases of the crassulacean acid metabolism plant *Kalanchoë daigremontiana*. Plant Physiol 121:849–856

Milburn TR, Pearson DJ, Ndegwe NA (1968) Crassulacean acid metabolism under natural tropical conditions. New Phytol 67:883–897

Moore BD, Cheng SH, Sims D, Seemann JR (1999) The biochemical and molecular basis for photosynthetic acclimation to elevated atmospheric CO_2. Plant Cell Environ 22:567–582

Mortensen LM, Moe R (1992) Effects of CO_2 enrichment and different day/night temperature combinations on growth and flowering of *Rosa* L. & *Kalanchoë blossfeldiana* v. Poelln. Sci Hortic 51:145–153

Nelson EA, Sage RF (2008) Functional constraints of CAM leaf anatomy: tight cell packing is associated with increased CAM function across a gradient of CAM expression. J Exp Bot 59:1841–1850

Nobel PS (1988) Environmental biology of agaves and cacti. Cambridge University Press, Cambridge

Nobel PS (1991) Achievable productivities of CAM plants: basis for high values compared with C_3 and C_4 plants. New Phytol 119:183–205

Nobel PS (1996) High productivity of certain agronomic CAM species. In: Winter K, Smith JAC (eds) Crassulacean acid metabolism: biochemistry. Ecophysiology and evolution. Springer-Verlag, Berlin, pp 255–265

Nobel PS (2000) Crop ecosystem responses of climatic change. Crassulacean acid metabolism crops. In: Reddy KR, Hodges HF (eds) Climate change and global crop productivity. CAB international, Oxford, pp 315–331

Nobel PS, Hartsock TL (1986) Short-term and long-term responses of crassulacean acid metabolism plants to elevated CO_2. Plant Physiol 82:604–606

Nobel PS, Israel AA (1994) Cladode development, environmental responses of CO_2 uptake, and productivity for *Opuntia ficus-indica* under elevated CO_2. J Exp Bot 45:295–303

Nobel PS, Cui M, Israel AA (1994) Light, chlorophyll, carboxylase activity and CO_2 fixation at various depths in the chlorenchyma of *Opuntia ficus-indica* (L.) Miller under current and elevated CO_2. New Phytol 128:315–322
North GB, Moore TL, Nobel PS (1995) Cladode development for *Opuntia ficus-indica* (Cactaceae) under current and doubled CO_2 concentrations. Am J Bot 82:159–166
Nowak EJ, Martin CE (1995) Effect of elevated CO_2 on nocturnal malate accumulation in the CAM species *Tillandsia ionantha* and *Crassula arborescens*. Photosynthetica 31:441–444
Osmond CB (1978) Crassulacean acid metabolism: a curiosity in context. Annu Rev Plant Physiol 29:379–414
Osmond CB, Björkman O (1975) Pathways of CO_2 fixation in the CAM plant *Kalanchoë daigremontiana* II. Effects of O_2 and CO_2 concentration on light and dark CO_2 fixation. Aust J Plant Physiol 2:155–162
Pagani M, Zachos JC, Freeman KH, Tipple B, Bohaty S (2005) Marked decline in atmospheric carbon dioxide concentrations during the Paleogene. Science 309:600–603
Poorter H, Navas ML (2003) Plant growth and competition at elevated CO_2: on winners, losers and functional groups. New Phytol 157:175–198
Popp M, Kramer D, Lee H, Diaz M, Ziegler H, Lüttge U (1987) Crassulacean acid metabolism in tropical dycotyledonous trees of the genus *Clusia*. Trees 1:138–247
Powles SB, Chapman KSR, Osmond CB (1980) Photoinhibition of intact attached leaves of C_4 plants: dependence on CO_2 and O_2 partial pressures. Aust J Plant Physiol 7:737–747
Pritchard SG, Rogers HH, Prior SA, Peterson CM (1999) Elevated CO_2 and plant structure: a review. Glob Change Biol 5:807–837
Raveh E, Gersani M, Nobel PS (1995) CO_2 uptake and fluorescence for a shade-tolerant cactus *Hylocereus undatus* under current and doubled CO_2 concentrations. Physiol Plant 93:505–511
Sage RF (2001) Environmental and evolutionary preconditions for the origin and diversification of the C4 photosynthetic syndrome. Plant Biol 3:202–212
Szarek SR, Holthe PA, Ting IP (1987) Minor physiological response to elevated CO_2 by the CAM plant *Agave vilmoriniana*. Plant Physiol 83:938–940
Ting IP (1994) CO_2 and Crassulacean acid metabolism plants: a review. In: Tolbert NE, Preiss J (eds) Regulation of atmospheric CO_2 and O_2 by photosynthetic carbon metabolism. Oxford University Press, New York, pp 176–183
Ting IP, Lord EM, da SL SL, DeNiro MJ (1985) Crassulacean acid metabolism in the strangler *Clusia rosea* Jacq. Science 229:969–971
Von Caemmerer S, Griffiths H (2009) Stomatal responses to CO_2 during a diel CAM cycle in *Kalanchoë daigremontiana* and *Kalanchoë pinnata*. Plant Cell Environ 32:567–576
von Caemmerer S, Furbank RT (2003) The C-4 pathway: an efficient CO_2 pump. Photosynth Res 77:191–207
Wang N, Nobel PS (1996) Doubling the CO_2 concentration enhanced the activity of carbohydrate-metabolism enzymes, source carbohydrate production, photoassimilate transport and sink strength for *Opuntia ficus-indica*. Plant Physiol 110:893–902
Ward JK, Tissue DT, Thomas RB, Strain BR (1999) Comparative responses of model C_3 and C_4 plants to drought in low and elevated CO_2. Glob Change Biol 5:857–867
Winter K, Engelbrecht B (1994) Short-term CO_2 responses of light and dark CO_2 fixation in the crassulacean acid metabolism plant *Kalanchoë pinnata*. J Plant Physiol 144:462–467
Winter K, Smith JAC (1996) Crassulacean acid metabolism: current status and perspectives. In: Winter K, Smith JAC (eds) Crassulacean acid metabolism: biochemistry. Ecophysiology and evolution. Springer-Verlag, Berlin, pp 230–246
Winter K, Zotz G, Baur B, Dietz KJ (1992) Light and dark CO_2 fixation in *Clusia uvitana* and the effects of plant water status and CO_2 availability. Oecologia 91:47–51
Winter K, Richter A, Engelbrecht B, Posada J, Virgo A, Popp M (1997) Effect of elevated CO_2 on growth and crassulacean acid metabolism activity of *Kalanchoë pinnata* under tropical conditions. Planta 201:389–396

Würth MKR, Winter K, Körner C (1998) *In situ* responses to elevated CO_2 in tropical forest understory plants. Funct Ecol 12:886–895
Zhu J, Bartholomew DP, Goldstein G (1997) Effect of elevated carbon dioxide on the growth and physiological responses of pineapple, a species with crassulacean acid metabolism. J Am Soc Hortic Sci 122:233–237
Zhu J, Goldstein G, Bartholomew DP (1999) Gas exchange and carbon isotope composition of *Ananas comosus* in response to elevated CO_2 and temperature. Plant Cell Environ 22:999–1007
Ziska LH, Hogan KP, Smith AP, Drake BG (1991) Growth and photosynthetic response of nine tropical species with long-term exposure to doubled carbon dioxide. Oecologia 86:383–389

Nuclear Magnetic Resonance Spectroscopic Analysis of Enzyme Products

Bernd Schneider

Contents

Abstract NMR spectroscopy has been increasingly used as an analytical tool to determine the structures of small molecules obtained as products of enzymes; these enzymes were isolated from plants or expressed in heterologous systems. The review covers several examples of products of type III polyketide synthases, oxydases, methyltransferases, acyltransferases, glycosyltransferases, and terpene cyclases. The examples reveal the advantages and limitations of NMR spectroscopic analysis and demonstrate that NMR is able to not only unambiguously identify products by comparing their spectra with those of authentic references but also elucidate structures of enzyme product de novo without using reference data. Recent enhancements of NMR sensitivity enable the analysis of mass-limited samples, including enzyme products. Coupling of NMR with liquid chromatography directly combines separation with structure analysis and extends the application of NMR to mixtures of enzyme products.

B. Schneider
Max-Planck-Institute for Chemical Ecology, Beutenberg Campus, Hans-Knöll-Str. 8, 07745 Jena, Germany
e-mail: schneider@ice.mpg.de

U. Lüttge et al. (eds.), *Progress in Botany 72*,
DOI 10.1007/978-3-642-13145-5_7,

1 Introduction

The identification of small molecules, which are produced by recombinant proteins or by enzymes isolated (or partially purified) from plant tissue, is an important objective of biochemical research. Product analysis is also necessary to characterize enzymes of interest from any other source, especially microorganisms and animal tissue. Chromatographic methods such as thin layer chromatography (TLC), high performance liquid chromatography (HPLC), ultra-performance liquid chromatography (UPLC), and gas chromatography (GC) are suitable for identifying enzyme products of known chemical structure. The analysis of enzyme products of unknown structure, however, requires spectroscopic methods, among which mass spectrometry (MS) and nuclear magnetic resonance (NMR) spectroscopy are the most powerful. Both MS and NMR can be used to identify isolated compounds and provide spectroscopic information for components of a mixture. In combination with a separation method, e.g., liquid chromatography (LC), structure elucidation is also possible for multicomponent mixtures.

NMR spectroscopy is the most informative analytical method for structure elucidation of small molecules. One of the special advantages of NMR is its ability to distinguish various types of isomeric compounds, including stereoisomers such as *E*/*Z*-isomers and diastereomers, which sometimes are difficult to discriminate by mass spectrometric methods. Furthermore, NMR is nondestructive, i.e., after NMR measurement, the sample is still available for other purposes. It may be subjected to another NMR spectroscopic measurement, mass spectrometric analysis, or any other analytical method. However, previously, the moderate sensitivity of NMR did not allow it to be used to analyze trace metabolites. Because molecules produced by isolated or recombinant proteins are usually formed on a small scale, they can be considered trace compounds and are subject to the limits mentioned above. This situation has changed in recent years because methods of enzyme purification have been improved, recombinant enzymes have become routinely available, and the sensitivity of NMR spectroscopy has been substantially increased by technical innovations such as the introduction of microprobes and cryogenically cooled probes (Russel et al. 2000; Kovacs et al. 2005). Hence, NMR spectroscopy is being increasingly used for analyzing enzyme products.

Here, examples of the structural identification of plant secondary metabolites formed by recombinant proteins or enzymes, which were isolated from plant material using conventional protein purification procedures, are reviewed. The examples discussed in this review have been arranged according to the enzymes to be characterized. 1D and 2D NMR spectra or partial spectra, mostly from samples measured in our laboratory, are shown in order to illustrate how they have been used to identify enzyme products. The purpose of the review is to recommend NMR as a useful tool for enzyme product analysis.

2 Polyketide Synthase Products

Structure identification of enzyme products, i.e., compounds formed by incubating an isolated or recombinant enzyme with putative substrates, often takes advantage of the fact that some structural information are already available. This is because the enzyme type is known from the isolation or cloning procedure and the catalyzed reactions mostly parallel known transformations. However, although the structure of the enzyme products is sometimes predictable, it needs to be confirmed experimentally by NMR or other analytical methods. If the authentic references compound is available, the product can be identified simply by comparing the ^{1}H NMR spectrum of the substrate with that of the product. This has been demonstrated, for example, for products formed by a recombinant type III polyketide synthase (Fig. 1), WtPKS1, from *Wachendorfia thyrsiflora*, a phenylphenalenone-producing plant from the Haemodoraceae family (Brand et al. 2006). The protein was cloned, heterologously expressed in *Escherichia coli*, and incubated with various phenylpropanoyl-CoAs (e.g., 4-coumaroyl-CoA) and malonyl-CoA. To produce samples sufficient for NMR analysis, the standard assays were scaled up 15-fold. The products were extracted from the assay mixture with ethyl acetate and separated by reverse-phase HPLC. The ^{1}H NMR spectra were then measured at 500 MHz using a cryogenically cooled probe. Comparing the spectra of products obtained from 4-coumaroyl-CoA and malonyl-CoA with those of authentic references identified the products as 4-hydroxy-benzalacetone and bisnoryangonin (Fig. 2). Although the expected in vivo products (diarylheptanoids) of WtPKS1 were not obtained, the investigation showed that the protein catalyzes condensation of the phenylpropanoyl-CoA (starter substrate) with one or two malonyl-CoA units (extender substrate) (Fig. 1).

Replacing the 4-coumaroyl-CoA with 4-hydroxyphenyl-propionyl-CoA (dihydro-4-coumaroyl-CoA) in the WtPKS1 assay resulted in two condensations with malonyl-CoA. Dihydrobisnoryangonin instead of bisnoryangonin was found as a product in vitro. Since a reference was not available, the structure of dihydrobisnoryangonin was identified by ^{1}H NMR and ^{1}H,^{1}H-lrCOSY, a homocorrelation spectrum optimized for long-range ^{1}H,^{1}H-couplings. The ^{1}H,^{1}H-lrCOSY (Fig. 3) not only displayed $^3J_{H,H}$-couplings ($^3J_{H\text{-}2''/6''\text{–}H\text{-}3''/5''}$; $^3J_{H\text{-}1\text{–}H\text{-}2}$) but also $^4J_{H,H}$-couplings between H-2″/6″ of the aromatic ring and H-2 ($^4J_{H\text{-}2''/6''\text{–}H\text{-}2}$), between H-1 and H-5′ ($^4J_{H\text{-}1\text{–}H\text{-}5'}$) of the 4-hydroxy-2-pyrone ring, and within the pyrone moiety ($^4J_{H\text{-}3'\text{–}H\text{-}5'}$). In the 1D ^{1}H NMR spectrum, the small ^4J coupling constants are not resolved (Fig. 3, projection), and the methylene signals of H-1 and H-2 overlap with the large HDO signal.

Dihydrobisnoryangonin was also formed in vitro by incubating stilbenecarboxylate synthase (STCS) from *Hydrangea macrophylla* with dihydro-4-coumaroyl-CoA (4-hydroxyphenyl-propionyl-CoA) and malonyl-CoA. Dihydrobisnoryangonin was identified as a product of two condensation steps, together with two products of three condensations, 5-hydroxylunularic acid and dihydro-4-coumaroyltriacetic acid lactone (dihydro-CTAL) (Fig. 4) (Eckermann et al. 2003). In addition to MS analysis,

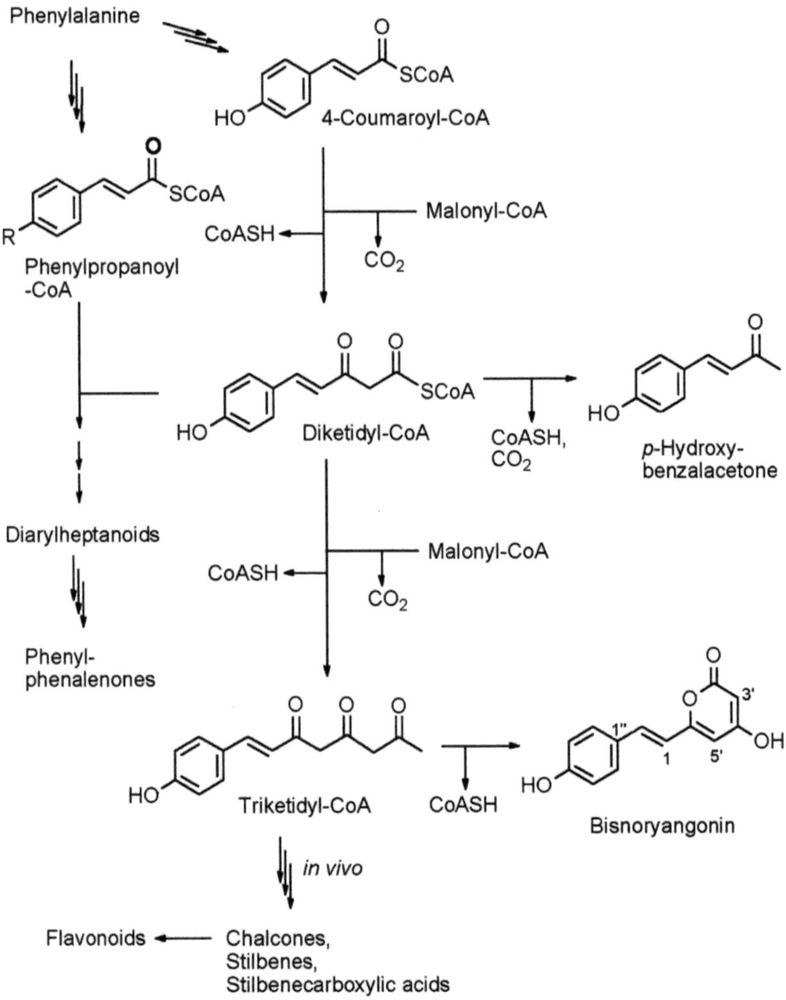

Fig. 1 Polyketide pathway in plants. 4-Coumaroyl-CoA is the most common starter substrate, but the pathway also operates analogously with other phenylpropanoyl-CoAs. Benzalacetones (e.g., *p*-hydroxybenzalacetone) and pyrones such as bisnoryangonin are byproducts, which have frequently been found in vitro (Schröder 2000)

the ^{1}H NMR measurement of the ethyl acetate extract of the enzyme assay indicated that each of the three enzyme products contained a 4-substituted phenyl ring (δ 6.7–6.8 and δ 7.0–7.1, $J = 8$–9 Hz). Further signals of the spectrum were assigned to H-3′ and H-5′ of the 4-hydroxy-2-pyrone rings of dihydro-CTAL and dihydrobisnoryangonin. The third compound was further characterized by two doublets at δ 6.3 ($J = 2.2$ Hz) of the 2-carboxyl-3,5-dihydroxyphenyl ring system because of the similarity of the chemical shifts values with those of flavonoid ring A protons. Furthermore, the integration of signals assigned to each of the compounds made it

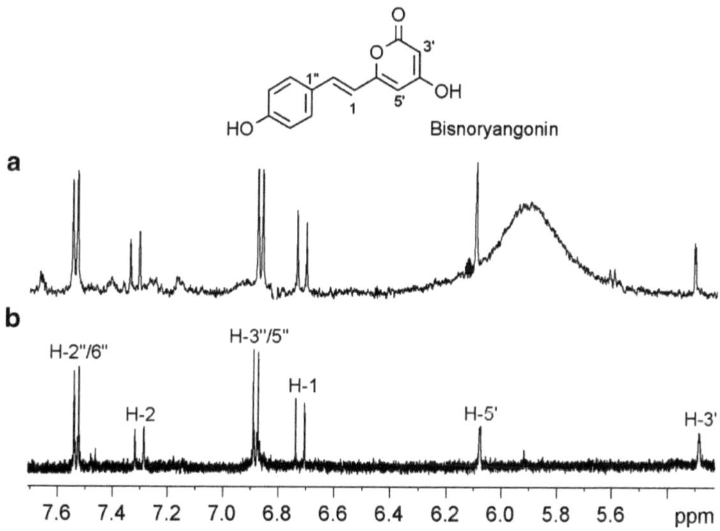

Fig. 2 ^{1}H NMR spectra (500 MHz, acetone-d_6) of bisnoryangonin (**a**) obtained from incubation of WtPKS1 with 4-coumaroyl-CoA and malonyl-CoA (Brand et al. 2006). (**b**) ^{1}H NMR spectrum of an authentic standard

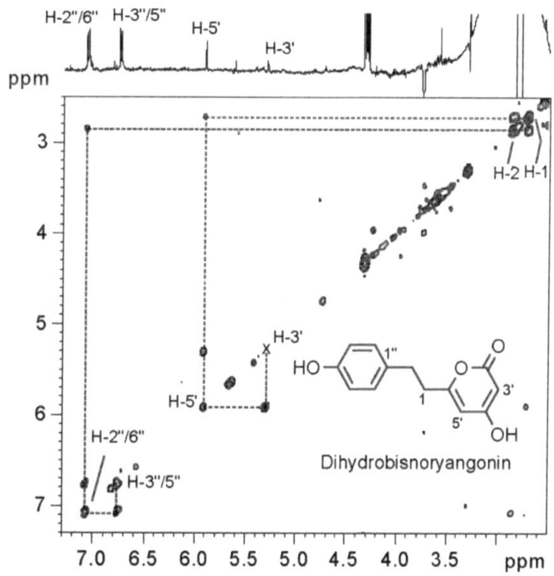

Fig. 3 ^{1}H,^{1}H-lrCOSY spectrum of dihydrobisnoryangonin obtained after incubating WtPKS1 with 4-hydroxyphenyl-propionyl-CoA and malonyl-CoA (Brand et al. 2006). The pyrone is formed in a process similar to the pathway outlined in Fig. 1. *Dotted lines* connect cross peaks, indicating $^3J_{H,H}$-couplings and $^4J_{H,H}$-couplings

possible to roughly determine their ratio in the extract. HPLC separation and further ^{1}H NMR analysis, especially of 5-hydroxylunularic acid, a compound containing a labile 2-carboxyl-3,5-dihydroxyphenyl residue (Gorham 1980), was difficult because of the instability of the compounds.

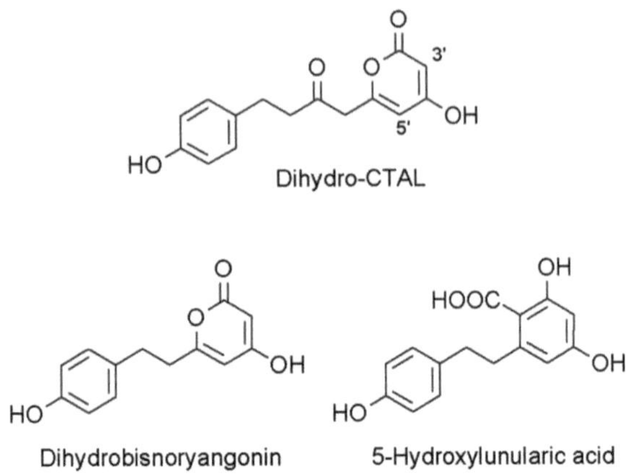

Fig. 4 Products formed by incubation of dihydrocoumaroyl-CoA and malonyl-CoA with STCS from *H. macrophylla* (Eckermann et al. 2003). Mass spectrometric determination of dihydrobisnoryangonin (a condensation product with two malonyl-CoA units), 5-hydroxylunularic acid, and dihydro-4-coumaroyltriacetic acid lactone (dihydro-CTAL) (products of three condensations) was supported by ^{1}H NMR spectroscopic signal assignment

Another drawback was that the NMR spectra were measured using a 2.5 mm inverse-detection room temperature probe since a cryoprobe was not available at that time. Nevertheless, two methylene groups with the same chemical shift (δ 2.75), and proton signals of a 2-carboxyl-3,5-dihydroxyphenyl ring (δ 6.15 and 6.16), consistent with the structure of 5-hydroxylunularic acid, were detected in the ^{1}H NMR spectrum of the isolated compound.

3 Oxygenase Products

A flavonoid 6-hydroxylase (F6H; CYP71D9) cDNA that was cloned earlier from elicited soybean (*Glycine max*) cell cultures (Schopfer and Ebel 1998) was heterologously expressed in yeast (Latunde-Dada et al. 2001). To characterize the metabolite formed by CYP71D9, microsomes of recombinant yeast were incubated with liquiritigenin, and the ethyl acetate extract of the reaction mixture was analyzed by HPLC, MS, and NMR spectroscopy. ^{1}H NMR spectra (500 MHz, acetone-d_6) showed that the proton signal of H-6 was absent in the reaction product, whereas H-5 and H-8 appeared as singlets and all other signals were very similar to those observed for the substrate, liquiritigenin. Hence, the hydroxylation was established as occurring at C-6 of ring A and the product was identified as 6,7,4′-trihydroxyflavanone (Fig. 5). As soybean produces isoflavonoid constituents possessing 6,7-dihydroxy substitution patterns on ring A, the biosynthetic relationship of flavonoid 6-hydroxylase to isoflavonoid biosynthesis was investigated by the same authors (Latunde-Dada et al. 2001). It was shown that recombinant 2-hydroxyisoflavanone synthase (2HIS; CYP93C1v2) efficiently used the product of F6H, 6,7,4′-trihydroxyflavanone, as a substrate. Therefore, the combined catalytic action of the recombinant proteins F6H and 2HIS was assayed. The large-scale reaction mixtures were

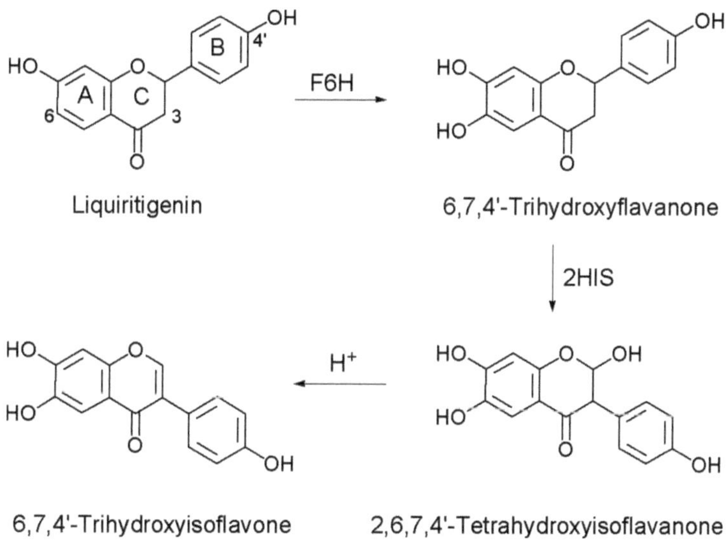

Fig. 5 Hydroxylation of liquiritigenin to 6,7,4′-trihydroxyflavanone and rearrangement to 2,6,7,4′-tetrahydroxyisoflavanone catalyzed by recombinant F6H and 2HIS, respectively, from soybean (Latunde-Dada et al. 2001). 6,7,4′-Trihydroxyflavanone and 6,7,4′-trihydroxyisoflavone, which was obtained by acid treatment of the labile enzyme product 2,6,7,4′-tetrahydroxyisoflavanone, were identified by spectroscopic methods including NMR

again extracted with ethyl acetate; the extract was subjected to acid treatment in order to dehydrate the chemically labile suspected reaction product, 2,6,7,4′-tetrahydroxyisoflavanone; and, after separation by HPLC, the product was analyzed by UV and NMR spectroscopy. Traces of acids, which strongly affected the chemical shifts ($\Delta\delta$ up to 0.3 ppm) but not the coupling constants of individual protons, were eliminated by high vacuum treatment. The ^{13}C chemical shifts obtained from a heteronuclear single quantum coherence (HSQC) spectrum measured before high vacuum treatment were consistent with those of the reference compound, 6,7,4′-trihydroxyisoflavone. After high vacuum treatment, also all proton chemical shifts agreed within 0.02 ppm with those of the reference. The results indicated that in the biosynthesis of polyhydroxylated isoflavones in soybean, 6-hydroxylation occurs before 1,2-aryl migration during isoflavanone formation (Fig. 5) (Latunde-Dada et al. 2001).

A cytochrome P450 has been shown to be involved in the biosynthetic hydroxylation of the arylnaphthalene lignan justicidin B to diphyllin (Fig. 6) (Hemmati et al. 2007). A ribosomal protein fraction from cell cultures of *Linum perenne* was incubated with justicidin B and the assay extracted with ethyl acetate. The organic layer was subjected to LC–SPE–^{1}H NMR. Postcolumn solid-phase extraction (SPE) was used to collect the desired diode array-detected peak, which showed the same retention time as the authentic standard, diphyllin. The SPE interface was connected by a capillary to an NMR cryoprobe equipped with a Cryofit™ flow

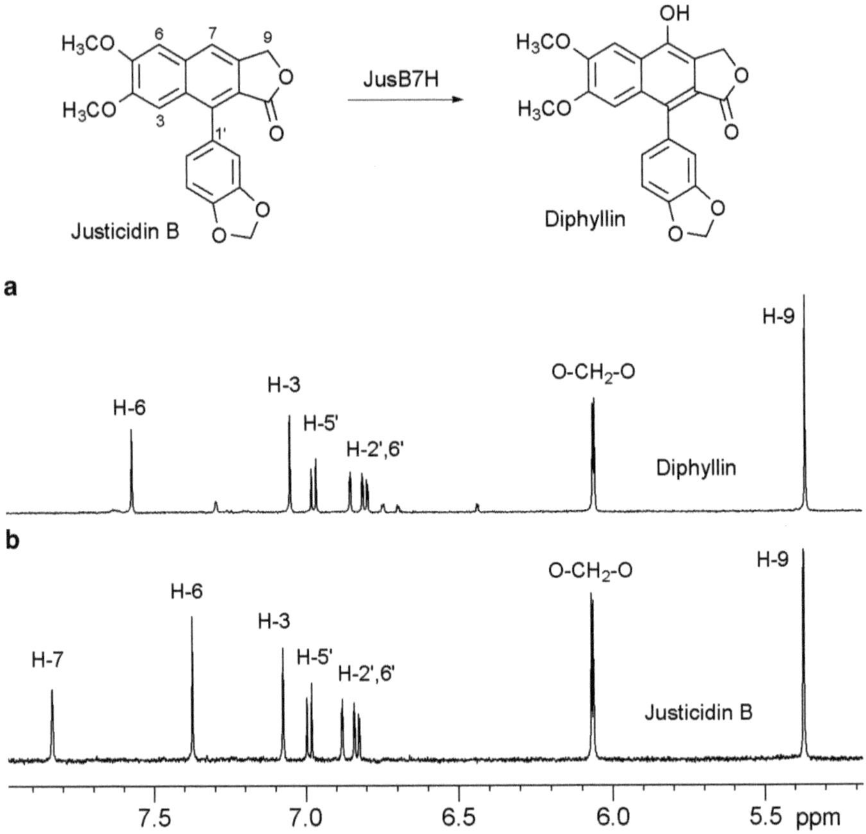

Fig. 6 Conversion of diphylline to justicidine B by a ribosomal protein fraction from cell cultures of *Linum perenne* (Hemmati et al. 2007). ^{1}H NMR spectra (500 MHz, MeCN-d_3) of the substrate (**b**) and the product (**a**) were obtained in the LC–SPE–NMR mode. The missing signal of H-7 and a shift of the signal of H-6 to a lower field in the spectrum of diphyllin compared to H-6 of justicidin B indicated hydroxylation at C-7

insert. From the nitrogen gas-dried SPE cartridge, the trapped enzyme product was eluted with deuterated acetonitrile into the NMR cryoprobe and the ^{1}H NMR spectrum was measured. The spectrum matched that of an authentic sample of diphyllin but differed from that of the substrate, justicidin B, which was obtained under identical LC–SPE–NMR conditions, because of the absence of the signal of H-7. In addition, H-6 of diphyllin appeared at lower field compared to H-6 of justicidin B, confirming that hydroxylation had occurred at C-7. The unambiguous identification of diphyllin helped to characterize the enzyme as a justicidin-7-hydroxylase (JusB7H) (Hemmati et al. 2007).

Unlike in the cytochrome P450-dependent monooxygenase-catalyzed hydroxylation of the *Linum* lignan justicidin B to diphyllin, a polyphenol oxidase (PPO) is involved in aromatic ring hydroxylation in the lignan biosynthetic pathway to nordihydroguaiaretic acid (Cho et al. 2003). The PPO protein was purified to apparent homogeneity from the creosote bush (*Larrea tridentata*) and the encoding gene was cloned. The ethyl acetate extract of a large-scale enzymatic incubation with 9.1 mg (±)-larreatricin was subjected to HPLC. ^{1}H NMR (500 MHz), ^{13}C NMR (125 MHz), and heteronuclear multiple-bond correlation (HMBC) spectra, which were recorded in a microcell tube, together with MS data, established the major product to be 3′-hydroxylarreatricin (Fig. 7). Because they were measured in dry DMSO-d_6, the three phenolic hydroxyl signals were visible in the ^{1}H NMR spectrum. Likewise, the minor product was shown to be the 3-hydroxy isomer. Enantioselectivity of the enzymatic reaction was substantiated by chiral HPLC analysis, comparing synthetic 3′-hydroxylarreatricin with the enzymatically generated product.

Recently, angelicin synthase (CYP71AJ4) has been isolated from *Pastinaca sativa* (parsnip) and expressed in yeast (Larbat et al. 2009). Activity screening of the crude yeast microsomal fractions using (+)-columbianetin as a substrate, according to UV spectra and MS data indicated the formation of angelicin as a major product and an unidentified hydroxycolumbianetin derivative as a minor product. In order to study the mechanism of the conversion of (+)-columbianetin to angelicin, specifically deuterated *syn*-[3′-^{2}H]columbianetin was incubated with the recombinant CYP71AJ4. A new product with the molecular mass of another hydroxycolumbianetin was formed at the expense of a relatively large decrease in the formation of angelicin and the minor hydroxycolumbianetin byproduct. LC–SPE–^{1}H NMR analysis of roughly 2 μg of the new product in a cryoprobe at 500 MHz showed two pairs of aromatic AB doublets (Fig. 8) and two methyl

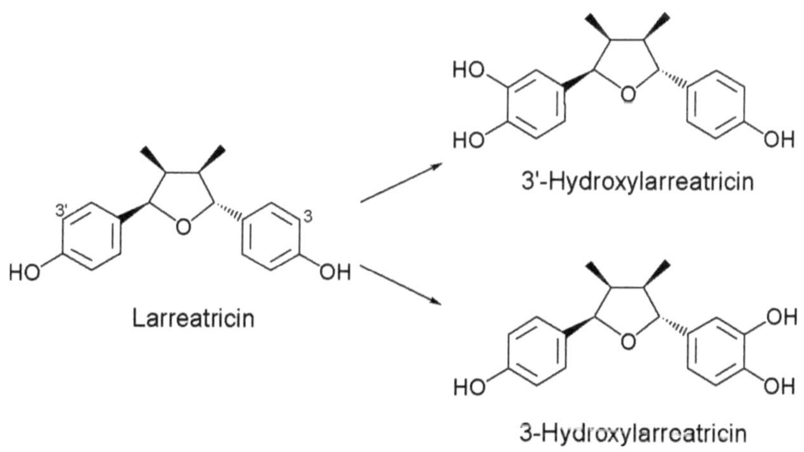

Fig. 7 Enantiospecific conversion of larreatricin to 3′-hydroxylarreatricin as a major and 3-hydroxylarreatricin as a minor product. The reaction is catalyzed by a polyphenol oxidase from the creosote bush (*Larrea tridentata*) (Cho et al. 2003). The structures of the products have been identified by 1D and 2D NMR spectroscopy and mass spectrometry

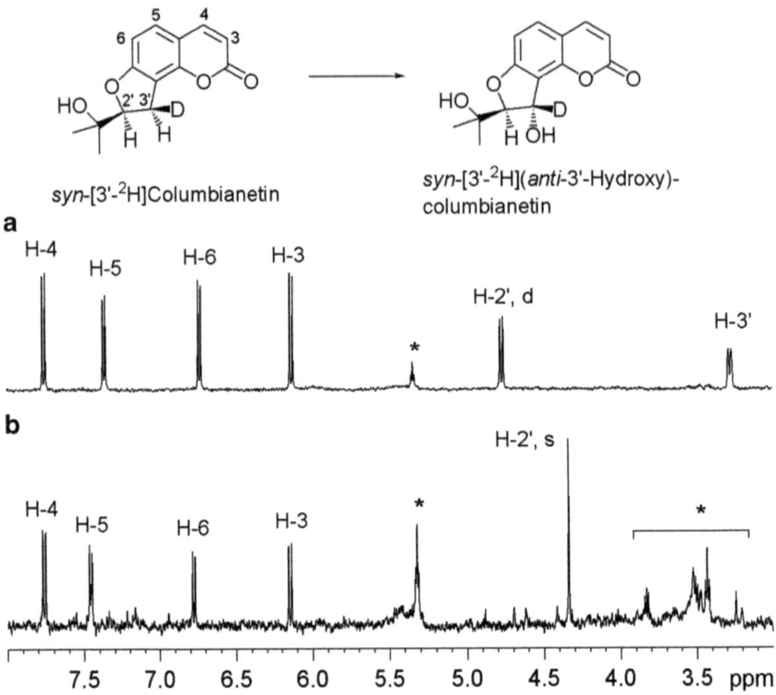

Fig. 8 Hydroxylation of deuterated columbianetin at C-3′ by a recombinant angelicin synthase from parsnip (Larbat et al. 2009). The product was identified by comparing its LC–SPE–^{1}H NMR spectrum (**b**) with that of the substrate, *syn*-[3′-^{2}H]columbianetin (**a**). The chemical shift and multiplicity of the remaining side chain proton signal (H-2′) indicated the deuterium had been retained at C-3′ in *syn*-position. Signals of the methyl groups are not shown. Signals marked with an *asterisk* are due to contamination

signals, which matched the chemical shifts and coupling constants of corresponding protons in the spectrum of columbianetin. The presence of these proton signals suggests that the CYP71AJ4-catalyzed hydroxylation must have occurred at C-2′ or C-3′. A missing signal of H-3′ and a singlet instead of a doublet of H-2′ in the ^{1}H NMR spectrum (Fig. 8) excluded the possibility that hydroxylation had occurred at C-2′ and indicated that the deuterium at C-3′ must be retained. These data were consistent with the new product as *syn*-[3′-^{2}H](*anti*-3′-hydroxycolumbianetin. The different product pattern, together with kinetic data observed after incubation of (+)-columbianetin and *syn*-[3′-^{2}H]columbianetin, clearly indicated that angelicin synthase from parsnip attacks the substrate by abstracting the *syn*-hydrogen from C-3′. However, CYP71AJ4 was also capable of abstracting the *anti*-hydrogen from C-3′, which was then followed by rebound hydroxylation at C-3′, and this reaction was strongly favored if the *syn*-3′-hydrogen was replaced by deuterium (Larbat et al. 2009).

A flavin monooxygenase enzyme, $FMO_{GS\text{-}OX1}$, from *Arabidopsis* was expressed in *E. coli* spheroplasts. The recombinant protein oxidized the methylsulfide group

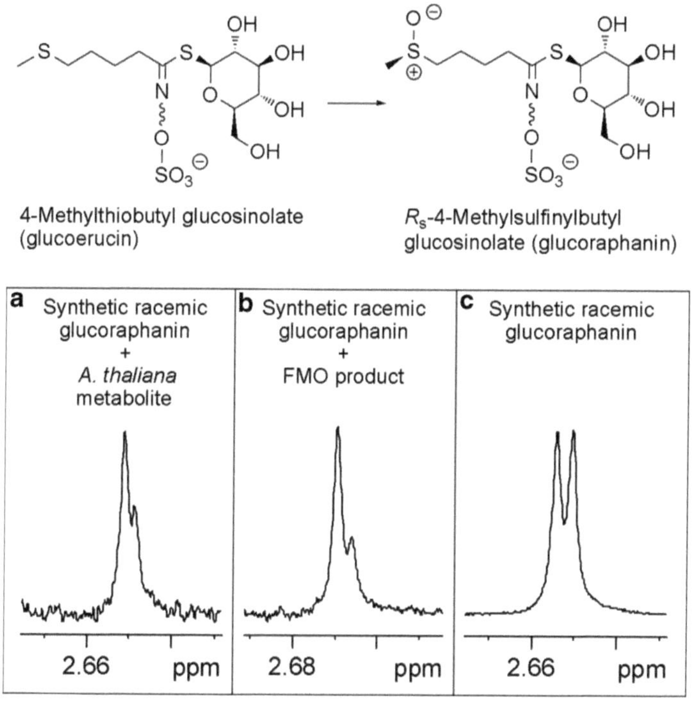

Fig. 9 Conversion of 4-methylthiobutyl glucosinolate (glucoerucin) to *R*s-4-methylsulfinylbutyl glucosinolate (glucoraphanin). Partial ^{1}H NMR spectra (400 MHz, MeCN-d_3, 50°C) of the product, glucoraphanin, displaying the signal of the *S*-methyl group are shown. Equimolar mixtures of (**a**) glucoraphanin isolated from *A. thaliana* and (**b**) glucoraphanin produced by *A. thaliana* flavin monooxygenase (FMO) with (**c**) synthetic racemic glucoraphanin were measured. Enhancement of the downfield signal in both cases indicated *R*s absolute configuration of glucoraphanin (Vergara et al. 2008). Slightly different chemical shifts are due to differences in the relative water contents of the samples

of 4-methylthiobutyl glucosinolate (glucoerucin) to produce 4-methylsulfinylbutyl glucosinolate (glucoraphanin, Fig. 9) (Hansen et al. 2007). The stereochemistry of the sulfoxide moiety of glucoraphanin was not determined by these authors. However, as determined for glucosinolate-derived isothiocyanates by optical rotary dispersion (Cheung et al. 1965), R_S absolute configuration could be suspected for glucoraphanin and other 4-methylsulfinylalkyl glucosinolates as well. This was confirmed by ^{1}H NMR spectroscopic analysis (400 MHz, MeCN-d_3, 50°C) for glucoraphanin produced by the recombinant FMO protein and glucoraphanin isolated from *Arabidopsis thaliana* (Vergara et al. 2008). Both spectra of equimolar mixtures of synthetic R_SS_S-glucoraphanin and glucoraphanin from *A. thaliana* and synthetic R_SS_S-glucoraphanin and glucoraphanin produced by $FMO_{GS\text{-}OX1}$ showed enhancement of the downfield signal while the upfield signal remained unaffected (Fig. 9). Therefore, both samples are enantiomerically pure R_S-glucoraphanin.

4 Methyltransferase Products

O-Methyltransferases (OMT's) play a crucial role in plant secondary metabolism (Roje 2006). An OMT from rice (*Oryza sativa*), ROMT-9, was cloned and characterized using RT-PCR. After heterologous expression in *E. coli*, the recombinant protein was purified and the product subjected to NMR analysis (400 MHz, DMSO-d_6) (Kim et al. 2006). ROMT-9 accepted quercetin as a substrate. The ^{1}H and ^{13}C NMR spectra of the reaction product gave new signals with chemical shifts of δ_H 3.83 and δ_C 55.9, respectively, which are characteristic of phenolic *O*-methyl groups and were not observed in the substrate, quercetin. These data indicate that a methoxy derivative of quercetin was formed by the action of ROMT-9. However, although all of the ^{1}H and ^{13}C chemical shifts of the product could be readily assigned by comparing the 1D spectra with those of the ^{1}H and ^{13}C spectra of the substrate, these data do not identify the methylation position. Therefore, 2D NMR, especially nuclear Overhauser effect spectroscopy (NOESY) and HMBC spectra, were recorded and cross peaks used to verify the position of the *O*-methyl group and the regioselectivity of ROMT-9. The cross peak between H-2′ (δ 7.75) and the *O*-methyl proton (δ 3.83) in the NOESY spectrum indicated close spatial proximity between these two groups and thereby identified the hydroxyl at C-3′ as the methylated position and the formed product as isorhamnetin. A ^{1}H,^{13}C long-range (HMBC) correlation between the *O*-methyl signal and C-3′ (δ 147.5) confirmed this result. The data show that ROMT-9 transfers a methyl group regioselectively to the 3′-OH of quercetin (Fig. 10) and other flavonoids, and characterizes the protein as a flavonoid 3′-*O*-methyltransferase (Kim et al. 2006).

A "small molecule" (SM)-OMT, accepting coniferyl alcohol and related phenolic alcohols as substrates, was cloned from suspension cultures of *Linum* species and expressed in *E. coli* (Berim et al. 2007). The occurrence of a phenolic and an alcoholic (allylic) hydroxyl group in the best substrate, coniferyl alcohol, required the acceptor hydroxyl group to be determined and thus the product structure to be elucidated. Because of its unique ability to distinguish isomers, NMR spectroscopy was used to find out whether the recombinant OMT produces the 3-*O*-methylconiferyl alcohol or the coniferyl alcohol 9-methyl ether (γ-methoxy-isoeugenol)

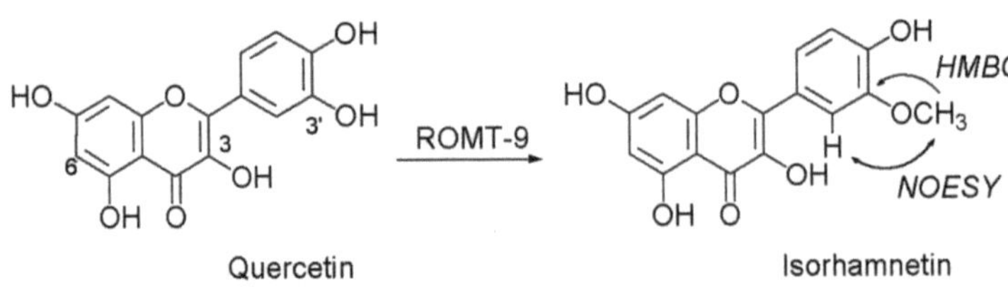

Fig. 10 3′-*O*-Methylation of quercetin, catalyzed by recombinant ROMT9 from rice (Kim et al. 2006), to form isorhamnetin. Nuclear Overhauser correlations (NOESY) and long-range ^{1}H–^{13}C heterocorrelations (HMBC) indicate the position of the methoxy group and the regiospecificity of the enzyme

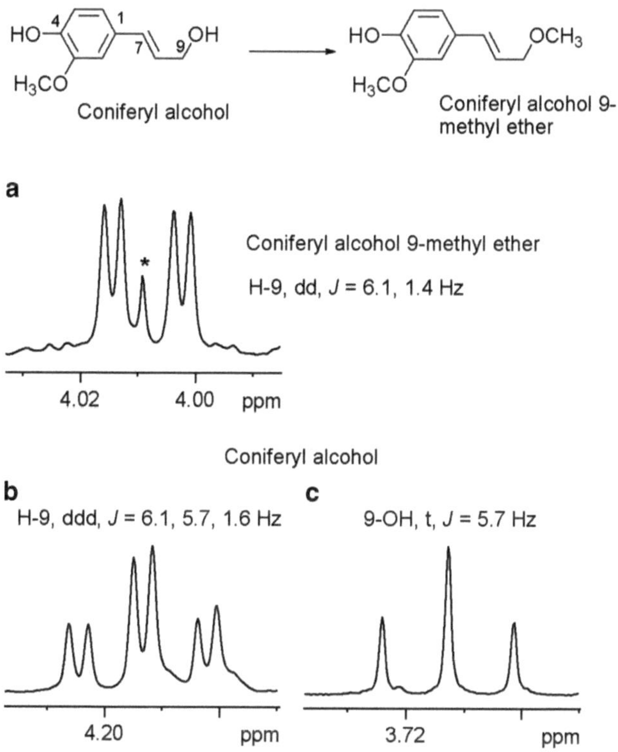

Fig. 11 OMT-catalyzed regiospecific *O*-methylation of coniferyl alcohol using a recombinant protein from *Linum* (Berim et al. 2007) and partial ^{1}H NMR spectra of the substrate (**b**, **c**) and the product, coniferyl alcohol 9-methyl ether (**a**). The *asterisk* in spectrum (**a**) designates an impurity signal. Because of the *O*-methylation occurring in the side chain, the aliphatic signal of the alcoholic OH proton is missing and the signal of H-9 is collapsed to a doublet of doublets

(Fig. 11). The product was first analyzed by ^{1}H NMR (500 MHz, cryoprobe, water-free acetone-d_6) and the spectrum compared to that of the substrate, coniferyl alcohol. The occurrence of a new *O*-methyl signal (δ 3.29), in addition to the *O*-methyl signals at δ 3.87, confirmed the dimethoxy structure of the product. The proton chemical shift of this new methyl singlet suggested methylation at the side chain hydroxyl rather than of the phenolic hydroxyl. Because dry acetone-d_6 was used as a solvent, the exchangeable protons of the phenolic and alcoholic hydroxyl groups became visible. In the spectrum of coniferyl alcohol, the singlet of 4-OH appeared at δ 7.56 and the triplet-signal of the allylic 9-OH at δ 3.71 ($^{3}J_{\text{OH-9–H-9}}$ = 5.7 Hz, Fig. 11, spectrum c). The latter signal is missing in the spectrum of the product and the methylene signal of H-9 (δ 4.01) is collapsed to a doublet of doublets, displaying couplings with H-8 ($^{2}J_{\text{H-9–H-8}}$ = 6.1 Hz) and H-7 ($^{3}J_{\text{H-9–H-7}}$ = 1.4 Hz) (Fig. 11, spectrum a) but did not show coupling with 9-OH, again indicating methylation of the side chain-hydroxyl group. HMBC experiments were used to further confirm the position of the *O*-methylation. The above-mentioned new methoxyl signal at δ 3.29 showed a HMBC correlation with C-9 (δ 73.8), unambiguously indicating attachment to the alcoholic hydroxyl group. Furthermore, the HMBC correlations of 4-OH (δ 7.61) with C-3 (δ 148.6), C-4 (δ 147.8), and C-5 (δ 115.9) indicated that the aromatic hydroxyl group remained unsubstituted. The reaction product therefore was identified as coniferyl alcohol 9-methyl ether.

5 Glycosyltransferase Products

The transfer of glycosyl units to secondary metabolites often occurs as the final biosynthetic step and results in hydrophilic conjugates, which are deposited in the vacuole, awaiting hydrolysis and release of the aglycon on demand for ecological or physiological purposes. Glycosyl transfer is also involved in the formation of di-, oligo-, and polysaccharides with various functions in cell constitution and metabolism. The identification of glycosyltransferase products belonging to three different classes of natural products, namely lignane glucosides, chalcone glucosides, and cell wall carbohydrates, will be discussed in this section.

Considerable levels of aryltetralin lignans such as podophyllotoxin and its congeners accumulate in the vacuoles of cultured *Linum* cells. Incubating a crude protein extract from suspension-cultured cells of *Linum nodiflorum* with podophyllotoxin (and other lignanes) as substrates resulted in a UDP-glucose-dependent formation of hydrophilic substances (Berim et al. 2008). The products were extracted with ethyl acetate and the residue of the organic phases subjected to LC–SPE–NMR (500 MHz, cryoprobe equipped with a Cryofit™ flow insert; MeCN-d_3). The identity of the major product was confirmed by comparing its ^{1}H NMR spectrum with that of the substrate, podophyllotoxin (Fig. 12). The ^{1}H NMR spectrum of the enzyme product displayed additional signals assignable to a glucose moiety, marked by an asterisk in Fig. 12). The doublet of H-1″ (δ 4.38, $J = 6.7$ Hz) indicated β-configuration of the anomeric center of the glucose. Downfield-shifted signals of H-6, H-7, and H-8 established the position of the β-glucose at the aglycon at C-7. This result was further confirmed by a change of the multiplicity of the signal of H-7, which appears as a doublet of doublets in the spectrum of podophyllotoxin and, as a result of the missing OH protons, is collapsed to a doublet in the spectrum of the glucoside. In addition, the doublet of OH-7 (δ 3.84) of the aglycone disappears in the spectrum of the glucoside for the same reason. Hence, the product of incubating the protein preparation from *L. nodiflorum* cell cultures with podophyllotoxin was identified as podophyllotoxin-7-*O*-β-glucoside (Berim et al. 2008).

Glycosylation of the hydroxyl group in position 2′ of phloretin, a characteristic dihydrochalcone in apple (*Malus* x *domestica*), is the final step in the biosynthesis of phloridzin. Several glycosyltransferases from apple and pear (*Pyrus communis*) were cloned and functionally expressed in yeast (Gosch et al. 2010). The recombinant glycosyltransferases were assayed regarding their substrate- and regiospecificity. The products were pooled from 50 assays, separated by HPLC, and the major

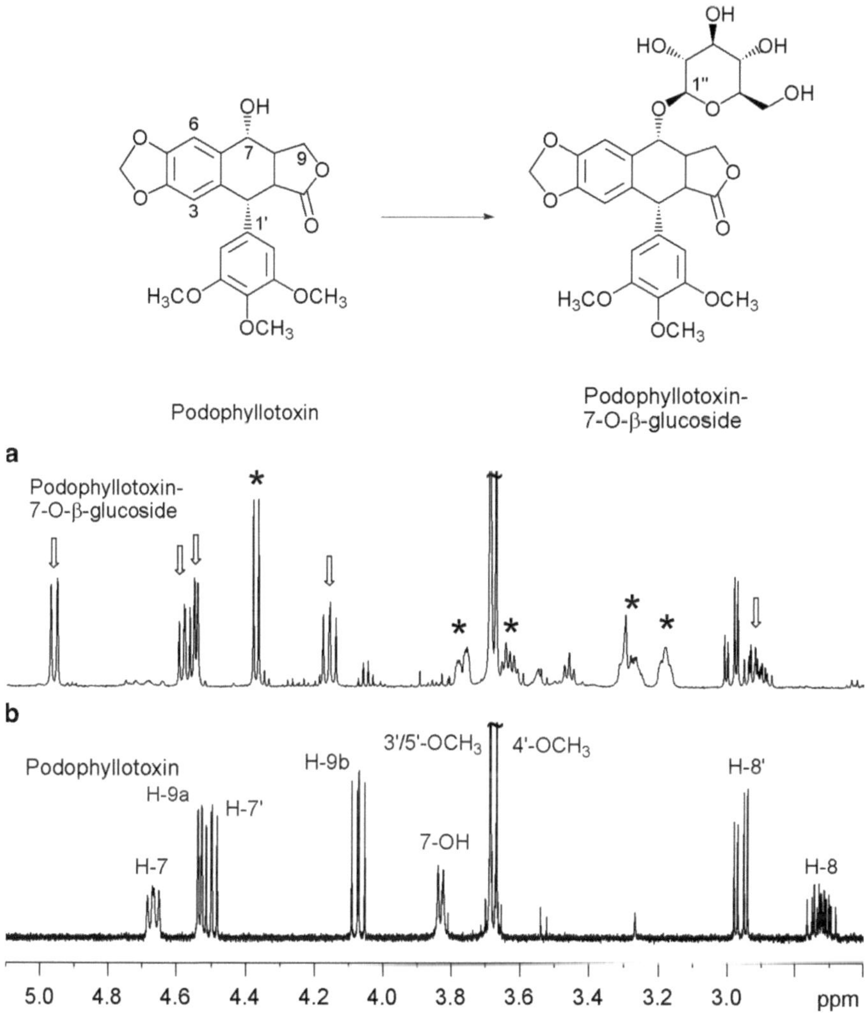

Fig. 12 7-*O*-Glycosidation of podophyllotoxin (Berim et al. 2008) and partial ^{1}H NMR spectra of the substrate (**b**) and the product, podophyllotoxin-7-*O*-β-glucoside (**a**). *Asterisks* denote signals of the β-glucosyl moiety and *open arrows* denote signal of H-7′, H-8, and H-9 which, because of the deshielding effect of the 7-*O*-β-glucosyl moiety, appear at a significantly lower field in comparison with the spectrum of the substrate (**b**). The signal of H-6 (not shown) is also shifted to a lower field. In addition, in the spectrum of the glucoside (**a**), the signal of 7-OH is missing and that of H-7 appears as a doublet at a low field

product, phloridzin, was identified by ^{1}H NMR, ^{1}H,^{1}H COSY, HMBC, and HMQC experiments using a cryoprobe at 500 MHz. Because of the relatively low substrate specificity of the recombinant proteins from *M.* x *domestica* reported by Gosch et al. 2010, phloretin-4-*O*-β-glucoside and phloretin-4′-*O*-β-glucoside (trilobatin) were formed as minor products, in addition to the major 2′-*O*-β-glucoside. The

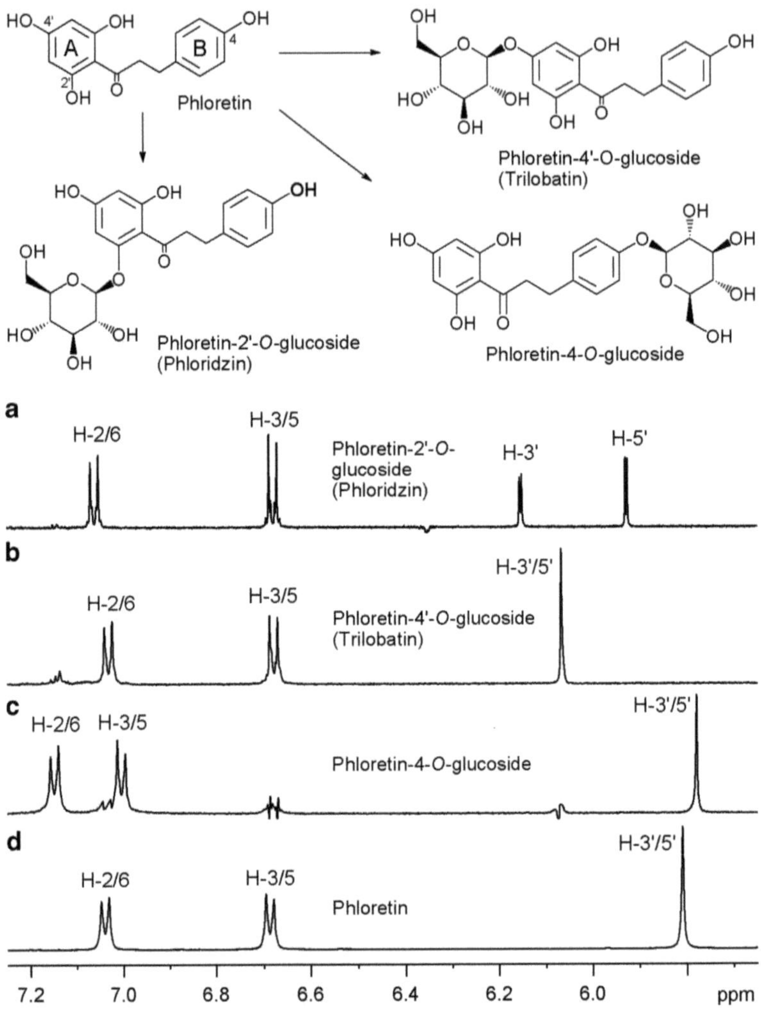

Fig. 13 Glucosidation of the dihydrochalcone phloretin by recombinant proteins from *Malus* x *domestica* and *Pyrus communis* (Gosch et al. 2010). The major product, phloridzin, and the minor products, phloretin-4-*O*-β-glucoside and phloretin-4′-*O*-β-glucoside, are distinguishable by their one-dimensional ^{1}H NMR spectra. Partial ^{1}H NMR spectra (500 MHz, MeOH-d_4) of phloretin-2′-*O*-β-glucoside (phloridzin) (**a**), phloretin-4′-*O*-β-glucoside (trilobatin) (**b**), phloretin-4-*O*-β-glucoside (**c**), and phloretin (substrate) (**d**) are shown

three products are readily distinguishable just from the chemical shifts and multiplicities of the ^{1}H NMR signals of the aromatic protons. The signals of H-3′ and H-5′ (Fig. 13, spectrum a) of phloretin-2′-*O*-β-glucoside appear as two separated doublets ($J = 2.3$ Hz), indicating asymmetric substitution of ring A, which is consistent only with a glucose attached to the 2′-hydroxyl group. Downfield shifts of the signals of H-3′ and H-5′ of phloretin-2′-*O*-β-glucoside in comparison to the signals of H-3′/5′ of phloretin (Fig. 13, spectrum d) support this finding. A similar downfield shift of H-3′/5′ but lacking a symmetry break (the signal of H-3′/5′ still appears as a singlet) in the spectrum of trilobatin confirms a glucose at the 4′-hydroxyl group (Fig. 13, spectrum d). An almost unchanged signal of H-3′/5′ and a downfield shift of B-ring signals (H-2/6 and H-3/5) in spectrum c (Fig. 13) was consistent with a glucose attached to 4-OH, i.e., the structure of phloretin-4-*O*-β-glucoside. 2D NMR spectra were necessary only for chemical shift assignment and to further confirm the results of 1D ^{1}H NMR.

Glycosyltransferases involved in the formation of a complex polysaccharide, pectic rhamnogalacturonan-II, which is essential to vascular plants, were reported by Egelund et al. (2006, 2008). Two rhamnogalacturonan-II xylosyltransferases, RGXT1 and RGXT2, from *A. thaliana* were expressed in baculovirus-transfected insect cells. Although for NMR (400 MHz, D_2O) purposes, an upscaled version of the assay with the recombinant protein was used, the available amount of material was too low to allow 1D ^{13}C or DEPT spectra to be recorded. Therefore, chemical shifts were extracted from the 2D spectra (dqCOSY, JRES, HSQC, and HMBC). For example, to achieve an appropriate HMBC spectrum, samples were measured for 90 h. Although the isolated sample was contaminated with an unknown impurity, unambiguous identification of the product was achieved. The NMR data revealed the regio- and stereochemistry of the methyl xylosyl-fucoside to be an α-(1,3)-linkage, indicating that RGXT1 and RGXT2 encode Golgi-localized (1,3)-α-D-xylosyltransferases involved in the biosynthesis of pectic rhamnogalacturonan-II (Egelund et al. 2008). Another rhamnogalacturonan-II xylosyltransferase (RGXT3) from *A. thaliana* was expressed in *Pichia pastoris* (Egelund et al. 2008). The recombinant protein was assayed and NMR analysis was conducted under similar conditions as those reported for RGXT1 and RGXT2. The data reported by Egelund et al. (2006, 2008) indicated that a small gene-family is responsible for the xylosylation of the internal fucose moiety of the RG II A-chain in *A. thaliana.*

6 Acyltransferase Products

Flavonol 3-*O*-β-glucosides esterified with ferulic or *p*-coumaric acid at the glucose moiety are the major UV-B screening pigments of the epidermal layer of Scots pine (*Pinus sylvestris*) and Norway spruce (*Picea abies*) needles. The final steps in the biosynthesis of these compounds are catalyzed by enzymes that transfer the acyl part of hydroxycinnamoyl-CoA esters to flavonol 3-*O*-β-glucosides (Kaffarnik et al. 2005). The hydroxycinnamoyltransferase activities of Scots pine needles were separated by anion exchange chromatography. Appropriate protein fractions were incubated with kaempferol 3-*O*-β-glucoside and *p*-coumaroyl-CoA as substrates. The products were extracted with ethyl acetate. The organic phases were pooled from 90 upgraded assays, dried under a stream of N_2 and separated by HPLC. Structures of the enzyme products were determined by ^{1}H NMR and an absolute value 2Q-^{1}H,^{1}H COSY spectrum (500 MHz) in MeCN-d_3. Differences in

^{1}H NMR chemical shifts were used to establish the acylated positions of the glucose moiety. 4″-*p*-Coumaroyl kaempferol 3-*O*-β-glucoside showed a chemical shift for H-4″ of δ 4.79 compared to δ 3.23 of the nonacylated kaempferol 3-*O*-β-glucoside, indicating that the acyl group was at position 4 of the glucose. 3″-*p*-Coumaroyl kaempferol 3-*O*-β-glucoside showed a chemical shift for H-3″ of δ 5.02 compared to δ 3.34 of kaempferol 3-*O*-β-glucoside, and was therefore acylated at position 3 of the glucose molecule. The NMR data together with the results of cochromatography thus proved the existence of position-specific 3″- and 4″-hydroxycinnamoyltransferases. In addition, a 6″-hydroxycinnamoyltransferase was found. These enzymes converted nonacylated flavonol 3-*O*-β-glucosides to the monoacylated 3″-*p*-coumaroyl flavonol 3-*O*-β-glucoside (a major product) and 4″-*p*-coumaroyl flavonol 3-*O*-β-glucoside (a minor product), respectively (Fig. 14). As substantiated by HPLC and diode array spectra using authentic reference compounds, both enzymes also convert the 6″-monoacylated flavonol 3-*O*-β-glucoside to the respective diacylated 3″,6″- and 4″,6″-di-*p*-coumaroyl kaempferol 3-*O*-β-glucosides (not shown in Fig. 14). Enzyme kinetic parameters indicated that acylation takes place in a well-defined order, beginning at position 6″ followed by acylation at position 3″ (Kaffarnik et al. 2005).

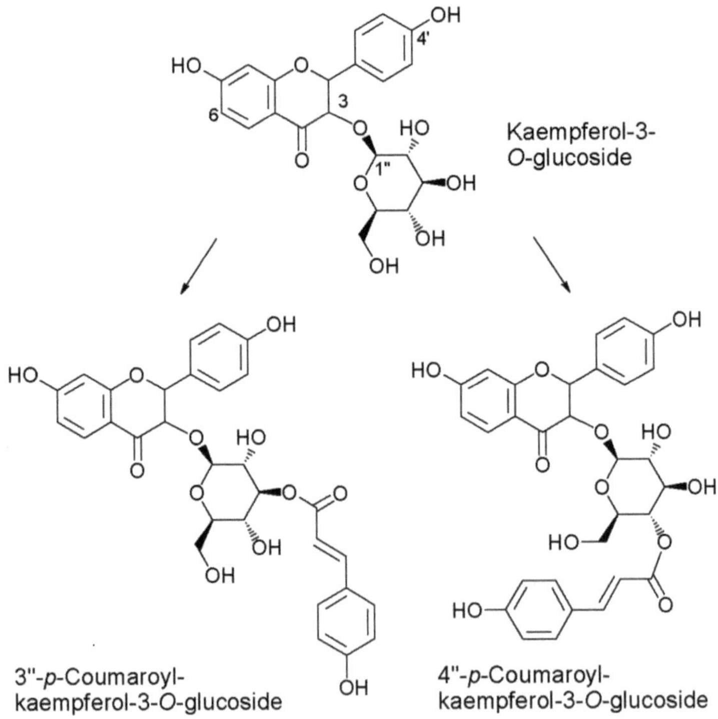

Fig. 14 Conversion of kaempferol-3-*O*-β-glucoside to 3″- and 4″-*p*-coumaroyl kaempferol 3-*O*-β-glucosides by position-specific hydroxycinnamoyltransferases from Scots pine needles. The shown products were identified by NMR spectroscopic methods (Kaffarnik et al. 2005)

Taxol biosynthesis has long been of special interest because of its function in tubulin polymerization and stabilization and the use of this polysubstituted diterpene as an anticancer drug (Kingston 2001). Benzoylation of the 3′-amino group in the C-13-side chain of 3′-debenzoyltaxol is the final step in the taxol biosynthetic pathway (Floss and Mocek 1995). In order to identify the responsible enzyme, a number of acyltransferases, including 3′-*N*-debenzoyl-2′-deoxytaxol *N*-benzoyltransferase (TAX 10), were isolated from jasmonate-induced *Taxus brevifolia* cells and overexpressed in *E. coli* (Walker et al. 2002). Recombinant TAX10 was partially purified by anion exchange and ceramic hydroxyapatite chromatography. The protein preparation was then incubated with the surrogate substrate *N*-debenzoyl-(3′*RS*)-2′-deoxytaxol and benzoyl-CoA as an acyl donor. In a preparative-scale incubation, approximately 1 mg of a TLC-purified product was obtained and analyzed by comparing the ^{1}H NMR spectrum (300 MHz, $CDCl_3$; Fig. 15) with that

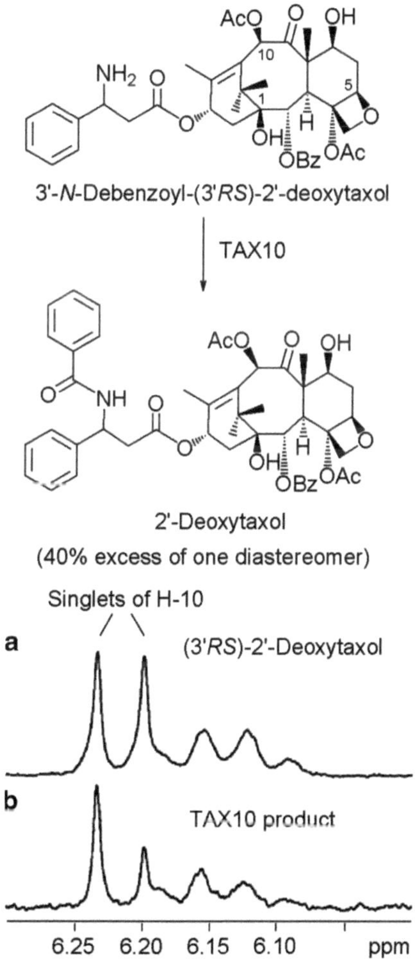

Fig. 15 3′-*N*-Debenzoyl-2′-deoxytaxol *N*-benzoyltransferase (TAX10)-catalyzed conversion of *N*-debenzoyl-2′-deoxytaxol to 2′-deoxytaxol and partial ^{1}H NMR spectra of 2′-deoxytaxol (300 MHz, $CDCl_3$) (Walker et al. 2002). The singlets of H-10 at δ 6.196 and 6.232 indicate a diastereoisomeric ratio of 1:1 of authentic (3′*RS*)-2′-deoxytaxol (spectrum a). The biosynthetic product obtained from incubating recombinant TAX10 with *N*-debenzoyl-(3′*RS*)-2′-deoxytaxol and benzoyl-CoA as cosubstrates shows a 40% excess of one diastereomer (spectrum b). Courtesy of Walker et al. Adapted from Walker et al. 2002 with their kind permission, copyright (2002) National Academy of Sciences, USA

of authentic 2′-deoxytaxol. The H-10 singlets at δ 6.196 and 6.232, respectively, were used as diagnostic signals to resolve diastereomers. The standard, 2′-deoxytaxol showed the expected 1:1 diasteromeric ratio of these signals (Fig. 15), while the product of TAX10 displayed one diastereomer in 40% excess. Hence, TAX10 catalyzes the stereoselective coupling of *N*-debenzoyl-(3′*RS*)-2′-deoxytaxol with benzoyl-CoA to form predominantly one 3′-epimer of 2′-deoxytaxol (Walker et al. 2002).

7 Terpene Synthase Products

Two genes isolated from maize encode very similar sesquiterpene synthases, TPS 6 and TPS 11 (Köllner et al. 2008). The proteins were overexpressed in *E. coli*. Incubation with (*E*,*E*)-farnesyldiphosphate (FPP) and GC-MS analysis showed that both enzymes produced an uncommon sesquiterpene hydrocarbon as the major product along with minor amounts of β-bisabolene and (*E*)-farnesene. For NMR analysis of the major product, *E. coli* extracts containing the recombinant protein were incubated with (*E*,*E*)-farnesyldiphosphate in an upscaled assay mixture, which was overlaid with *n*-pentane to avoid a loss of volatile products by evaporation. The assay was then extracted three times with *n*-pentane, and the pooled organic phases passed through a silica column to remove traces of farnesol and other oxygenated extraction products. Approximately 1 mg of the reaction product was obtained for ^{1}H NMR, ^{1}H–^{1}H COSY, HMQC and HMBC analysis (500 MHz; benzene-d_6) using a cryogenically cooled 5 mm-probe. Assignment of the structure started from ^{1}H NMR signals characteristic of two geminal methyl groups (δ 0.915 and 0.919), a broad signal (δ 1.65) of a methyl group attached to a double bond and broad singlets at δ 5.46 and 5.42 (Fig. 16a and c) attributable to two olefinic protons. Homo- and heterocorrelation 2D NMR experiments were used to assign the structure as 4′,5,5-trimethyl-1,1′-bi(cyclohexane)-1,3′-diene, an unusual, bicyclic sesquiterpene reported as (*S*)-β-macrocarpene (Cool 2005). The identity of the product of TPS 6 and TPS11 from maize was finally provided by comparing its ^{1}H NMR and HSQC spectra with those of authentic β-macrocarpene (Fig. 16). Gas chromatographic analysis on a chiral column revealed that the (*S*)-enantiomer was formed by TPS6. The mechanism of (*S*)-β-macrocarpene formation via a neutral bisabolene intermediate has been discussed (Köllner et al. 2008).

NMR spectroscopy has been extensively used to study mechanistic aspects of the complex reaction cascade involved in geranylgeranyldiphosphate cyclization to afford taxadiene (Williams et al. 2000; Jin et al. 2005a; Jin et al. 2005b). In part, these studies were reviewed in the context of isotopic labeling in biosynthetic studies, especially deuterium (Schneider 2007), and thus, although recombinant enzyme preparations were used, they shall not be discussed here in detail.

In addition to deuterated precursors, fluorinated analogs of native enzyme substrates are useful in mechanistic studies. [6-^{19}F]GGPP was used by Jin et al. 2005b to block the proton transfer from C-11 to C-7 that normally occurs in the course of

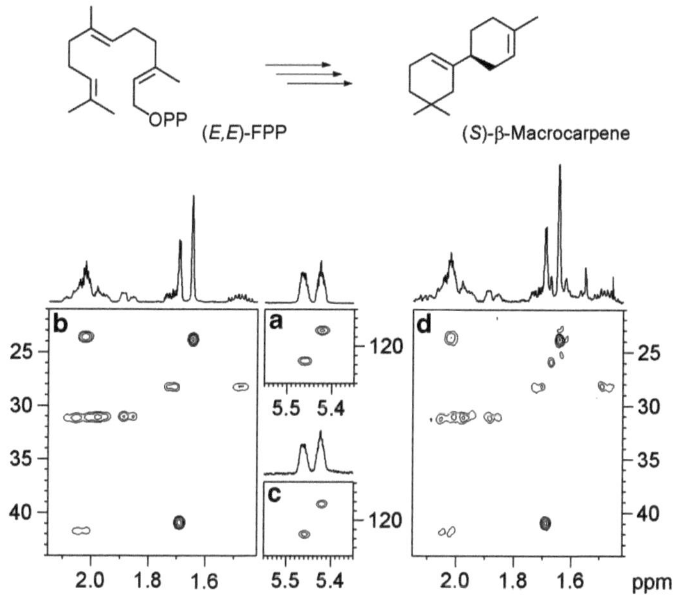

Fig. 16 Formation of (*S*)-β-macrocarpene from (*E*,*E*)-farnesyldiphosphate catalyzed by terpene synthases TPS6 and TPS11 (Köllner et al. 2008) and partial ^{1}H NMR and HSQC spectra of (*S*)-β-macrocarpene from maize (**c**, **d**) and an authentic standard (**a**, **b**) (Cool 2005)

taxadiene synthase reaction; to terminate the multistep cascade at the verticillene step to identify the resulting product; and to infer the enzymatic reaction mechanism. Indeed, three major and two minor products were observed upon the incubation of recombinant taxadiene synthase with [6-^{19}F]GGPP. The three major products were separated or enriched, and, on the basis of extensive NMR analysis and optical data, their structures were elucidated as *exo*-7-fluoroverticillene, *endo*-7-fluoroverticillene, and 7-fluoroverticilla-4(20),7(8),11(12)-triene (Fig. 17). NMR coupling and NOE data, especially of *exo*-7-fluoroverticillene, provided evidence for a verticillen-12-yl carbocation intermediate with 11*R* configuration (Jin et al. 2005b).

Another example of how the stereochemical mechanism of terpene cyclization has been established by NMR spectroscopic analysis is cyclic ether formation in 1,8-cineol biosynthesis (Wise et al. 2002). Specifically, pentadeuterated substrates, (2*E*,6*E*)-[1,1,8,8,8-$^{2}H_5$]- and (2*E*,6Z)-[1,1,9,9,9-$^{2}H_5$]geranyldiphosphate and unlabeled GDP, have been incubated in parallel with a recombinant 1,8-cineol synthase from *Salvia officinalis* on a large scale in order to get sufficient product for NMR spectroscopic analysis. ^{1}H NMR and, using special pulse sequences, 1D NOE spectra of chromatographically purified cineol-d_0 and cineol-d_5 samples were measured at 500 MHz ($CDCl_3$). In addition, upon incubating the recombinant cyclase with GDP in deuterated buffer, incorporation of a solvent-derived deuterium in 6-*exo*-position of 1,8-cineol-d_1 was demonstrated by ^{2}H NMR spectroscopy. The data showed that the *syn* product was formed in the 1,8-cineol synthase-catalyzed bicyclization reaction.

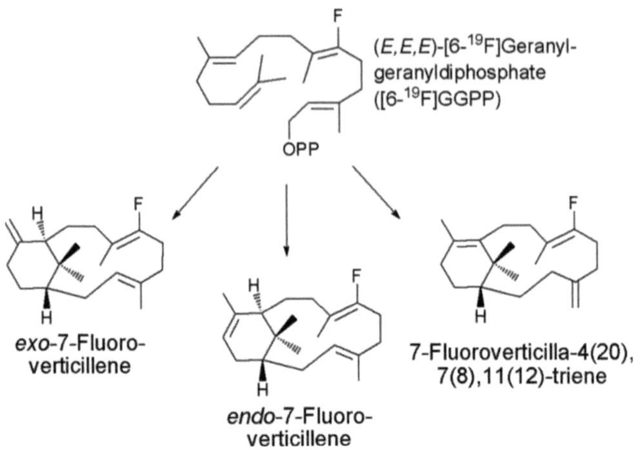

Fig. 17 Conversion of [6-^{19}F]geranylgeranyldiphosphate to *exo*-7-fluoroverticillene, *endo*-7-fluoroverticillene, and 7-fluoroverticilla-4(20),7(8),11(12)-triene as the major products obtained upon incubation with recombinant taxadiene synthase (Jin et al. 2005b)

8 Conclusions

Since NMR spectroscopy is still considered a relatively insensitive analytical method, and indeed is much less sensitive than MS, a standard enzyme assay is rarely able to produce sufficient material for structure identification by NMR spectroscopic methods; upscaled or multiple parallel assays are required. However, because of the introduction of cryogenic probes and microprobes, considerable progress has been made in NMR sensitivity. As demonstrated by the examples of this review, an NMR spectrometer is no longer an unusual tool for use in enzyme assays. Extraction of the aqueous assay mixture with ethyl acetate is the standard procedure for separating the medium-polar small molecules from the proteins, cofactors, and buffer substances, and then measuring them in deuterated organic solvent. Highly hydrophilic products can be measured in deuterated water (Egelund et al. 2008) and volatile samples (e.g., the β-macrocarpene) require special sample preparation (Köllner et al. 2008).

Although the purity of the analyte sample is an important issue, contaminants may be tolerated if NMR signals do not overlap much if at all with those of the enzyme product of interest (Figs. 8b, 11a, and 16d). Mixtures containing multiple samples or contaminants should be separated prior to NMR analysis or subjected to LC–NMR or LC–SPE–NMR coupling (Hemmati et al. 2007; Berim et al. 2008; Larbat et al. 2009).

Most of the enzymatic reactions discussed in this review modify the structure of the substrate in a predictable manner. Hence, the structures of enzyme products are often closely related to those of the substrates, and interpretation of changes of the chemical shift and coupling constants according to the established rules is sufficient for structure identification. Such examples are hydroxylation of diphylline to justicidine B (Hemmati et al. 2007), regiospecific *O*-methylation of coniferyl alcohol (Berim et al. 2007), and the conversion of to 3′-*N*-debenzoyl-2′-deoxytaxol to 2′-deoxytaxol (Walker et al. 2002). In other cases, such as terpene synthese-catalyzed formation of macrocarpene (Köllner et al. 2008), multiple cyclization possibilities exist and either an authentic reference or even de novo structure elucidation by 2D NMR methods may be necessary. In conclusion, the examples reviewed here demonstrate that NMR spectroscopy is an established analytical method for identifying and elucidating structures of enzyme products, for unraveling stereochemical aspects of enzymatic reactions, and for providing information about the use of isotopically labeled substrates by isolated or recombinant proteins.

Acknowledgments The author thanks Wilhelm Alfermann, Frédéric Bourgaud, Jörg Degenhardt, Elisabeth Fuss, Jonathan Gershenzon, Ulrich Matern, Maike Petersen, Joachim Schröder, Karl Stich, and their collaborators for agreeing to present partly unpublished results, the Max Planck Institute for Chemical Ecology for financial support, and Emily Wheeler for editorial assistance.

References

Berim A, Schneider B, Petersen M (2007) Methyl allyl ether formation in plants: novel *S*-adenosyl L-methionine:coniferyl alcohol 9-*O*-methyltransferase from suspension cultures of three *Linum* species. Plant Mol Biol 64:279–291

Berim A, Ebel R, Schneider B, Petersen M (2008) UDP-glucose:(6-methoxy)podophyllotoxin 7-O-glucosyltransferase from suspension cultures of *Linum nodiflorum*. Phytochemistry 69:374–381

Brand S, Hölscher D, Schierhorn A, Svatoš A, Schröder J, Schneider B (2006) A type III polyketide synthase from *Wachendorfia thyrsiflora* and its role in diarylheptanoid and phenylphenalenone biosynthesis. Planta 224:413–428

Cheung K, Kjær A, Sim G (1965) The absolute configuration of sulphoxide mustard oils. Chem Commun 6:100–102

Cho MH, Moinuddin SGA, Helms GL, Hishiyama S, Eichinger D, Davin LB, Lewis NG (2003) +)-Larreatricin hydroxylase, an enantio-specific polyphenol oxidase from the creosote bush (*Larrea tridentata*. Proc Natl Acad Sci USA 100:10641–10646

Cool LG (2005) Sesquiterpenes from *Cupressus macrocarpa* foliage. Phytochemistry 66: 249–260

Eckermann C, Schröder G, Eckermann S, Strack D, Schmidt J, Schneider B, Schröder J (2003) Stilbenecarboxylate biosynthesis: a new function in the family of chalcone synthase related proteins. Phytochemistry 62:271–286

Egelund J, Petersen BL, Motawia MS, Damager I, Faik A, Olsen CE, Ishii T, Clausen H, Ulvskov P, Geshi N (2006) *Arabidopsis thaliana RGXT1* and *RGXT2* encode golgi-localized (1,3)-α-D-xylosyltransferases involved in the synthesis of pectic rhamnogalacturonan-II. Plant Cell 18:2593–2607

Egelund J, Damager I, Faber K, Olsen CE, Ulvskov P, Petersen BL (2008) Functional characterisation of a putative rhamnogalacturonan II specific xylosyltransferase. FEBS Lett 582: 3217–3222

Floss HG, Mocek U (1995) Biosynthesis of taxol. In: Suffness M (ed) Taxol – science and applications. CRC, Boca Raton, FL, USA, pp 191–208

Gorham J (1980) The stilbenoids. Prog Phytochem 6:203–252

Gosch C, Halbwirth H, Schneider B, Hölscher D, Stich K (2010) Cloning and heterologous expression of glycosyltransferases from *Malus* x *domestica* and *Pyrus communis*, which convert phloretin to phloretin 2′-O-glucoside (phloridzin). Plant Sci 178:299–306

Hansen B, Kliebenstein D, Halkier B (2007) Identification of a flavinmonooxygenase as the *S*-oxygenating enzyme in aliphatic glucosinolate biosynthesis in *Arabidopsis*. Plant J 50:902–910

Hemmati S, Schneider B, Schmidt T, Federolf K, Alfermann AW, Fuss E (2007) Justicidin B 7-hydroxylase, a cytochrome P450 monooxygenase from cell cultures of *Linum perenne* Himmelszelt involved in the biosynthesis of diphyllin. Phytochemistry 68:2736–2743

Jin Q, Williams DC, Hezari M, Croteau R, Coates RM (2005a) Stereochemistry of the macrocyclization and elimination steps in taxadiene biosynthesis through deuterium labeling. J Org Chem 70:4667–4675

Jin Y, Williams DC, Croteau R, Coates RM (2005b) Taxadiene synthase-catalyzed cyclization of 6-fluorogeranylgeranyl diphosphate to 7-fluoroverticillenes. J Am Chem Soc 127:7834–7842

Kaffarnik F, Heller W, Hertkorn N, Sandermann H (2005) Flavonol 3-*O*-glycoside hydroxycinnamoyltransferases from Scots pine (*Pinus sylvestris* L.). FEBS J 272:1415–1424

Kim BG, Lee Y, Hur HG, Lim Y, Ahn JH (2006) Flavonoid 3′-*O*-methyltransferase from rice: cDNA cloning, characterization and functional expression. Phytochemistry 67:387–394

Kingston DGI (2001) Taxol, a molecule for all seasons. Chem Commun 10:867–880

Köllner TG, Schnee C, Li SH, Svatoš A, Schneider B, Gershenzon J, Degenhardt J (2008) Protonation of a neutral (*S*)-beta-bisabolene intermediate is involved in (*S*)-beta-macrocarpene formation by the maize sesquiterpene synthases TPS6 and TPS11. J Biol Chem 283: 20779–20788

Kovacs H, Moskau D, Spraul M (2005) Cryogenically cooled probes – a leap in NMR technology. Prog Nucl Magn Reson Spectrosc 46:131–155

Larbat R, Hehn A, Hans J, Schneider S, Jugdé H, Schneider B, Matern U, Bourgaud F (2009) Isolation and functional characterization of CYP71AJ4 encoding for the first P450 monooxygenase of angular furanocoumarin biosynthesis. J Biol Chem 284:4776–4785

Latunde-Dada AO, Cabello-Hurtado F, Czittrich N, Didierjean L, Schopfer C, Hertkorn N, Werck-Reichhart D, Ebel J (2001) Flavonoid 6-hydroxylase from soybean (*Glycine max* L.), a novel plant P-450 monooxygenase. J Biol Chem 276:1688–1695

Roje S (2006) *S*-Adenosyl-L-methionine: beyond a universal methyl group donor. Phytochemistry 67:1686–1698

Russel DJ, Hadden CE, Martin GE, Gibson AA, Zens AP, Carolan JL (2000) A comparison of inverse-detected heteronuclear NMR performance: Conventional vs cryogenic microprobe performance. J Nat Prod 63:1047–1049

Schneider B (2007) Nuclear magnetic resonance spectroscopy in biosynthetic studies. Prog Nucl Magn Reson Spectrosc 51:155–198

Schröder J (2000) The family of chalcone-synthase related proteins: functional diversity and evolution. Recent Adv Phytochem 34:55–89

Schopfer CR, Ebel J (1998) Identification of elicitor-induced cytochrome P450s of soybean (*Glycine max* L.) using differential display of mRNA. Mol Gen Genet 258:315–322

Williams DJ, Carroll BC, Jin Q, Rithner CD, Lenger SR, Floss HG, Coates RM, Williams RM, Croteau R (2000) Intramolecular proton transfer in the cyclization of geranylgeranyl diphosphate to the taxadiene precursor of taxol catalyzed by recombinant taxadiene synthase. Chem Biol 7:969–977

Walker K, Long R, Croteau R (2002) The final acylation step in Taxol biosynthesis: cloning of the taxoid C13-side-chain *N*-benzoyltransferase from *Taxus*. Proc Natl Acad Sci USA 99:9166–9171

Wise ML, Urbansky M, Helms GL, Coates RM, Croteau R (2002) Syn stereochemistry of cyclic ether formation in 1,8-cineole biosynthesis catalyzed by recombinant synthase from *Salvia officinalis*. J Am Chem Soc 124:8564–8547

Vergara F, Wenzler M, Hansen BG, Kliebenstein DJ, Halkier BA, Gershenzon J, Schneider B (2008) Determination of the absolute configuration of the glucosinolate methyl sulfoxide group reveals a stereospecific biosynthesis of the side chain. Phytochemistry 69:2737–2742

Part IV
Systematics

Phylogeny of Cyanobacteria: An Overview

Frank Kauff and Burkhard Büdel

Contents

Abstract Cyanobacteria are the oldest microorganisms performing oxygenic photosynthesis. Despite their morphological diversity, establishing a taxonomy that reflects evolutionary relationships has always been challenging. Molecular data have helped us to understand some aspects of the complex evolution of the cyanobacteria, but many of the published phylogenetic trees still lack the statistical support on internal branches necessary for reliable conclusions. Recent analyses involving whole genome data have provided us with new insights into cyanobacterial evolution, but taxon numbers are still comparatively small. Ultrastructural

F. Kauff (✉)
Molecular Phylogenetics, Department of Biology, University of Kaiserslautern, Postfach 3049, 67653 Kaiserslautern, Germany
e-mail: fkauff@biologie.uni-kl.de

B. Büdel
Plant Ecology and Systematics, Department of Biology, University of Kaiserslautern, Postfach 3049, 67653 Kaiserslautern, Germany
e-mail: buedel@rhrk.uni-kl.de

U. Lüttge et al. (eds.), *Progress in Botany 72*,
DOI 10.1007/978-3-642-13145-5_8,

characters have the potential to complement molecular data and can help to resolve phylogenetic relationships where sequence data alone are not sufficient. Despite all uncertainties involved, several well supported natural groups started to emerge, whereas other long-standing classical cyanobacterial taxa were shown to be non-monophyletic assemblages. This article summarizes the current knowledge of phylogenetic research in the cyanobacteria, focusing on the main evolutionary lineages at ordinal level and above.

1 Introduction

Cyanobacteria are prokaryotes that expose a lifestyle quite similar to that of eukaryotic algae. As organisms performing oxygenic photosynthesis, cyanobacteria and eukaryotic algae share the same habitats. Beyond that, cyanobacteria are able to live in some of the most extreme habitats on earth (e.g., Seckbach 2007 for an overview). Inferred from morphological similarities, the rare fossil record suggests an age of about 3.5 billion years for the cyanobacterial lineage (Schopf 2000). There is strong evidence that cyanobacteria are the oldest microorganisms performing oxygenic photosynthesis, resulting in a sharp rise in atmospheric oxygen about 2.45–2.32 billion years ago (Rasmussen et al. 2008).

Many cyanobacteria have large cells that are easy to observe under a light microscope. Some even form macroscopic colonies, exhibiting sizes from a few millimeters to up to 30 cm, with a fresh weight of almost 3 kg (Dodds and Castenholz 1988). Furthermore, cyanobacteria are morphologically diverse, including simple, unicellular forms, unicellular colony forming taxa, simple filamentous forms, or complex truly branched forms with up to four different cell types (e.g., Geitler 1932). Therefore, and due to their algal way of life – cyanobacteria "usually behave like algae" (Wilmotte 1994) – they were treated as plants under the botanical code of nomenclature until the late seventies of the last century (e.g., Stanier et al. 1978; Anagnostidis and Komárek 1985). However, because cyanobacteria are essentially prokaryotes, their treatment under the bacteriological code of nomenclature seems appropriate (e.g., Stanier et al. 1978; Rippka et al. 1979; Castenholz 2001).

Taxonomy and classification have always been a challenge in the cyanobacteria. For example, Geitler (1932) introduced ca. 1,500 species and 150 genera, many of them distinguished by only a single character, whereas Drouet (1981) accepted only nine genera based on ecophysiological criteria. Classification still follows mostly morphological traits at the higher taxonomic ranks, often combined with ecology in the lower taxonomic ranks (e.g., Geitler 1932; Castenholz 2001; Anagnostidis and Komárek 1985, 1988, 1990; Komárek and Anagnostidis 1986, 1989). Originally, the main difference in the cyanobacterial lineage was thought to be that of unicellular and multicellular-filamentous taxa. Geitler (1932) proposed two unicellular orders, Chroococcales and Chamaesiphonales, and one order of filamentous cyanobacteria, the Hormogonales. The latter were subdivided into 14 families according to the presence or absence of false branching, true branching, heterocytes (see Komárek and Anagnostidis 1999 for a discussion on the use of "heterocytes" instead of "heterocysts"), and akinetes (resting cells).

Despite their differences, both the botanical and the bacteriological systems of nomenclature essentially established five largely corresponding groups: Section I (Chroococcales) for unicellular taxa, Section II (Pleurocapsales) for taxa with reproduction by multiple fission and baeocytes, Section III (Oscillatoriales) for filamentous taxa without heterocytes, Section IV (Nostocales) for heterocyte-forming filamentous taxa with no branching or false branching, and Section V (Stigonematales) for heterocyte-forming filamentous taxa with false or true branching (for a more comprehensive overview of the development of classification and nomenclature of the cyanobacteria see Anagnostidis and Komárek 1985).

Using morphological characters, the next step – establishing a natural system composed of monophyletic entities – has been problematic. Although cyanobacteria are comparatively rich in features when compared to other prokaryotes, many morphological features are highly variable, and often dependent of environmental factors or culture conditions (Pearson and Kingsbury 1966). Shared morphological or ecological traits might also be misleading. For example, the close relationship between *Synechocystis* and *Crocosphaera* (Dvornyk 2006; Dyhrman and Haley 2006) might indicate that genetically related taxa do not necessarily share ecological features. Initial morphological taxonomies might thus have overestimated species numbers because in many cases species were distinguished only by features like cell size, cell shape, or other cellular structures (Wilmotte 1994). Consequently, Woese (1987) dismissed any phylogenetic hypothesis in the cyanobacteria as purely speculative because of the small amount of observable morphological and ultrastructural characters.

Classification above the level of species and genera based on cell organization (filamentous, colonial, unicellular), ability and strategies of nitrogen fixation, and modes of propagation (e.g., bacocytes) often does not reflect evolutionary relationships (e.g., Turner et al. 1999; Urbach et al. 1992; Wilmotte 1994; Fewer et al. 2002). Only apomorphic characters can potentially be useful for delimiting monophyletic groups (Giovannoni et al. 1988). The example of *Chroococcidiopsis* (traditionally placed in the Pleurocapsales, but assumed a distinct lineage of its own in 16S phylogenies (Fewer et al. 2002)) and the Pleurocapsales shows that finding such apomorphic morphological traits is not a trivial task.

Recent results from molecular phylogenies, however, indicate a more complex classification scheme, not following the simple division between uni- and multicellular forms (e.g., Wilmotte and Herdman 2001; Fewer et al. 2002; Gugger and Hoffman 2004). Conflicting results of phylogenies based only on 16S rDNA sequences suggest that phylogenies based on only a single gene are not sufficient. Litvaitis (2002) expressed this challenge of systematics and phylogeny as follows: "Thus, an ever-changing classification system and a lack of a consensus phylogeny are the imminent proofs of the unresolved evolutionary relationships among cyanobacteria."

2 Molecular Studies: Problems and Limitations

As for any group of organisms, molecular sequence data provide a unprecedented amount of information for phylogenetic analyses. The assumption that previously established taxonomic units in many cases do not reflect natural – monophyletic – groups was already confirmed by the first molecular analyses (e.g., Giovannoni et al. 1988). For example, the use of unicellular or filamentous growth as the two principal morphological features to distinguish between Chroococcales (Section I) and Oscillatoriales (Section III), was shown to be of minor value to define monophyletic entities, as taxa from both groups were scattered all over the tree, and the nonmonophyly of unicellular and filamentous groups has been supported by all subsequent molecular analyses (e.g., Honda et al. 1999; Wilmotte and Herdman 2001; Litvaitis 2002; Hoffmann et al. 2004; Sanchez-Baracaldo et al. 2005). Recent analyses (Fewer et al. 2002; Litvaitis 2002; Tomitani et al. 2006) indicate the possibility that from the initial five major groups mentioned above, only the Stigonematales (seen as the most derived group) may actually be a natural monophyletic entity.

Owing to the few readily observable features in cyanobacterial morphology, misidentification of strains is not uncommon, and some authors expect up to 50% of all strains available in culture collections to be misidentified or incorrectly assigned to taxonomic groups (Komárek and Anagnostidis 1989). Observed topological discrepancies and unexpected polytomies within species, genera, or even at higher levels, may be attributed to both analytical limitations and incorrectly named specimen.

Various genes have been used for molecular phylogenetics in the cyanobacteria, and the 16S small subunit ribosomal RNA gene (16S rDNA) is by far the most common (e.g., Neilan et al. 1997; Giovannoni et al. 1988; Honda et al. 1999; Turner et al. 1999; Iteman et al. 2000; Ishida et al. 2001; Litvaitis 2002; Gugger and Hoffman 2004). The phylogenetic resolution of 16S rDNA, however, is limited, and often not sufficient to resolve relationships neither among very closely nor very distinctly related organisms (Fox et al. 1992; Casamatta et al. 2005), resulting in unresolved trees and low bootstrap support. A possible rapid radiation of the cyanobacteria after the invention of oxygenic photosynthesis might further exacerbate the problem. Although a varying number of statistically supported groups can be observed across most published papers, the connections among these groups (the "backbone" of the tree) lack support, which is often interpreted as the result of a quick exploitation of the novel niche (Giovannoni et al. 1988; Wilmotte and Golubic 1991; Ferris et al. 1996; Nelissen et al. 1996; Ishida et al. 2001; Garcia-Pichel et al. 1998).

The ITS (internal transcribed spacer), situated between the 16S and the large subunit rDNA (23S), is more variable in length and nucleotide composition, and thus generally more suitable for differentiating relationships at species and subspecies level. Together with the high frequency of indels (insertions and deletions), ITS alignments can become difficult. An increasing amount of necessary manual adjustments can also lead to possible biases or errors in the analyses (Boyer et al. 2001; Rocap et al. 2002; Brown et al. 2005), and multiple presumably independent copies of ITS regions in cyanobacteria have been reported (Lu et al. 1997; Neilan et al. 1997; Iteman et al. 2000; Boyer et al. 2001).

Various protein coding sequences have also been used for inferring phylogenies in the cyanobacteria (e.g., *rbcL/S*: Morden and Golden (1991); Shimada et al. 1995; Watson and Tabita (1997); *psbA*: Zeidner et al. (2005); Morden and Golden (1989); Lockhart et al. 1992; Hess et al. 1995; *nifD/H/E/K/N*: Henson et al. (2004); Raymond et al. (2004); *tufA*: Delwiche et al. (1995); *rpoD1*, *gyrB*, *rpoC1*: Seo and Yokota (2003); *cpmA*: Dvornyk (2006); *pcb* gene familiy: Garczarek et al. (2001); phycocyanin operon: Robertson et al. (2001); Crosbie et al. (2003)). Multilocus sequence typing (Maiden et al. 1998) with a set of housekeeping genes shared in *Microcystis* (*ftsZ*, *glnA*, *gltX*, *gyrB*, *pgi*, *recA*, *tpi*) has been implemented by Tanabe et al. (2007). However, the number of sequences of nonribosomal DNA genes is comparatively small and/or restricted to smaller groups of taxa, and the suitability of these loci for cyanobacterial phylogenetics on a larger taxonomic scale has yet to be demonstrated.

Single gene analyses are known to provide insufficient support of basal branches in prokaryotes and in other organisms, but their resolution can be increased with multigene studies (Brown et al. 2005; Blank 2004). Combined analyses (analyses of several single gene data sets concatenated into one larger data set) are rare in the cyanobacteria. The available single locus gene data vary greatly across taxa and species, with relatively few taxa sharing multiple genes (except for the 16S rDNA). Robertson et al. (2001) used two genes, Seo and Yokota (2003) used four, but neither set of authors combined their data into a single data set. Sanchez-Baracaldo et al. (2005) combined data from several cyanobacterial genomes and from several single loci for a comprehensive analysis. Other publications that involve larger numbers of loci either sample within a rather broad taxonomic context and include only few cyanobacterial taxa (Martin et al. 2002; Ciccarelli et al. 2006), or focus on topics outside taxonomy and phylogeny (Mulkidjanian et al. 2006).

Consequently, unresolved, poorly resolved, and statistically weakly supported or unsupported trees make the reconstruction of ancestral states at internal nodes highly ambiguous, and thus many hypotheses about the origin and the evolution of cyanobacteria may remain purely speculative. Sanchez-Baracaldo et al. (2005) hypothesize the common ancestor of cyanobacteria to be a unicellular organism living in a freshwater habitat. Most other morphological and physiological traits might have been gained secondarily and/or multiple times in parallel; however, more complex features such as heterocytes, akinetes, and hormogonia have most likely developed only once (Sanchez-Baracaldo et al. 2005).

Despite the fact that morphological characters alone are not sufficient to establish a reliable taxonomic system in the cyanobacteria, it is important to note that morphological and molecular characters are not necessarily contradicting or mutually exclusive. Unfortunately, according to Sanchez-Baracaldo et al. (2005), fundamental morphological and physiological characters are still unknown for a large number of strains for which sequence data are already available. Many physiological traits, such as the potential for N_2-fixation, cannot be readily observed using a light microscope. Molecular analyses have shown many times that some of the traditional genera are poly- or paraphyletic (e.g., *Synechococcus*, *Oscillatoria*, *Leptolyngbya*). Komárek and Kaštovský (2003) demonstrate how ultrastructural characters, such as thylakoid arrangement, can be used to characterize and delimit taxa for which molecular data are yet insufficient. Based on the evidence from molecular and morphological data (presence of thylakoids, types of thylakoids, heterocytes), Hoffmann et al. (2004) propose four subclasses as the main cyanobacterial lineages: Gloeobacterophycidae, Synechococcophycidae, Oscillatoriophycidae, and Nostocophycidae.

3 Genomic Projects

The steadily growing number of sequenced genomes provides taxonomy and phylogeny with new possibilities, but also with new challenges. Currently (April 2010), 38 completely sequenced cyanobacterial genomes are available, and an additional 55 genomes are currently being sequenced (NCBI (National Center for Biotechnology Information)). Estimated genome sizes vary from 1.6×10^6 base pairs in unicellular taxa to 8.6×10^6 bp in filamentous forms (Castenholz 1992) and up to 13.2×10^6 bp (Iteman et al. 2000). In *Prochlorococcus*, genome sizes of different strains vary with light adaptation, and can exhibit substantial differences, from 1.7×10^6 in high-light-adapted to 2.4×10^6 bp in low-light-adapted strains (Rocap et al. 2003).

With the increasing availability of genomic data and the computational power to easily compute phylogenetic trees for thousands of genes and thousands of taxa (Stamatakis et al. 2007), choosing suitable loci for phylogenetic analyses is no longer limited by an artificial number of loci feasible for an analysis. Identifying suitable genes, which are present across all taxa and which represent the underlying organismic evolution, is a crucial step prior to an analysis. Horizontal gene transfer (HGT), gene duplication or loss, and recombination (Tanabe et al. 2007) are important events during evolution, and can severely obscure the underlying phylogenetic pattern (Gogarten et al. 2002; Gogarten and Townsend 2005). Different methods for detecting HGT (e.g., phylogenetic trees, nucleotide composition) may result in different estimates for the amount of HGT present in a given data set. A comparison of nucleotide composition indicates that some cyanobacterial genomes might have acquired between 10 and 17% of their genes due to HGT (Ochman et al. 2000; Ragan 2001); however, these numbers might underestimate the true amount of HGT (Zhaxybayeva et al. 2006). HGT among cyanobacteria is more frequent than HGT between cyanobacteria and other phyla, and consequently, cyanobacterial populations can show a high degree of HGT (Lodders et al. 2005). Cyanophages may also transfer photosynthetic genes involved in photosynthesis among unrelated cyanobacteria (Mann et al. 2003; Zeidner et al. 2005; Lindell et al. 2004; Millard et al. 2004; Sullivan et al. 2005).

Thus, the phylogenetic evolution of entire genomes may not be adequately represented by only one single phylogenetic tree. However, when analyzing a large enough number of loci, a plurality signal of vertically transmitted genes may reveal the underlying phylogenetic history of the organism (Zhaxybayeva et al. 2006), whereas attempts to infer the phylogeny of a set of taxa from a single locus or a set of arbitrarily selected loci are likely to produce erroneous results. Depending on the degree of relation among the investigated organisms, at least a subset of genes may be expected to have congruent phylogenies and share a common evolution (Swingley et al. 2008b), and it is obvious that the closer related a set of taxa is, the more orthologous loci can be found, and the less HGT events will hinder phylogenetic reconstruction. Suitable gene loci can be found by blasting (tblastn) known protein sequences against genome data (e.g., Sanchez-Baracaldo et al. 2005), or by employing statistical methods, such as Markov clustering (Enright et al. 2002; Swingley et al. 2008a,b) or Maximum Likelihood mapping (Zhaxybayeva and Gogarten 2002; Zhaxybayeva et al. 2004).

In most cases, analysis of concatenated single locus data produces better results compared to averaging tree topologies derived from individual single gene data sets, and trees will converge faster on the true topology (Gadagkar et al. 2005) with increasing numbers of loci. Single gene topologies even from perfectly orthologous loci are likely to differ to some degree (e.g., Seo and Yokota 2003). Especially, when lacking statistical support, this does not necessarily indicate incongruent signal, but can be attributed to the "usual" variations among different methods of phylogenetic reconstruction.

4 Main Lineages

4.1 Gloeobacter *and the Origin of Cyanobacteria*

The monophyly of the cyanobacteria has so far been supported by every molecular analysis (e.g., Giovannoni et al. 1988; Brochier et al. 2002; Ciccarelli et al. 2006), including an analysis of indel sequences of ten proteins (Gupta et al. 2003). Cyanobacteria have originated from within the bacteria, and might be among the latest bacterial groups to develop, with the Proteobacteria as their closest bacterial relatives (Woese 1987; Schleifer and Ludwig 1989; Blank 2004; Ciccarelli et al. 2006).

Gloeobacter violaceus consistently emerges as the earliest branch within the cyanobacteria (e.g., Nelissen et al. 1996; Honda et al. 1999; Fewer et al. 2002; Nakamura et al. 2003; Sanchez-Baracaldo et al. 2005; Taton et al. 2006; Swingley et al. 2008a,b), and thus its unique features, such as the lack of photosynthetic thylakoids, are likely to constitute plesiomorphic traits.

4.2 Prochlorococcus, Prochlorothrix, *and* Prochloron

The genera *Prochlorococcus*, *Prochlorothrix*, and *Prochloron* are unique among the cyanobacteria due to the possession of chlorophyll$_b$ in addition to chlorophyll$_a$, the lack of phycobilisomes, and a distinctive thylakoid organization (Giddings et al. 1980; Burger-Wiersma et al. 1986; Miller et al. 1988; Chisholm et al. 1992; Hess et al. 1996). Based on these shared features, the three genera were initially placed together into the "Prochlorales," or "Prochlorophyta." A relationship to the chloroplasts of rhodophytes, which exhibit the same set of pigments, has been suggested (Raven 1970), and analyses of *psbA* indicated a close connection to the chloroplasts of green plants (Morden and Golden 1989). Subsequent phylogenetic analyses supported the placement of these taxa in the cyanobacteria, but at the same time showed the polyphyly of the "Prochlorales" (Palenik and Haselkorn 1992; Urbach et al. 1992; Palenik and Swift 1996; Turner et al. 1999; Litvaitis 2002). Tomitani et al. (1999) propose a common chlorohyll$_b$ containing ancestor of the cyanobacteria, "Prochlorales," and chloroplasts of green plants based on a shared ancestry of chlorophyll$_a$ oxygenase, the key enzyme involved in chlorophyll$_b$ synthesis. Swingley et al. (2008b) report the chlorophyll$_a$ oxygenase of *Prochlorococcus* as being only distantly related to that of *Prochlorothrix* and *Prochloron*. If the features shared by *Prochlorococcus*, *Prochlorothrix*, and *Prochloron* have in fact evolved multiple times independently (Urbach et al. 1992; Palenik and Haselkorn 1992), it might indicate a strong connection between chlorophyll$_b$ synthesis and the structural elements of the photosynthetic apparatus (Swingley et al. 2008b).

Strains of *Prochlorococcus* compose a monophyletic group nested within a paraphyletic assemblage of marine strains of *Synechococcus* (Litvaitis 2002; Rocap et al. 2002; Scanlan and West 2002; Scanlan 2003; Swingley et al. 2008a, b). Using an orthologous protein library, Swingley et al. (2008b) identify only nine genes as unique to *Prochlorococcus*. *Prochloron* might be closely related to the Pleurocapsales (Litvaitis 2002; Sanchez-Baracaldo et al. 2005), and the position of *Prochlorothrix* is not yet established with sufficient degree of certainty (Turner et al. 1989; Wilmotte 1994; Palenik and Swift 1996; Nelissen et al. 1996; Nadeau et al. 2001; Taton et al. 2006).

4.3 Basal Clades

The branching patterns of basal cyanobacterial taxa are poorly resolved, and bootstrap support is in most cases missing or low (e.g., Giovannoni et al. 1988; Honda et al. 1999; Robertson et al. 2001; Wilmotte and Herdman 2001). Depending on the number of taxa and method of analysis, the number of reported "major lineages" varies greatly (Wilmotte 1994: eight clades, Turner 1997: ten clades, Honda et al. 1999: seven groups).

Some groups, however, reoccur frequently in several analyses, although their exact circumscription is not yet established:

- *Synechococcus*, originally classified based upon a few shared morphological and physiological traits, is not a monophyletic genus. Thermophilic basal strains, freshwater strains and marine strains of *Synechococcus* form distinct clades, and marine forms of *Synechococcus* are paraphyletic with *Prochlorococcus* (Litvaitis 2002; Scanlan and West 2002; Scanlan 2003; Rocap et al. 2002; Swingley et al. 2008a,b). The basal position of freshwater forms of *Synechococcus* PCC6301 and PCC 7942, the intermediate position of freshwater-tolerant *Synechococcus* WH 5701, and the terminal clade with marine forms could be interpreted as a transition from freshwater to marine environment (Fuller et al. 2003; Rocap et al. 2003; Swingley et al. 2008b).

  In contrast to their different physiology, *Synechocystis* PCC 6803 and *Crocosphaera watsonii* are probably closely related (Dvornyk 2006; Dyhrman and Haley 2006; Swingley et al. 2008a, b).

  Closely related to the *Synechococcus–Prochlorococcus* clade is a monophyletic clade composed of strains of *Synechococcus elongatus*, *Microcystis elabens*, and *M. holsatica* (Urbach et al. 1998; Honda et al. 1999; Turner et al. 1999; Robertson et al. 2001; Wilmotte and Herdman 2001; Sanchez-Baracaldo et al. 2005).
- *Thermosynechococcus elongatus* and *Synechococcus lividus* PCC 6716 branch together early near *Gloeobacter* (Turner et al. 1999; Honda et al. 1999; Robertson et al. 2001). Sanchez-Baracaldo et al. (2005), however, interpret this as a result of a possible long-branch-attraction, with *T. elongatus* branching higher up in the tree after the exclusion of *S. lividus*.
- The nonmonophyly of the LPP-group (Rippka 1988; Honda et al. 1999; Ishida et al. 2001; originally composed of *Lyngbya*, *Phormidium*, *Plectonema*, *and others*, taxa susceptible to a cyanophage called LPP-1) is ascribed to a long branch attraction by Sanchez-Baracaldo et al. (2005).
- The SPM-clade (*Synechocystis/Pleurocapsa/Microcystis*, Turner et al. 1999), which is usually not exactly circumscribed and poorly supported, possibly includes halotolerant taxa and *Dermocarpella* (Ishida et al. 2001; Litvaitis 2002; Sanchez-Baracaldo et al. 2005).

4.4 Chroococcidiopsis *and the Pleurocapsales*

Although the heterocyte-forming cyanobacteria are nested within filamentous strains (Honda et al. 1999; Swingley et al. 2008a,b), their sister group is the unicellular cyanobacterium *Chroococcidiopsis* (Wilmotte 1994; Turner 1997; Priscu et al. 1998; Bhattacharya et al. 1999; Fewer et al. 2002). *Chroococcidiopsis* has been traditionally classified within the Pleurocapsales (group II) owing to its shared mode of reproduction through multiple fission and the formation of non-motile baeocytes. The remaining taxa of the Pleurocapsales are resolved as monophyletic (but see also Ishida et al. 2001) and phylogenetically distinct from *Chroococcidiopsis*, indicating that reproduction by multiple fission and the production of baeocytes has been developed multiple times (Reeves 1996; Turner 1997; Rudi et al. 1997; Garcia-Pichel et al. 1998; Billi and Grilli-Caiola 1996; Ishida et al. 2001; Fewer et al. 2002). Consequently, the striking morphological similarity between *Myxosarcina* (Pleurocapsales) and *Chroococcidiopsis* is likely to be the result of convergent evolution (Fewer et al. 2002). *Chroococcidiopsis* can produce "survival cells" under N_2-limiting conditions (Billi and Grilli-Caiola 1996), and many heterocyte-forming taxa can produce akinetes (Rippka et al. 1979) under stress. Both the survival cells and the akinete differentiation may be ancestral to the heterocytes of Nostocales and Stigonematales (Wolk et al. 1994; Fewer et al. 2002).

4.5 Heterocyte-Forming Cyanobacteria

The filamentous cyanobacteria originally classified in the Nostocales and Stigonematales are characterized by the ability to produce heterocytes. Heterocytes are specialized cells that provide a low-oxygen environment required for nitrogen fixation. A great number of phylogenetic analyses, based on a variety of genes and gene combinations, support the monophyly of heterocyte-forming cyanobacteria (e.g., Giovannoni et al. 1988; Urbach et al. 1992; Wilmotte 1994; Nelissen et al. 1996; Turner 1997; Zehr et al. 1997; Garcia-Pichel et al. 1998; Turner et al. 1989, 1999; Honda et al. 1999; Wilmotte and Herdman 2001; Litvaitis 2002; Seo and Yokota 2003; Henson et al. 2004; Swingley et al. 2008a,b). The two orders are mainly separated by their type of branching: in contrast to the Nostocales, the Stigonematales are characterized by the apomorphy of true branching with multiseriate trichomes (Komárek and Anagnostidis 1989; Anagnostidis and Komárek 1990). Whereas the Stigonematales are usually resolved as a monophyletic group, the Nostocales often appear paraphyletic (Fewer et al. 2002; Litvaitis 2002[1]; Tomitani et al. 2006). Gugger and Hoffman (2004) report both the Nostocales and Stigonematales as paraphyletic.

5 Conclusions

Generating genomic data for cyanobacterial taxonomy will become faster and less expensive, and an increasing amount of data will soon be available for molecular phylogenetics. Nonetheless, with regard to the great diversity and the large numbers

[1]Although referred to as monophyletic in the text, both figures in Litvaitis (2002) show the Nostocales as paraphyletic.

of strains and species of cyanobacterial taxa, single gene sequencing is likely to remain the method of choice for the foreseeable future. Full genome sequencing for a larger number of taxa solely for phylogenetic purposes is not yet a reasonable option, because sequencing of a complete genome is still prohibitively expensive when compared to direct sequencing of even 10 or 20 or more single gene loci. But genomic data can help to identify genes that are not subjected to HGT or other processes that interfere with phylogenetic reconstruction, and a combined analysis of several such genes will help us to improve resolution and statistical support of internal and basal branches of the cyanobacterial tree of life.

Despite the undisputed importance of molecular data, morphology continues to play an important role in cyanobacterial taxonomy. First, Komárek and Kaštovský (2003) and Hoffmann et al. (2004) have shown that ultrastructural characters are extremely valuable to either morphologically characterize clades derived from molecular analyses or delimit groups that cannot be sufficiently resolved by molecular data alone. Second, generating phylogenetic trees and even full genome sequences without a proper morphological characterization of the actual organisms might give us a lot of information about genetic or genomic evolution, but is yet of only very limited use for a truly profound understanding of the evolutionary development of physiological or ecological key features. Third, any interpretation of phylogenetic trees is only meaningful with a proper identification of the taxa at the tips, especially in groups like the cyanobacteria with a comparatively large number of strains in public culture collections being incorrectly identified (Komárek and Anagnostidis 1989). Both the morphological and the molecular aspects are essential factors in order to achieve progress in cyanobacterial taxonomy and phylogenetics that truly helps us to understand the evolution of these important organisms.

References

Anagnostidis K, Komárek J (1985) Modern approach to the classification system of cyanophytes. 1 – Introduction. Arch Hydrobiol Suppl 71(Algological Studies 38/39):291–302

Anagnostidis K, Komárek J (1988) Modern approach to the classification system of cyanophytes. 3 – Oscillatoriales. Arch Hydrobiol Suppl 80(Algological Studies 50/53):327–472

Anagnostidis K, Komárek J (1990) Modern approach to the classification system of cyanophytes. 5 – Stigonematales. Arch Hydrobiol Suppl 86(Algological Studies 59):1–73

Bhattacharya D, Friedl T, Schmidt H (1999) The phylogeny of thermophiles and hyperthermophiles and the three domains of life. In: Seckbach J (ed) Enigmatic microorganisms and life in extreme environments. Kluwer Academic, Dordrecht, Netherlands, pp 291–304

Billi D, Grilli-Caiola M (1996) Effect of nitrogen limitation and starvation on *Chroococcidiopsis* sp. (Chroococcales). New Phytol 133:563–571

Blank CE (2004) Evolutionary timing of the origins of mesophilic sulfate reduction and oxygenic photosynthesis: a phylogenomic dating approach. Geobiology 2:1–20

Burger-Wiersma T, Veenhuis M, Korthals HJ, Van de Wiel CCM, Mur LR (1986) A new prokaryote containing Chls *a* and *b*. Nature 320:262–264

Boyer SL, Flechtner VR, Johansen JR (2001) Is the 16S–23S rRNA internal transcribed spacer region a good tool for use in molecular systematics and population genetics? A case study in cyanobacteria. Mol Biol Evol 18:1057–1069

Brochier C, Bapteste E, Moreira D, Philippe H (2002) Eubacterial phylogeny based on translational apparatus proteins. Trends Genet 18:1–5

Brown MV, Schwalbach MS, Hewson I, Fuhrman JA (2005) Coupling 16S-ITS rDNA clone libraries and automated ribosomal intergenic spacer analysis to show marine microbial diversity: development and application to a time series. Eviron Microbial 7:1466–1479

Casamatta DA, Johansen JR, Vis ML, Broadwater ST (2005) Molecular and morphological characterization of ten polar and near-polar strains within the Oscillatoriales (cyanobacteria). J Phycol 41:421–438

Castenholz RW (1992) Species usage, concept and evolution in the cyanobacteria (blue green algae). J Phycol 28:737–745

Castenholz RW (2001) Phylum BX. Cyanobacteria. Oxygenic photosynthetic bacteria. In: Boode DR, Castenholz RW (eds) Bergey's manual of systematic bacteriology, 2nd edn. New York, Springer

Chisholm SW, Frankel SL, Goericke R, Olson RJ, Palenik B, Waterbury JB, West-Johnsrud L, Zettler ER (1992) *Prochlorococcus marinus* gen. nov.: an oxyphototrophic marine prokaryote containing divinyl chlorophyll a and b. Arc. Microbiology 157:297–300

Ciccarelli FD, Doerks T, von Mering C, Creevey CJ, Snel B, Bork P (2006) Toward automatic reconstruction of a highly resolved tree of life. Science 311:1283–1287

Crosbie ND, Pöckl M, Weisse T (2003) Dispersal and phylogenetic diversity of nonmarine picocyanobacteria, inferred from 16S rRNA gene and cpcBA-intergenic spacer sequence analyses. Appl Environ Microbiol 69:5716–5721

Delwiche CF, Kuhsel M, Palmer JD (1995) Phylogenetic analysis of tufA sequences indicates a cyanobacterial origin of all plastids. Mol Phylogenet Evol 4:110–128

Dodds WK, Castenholz RW (1988) Effects of grazing and light on the growth of *Nostoc pruniforme* (cyanobacteria). Br Phycol J 23:219–227

Drouet F (1981) Summary of the classification of blue-green alga. Beih Nova Hedwigia 66:135–209

Dvornyk V (2006) Subfamilies of cpmA, a gene involved in circadian output, have different evolutionary histories in cyanobacteria. Microbiology 152:75–84

Dyhrman ST, Haley ST (2006) Phosphorus scavenging in the unicellular marine diazotroph *Crocosphaera watsonii*. Appl Environ Microbiol 72:1452–1458

Enright AJ, Van Dongen S, Ouzounis CA (2002) An efficient algorithm for large-scale detection of protein families. Nucleic Acids Res 30:1575–1584

Ferris MJ, Ruff-Roberts AL, Kopczynski ED, Bateson MM, Ward MD (1996) Enrichment culture and microscopy conceal diverse thermophilic *Synechococcus* populations in a single hot spring microbial mat habitat. Appl Environ Microbiol 62:1045–1050

Fewer D, Friedl T, Büdel B (2002) *Chroococcidiopsis* and heterocyst-differentiating cyanobacteria are each other's closest relatives. Mol Phylogenet Evol 23:82–90

Fox GE, Wisotzkey JD, Jurtshuk P (1992) How close is close: 16S rRNA sequence identity may not be sufficient to guarantee species identity. Int J Syst Bacteriol 42:166–170

Fuller NJ, Marie D, Partensky F, Vaulot D, Post AF, Scanlan DJ (2003) Clade-specific 16S ribosomal DNA oligonucleotides reveal the predominance of a single marine Synechococcus clade throughout a stratified water column in the Red Sea. Appl Environ Microbiol 69:2430–2443

Geitler L (1932) Rabenhorst's Kryptogamenflora von Deutschland, Österreich und der Schweiz. Vierzehnter Band: Cyanophyceae. Akademische Verlagsgesellschaft, Leipzig, Germany, p 1196

Gadagkar SR, Rosenberg MS, Kumar S (2005) Inferring species phylogenies from multiple genes: concatenated sequence tree versus consensus gene tree. J Exp Zool B Mol Dev Evol 304:64–74

Garcia-Pichel F, Nubel U, Muyzer G (1998) The phylogeny of unicellular, extremely halotolerant cyanobacteria. Arch Microbiol 169:469–482

Garczarek L, Van der Staay GWM, Hess WR, Le Gall F, Partensky F (2001) Expression and phylogeny of the multiple antenna genes of the low-light-adapted strain *Prochlorococcus marinus* SS120 (Oxyphotobacteria). Plant Mol Biol 46:683–693

Giddings TH, Withers NW, Staehelin LA (1980) Supramolecular structure of stacked andunstacked regions of the photosynthetic membranes of *Prochloron* sp., a prokaryote. Proc Natl Acad Sci U S A 77:352–356

Giovannoni SJ, Turner S, Olsen GJ, Barns S, Lane DJ, Pace NR (1988) Evolutionary relationships among cyanobacteria and green chloroplasts. J Bacteriol 170:3584–3592

Gogarten JP, Doolittle WF, Lawrence JG (2002) Prokaryotic evolution in light of gene transfer. Mol Biol Evol 19:2226–2238

Gogarten JP, Townsend JP (2005) Horizontal gene transfer, genome innovation and evolution. Nat Rev Microbiol 3:679–687

Gugger MF, Hoffman L (2004) Polyphyly of true branching cyanobacteria (Stigonematales). Int J Syst Evol Microbiol 54:349–357

Gupta RS, Pereira M, Chandrasekera C, Johari V (2003) Molecular signatures in protein sequences that are characteristic of cyanobacteria and plastid homologues. Int J Syst Evol Microbiol 53:1833–1842

Henson BJ, Watson LE, Barnum SR (2004) The evolutionary history of nitrogen fixation, as assessed by NifD. J Mol Evol 58:390–399

Hess WR, Weihe A, Loiseaux-de Goër S, Partensky F, Vaulot D (1995) Characterization of the single psbA gene of *Prochlorococcus marinus* CCMP 1375 (Prochlorophyta). Plant Mol Biol 27:1189–1196

Hess WR, Partensky F, van der Staay GWM, Garcia-Fernandez JM, Borner T, Vaulot D (1996) Coexistence of phycoerythrin and a chlorophyll a/b antenna in a marine prokaryote. Proc Natl Acad Sci U S A 93:11126–11130

Hoffmann L, Komárek J, Kaštovský J (2004) System of cyanoprocaryotes (cyanobacteria) – state in 2004. Algol Stud 117:95–115

Honda D, Yokota A, Sugiyama J (1999) Detection of seven major evolutionary lineages in cyanobacteria based on the 16S rRNA gene sequence analysis with new sequences of five marine *Synechococcus* strains. J Mol Evol 48:723–739

Ishida T, Watanabe MM, Sugiyama J, Yokota A (2001) Evidence for polyphyletic origin of the members of the orders of oscillatoriales and pleurocapsales as determined by 16S rDNA analysis. FEMS Microbiol Lett 201:79–82

Iteman I, Rippka R, Tandeau de Marsac N, Herdman M (2000) Comparison of conserved structural and regulatory domains within divergent 16S rRNA-23S rRNA spacer sequences of cyanobacteria. Microbiology 146:1275–1286

Komárek J, Anagnostidis K (1986) Modern approach to the classification system of cyanophytes. 2 – Chroococcales. Arch Hydrobiol Suppl 73(Algological Studies 43):157–226

Komárek J, Anagnostidis K (1989) Modern approach to the classification system of cyanophytes. 4 – Nostocales. Arch Hydrobiol Suppl 82(Algological Studies 56):247–345

Komárek J, Anagnostidis K (1999) Cyanoprokaryota. 1. Teil Chroococcales. In: Ettl H, Gärtner G, Heynig H, Mollenhauer D (eds) Süßwasserflora von Mitteleuropa 19/1. Gustav Fischer Verlag, Jena, pp 1–548

Komárek J, Kaštovský J (2003) Coincidences of structural and molecular characters in evolutionary lines of cyanobacteria. Algol Stud 109:305–325

Lindell D, Sullivan MB, Johnson ZI, Tolonen AC, Rohwer F, Chisholm SW (2004) Transfer of photosynthesis genes to and from *Prochlorococcus* viruses. Proc Natl Acad Sci 101:11013–11018

Litvaitis MK (2002) A molecular test of cyanobacterial phylogeny: inferences from constraint analyses. Hydrobiology 468:135–145

Lockhart PJ, Penny D, Hendy MD, Howe CJ, Beanland TJ, Larkum AWD (1992) Controversy on chloroplast origins. FEBS Lett 301:127–131

Lodders N, Stackebrandt E, Nubel U (2005) Frequent genetic recombination in natural populations of the marine cyanobacterium *Microcoleus chthonoplastes*. Environ Microbiol 7:434–442

Lu W, Evans HE, McColl M, Saunders VA (1997) Identification of cyanobacteria by polymorphism of PCR-amplified ribosomal DNA spacer region. FEMS Microbiol Lett 153:141–149

Maiden MCJ, Bygraves JA, Feil E, Morelli G, Russell JE, Urwin R, Zhang Q, Zhou J, Zurth K, Caugant DA, Feavers IM, Achtman M, Spratt BG (1998) Multilocus sequence typing: a portable approach to the identification of clones within populations of pathogenic microorganisms. Proc Natl Acad Sci 95:3140–3145

Mann NH, Cook A, Millard A, Bailey S, Clokie M (2003) Marine ecosystems: bacterial photosynthesis genes in a virus. Nature 424:741

Martin W, Rujan T, Richly E, Hansen A, Cornelsen S, Lins T, Leister D, Stoebe B, Hasegawa M, Penny D (2002) Evolutionary analysis of *Arabidopsis*, cyanobacterial, and chloroplast genomes reveals plastid phylogeny and thousands of cyanobacterial genes in the nucleus. Proc Natl Acad Sci U S A 99:12246–12251

Millard A, Clokie MR, Shub DA, Mann NH (2004) Genetic organization of the psbAD region in phages infecting marine *Synechococcus* strains. Proc Natl Acad Sci 101:11007–11012

Miller KR, Jacob JS, Burger-Wiersma T, Matthijs HC (1988) Supramolecular structure of the thylakoid membrane of *Prochlorothrix hollandica*: a chlorophyll b-containing prokaryote. J Cell Sci 91:577–586

Morden CW, Golden SS (1989) psbA genes indicate common ancestry of prochlorophytes and chloroplasts. Nature 337:382–385

Morden CW, Golden SS (1991) Sequence analysis and phylogenetic reconstruction of the genes encoding the large and small subunits of ribulose-1,5-bisphosphate carboxylase/oxygenase from the chlorophyll b containing prokaryote *Prochlorothrix hollandica*. J Mol Evol 32:379–395

Mulkidjanian AY, Koonin EV, Makarova KS, Mekhedov SL, Sorokin A, Wolf YI, Dufresne A, Partensky F, Burd H, Kaznadzey D, Haselkorn R, Galperin MY (2006) The cyanobacterial genome core and the origin of photosynthesis. Proc Natl Acad Sci 103: 13126–13131

Nadeau TL, Milbrandt EC, Castenholz RW (2001) Evolutionary relationships of cultivated *Antarctic oscillatorians* (cyanobacteria). J Phycol 37:650–654

Nakamura Y, Kaneko T, Sato S, Mimuro M, Myashita H, Tsuchiya T, Sasamoto S, Watanabe A, Kawashima K, Kishida Y, Kiyokawa C, Kohara M, Matsumoto M, Matsuno A, Nakazaki N, Shimpo S, Takeuchi C, Yamada M, Tabata S (2003) Complete genome structure of *Gloeobacter violaceus* PCC 7421, a cyanobacterium that lacks thylakoids (suppl.). DNA Res 10:181–201

Neilan BA, Stuart JL, Goodman AE, Cox PT, Hawkins PR (1997) Specific amplification and restriction polymorphisms of the cyanobacterial rRNA operon spacer region. Syst Appl Microbiol 20:612–621

Nelissen B, De Baere R, Wilmotte A, De Waechter R (1996) Phylogenetic relationships of nonaxenic filamentous cyanobacteria strains based on 16S rDNA sequence analysis. J Mol Evol 42:194–200

Ochman H, Lawrence JG, Groisman EA (2000) Lateral gene transfer and the nature of bacterial innovation. Nature 405:299–304

Palenik B, Haselkorn R (1992) Multiple evolutionary origins of prochlorophytes, the chlorophyll b-containing prokaryotes. Nature 355:265–267

Palenik B, Swift H (1996) Cyanobacterial evolution and prochlorophyte diversity as seen in DNA-dependent RNA polymerase gene sequences. J Phycol 32:638–646

Pearson JE, Kingsbury JM (1966) Culturally induced variation in four morphologically diverse blue-green algae. Am J Bot 53:192–200

Priscu JC, Fritsen CH, Adams EE, Giovannoni SJ, Paerl HW, McKay CP, Doran PT, Gordon DA, Lanoil BD, Pinckney JL (1998) Perennial Antarctic lake ice: an oasis for life in a polar desert. Science 280:2095–2098

Ragan MA (2001) Detection of lateral gene transfer among microbial genomes. Curr Opin Genet Dev 11:620–626

Rasmussen B, Fletcher IR, Brocks JJ, Kilburn MR (2008) Reassessing the first appearance of eukaryotes and cyanobacteria. Nature 455:1101–1104

Raymond J, Siefert JL, Staples CR, Blankenship RE (2004) The natural history of nitrogen fixation. Mol Biol Evol 21:541–554

Raven PH (1970) A multiple origin for plastids and mitochondria. Nature 169:641–646

Reeves RH (1996) 16S ribosomal RNA and the molecular phylogeny of the cyanobacteria. Nova Hedwigia 112:55–67

Rippka R, Deruelles J, Waterbury JB, Herdman M, Stanier RY (1979) Generic assignments, strain histories and properties of pure cultures of cyanobacteria. J Gen Microbiol 111:1–61

Rippka R (1988) Recognition and identification of cyanobacteria. Methods Enzymol 167:28–67

Rocap G, Distel DL, Waterbury JB, Chisholm SW (2002) Resolution of *Prochlorococcus* and *Synechococcus* ecotypes by using 16S-23S ribosomal DNA internal transcribed spacer sequences. Appl Eviron Microbiol 68:1180–1191

Rocap G, Larimer FW, Lamerdin J, Malfatti S, Chain P, Ahlgren NA, Arellano A, Coleman M, Hauser L, Hess WR, Johnson ZI, Land M, Lindell D, Post AF, Regala W, Shah M, Shaw SL, Steglich C, Sullivan MB, Ting CS, Tolonen A, Webb EA, Zinser ER, Chisholm SW (2003) Genome divergence in two *Prochlorococcus* ecotypes reflects oceanic niche differentiation. Nature 424:1042–1047

Robertson BR, Tezuka N, Watanabe MM (2001) Phylogenetic analyses of *Synechococcus* strains (cyanobacteria) using sequences of 16S rDNA and part of the phycocyanin operon reveal multiple evolutionary lines and reflect phycobilin content. J Syst Evol Microbiol 51:861–871

Rudi K, Skulberg OM, Larsen F, Jakobsen KS (1997) Strain characterization and classification of oxyphotobacteria in clone cultures on the basis of 16S rRNA sequences from the variable regions V6, V7, and V8. Appl Environ Microbiol 63:2593–2599

Sanchez-Baracaldo P, Hayes PK, Blank CE (2005) Morphological and habitat evolution in the cyanobacteria using a compartimental approach. Geobiology 3:145–165

Schleifer KH, Ludwig W (1989) Phylogenetic relationships of bacteria. In: Fernholm B, Bremer K, Jörnvall H (eds) The hierarchy of life. Elsevier Science Publishers B. V, Amsterdam, pp 103–117

Scanlan DJ (2003) Physiological diversity and niche adaptation in marine *Synechococcus*. Adv Microb Physiol 47:1–64

Scanlan DJ, West NJ (2002) Molecular ecology of the marine cyanobacterial genera *Prochlorococcus* and *Synechococcus*. FEMS Microb Lett 40:1–12

Schopf JW (2000) The fosil record: tracing the roots of the cyanobacterial lineage. In: Whitton BA, Potts M (eds) The ecology of cyanobacteria. Their diversity in time and space. Kluwer Academic Publishers, Dordrecht, pp 13–35

Seckbach J (2007) Algae and cyanobacteria in extreme environments. In: Seckbach J (ed) Cellular origin, life in extreme habitats and astrobiology, vol 11., pp 1–811

Seo PS, Yokota A (2003) The phylogenetic relationships of cyanobacteria inferred from 16S rRNA, gyrB, rpoC1 and rpoD1 gene sequences. J Gen Appl Microbiol 49:191–203

Shimada A, Kanai S, Maruyama T (1995) Partial sequence of ribulose-l, 5-bisphosphate carboxylase/oxygenase and the phylogeny of Prochloron and Prochlorococcus (Prochlorales). J Mol Evol 40:671–677

Stamatakis A, Auch AF, Meier-Kolthoff JP, Göker M (2007) AxPcoords & parallel AxParafit: statistical co-phylogenetic analyses on thousands of taxa. BMC Bioinformatics 8:405–414

Stanier RY, Sistrom WR, Hansen TA, Whitton BA, Castenholz RW, Pfennig N, Gorlenko VN, Kondratieva EN, Eimhjellen KE, Whittenbury R, Gherna RL, Trüper HB (1978) Proposal to place the nomenclature of the cyanobacteria (blue-green algae) under the rules of the international code of nomenclature of bacteria. Int J Syst Bacteriol 28:335–336

Sullivan MB, Coleman ML, Weigele P, Rohwer F, Chisholm SW (2005) Three *Prochlorococcus* cyanophage genomes: signature features and ecological interpretations. PLoS Biol 3:e144

Swingley WD, Blankenship RE, Raymond J (2008a) Integrating Markov clustering and molecular phylogenetics to reconstruct the cyanobacterial species tree from conserved protein families. Mol Biol Evol 25:643–654

Swingley WD, Blankenship RE, Raymond J (2008b) Insights into cyanobacterial evolution from comparative genomics. In: Herrero A, Flores E (eds) The cyanobacteria. Caister Academic, Norfolk, UK, pp 21–44

Tanabe Y, Kasai F, Watanabe MM (2007) Multilocus sequence typing reveals high genetic diversity and clonal population structure of thetoxic cyanobacterium *Microcystis aeruginosa*. Microbiology 153:3695–3703

Taton A, Grubisic S, Ertz D, Hodgson DA, Piccardi R, Biondi N, Tredici MR, Mainini M, Losi D, Marinelli F, Wilmotte A (2006) Polyphasic study of antarctic cyanobacterial strains. J Phycol 42:1257–1270

Tomitani A, Okada K, Miyashita H, Matthijsk HCP, Ohno T, Tanaka A (1999) Chlorophyll b and phycobilins in the common ancestor of cyanobacteria and chloroplasts. Nature 400:159–162

Tomitani A, Knoll AH, Cavanaugh CM, Ohno T (2006) The evolutionary diversification of cyanobacteria: molecular-phalogenetic and paleontological perspectives. Proc Natl Acad Sci U S A 103:5442–5447

Turner S (1997) Molecular systematics of oxygenic photosynthetic bacteria. Plant Syst Evol 11 (Suppl):13–52

Turner S, Burger-Wiersma T, Giovannoni SJ, Mur LR, Pace NR (1989) The relationship of a prochlorophyte *Prochlorothrix hollandica* to green chloroplasts. Nature 337:380–382

Turner S, Pryer KM, Miao VPW, Palmer JD (1999) Investigating deep phylogenetic relationships among cyanobacteria and plastids by small subunit rRNA sequence analysis. J Eukaryot Microbiol 46:327–338

Urbach E, Robertson DL, Chisholm SW (1992) Multiple evolutionary origins of prochlorophytes within the cyanobacterial radiation. Nature 355:267–270

Urbach E, Scanlan DJ, Diestel DL, Waterbury JB, Chisholm SW (1998) Rapid diversification of marine phytoplankton with dissimilar light-harvesting structures inferred from sequences of *Prochlorococcus* and *Synechococcus* (cyanobacteria). J Mol Evol 46:188–201

Watson GMF, Tabita FR (1997) Microbial ribulose 1,5-biphosphate carboxylase/oxygenase: a molecule for phylogenetic and enzymological investigation. FEMS Microbiol Lett 146:13–22

Wilmotte A (1994) Molecular evolution and taxonomy of cyanobacteria. In: Bryant DA (ed) The molecular biology of cyanobacteria. Kluwer Academic Publishers, Dordrecht, The Netherlands, pp 1–25

Wilmotte A, Golubic (1991) Morphological and genetic criteria in the taxonomy of cyanophyta/ cyanobacteria. Arch Hydrobiol 64(Suppl):1–24

Wilmotte A, Herdman M (2001) Phylogenetic relationships among the cyanobacteria based on 16S rRNA sequences. In: Garrity GM (ed) Bergey's manual of systematic bacteriology. Springer, New York, pp 487–493

Woese CR (1987) Bacterial evolution. Microbiol Rev 51:221–271

Wolk CP, Ernst A, Ehlai J (1994) Heterocyst metabolism and development. In: Bryant DA (ed) The molecular biology of cyanobacteria. Kluwer Academic, Dordrecht, pp 769–823

Zehr JP, Mellon MT, Hiorns WD (1997) Phylogeny of cyanobacterial nifH genes: evolutionary implications and potential applications to natural assemblages. Microbiology 143:1443–1450

Zeidner G, Bielawski JP, Shmoish M, Scanlan DJ, Sabehi G, Beja O (2005) Potential photosynthesis gene recombination between *Prochlorococcus* and *Synechococcus* via viral intermediates. Environ Microbiol 7:1505–1513

Zhaxybayeva O, Gogarten JP (2002) Bootstrap, Bayesian probability and maximum likelihood mapping: exploring new tools for comparative genome analyses. BMC Genomics 3:4

Zhaxybayeva O, Lapierre P, Gogarten JP (2004) Genome mosaicism and organismal lineages. Trends Genet 20:254–260

Zhaxybayeva O, Gogarten JP, Charlebois RL, Doolittle WF, Papke RT (2006) Phylogenetic analyses of cyanobacterial genomes: quantification of horizontal gene transfer events. Genome Res 16:1099–1108

Part V
Ecology

Carbon and Oxygen Isotopes in Trees: Tools to Study Assimilate Transport and Partitioning and to Assess Physiological Responses Towards the Environment

Arthur Gessler

Contents

Abstract At present, there is lack of knowledge on how plant physiological processes, the transfer of carbon within the plant, carbon storage, and remobilization in the plant tissues as well as the release of carbon from the roots to the soil interact with ecosystem-scale processes. On the background of global climate change, we need to mechanistically link plant physiology, CO_2 net exchange between ecosystems and the atmosphere and plant biomass accumulation. This is the basis for predicting productivity of forests as well as their carbon sequestration potential in future. This chapter will give an overview on how stable isotope studies can give insights into the fate of newly assimilated carbon transported within trees and transferred to the soil and atmosphere. The review includes studies either characterizing temporal and spatial variation in the natural abundance of carbon and oxygen isotopes or applying isotopically enriched tracers. In addition, it highlights the fact that the stable isotope composition of assimilates transported within

A. Gessler
Institute for Landscape Biogeochemistry, Leibniz-Zentrum für Agrarlandschaftsforschung (ZALF) e.V., Eberswalderstr. 84, 15374 Müncheberg, Germany
Professorship for Landscape Biogeochemistry, Humboldt-University Berlin, Lentze-Allee 75, 14195 Berlin Germany
e-mail: gessler@zalf.de

U. Lüttge et al. (eds.), *Progress in Botany 72*,
DOI 10.1007/978-3-642-13145-5_9, © Springer-Verlag Berlin Heidelberg 2011

the plant contains important time integrated information on environmental conditions, leaf physiology, and postphotosynthetic metabolism.

This review will on the one hand focus on the fast turn over carbon pools, which fuel plant respiration and soil microbial activity and on the other hand explore the transfer of the isotope information to long-lived compounds in the tree-ring archive.

1 Introduction

The anthropogenic emission of increased amounts of CO_2 to the atmosphere mainly due to the combustion of fossil fuel and land use changes (IPCC 2007) has various effects on plants and ecosystems. Mainly not only the carbon but also the water balance are affected, directly by the change in atmospheric CO_2 concentration and indirectly by the effects of CO_2 on climate.

There are numerous studies, which show that the water use efficiency (WUE; given as the ratio between net assimilation rate (A) and transpiration rate (E)) increases with rising atmospheric CO_2 concentration (Drake et al. 1997; McCarroll and Loader 2004). In dry regions, this effect might in principle lead to decreased drought stress for plants. On the other hand, the anthropogenic increase in CO_2 concentration does not only affect plant physiology but is also the main reason for global climate change (IPCC 2007). In the last 10 years, the air temperature in the troposphere has been increasing by approximately 0.6°C and within the next 40 years an additional increase by 2°C is expected, for example, in the southern part of Central Europe (Forkel and Knoche 2006; Gessler et al. 2007b). One of the potential consequences of global warming is the intensification of the hydrological cycle with increased precipitation in early spring but with decreased amounts of rainfall in summer. This will most probably lead to an increase in duration and intensity of drought periods during the vegetation period in Central Europe (Angert et al. 2005). As a consequence, extremely dry summers like in 2003 are expected to become more frequent (Gessler et al. 2007b). In order to better characterize the effects of global climate change on the carbon balance of ecosystems, intensive research on CO_2 net exchange (NEE) between plant canopies and the atmosphere as affected by environmental conditions has been carried out within the last decade (e.g., Buchmann 2002; www.fluxnet.ornl.gov/fluxnet/index.cfm). During the dry summer in 2003, a strong decrease in the CO_2 sink strength of forests was observed all over Europe; particular ecosystems that were normally CO_2 sinks during the growing season even turned into CO_2 sources for longer periods (Ciais et al. 2005). These results have been obtained from studies on the ecosystem level, where the net CO_2 flux from or to the atmosphere is determined by applying micrometeorological techniques, and upscaling to the landscape level has been achieved by modeling (Bowling et al. 2003; Friend et al. 2007).

Until now, there is, however, a lack of knowledge on how plant physiological processes, such as the transfer of carbon within the plant, carbon storage, remobilization, and respiration in the plant tissues as well as the release of carbon from the roots to the soil interact with and affect ecosystem processes related to primary production and carbon sequestration. It becomes more and more obvious that the vegetation is the central and important driver also of soil carbon dynamics and the carbon sequestration potential of terrestrial ecosystems. This is because the vegetation is not only the original source for soil organic matter thus contributing to long-term carbon accumulation in the organic soil layers but also it determines below-ground processes such as soil respiration over the short term. As an example, Bhupinderpal et al. (2003) found that more than 60% of soil respiration originates from recent assimilates. If we want to be able to consider the adaptability of stands and ecosystems to develop strategies for forest management aiming at minimizing the negative effects of the predicted climate change and maintaining the carbon sequestration potential, we have to deepen our knowledge on the processes driving plant physiology. This has to include studies on carbon exchange, and assessments on how plant physiology is influenced by external factors like drought events and how water and carbon balance interact on the plant level.

Stable isotope studies can give such novel insights into the fate of newly assimilated carbon transported and deposited within the plant and transferred to the soil and atmosphere. Such processes can be explored by either assessing the natural abundance of carbon and oxygen isotopes or applying isotopically enriched tracers. Due to isotope fractionation processes associated with phase transition, diffusion, and with enzyme activities in leaves and heterotrophic plant parts, the natural abundance of carbon and oxygen isotopes in different organs and chemical compounds can give additional information on the physiology of the carbon and water balance as affected by (changing) environmental conditions.

2 Assimilate Fluxes Within Trees and Transfer of Carbon to the Soil: Short-Term Dynamics

The flux of recently assimilated carbon through trees and the transfer to soil microbiota have been assessed by manipulation experiments (e.g., girdling; Högberg et al. 2001), by tracking the natural variation of stable isotope signatures through the different organic matter pools of a tree and to CO_2 emitted from plants and soil (e.g., (Brandes et al. 2006; Gessler et al. 2007a; Kodama et al. 2008), and by the application of highly enriched isotope tracers (e.g., Högberg et al. 2008) in pulse labeling experiments.

The field scale girdling experiment in boreal forests in Sweden (Högberg et al. 2001) as well as trenching studies in Canada (Bond-Lamberty et al. 2004) have shown that the supply of new assimilates to tree roots is crucial for fuelling soil respiration. Cutting off this supply results in a rapid decline in below ground respiration within days to weeks. These experiments clearly show the close and immediate link between canopy assimilation on the one hand and heterotrophic energy gain and below ground metabolism on the other hand. There is, however, also information that other carbon pools with longer residence time are used as substrates for respiration. The increased $\delta^{13}C$ value of soil respired CO_2 in the longer term after girdling indicates that ^{13}C enriched ectomycorrhizal mycelium and root starch and sugar reserves may also supply substrate for respiration and thus energy gain of roots and microorganisms (Bhupinderpal et al. 2003).

2.1 Tracer Experiments

Applying a $^{13}CO_2$ pulse to the canopy of a *Pinus sylvestris* forest, Högberg et al. (2008) observed a label peak in phloem organic matter after 24 h and in soil microbial biomass after 4–7 days. For black spruce seedlings, Carbone et al. (2007) determined a mean residence time of ^{14}C-labeled assimilates of 6 days in the canopy and a maximum ^{14}C peak in the root and rhizosphere respiration after 4 days. A very fast transfer of recently fixed ^{13}C-labeled carbon to soil microbial organisms, within less than 24 h was observed under well watered conditions for beech seedlings, highlighting the close coupling between tree photosynthesis and soil activity. Mean residence time of the assimilates in the leaves was 2.4 days (Rühr et al. 2009). In almost 10-m tall beech trees, a maximum ^{13}C abundance was observed in the CO_2 emitted from trunks and from the soil 2–3 and 3–4 days, respectively, after the crown has been exposed to a $^{13}CO_2$ pulse (Plain et al. 2009).

The CO_2 pulse labeling experiments thus support the findings from the girdling approaches that soil and rhizosphere processes are closely coupled to canopy photosynthesis. Högberg et al. (2008) postulated that half or more of the soil activity is driven by recent assimilates.

The field experiments with trees mentioned above, did, however, not include the effects of changing environmental conditions (e.g., reduction of water availability, temperature increase) on assimilate transport in the tree or on transfer to the soil and to soil microorganisms. There are many studies published, which examine assimilate transport as affected by water supply under controlled and field conditions. However, the majority of the papers focuses on herbs or grasses (e.g., Plaut and Reinhold 1965; Wardlaw 1969; Palta and Gregory 1997) and on particular processes of carbon allocation such as leaf assimilate export (e.g., Deng et al. 1990; Li et al. 2003). Whereas drought-induced regulation of carbon assimilation in the leaves is reasonably well understood (Flexas and Medrano 2002), there is not much integrated information available on carbon export from the leaf connected to the transport within the plant and to the export to the rhizosphere under drought conditions. To understand the impact of summer droughts, which will increase in intensity and duration in the temperate regions in future (IPCC 2007), on the carbon balance of ecosystems, we need, however, clear-cut information on how tree–water relations affect carbon flux and partitioning.

In a ^{13}C pulse labeling study with wheat, the assimilate transfer to roots and the recovery of ^{13}C in soil respiratory CO_2 was higher under drought conditions compared to controls 2 days after the label application (Palta and Gregory 1997).

In contrast, lower transfer of recently assimilated carbon to soil and microorganisms was observed in shrublands (Gorissen et al. 2004) a few days after a ^{14}C pulse label application. The main shortcoming of these two studies is their lack in temporal resolution as achieved in the more recent studies by Högberg et al. (2008) and Carbone et al. (2007).

Only recently, Rühr et al. (2009) have investigated the effects of drought on the translocation of recently assimilated carbon by pulse labeling 1.5-year-old beech tree microcosms with $^{13}CO_2$. The ^{13}C tracer signals in different organic matter pools of trees and soil microbes as well as in soil respired CO_2 were determined up to 10 days with a daily resolution. Drought not only reduced C assimilation but also doubled the residence time of recently assimilated C in leaves from 2.4 to 5 days.

In phloem organic matter, the ^{13}C label peaked immediately after labeling, then decayed exponentially in the control treatment, while under drought the maximum label recovery was observed 4 days after the pulse labeling (Fig. 1). The label peaked in soil microbial biomass 1 day after labeling in the control treatment, whereas under drought no peak was measured in soil microbial biomass within the 10 days of observation. The study with beech seedlings showed that drought can reduce the coupling between canopy photosynthesis and belowground processes. It is likely that the reduced inputs of labile carbon to the soil under drought as observed in the study of Rühr et al. (2009) results in a shift of the soil microbial community toward increased decomposition of soil organic carbon (Bradford et al. 2008). Together with a reduction in the overall root biomass, this may decrease soil C sequestration potential under water-limited conditions. It has to be examined carefully how shifts in soil carbon pools used for soil respiration during short extreme periods affect long-term soil quality and functioning of the soil microbial communities. The reduced carbon transfer to the roots is also likely to constrain the supply with energy and carbon skeletons for plant nutrient uptake and assimilation. This assumption is supported by the finding that the maximum nitrate uptake rate is reduced in adult beech trees under drier climatic conditions (Gessler et al. 2005). We have also to assume that at least part of the energy-consuming processes in heterotrophic tree tissues have to be fuelled by remobilized storage carbon originating from stems and roots in times when recent assimilates are not sufficiently available due to transport retardation. Whether and how such temporal switches to storage carbon will affect growth and development of below and above ground tissues at the beginning of the subsequent growing season is not yet resolved.

Besides water availability, temperature also affects the carbon transfer from plants to mycorrhizae and rhizosphere. Hawkes et al. (2008) observed that at higher soil temperatures, the rate of transfer of ^{13}C-labeled carbon from the roots to the fungus increased. The carbon allocation within the fungus changed from storage to growth and the fungal respiration rates (per unit of hyphal length) increased. The authors hypothesize that higher temperatures result in an increased carbon loss to the atmosphere due to higher fungal respiration. It remains to be assessed how the combined variation of temperature and water availability affects assimilate transport within the plant and transfer to soil and rhizosphere microorganisms. It can be (1) assumed that the reduced carbon export by the roots under drought might be

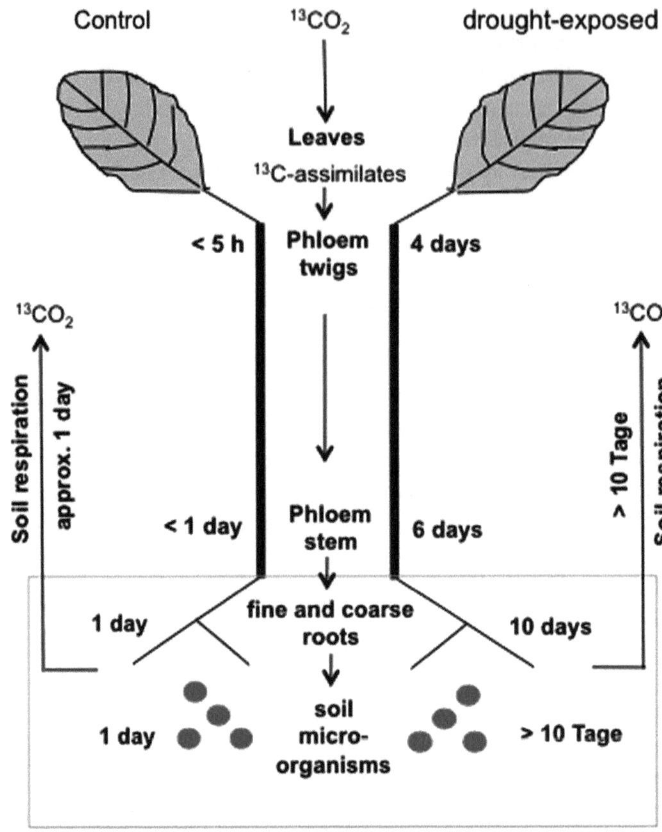

Fig. 1 Carbon transport times in control and drought-treated beech seedlings. The times indicate the maximum (peak) occurrence of the ^{13}C label in either water soluble organic matter or CO_2. The data are taken from Rühr et al. (2009)

(partially) compensated with increased temperatures. However, (2) the higher rates of respiration in plant heterotrophic tissues and the related higher demand of respiratory substrates at higher temperatures might even intensify the shortage of recent assimilates for mycorrhizal fungi and soil microorganisms.

2.2 *Natural Abundance Techniques*

The natural abundance technique for assessing carbon transfer within trees uses the natural variation of the carbon isotope composition of assimilates for calculating transport and turnover times (cf. Brandes et al. 2006). This approach has the advantage of being nondestructive and of not introducing highly enriched isotope tracers into an ecosystem. Besides tracking the transport of carbon through different

C-pools in the tree, physiological information can be deciphered: e.g., the direct impact of air humidity (Ekblad and Högberg 2001; Bowling et al. 2002) using well-established fractionation models (e.g., Farquhar et al. 1982; Cernusak et al. 2005). As a consequence, a combined analysis of carbon flux as affected by leaf level physiological processes (assimilation rate, stomatal conductance, and transpiration rate) is possible.

Due to the relationship between photosynthetic carbon isotope fractionation and the ratio of leaf intercellular to ambient CO_2 concentration (c_i/c_a), the $\delta^{13}C$ of newly assimilated organic matter can be generally used to characterize environmental effects on the physiology of photosynthesis. Stomatal closure due to water deficits generally reduces c_i, leading to an increase in $\delta^{13}C$ (e.g., Farquhar et al. 1982; Korol et al. 1999; Brugnoli and Farquhar 2000). As light limitation of photosynthesis increases c_i, $\delta^{13}C$ can also depend on radiation (Leavitt and Long 1986; McCarroll and Pawellek 2001) under particular conditions, but also combined influences of water and light availability have been observed (Gessler et al. 2001).

The oxygen stable isotope composition ($\delta^{18}O$) or the oxygen isotopic enrichment above source water ($\Delta^{18}O$) of plant organic matter, has been shown to provide additional information to separate effects of stomatal conductance (g_s) from the influence of changes in photosynthetic capacity on $\delta^{13}C$ (Farquhar et al. 1998; Adams and Grierson 2001). This can be explained by the fact that oxygen isotope patterns are related to stomatal conductance, when g_s is controlled by the water vapor pressure of the ambient air. As a consequence, oxygen isotope patterns share the dependence on stomatal conductance with the $\delta^{13}C$ signature, but they are not dependent on Rubisco activity (Barbour and Farquhar 2000; Scheidegger et al. 2000). In addition, the evaporative enrichment of leaf water, which is imprinted on the newly assimilated organic matter (i.e., the carbonyl group) with an equilibrium fractionation factor (ε_{wc}) of 27‰, has been shown to provide information on transpiration rate (Wang and Yakir 2000; Barnard et al. 2007).

Changes in environmental conditions and as a consequence in leaf physiology alter the photosynthetic carbon isotope fractionation and the evaporative ^{18}O enrichment. Newly assimilated organic matter, which is distributed within the tree can then be tracked by its altered isotope composition (Fig. 2).

Until present, several studies have assessed the temporal variation in the oxygen and/or carbon isotope composition in different tree organs and tissues (for an overview see Gessler et al. (2009). However, most of these studies (e.g., Damesin et al. 1998; Barbour et al. 2002; West et al. 2007) have concentrated on longer term seasonal patterns – which do not allow to characterize the transport of recent assimilates with high temporal resolution. Only recently, however, Brandes et al. (2006) and Barnard et al. (2007) have assessed the carbon and oxygen isotope composition in needles and phloem (twig phloem and phloem of the trunk in different heights) in *Pinus sylvestris* over several diel courses with a resolution in time of approximately 6 h.

The measured evaporative enrichment above source water ($\Delta^{18}O_L$) of leaf water and organic matter under field conditions could be readily explained (cf. Barnard et al. 2007) when applying mechanistic models according to Farquhar and Cernusak

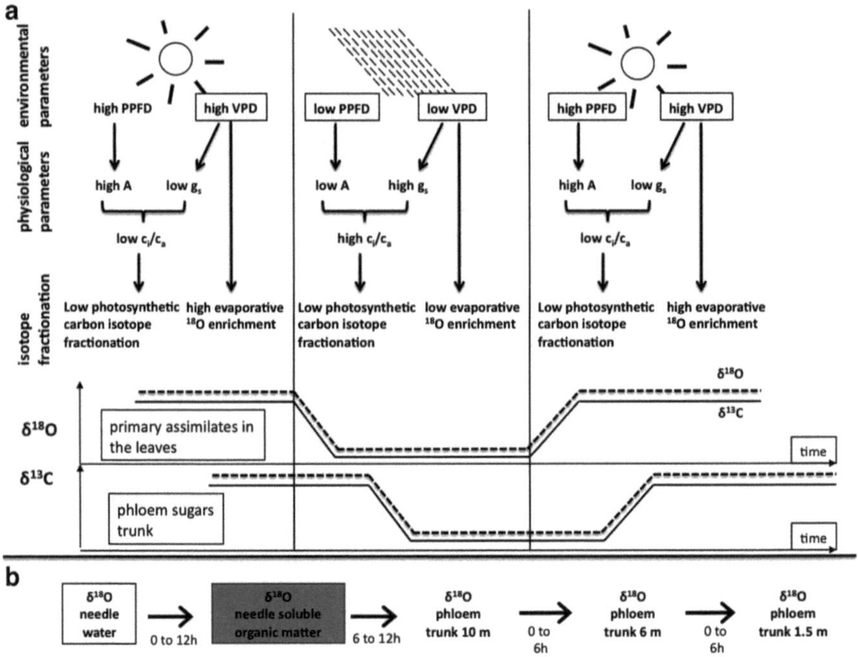

Fig. 2 Tracing natural abundance isotope signals within the tree. (**a**) Changes in environmental conditions and as a consequence in leaf physiology alter the photosynthetic carbon isotope fractionation and the evaporative ^{18}O enrichment. Newly assimilated organic matter, which is distributed within the tree can be tracked by its altered isotope composition (**b**) Example from Barnard et al. 2007: By applying cross-correlation analysis the time lags (in hours) among the oxygen isotope composition of sugars in different tissues have been characterized. These time lags indicate transport and turn-over times of recently assimilated organic matter (Gessler et al. 2009). *a*: assimilation rate, g_s: stomatal conductance; c_i/c_a: ratio of the leaf internal (intercellular) and the ambient air CO_2 concentration; PPFD: Photosynthetic photon flux density; VPD: Water pressure deficit of the air

(2005), which take into account ^{18}O fractionation during phase transition of water and diffusion of water vapor through the stomata and the boundary layer, isotopic nonsteady state, and isotopic inhomogeneity of leaf water. Brandes et al. (2006) could explain at least part of the variation of $\delta^{13}C$ in newly assimilated organic carbon from the variations in c_i/c_a.

By applying cross-correlation analyses, the authors assessed the time lags between the temporal variations of the stable isotope composition from the leaf to the phloem at the stem base. They found significant time lags for $\delta^{13}C$ (Brandes et al. 2006) and $\delta^{18}O$ (Barnard et al. 2007) in different water and organic matter pools (needle water, needle water-soluble organic matter, phloem sugars at different positions along the tree axis), which represent residence and transport times of newly produced assimilates (Gessler et al. 2009). In total, it took 1–2 days for the oxygen isotope signal to be transferred from the leaf water to the phloem-transported sugars at the base of the trunk (Fig. 2b). Comparable values were estimated with $\delta^{13}C$ (Brandes et al. 2006). With the trees being 15 m high, phloem transport velocities thus amounted to approximately 0.8 m h^{-1}, a value, which is within the range for transport velocities in other species determined with different techniques (poplar: 0.9–1.2 m h^{-1}, magnetic resonance imaging (MRI) (Windt et al. 2006); beech: 1 m h^{-1}, ^{13}C pulse labeling (Plain et al. 2009).

Keitel et al. (2003, 2006); Barbour et al. (2005), and Brandes et al. (2007) applied time lagged correlation analyses between environmental conditions and canopy physiological parameters (e.g., canopy stomatal conductance) and $\delta^{13}C$ or $\delta^{18}O$ of trunk phloem organic matter. With these approaches, no samples from different heights were taken and compared, but the changes in carbon isotope fractionation or evaporative oxygen isotope enrichment over time due to changes in meteorological or physiological parameters were tracked at given position at the stem. Barbour et al. (2005) found the best correlation between modeled photosynthetic carbon isotope fractionation in the canopy of *Nothofagus solandri* (tree height 15–25 m) and trunk base phloem sap (honeydew from phloem tapping insects) when a time lag of 3 days was included. From these results, the authors concluded that $\delta^{13}C$ represents a photosynthesis-weighted, integrative record of canopy photosynthesis and conductance. Consistent with the work on *N. solandri*, Brandes et al. (2007) found 3-day time lags between modeled canopy leaf water evaporative enrichment in *Pinus sylvestris* (tree height approx 15 m) and $\delta^{18}O$ in the phloem at the trunk base over the whole growing season. One major constraint of the two above-mentioned approaches is that no phloem transport velocities can be calculated as the time lag observed is a result of assimilate turnover time in the leaves plus phloem transport time. Keitel et al. (2003) found best correlation between canopy stomatal conductance and $\delta^{13}C$ or $\delta^{18}O$ in the trunk phloem of beech when mean stomatal conductance values of the 2 days before phloem sampling were used for analysis. This temporal integration of canopy processes suggests that the stem phloem comprises a mixture of carbon with different metabolic history and residence time. In other words, we have to assume that part of the assimilates found in the phloem at the trunk base might be transported directly down the trunk, whereas other parts undergo storage and remobilization processes and/or are unloaded from the sieve tubes and retrieved again (e.g., Van Bel 2003).

Time lags between changes in environmental conditions and $\delta^{13}C$ in ecosystem respired CO_2 ($\delta^{13}C_R$) have also been used to estimate the (temporal) coupling between photosynthesis and respiration. Environmental factors, varying at different time scales, from single extreme weather events to daily and seasonal changes, showed a link with $\delta^{13}C$ of ecosystem respired CO_2 with a consistent time lag of 4–5 days in a beech dominated temperate forest (Knohl et al. 2005). This lag is in the same range as given for the $^{13}CO_2$ (Högberg et al. 2008) and $^{14}CO_2$ (Carbone et al. 2007) labeling experiments. When studying the carbon isotope composition of ecosystem respiration along a precipitation gradient, Bowling et al. (2002) observed that there was a strong link between $\delta^{13}C_R$ and the water vapor saturation deficit of the air 5–10 days earlier. This relationship was found both across and within sites, which were dominated by *Picea sitchensis*, *Tsuga heterophylla*, *Pseudotsuga menziesii*, *Pinus ponderosa*, and *Juniperus occidentalis*. The relationship observed was consistent with the expected stomatal response to changes in vapor pressure deficit and thus with changes in photosynthetic carbon isotope discrimination.

As illustrated by the $^{13}CO_2$ pulse labeling experiment by Rühr et al. (2009), environmental conditions might strongly affect the transfer of assimilates within the trees as well as the export to the soil. As a consequence, it must be assumed that the isotopic time lag between assimilation (and photosynthetic carbon isotope fractionation) and respiration is not always constant over time (Werner et al. 2006; Steinmann et al. 2004). The degree to which changes in environmental conditions are reflected by changes in $\delta^{13}C_R$ should also depend on assimilation rates and on the amount of assimilates being produced during these conditions. In agreement with this hypothesis Knohl et al. (2005) observed different intensities of isotopic coupling between assimilation and respiration; briefly when photosynthesis was low, the isotopic signal induced by a low vapor pressure deficit was not as clearly recovered in respiratory CO_2 as under conditions with high assimilation.

In conclusion, both isotope pulse labeling approaches as well as natural abundance techniques can give deeper insights into the fate of newly assimilated organic matter within the tree and the whole ecosystem. Whereas pulse labeling allows very accurate calculations of turnover and residence times, the natural abundance approaches have the potential to characterize seasonal variations in the assimilate allocation and the coupling between assimilation and respiration without introducing artificial isotope label to an ecosystem.

However, one possible constraint for applying the natural abundance approaches is the fact that not only assimilation and leaf water evaporation induces isotope fractionation, the variation of which can be traced through the tree and ecosystem, but also processes in the downstream metabolism in autotrophic and heterotrophic tissues as well as isotope fractionation associated with transport can alter the original isotope signal. For oxygen isotopes, the exchange of organic oxygen with oxygen of the reaction water might also occur during chemical transformations. This is mainly important during the production of cellulose from sugars and thus will be dealt with in the chapter "isotope archives." Postphotosynthetic isotope fractionation might on the one hand blur the original (photosynthetic) isotope signal making it more difficult to follow the natural isotope signal through the plant. On the other hand, the understanding of the mechanisms behind might give new insights into carbon allocation between metabolic pathways as affected by environmental conditions.

3 Postphotosynthetic Isotope Fractionation

It was recognized only recently that postphotosynthetic isotope fractionation (sometimes also referred to as postcarboxylation fractionation) due to equilibrium and kinetic isotope effects beyond CO_2 diffusion and fixation by Rubisco is of importance in autotrophic and heterotrophic tissues (Ghashghaie et al. 2003; Badeck et al. 2005). The isotope effects result in differences in isotopic signatures among metabolites and in nonstatistical intramolecular isotope distributions

(Gleixner and Schmidt 1997; Schmidt 2003; Tcherkez and Farquhar 2005; Brandes et al. 2006), thereby potentially uncoupling the isotopic information reaching the heterotrophic tissues of a tree or the soil microbes from leaf physiological processes. Not only the assessment of assimilate transport based on natural stable isotope abundances requires understanding of postphotosynthetic carbon isotope fractionation as well as oxygen atom exchange between organic matter and water. In general, the proper interpretation of carbon and oxygen isotope signatures in plant chemical compounds beyond the primary assimilates (i.e., 3-phospho-glycerate) including phloem sugars, respired CO_2, and tree-ring cellulose needs to consider these issues. Figure 3 shows the compilation of carbon and oxygen isotope fractionation steps (plus an overview of potential oxygen isotope exchange) within a tree. Even though postphotosynthetic fractionation has been intensively studied within the last years (Ghashghaie et al. 2001; Gessler et al. 2009; Badeck et al. 2005) we lack a clear picture of what is happening in downstream metabolic processes (c.f. Brandes et al. 2006) and – which is of central importance – how potential postphotosynthetic fractionation processes are influenced by environmental and plant internal factors.

Only recently day–night variations in $\delta^{13}C$ of phloem sugars due to fractionation processes associated with the aldolase reaction in the chloroplast (Tcherkez et al. 2003; Tcherkez et al. 2004) have been characterized in trees (Gessler et al. 2007a). At a first glance, these diel variations in $\delta^{13}C$ might be also used – in addition the day-to-day variations imposed by the photosynthetic carbon isotope fractionation – to be traced and followed within the tree. However, Kodama et al. (2008) observed that the distinct diel variations in $\delta^{13}C$ observed in newly assimilated organic matter in the tree crown (leaf water soluble organic matter) of *Pinus sylvestris* were strongly damped during phloem transport to the trunk. The damping of the diel $\delta^{13}C$ variations in phloem-transported organic matter at the trunk base can be explained by a mixing of various C pools with different metabolic histories during phloem transport down the trunk (Brandes et al. 2006). This hypothesis is supported by findings of Keitel et al. (2003) and Keitel et al. (2006) (s. above) showing that $\delta^{13}C$ in trunk phloem sap organic matter integrates mean canopy c_i/c_a over several days in European beech. We have, on the one hand, to consider here that such mixing might also negatively affect the sensitivity of any of the natural abundance approaches for carbon phloem transport as described in the Chapter 2.2. On the other hand, exactly these isotope approaches gave the first important hints for the mixing of carbon pools likely to occur during the transport of recent assimilates within the tree and can be the basis for further studies on whole tree assimilate partitioning. More detailed highly temporally resolved and compound-specific isotope analysis in $^{13}CO_2$ pulse labeling experiments might be helpful to understand the allocation of newly produced assimilates to export, storage, transport, and respiration in auto- and heterotrophic tissues as affected by environmental conditions.

Badeck et al. (2005) reviewed more than 80 publications for differences in $\delta^{13}C$ signatures between organs caused by postphotosynthetic fractionation processes and showed that on average heterotrophic tissues were enriched in ^{13}C by

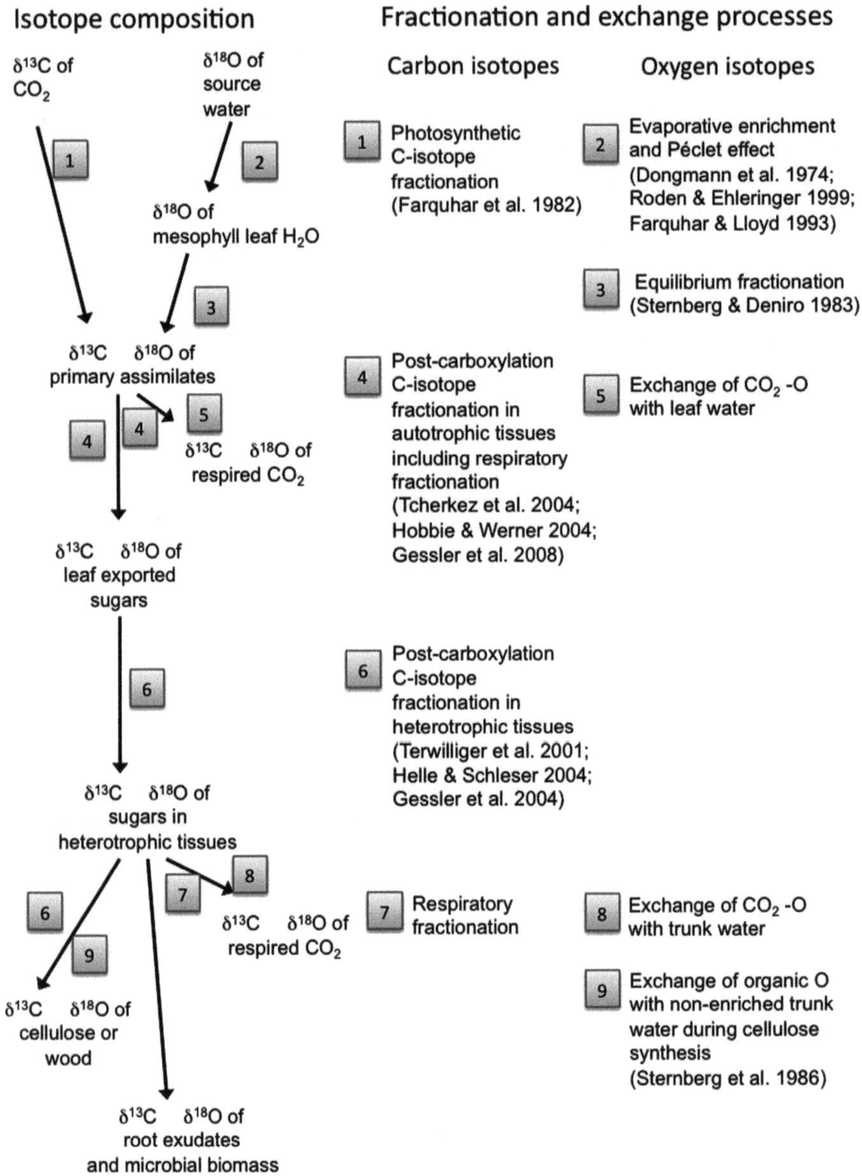

Fig. 3 Summary of the plant-related processes that potentially influence the carbon and oxygen isotopic composition of organic matter. Carbon and oxygen isotope fractionation or atom exchange processes are listed on the right side of the figure. In addition to the listed fractionation and exchange processes, the carbon isotope composition of atmospheric CO_2 and the oxygen isotope composition of the source water influence $\delta^{13}C$ and $\delta^{18}O$ of organic matter, respectively. The figure is adapted from Gessler et al. (2009)

approximately 1.3‰ compared to leaves. In particular in woody plants, the difference in carbon isotope signatures between leaves and stems showed a strong variation among species, ranging from a relative enrichment of the woody tissues of more than 8‰ to a relative depletion of 1.8‰. Preferential export of ^{13}C enriched sugars from the leaves to the phloem (Hobbie and Werner 2004), phloem unloading of sugars, partial metabolization, and preferential reloading of isotopically heavy sugars into the sieve tubes (Gessler et al. 2004), and fixation of organic matter in stems and roots by phosphoenolpyruvate carboxylase (PEPC) (Badeck et al. 2005) have been discussed as potential reasons for the ^{13}C enrichment in heterotrophic tissues. Any continuous and temporally as well as spatially constant ^{13}C enrichment of phloem-transported sugars will mainly challenge the calculation of intrinsic WUE or c_i/c_a values from phloem or tree-ring $\delta^{13}C$ values (for an detailed discussion see (Gessler et al. 2009) but should not impair assimilate transport estimates. This points to the fact that tree-ring isotope archives not necessarily conserve the absolute values of leaf intrinsic WUE but still allow comparisons among years and over the longer term.

However, Gessler et al. (2004, 2009) showed that postphotosynthetic fractionation processes that are related to sugar export from the leaf to the sieve tubes and associated with phloem transport can vary with time and might thus introduce additional variation in $\delta^{13}C$. This additional source of variation in the isotope composition might interfere with the variation caused by photosynthesis and can thus complicate or even prevent the tracing of assimilate transport through the plant by natural abundance approaches under particular conditions. There is, however, also the prospect that variations in postphotosynthetic isotope fractionation might give new and additional insights into the plant metabolism and shifts in metabolite turnover. As an example to illustrate this potential, we will have a closer look on respiratory carbon isotope fractionation in the next section.

Respiratory fractionation is due to nonstatistical intramolecular isotope distribution in respiratory substrates (see Rossmann et al. (1991) and Tcherkez et al. (2004), combined with fragmentation of the organic molecules during decarboxylation (Tcherkez et al. 2003) and can change the isotopic composition of respired CO_2 independently of the isotopic composition of the substrate. Respiratory carbon isotope fractionation on the one hand complicates the interpretation of the coupling between assimilation and respiration but provides on the other hand, valuable information on the biochemical processes underlying respiration under field conditions (Kodama et al. 2008): The CO_2 released by the decarboxylation of pyruvate (via the enzyme pyruvate dehydrogenase (PDH)) originates from the relatively ^{13}C-enriched C-3 and C-4 atoms of a glucose molecule, whereas the CO_2 emitted from Krebs cycle reactions originates from the C-1, C-2, C-5, and C-6 atoms, which are relatively ^{13}C depleted (Rossmann et al. 1991; Tcherkez et al. 2003). Any apparent ^{13}C enrichment of CO_2 compared to the organic respiratory substrate is thus due to incomplete oxidation of glucose molecules with a higher proportion of C-3 and C-4 atoms converted to CO_2. The simultaneous characterization of $\delta^{13}C$ in CO_2, and the potential substrates for respiration (including respiratory quotients; (Tcherkez et al. 2003) allows to quantify respiratory fractionation and to characterize shifts between PDH and Krebs cycle-mediated CO_2 production. Kodama et al. (2008) showed that respiratory fractionation was related to both temperature and respiration rate. At low temperatures (and consequently lower respiration rates), respiratory CO_2 was highly enriched compared to the organic respiratory substrate. Glycolysis (and thus decarboxylation of pyruvate) is less temperature-dependent than mitochondrial oxidation (citrate cycle) (Berry and Raison 1981; Atkin et al. 2000) and, thus, the relative contribution of CO_2 from glycolysis to total respiration is likely to increase at lower temperatures resulting in higher apparent isotope fractionation. In contrast, at higher temperatures, a relatively higher contribution of citrate cycle-derived CO_2 to total respiration would not only result in increased respiration rates but also in apparent fractionation approaching zero when complete oxidation of the substrate is assumed. Hymus et al. (2005) postulated that from the difference in $\delta^{13}C$ between sugars and CO_2 from leaf respiration, partitioning of carbon between respiration and secondary metabolites derived from acetyl-CoA could be characterized. Only recently, this hypothesis was further supported with positional labeling experiments by Priault et al. (2009).

4 Isotope Archives

A large portion of the assimilated carbon is transferred from short-lived high turnover pools (i.e., carbohydrates) not only to long-lived compounds like cellulose, lignin but also to transient storage compounds like starch. The information archived in growth parameters like tree-ring width or maximum latewood density and also in cell structure data can be used to retrospectively record growth and vitality of trees subjected to various environmental conditions over different climate zones. It also allows to reconstruct past climatic conditions on various temporal and spatial scales (Schweingruber 1996).

Besides the examination of the classical growth parameters, the assessment of stable isotope ratios in tree-ring cellulose or whole wood is gaining importance. The stable isotope composition in this tissue contains retrospective information on environmental conditions and the ecophysiological processes affected by these conditions (Leavitt 2002; McCarroll and Loader 2004; Poussart et al. 2004; Saurer et al. 1997; Schleser et al. 1999; Treydte et al. 2001; Treydte et al. 2006). However, there is still a considerable lack of information on the temporal dynamics of the transfer of the isotope signal from the sugar pools to the cellulose or wood pools as well as on potential postphotosynthetic fractionation and oxygen atom exchange during cellulose synthesis. Such fractionation and exchange processes might on the one hand partially uncouple the tree-ring signal from leaf physiology and the controlling environmental parameters. On the other hand, oxygen atom exchange between organic matter and nonenriched source water in the stem might pronounce the precipitation $\delta^{18}O$ signal in the tree ring and thus might provide important paleoclimatic information (Treydte et al. 2004;

Treydte et al. 2006). On average approximately 40–50% of the oxygen destined for the cellulose have been observed to exchange with not enriched stem water (Roden et al. 2000; Sternberg 2009). As described in detail by Farquhar et al. (1998), it is assumed that a portion of the hexose phosphates, which originate from sucrose, cycles through a triose phosphate pool before cellulose synthesis (Hill et al. 1995). The effect of this recycling will be that two out of three oxygen atoms that go through triose phosphates exchange with the nonenriched reaction water. In addition, from the ten oxygen atoms of a cellobiose unit, two atoms exchange with water already in glucose and fructose. Hill et al. (1995) calculated that 40–50% of hexose phosphate cycles trough triose phosphates during cellulose synthesis in oak stem tissue. From this calculation, 47% of the sucrose oxygen exchanges with stem water, thus being in accordance with the observed values (Farquhar et al. 1998).

A promising tool to characterize isotope fractionation and oxygen atom exchange during cellulose synthesis over the growing season is the high resolution measurement of intra-annual variations in the carbon and/or oxygen isotope composition of tree rings ideally combined with the temporally resolved assessments of sugar pools. When comparing high-resolution intra-annual tree-ring ^{13}C data with sugar pools it has, however, to be considered that the structural material in the tree ring is more intensively integrating isotope signals over time compared to the high turnover sugar pools.

By comparing the intra-annual variation of $\delta^{13}C$ in leaf material and tree-ring cellulose, Helle and Schleser (2004) applied $\delta^{13}C$ analyses to characterize the influence of stored carbon resources on the intra-annual production of tree-ring tissues and identified growth periods when newly assimilated carbon is readily transferred to cellulose and periods, when carbon derived from starch is used for radial growth. These authors were able to identify a seasonally recurrent triphase carbon isotope pattern in tree rings of broad leaf deciduous tree species. The most positive $\delta^{13}C$ values in wood or cellulose were observed at the very beginning of each growing season indicating the carbon to originate from remobilized starch (phase 1). Thereafter (phase 2), the decline in $\delta^{13}C$ represents the switch to newly assimilated carbon. In a third phase at the end of the growing season, $\delta^{13}C$ increases again, potentially due to tree-ring formation depending again on reserves. Owing to the seasonally varying role of stored carbon for the production of tree-ring material, the isotopic composition of the mid section (phase 2) of an annual ring is most likely to record the variations of environmental conditions and leaf physiology of the current year. Especially the early wood isotope signature might integrate climatic and physiological information also over the previous growing season or even longer, and is – at least in the case of $\delta^{13}C$ – affected by the carbon isotope fractionation during starch formation.

$^{13}CO_2$ pulse labeling experiments performed with saplings of the deciduous conifer *Larix gmelinii* (Kagawa et al. 2006) confirmed the natural abundance approaches as they showed that earlywood contained a large portion of nonlabeled carbon from the last growing season and that newly assimilated ^{13}C-labeled carbon was incorporated directly into the late wood.

In evergreen coniferous species, the coupling between the isotope composition of newly assimilated carbon and tree-ring organic matter seems more tight. Klein et al. (2005) showed in *Pinus halepensis* that subsections of tissue across annual rings and along needles, for which time of formation was resolved from growth rate analyses, showed rapid growth and $\delta^{13}C$ responses to changing environmental conditions. The $\delta^{13}C$ estimates calculated from gas exchange explained reasonably well the seasonal and interannual variations not only in needles but also in tree-ring tissues. Barbour et al. (2002) applied modeling approaches to mechanistically describe the $\delta^{13}C$ and $\delta^{18}O$ signatures of tree-ring cellulose in the coniferous species *Pinus radiata* and observed clear intra-annual responses of the isotope signatures in the archive to variations in water availability.

In order to fully understand the transfer of the isotopic signal from assimilation to the production of cellulose, we need, however, consistent information on the seasonal dynamics of $\delta^{13}C/\delta^{18}O$ in leaf sugars and phloem sugars as they are transported down the trunk and incorporated into cellulose. In a case study with *Pinus sylvestris*, Gessler et al. (2009) observed that the oxygen isotope signal was transferred in Scots pine from the leaf water to the tree ring with the expected enrichment of 27‰ from water to carbonyl oxygen, with time lags caused by metabolite turnover (in leaves and trunk) and transport time and with an approximately 40% exchange between organic oxygen and xylem water oxygen during cellulose synthesis. Accounting for these factors, it was possible to extract a transpiration signal from $\Delta^{18}O$ in tree-ring cellulose with an intra-annual resolution.

From this compilation of research related to the transfer of isotope information from the fast turnover assimilate pools in the canopy to the tree rings, the following might be concluded:

1. Especially in deciduous trees like beech, the seasonal variation of the carbon isotope composition is strongly influenced by the origin of the organic matter (storage pools vs. recent assimilates) used for wood/cellulose synthesis. Only particular parts of a tree ring, which are supplied with newly formed assimilates, carry the current year information on climate and physiology.
2. In coniferous species, a more direct coupling between leaf level carbon and oxygen isotope composition in recent assimilates and the isotope information in the tree ring seems to be present, even though the use of reserves might be also important for wood tissue produced at the beginning of the growing season.

For climate reconstruction and retrospective analyses of leaf and canopy physiology, early and late wood should be analyzed separately as they integrate information over different time scales. In addition, we need more information on postphotosynthetic carbon isotope fractionation associated with leaf export, phloem transport, and cellulose synthesis. There is the need for an interspecies comparison of such discrimination and we have to seek for a mechanistic understanding of the underlying processes. Moreover, pulse label experiments with $^{13}CO_2$ (or ^{18}O-labeled source water or water vapor) at different times during the growing season as well as in years with different environmental conditions are needed with larger trees under field conditions in order to understand the timing of cellulose deposition and partitioning of recent assimilates between storage pools, structural tissues, respiration, and transfer to soil microorganisms. With such a comprehensive work, a system understanding of the mechanisms and the temporal variation of the carbon cycling in forest ecosystems might be possible.

5 Conclusions

The assessment of natural abundance stable carbon and oxygen isotope signatures in different organic matter pools is a valuable tool to characterize the transport and distribution of new assimilates within the plant as well as the transfer to soil and microorganisms. The natural variation of the $\delta^{13}C$ and $\delta^{18}O$ values can on the one hand be followed through the whole ecosystem in order to characterize transport and turn-over times of assimilates in a temporal resolution from hours to years. On the other hand, the stable isotope composition contains important information on environmental conditions and leaf physiology, which can be used to reconstruct WUE (Farquhar et al. 1982), stomatal conductance (Keitel et al. 2003) and transpiration (Gessler et al. 2009) as well as climatic conditions. Depending on the tissue, organ, and chemical compound examined, different temporal integration related to the turn-over times of the organic matter pools are obtained. Pulse labeling experiments allow more accurate calculations of assimilate turn-over and residence times compared to the natural abundance technique and might be used in future to combine cross-scale metabolic flux analysis with tree and ecosystem carbon allocation assessments under field conditions. The combination of both approaches might help to elucidate postphotosynthetic isotope fractionation and thus deepen our understanding on the effects of metabolic flux partitioning on isotope fractionation and oxygen atom exchange. Integration of stable isotope approaches into ecosystem scale assessments of carbon and water fluxes and biomass accumulation will as a consequence provide novel insights into the balance of matter and the underlying mechanisms in forest stands. This is a further step to a mechanistic understanding of the carbon and water fluxes, which is a prerequisite for estimating the response of trees and ecosystem to the changing environmental conditions in future.

References

Adams MA, Grierson PF (2001) Stable isotopes at natural abundance in terrestrial plant ecology and ecophysiology: an update. Plant Biol 3:299–310

Angert A, Biraud S, Bonfils C, Henning CC, Buermann W, Pinzon J, Tucker CJ, Fung I (2005) Drier summers cancel out the CO_2 uptake enhancement induced by warmer springs. Proc Natl Acad Sci USA 102:10823–10827

Atkin OK, Edwards EJ, Loveys BR (2000) Response of root respiration to changes in temperature and its relevance to global warming. New Phytol 147:141–154

Badeck FW, Tcherkez G, Nogues S, Piel C, Ghashghaie J (2005) Post-photo synthetic fractionation of stable carbon isotopes between plant organs – a widespread phenomenon. Rapid Commun Mass Spectrom 19:1381–1391

Barbour MM, Farquhar GD (2000) Relative humidity- and ABA-induced variation in carbon and oxygen isotope ratios of cotton leaves. Plant Cell Environ 23:473–485

Barbour MM, Hunt JE, Dungan RJ, Turnbull MH, Brailsford GW, Farquhar GD, Whitehead D (2005) Variation in the degree of coupling between $\delta^{13}C$ of phloem sap and ecosystem respiration in two mature Nothofagus forests. New Phytol 166:497–512

Barbour MM, Walcroft AS, Farquhar GD (2002) Seasonal variation in $\delta^{13}C$ and $\delta^{18}O$ of cellulose from growth rings of *Pinus radiata*. Plant Cell Environ 25:1483–1499

Barnard RL, Salmon Y, Kodama N, Sorgel K, Holst J, Rennenberg H, Gessler A, Buchmann N (2007) Evaporative enrichment and time lags between $\delta^{18}O$ of leaf water and organic pools in a pine stand. Plant Cell Environ 30:539–550

Berry JA, Raison JK (1981) Responses of macrophytes to temperature. In: Lange OL, Nobel PS, Osmond B, Ziegler H (eds) Physiological plant ecology I. Responses to the physical environment. Springer, Berlin, pp 277–338

Bhupinderpal S, Nordgren A, Lofvenius MO, Hogberg MN, Mellander PE, Hogberg P (2003) Tree root and soil heterotrophic respiration as revealed by girdling of boreal Scots pine forest: extending observations beyond the first year. Plant Cell Environ 26:1287–1296

Bond-Lamberty B, Wang CK, Gower ST (2004) Contribution of root respiration to soil surface CO_2 flux in a boreal black spruce chronosequence. Tree Physiol 24:1387–1395

Bowling DR, Pataki DE, Ehleringer JR (2003) Critical evaluation of micrometeorological methods for measuring ecosystem–atmosphere isotopic exchange of CO_2. Agric For Meteorol 116:159–179

Bowling DR, McDowell NG, Bond BJ, Law BE, Ehleringer JR (2002) ^{13}C content of ecosystem respiration is linked to precipitation and vapor pressure deficit. Oecologia 131:113–124

Bradford MA, Fierer N, Reynolds JF (2008) Soil carbon stocks in experimental mesocosms are dependent on the rate of labile carbon, nitrogen and phosphorus inputs to soils. Funct Ecol 22:964–974

Brandes E, Kodama N, Whittaker K, Weston C, Rennenberg H, Keitel C, Adams MA, Gessler A (2006) Short-term variation in the isotopic composition of organic matter allocated from the leaves to the stem of *Pinus sylvestris*: effects of photosynthetic and postphotosynthetic carbon isotope fractionation. Global Change Biol 12:1922–1939

Brandes E, Wenninger J, Koeniger P, Schindler D, Rennenberg H, Leibundgut C, Mayer H, Gessler A (2007) Assessing environmental and physiological controls over water relations in a Scots pine (*Pinus sylvestris* L.) stand through analyses of stable isotope composition of water and organic matter. Plant Cell Environ 30:113–127

Brugnoli E, Farquhar GD (2000) Photosynthetic fractionation of carbon isotopes. In: Leegood RC, Sharkey TD, von Caemmerer S (eds) Photosynthesis: physiology and metabolism. Kluwer Akademic Publishers, Dordrecht, pp 399–434

Buchmann N (2002) Plant ecophysiology and forest response to global change. Tree Physiol 22:1177–1184

Carbone MS, Czimczik CI, McDuffee KE, Trumbore SE (2007) Allocation and residence time of photosynthetic products in a boreal forest using a low-level ^{14}C pulse-chase labeling technique. Global Change Biol 13:466–477

Cernusak LA, Farquhar GD, Pate JS (2005) Environmental and physiological controls over oxygen and carbon isotope composition of Tasmanian blue gum, *Eucalyptus globulus*. Tree Physiol 25:129–146

Ciais P, Reichstein M, Viovy N, Granier A, Ogee J, Allard V, Aubinet M, Buchmann N, Bernhofer C, Carrara A, Chevallier F, De Noblet N, Friend AD, Friedlingstein P, Grunwald T, Heinesch B, Keronen P, Knohl A, Krinner G, Loustau D, Manca G, Matteucci G, Miglietta F, Ourcival JM, Papale D, Pilegaard K, Rambal S, Seufert G, Soussana JF, Sanz MJ, Schulze ED, Vesala T,

Valentini R (2005) Europe-wide reduction in primary productivity caused by the heat and drought in 2003. Nature 437:529–533

Damesin C, Rambal S, Joffre R (1998) Seasonal and annual changes in leaf $\delta^{13}C$ in two co-occurring Mediterranean oaks: relations to leaf growth and drought progression. Funct Ecol 12:778–785

Deng XM, Joly RJ, Hahn DT (1990) The influence of plant water deficit on photosynthesis and translocation of ^{14}C-labeled assimilates in cacao seedlings. Physiol Plant 78:623–627

Drake BG, Gonzalez-Meler MA, Long SP (1997) More efficient plants: a consequence of rising atmospheric CO_2? Ann Rev Plant Physiol Plant Mol Biol 48:609–639

Ekblad A, Högberg P (2001) Natural abundance of ^{13}C in CO_2 respired from forest soils reveals speed of link between tree photosynthesis and root respiration. Oecologia 127:305–308

Farquhar GD, Barbour MM, Henry BK (1998) Interpretation of oxygen isotope composition of leaf material. In: Griffiths H (ed) Stable isotopes – integration of biological, ecological and geochemical processes. Oxford, Bios Scientific Publishers, pp 27–74

Farquhar GD, Cernusak LA (2005) On the isotopic composition of leaf water in the non-steady state. Funct Plant Biol 32:293–303

Farquhar GD, O' Leary MH, Berry JA (1982) On the relationship between carbon isotope discrimination and the intercellular carbon-dioxide concentration in leaves. Aust J Plant Physiol 9:121–137

Flexas J, Medrano H (2002) Drought-inhibition of photosynthesis in C_3 plants: stomatal and non-stomatal limitations revisited. Ann Bot 89:183–189

Forkel R, Knoche R (2006) Regional climate change and its impact on photooxidant concentrations in southern Germany: simulations with a coupled regional climate-chemistry model. J Geophys Res Atmosph 111:G02018

Friend AD, Arneth A, Kiang NY, Lomas M, Ogee J, Rodenbeckk C, Running SW, Santaren JD, Sitch S, Viovy N, Woodward FI, Zaehle S (2007) FLUXNET and modelling the global carbon cycle. Global Change Biol 13:610–633

Ghashghaie J, Badeck FW, Lanigan G, Nogues S, Tcherkez G, Deleens E, Cornic G, Griffiths H (2003) Carbon isotope fractionation during dark respiration and photorespiration in C_3 plants. Phytochem Rev 2:145–161

Gessler A, Brandes E, Buchmann N, Helle G, Rennenberg H, Barnard R (2009) Tracing carbon and oxygen isotope signals from newly assimilated sugars in the leaves to the tree ring archive. Plant Cell Environ 32:780–795

Gessler A, Jung K, Gasche R, Papen H, Heidenfelder A, Borner E, Metzler B, Augustin S, Hildebrand E, Rennenberg H (2005) Climate and forest management influence nitrogen balance of European beech forests: microbial N transformations and inorganic N net uptake capacity of mycorrhizal roots. Eur J For Res 124:95–111

Gessler A, Keitel C, Kodama N, Weston C, Winters AJ, Keith H, Grice K, Leuning R, Farquhar GD (2007a) $\delta^{13}C$ of organic matter transported from the leaves to the roots in *Eucalyptus delegatensis*: short-term variations and relation to respired CO_2. Funct Plant Biol 34:692–706

Gessler A, Keitel C, Kreuzwieser J, Matyssek R, Seiler W, Rennenberg H (2007b) Potential risks for European beech (*Fagus sylvatica* L.) in a changing climate. Trees Struct Funct 21:1–11

Gessler A, Rennenberg H, Keitel C (2004) Stable isotope composition of organic compounds transported in the phloem of European beech – evaluation of different methods of phloem sap collection and assessment of gradients in carbon isotope composition during leaf-to-stem transport. Plant Biol 6:721–729

Gessler A, Schrempp S, Matzarakis A, Mayer H, Rennenberg H, Adams MA (2001) Radiation modifies the effect of water availability on the carbon isotope composition of beach (*Fagus sylvatica*). New Phytol 150:653–664

Ghashghaie J, Duranceau M, Badeck FW, Cornic G, Adeline MT, Deleens E (2001) $\delta^{13}C$ of CO_2 respired in the dark in relation to $\delta^{13}C$ of leaf metabolites: comparison between *Nicotiana sylvestris* and *Helianthus annuus* under drought. Plant Cell Environ 24:505–515

Gleixner G, Schmidt HL (1997) Carbon isotope effects on the fructose-1,6-bisphosphate aldolase reaction, origin for non-statistical 13C distributions in carbohydrates. J Biol Chem 272:5382–5387

Gorissen A, Tietema A, Joosten NN, Estiarte M, Penuelas J, Sowerby A, Emmett BA, Beier C (2004) Climate change affects carbon allocation to the soil in shrublands. Ecosystems 7:650–661

Hawkes CV, Hartley IP, Ineson P, Fitter AH (2008) Soil temperature affects carbon allocation within arbuscular mycorrhizal networks and carbon transport from plant to fungus. Global Change Biol 14:1181–1190

Helle G, Schleser GH (2004) Beyond CO_2-fixation by Rubisco – an interpretation of $^{13}C/^{12}C$ variations in tree rings from novel intra-seasonal studies on broad-leaf trees. Plant Cell Environ 27:367–380

Hill SA, Waterhouse JS, Field EM, Switsur VR, Aprees T (1995) Rapid recycling of triose phosphates in oak stem tissue. Plant Cell Environ 18:931–936

Hobbie EA, Werner RA (2004) Intramolecular, compound-specific, and bulk carbon isotope patterns in C_3 and C_4 plants: a review and synthesis. New Phytol 161:371–385

Högberg P, Hogberg MN, Gottlicher SG, Betson NR, Keel SG, Metcalfe DB, Campbell C, Schindlbacher A, Hurry V, Lundmark T, Linder S, Nasholm T (2008) High temporal resolution tracing of photosynthate carbon from the tree canopy to forest soil microorganisms. New Phytol 177:220–228

Högberg P, Nordgren A, Buchmann N, Taylor AFS, Ekblad A, Hogberg MN, Nyberg G, Ottosson-Lofvenius M, Read DJ (2001) Large-scale forest girdling shows that current photosynthesis drives soil respiration. Nature 411:789–792

Hymus GJ, Maseyk K, Valentini R, Yakir D (2005) Large daily variation in ^{13}C-enrichment of leaf-respired CO_2 in two *Quercus* forest canopies. New Phytol 167:377–384

IPCC (2007) Climate change 2007: the physical science Basis. Contribution of Working Group I to the Fourth Assessment Report of the Intergovernmental Panel on climate change

Kagawa AKIR, Sugimoto ATSU, Maximov TC (2006) $^{13}CO_2$ pulse-labelling of photoassimilates reveals carbon allocation within and between tree rings. Plant Cell Environ 29: 1571–1584

Keitel C, Adams MA, Holst T, Matzarakis A, Mayer H, Rennenberg H, Gessler A (2003) Carbon and oxygen isotope composition of organic compounds in the phloem sap provides a short-term measure for stomatal conductance of European beech (*Fagus sylvatica* L.). Plant Cell Environ 26:1157–1168

Keitel C, Matzarakis A, Rennenberg H, Gessler A (2006) Carbon isotopic composition and oxygen isotopic enrichment in phloem and total leaf organic matter of European beech (*Fagus sylvatica* L.) along a climate gradient. Plant Cell Environ 29:1492–1507

Klein T, Hemming D, Lin TB, Grunzweig JM, Maseyk K, Rotenberg E, Yakir D (2005) Association between tree-ring and needle $\delta^{13}C$ and leaf gas exchange in *Pinus halepensis* under semi-arid conditions. Oecologia 144:45–54

Knohl A, Werner RA, Brand WA, Buchmann N (2005) Short-term variations in $\delta^{13}C$ of ecosystem respiration reveals link between assimilation and respiration in a deciduous forest. Oecologia 142:70–82

Kodama N, Barnard R, Salmon Y, Weston C, Ferrio JP, Holst J, Werner R, Sauer M, Eugster W, Buchmann N, Gessler A (2008) Temporal dynamics of the carbon isotope composition in a *Pinus sylvestris* stand – from newly assimilated organic carbon to respired CO_2. Oecologia 156:737–750

Korol RL, Kirschbaum MUF, Farquhar GD, Jeffreys M (1999) Effects of water status and soil fertility on the C-isotope signature in *Pinus radiata*. Tree Physiol 19:551–562

Leavitt SW (2002) Prospects for reconstruction of seasonal environment from tree-ring $\delta^{13}C$: baseline findings from the Great Lakes area, USA. Chem Geol 192:47–58

Leavitt SW, Long A (1986) Stable-carbon isotope variability in tree foliage and wood. Ecology 67:1002–1010

Li TH, Li SH, Wang J, Yu KS (2003) Effects of water stress at different deficit intensities on transport and distribution of ^{14}C-assimilates in micropropagated apple plants. Eur J Hort Sci 68:227–233

McCarroll D, Loader NJ (2004) Stable isotopes in tree rings. Quat Sci Rev 23:771–801

McCarroll D, Pawellek F (2001) Stable carbon isotope ratios of *Pinus sylvestris* from northern Finland and the potential for extracting a climate signal from long Fennoscandian chronologies. Holocene 11:517–526

Palta J, Gregory P (1997) Drought affects the fluxes of carbon to roots and soil in ^{13}C pulse labelled plants of wheat. Soil Biol Biochem 29:1395–1403

Plain C, Gerant D, Maillard P, Dannoura M, Dong Y, Zeller B, Priault P, Parent F, Epron D (2009) Tracing of recently assimilated carbon in respiration at high temporal resolution in the field with tuneable diode laser absorption spectrometer after in situ $^{13}CO_2$ pulse labelling of 20 year-old beech trees. Tree Physiol 29:1433–1445

Plaut Z, Reinhold L (1965) Effect of water stress on ^{14}C sucrose transport in bean plant. Aust J Biol Sci 18:1143–1155

Poussart PF, Evans MN, Schrag DP (2004) Resolving seasonality in tropical trees: multi-decade, high-resolution oxygen and carbon isotope records from Indonesia and Thailand. Earth Planet Sci Lett 218:301–316

Priault P, Wegener F, Werner C (2009) Pronounced differences in diurnal variation of carbon isotope composition of leaf respired CO_2 among functional groups. New Phytol 181:400–412

Roden JS, Lin GG, Ehleringer JR (2000) A mechanistic model for interpretation of hydrogen and oxygen isotope ratios in tree-ring cellulose. Geochim Cosmochim Acta 64:21–35

Rossmann A, Butzenlechner M, Schmidt HL (1991) Evidence for a nonstatistical carbon isotope distribution in natural glucose. Plant Physiol 96:609–614

Rühr N, Offermann C, Gessler A, Winkler JB, Ferrio JP, Buchmann N, Barnard RL (2009) Effects of drought on allocation of recent carbon: From beech leaves to soil respiration. New Phytol 184:950–961

Saurer M, Aellen K, Siegwolf R (1997) Correlating δ^{13}C and δ^{18}O in cellulose of trees. Plant Cell Environ 20:1543–1550

Scheidegger Y, Saurer M, Bahn M, Siegwolf R (2000) Linking stable oxygen and carbon isotopes with stomatal conductance and photosynthetic capacity: a conceptual model. Oecologia 125: 350–357

Schleser GH, Helle G, Lucke A, Vos H (1999) Isotope signals as climate proxies: the role of transfer functions in the study of terrestrial archives. Quat Sci Rev 18:927–943

Schmidt HL (2003) Fundamentals and systematics of the non-statistical distributions of isotopes in natural compounds. Naturwissenschaften 90:537–552

Schweingruber FH (1996) Dendrochronology – an extremely exact measuring tool for the study of environmental and human history. Naturwissenschaften 83:370–377

Steinmann K, Siegwolf RTW, Saurer M, Körner C (2004) Carbon fluxes to the soil in a mature temperate forest assessed by ^{13}C isotope tracing. Oecologia 141:489–501

Sternberg L (2009) Oxygen stable isotope ratios of tree-ring cellulose: the next phase of understanding. New Phytol 181:553–562

Tcherkez G, Farquhar GD, Badeck F, Ghashghaie J (2004) Theoretical considerations about carbon isotope distribution in glucose of C_3 plants. Funct Plant Biol 31:857–877

Tcherkez G, Farquhar GD (2005) Carbon isotope effect predictions for enzymes involved in the primary carbon metabolism of plant leaves. Funct Plant Biol 32:277–291

Tcherkez G, Nogues S, Bleton J, Cornic G, Badeck F, Ghashghaie J (2003) Metabolic origin of carbon isotope composition of leaf dark-respired CO_2 in French bean. Plant Physiol 131:237–244

Treydte K, Schleser GH, Schweingruber FH, Winiger M (2001) The climatic significance of δ^{13}C in subalpine spruces (Lotschental, Swiss Alps) – a case study with respect to altitude, exposure and soil moisture. Tellus Ser B-Chem Phys Meteorol 53:593–611

Treydte K, Welscher C, Schleser GH, Helle G, Esper J, Frank D, Buntgen U (2004) The climatic signal in oxygen isotopes of junipers at the lower timberline in the Karakorum, Pakistan. Tree Rings Archaeol 2:100–106

Treydte KS, Schleser GH, Helle G, Frank DC, Winiger M, Haug GH, Esper J (2006) The twentieth century was the wettest period in northern Pakistan over the past millennium. Nature 440:1179–1182

Van Bel AJE (2003) The phloem, a miracle of ingenuity. Plant Cell Environ 26:125–149

Wang XF, Yakir D (2000) Using stable isotopes of water in evapotranspiration studies. Hydrolog Process 14:1407–1421

Wardlaw I (1969) Effect of water stress on translocation in relation to photosynthesis and growth II. Effect during leaf development in Lolium temulentum. Aust J Biol Sci 22:1–16

Werner C, Unger S, Pereira JS, Maia R, David TS, Kurz-Besson C, David JS, Máguas C (2006) Importance of short-term dynamics in carbon isotope ratios of ecosystem respiration ($\delta^{13}C_R$) in a Mediterranean oak woodland and linkage to environmental factors. New Phytol 172:330–346

West AG, Hultine KR, Burtch KG, Ehleringer JR (2007) Seasonal variations in moisture use in a pinon-juniper woodland. Oecologia 153:787–798

Windt CW, Vergeldt FJ, De Jager PA, van As H (2006) MRI of long-distance water transport: a comparison of the phloem and xylem flow characteristics and dynamics in poplar, castor bean, tomato and tobacco. Plant Cell Environ 29:1715–1729

Appropriate Use of Genetic Manipulation for the Development of Restoration Plant Materials

T.A. Jones and J.G. Robins

Contents

Abstract The diversity of approaches for developing restoration plant material reflects a variety of philosophies that represent what can and should be accomplished by restoration. The "natural" approach emphasizes emulation of putative naturally occurring patterns of genetic variation. The "genetically manipulated" approach involves such techniques as artificial selection, hybridization, bulking, and chromosome doubling to create populations that are ostensibly as well or better equipped to restore ecosystem function than the extirpated natural populations that they are designed to replace. A number of caveats have been issued regarding manipulated plant materials, including concerns regarding improper genetic identity, outbreeding

T.A. Jones (✉) and J.G. Robins
USDA-Agricultural Research Service, Forage and Range Research Laboratory, Utah State University, Logan, UT 84322-6300, USA
e-mail: Thomas.Jones@ars.usda.gov

U. Lüttge et al. (eds.), *Progress in Botany 72*,
DOI 10.1007/978-3-642-13145-5_10, © Springer-Verlag Berlin Heidelberg 2011

depression, maladaptation, and inappropriate amounts of genetic variation. Here we detail (1) when these concerns are likely to be valid or inconsequential, (2) how precautions may be taken to minimize these concerns, and (3) how to respect, as much as possible, the principles cherished by proponents of natural plant materials, yet still take advantage of the benefits of genetic manipulation.

1 Use of Natural and Genetically Manipulated Plant Materials

In recent years, the use of prevariety germplasm (Fig. 1) has become popular for restoration applications in the USA. Prevariety germplasm consists of released (termed "selected" or "tested" class) or unreleased (termed "source identified" class) plant material that qualifies for seed certification but lacks the field testing across multiple locations or years required for cultivar release. In lieu of this testing, the native plant material is presumed to be adapted to the locale from which it originates. The prevariety germplasm scheme supports two "tracks" for native plant materials, "natural-track" and "manipulated-track" (Young et al. 2004). Materials that are an "unrestricted representation of the intact wildland population on the original site" qualify for natural-track status. In addition, intentional genetic manipulation must be avoided when the material is being increased or tested. On the other hand, manipulated-track materials have been "purposefully or inadvertently hybridized with other accessions or selected for distinctive traits."

Based on the objectives for the plant material, the plant-material developer must decide whether a plant material's track should be natural or manipulated. The relative merit of the two tracks is the subject of much debate. Currently, both approaches are widely used, and each has its merits. By addressing the common objections to genetic manipulation in restoration plant materials, we attempt to show how the use of such materials is legitimate when environmental degradation is severe. Indeed, using genetic manipulation to develop plant materials that are able to resist biotic and abiotic stresses may be the best hope for the greatest restoration challenges (Jones and Monaco 2009; Jones et al. 2010).

Nevertheless, manipulated materials are often deemed inappropriate for restoration practice. For example, in 2001 we released P-7 germplasm of bluebunch wheatgrass (*Pseudoroegneria spicata* [Pursh] A. Löve), a multiple-origin polycross of 25 populations, based on its high genetic diversity (Larson et al. 2000; Sect. 3.6). But despite P-7's genetic similarity to other released bluebunch wheatgrass materials (Larson et al. 2004) and despite support in the restoration ecology literature for such an approach (Millar and Libby 1989; Burton and Burton 2002; Rice and Emery 2003), this material has been roundly criticized for its lack of a single point-of-origin. While some are ready and willing to implement such materials onto the landscape as the best hope for countering what they perceive to be a ferocious and overwhelming threat, others consider their use to be unacceptable. We contend that the desired approach is the one that maximizes the probability of success for a particular situation. Our personal plant material development efforts have entailed

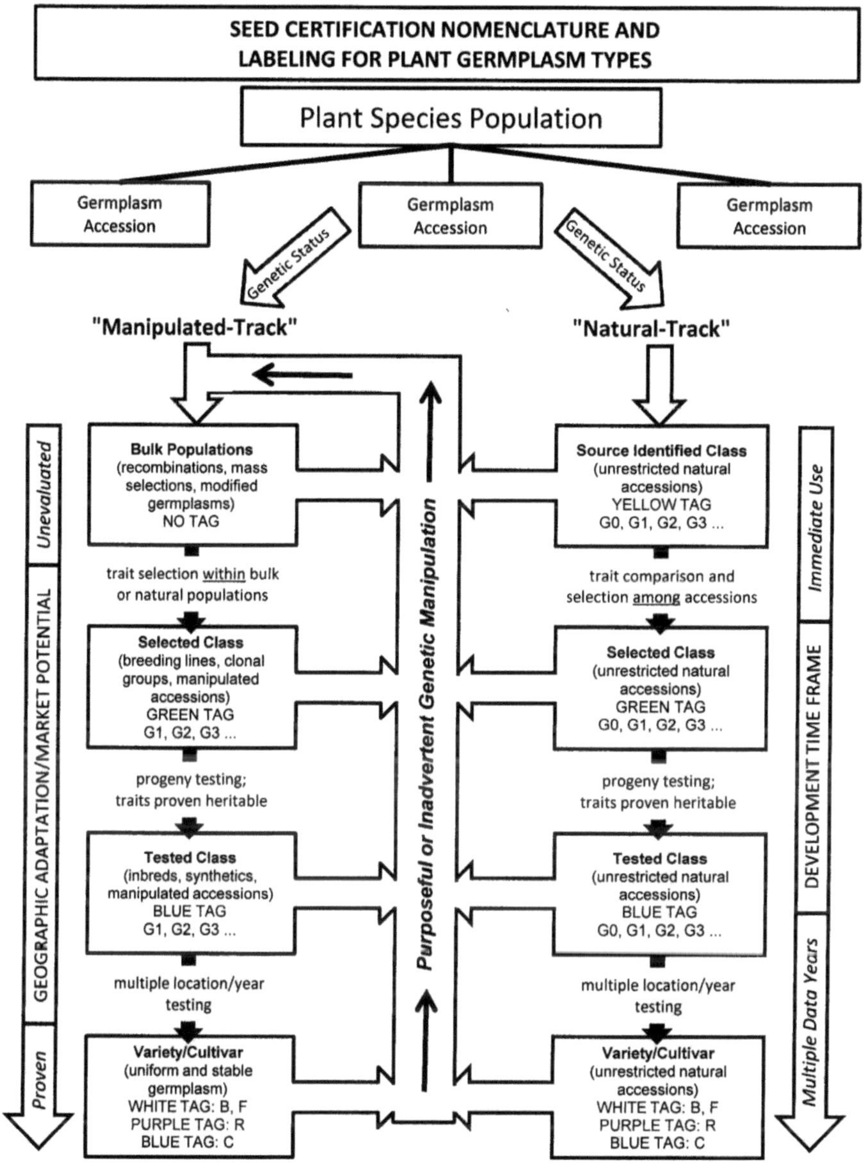

Fig. 1 Prevariety germplasm development flow chart as updated from Young et al. (2004). (With permission of the Association of Official Seed Certifying Agencies (AOSCA))

both approaches, and we see nothing to be gained by insistence on the exclusive use of one approach over the other regardless of the restoration challenge at hand.

The natural approach has been widely accepted because it appeals to restoration ecologists and practitioners trained in the evolutionary paradigm, which is preeminent in the field of conservation biology (Meffe and Carroll 1997). However, the manipulated approach may seem suspect because it does not resonate with the training and values of this clientele. The basic elements of the manipulated approach originate in the field of plant breeding, which adheres to the consumptive resource-use paradigm of agriculture. This paradigm may be considered inappropriate by the natural-resources community, which often emphasizes nonconsumptive uses. On the other hand, as a result of past failures, others have come to accept its utility as a viable alternative. Because the manipulated approach remains controversial, the remainder of this chapter deals with issues surrounding its application. This should not be construed as a bias on our part. Our contention is that, when used properly, manipulated plant material may provide viable and valuable options to assist evolutionary processes (Jones and Monaco 2009) in severely modified environments where nature has been fundamentally altered by man's intentional or inadvertent activities (Milton 2003; Hobbs et al. 2006; Kareiva et al. 2007).

To this end, we detail here the essential elements of the manipulated approach and respond to seven criticisms commonly aired against it. While these criticisms may have merit in individual circumstances, we do not believe they disqualify manipulated plant materials in general. Nor do they legitimately discredit an approach that recognizes situations and circumstances in which man's domestication of nature is inevitable (Wilkinson 2004; Kareiva et al. 2007).

2 Development of Genetically Manipulated Plant Materials

Four genetic-manipulation procedures, artificial selection, hybridization, bulking, and chromosome doubling, are highlighted below. The purpose of artificial selection is to increase the frequency of desirable genes for the trait in question. Artificial selection is much more likely to be employed for a cross-pollinated than for a self-pollinated or apomictic species. Cross-pollinated species are more responsive to this approach because (1) their naturally occurring populations feature much more heterozygosity and genetic variation than naturally occurring self-pollinated and apomictic populations and (2) their mating system makes genetic manipulation relatively easy. Nevertheless, hybridization between self-pollinated populations can generate considerable genetic variation, which can then be subjected to artificial selection. For many row crops, the objective has typically been to produce a uniform phenotype, which is typically desired for agriculture. For native plants, however, selection may be used to increase mean performance with the concurrent objective of maintaining as much genetic variation for the trait as possible.

Artificial selection is probably the plant breeding profession's oldest and most essential activity, but some believe that such material should be disqualified for restoration use. A common perception is that artificial selection alters the plant's response in unforeseen ways or that it has been modified such that it is productive only in a cultivated setting, e.g., a seed field. The emphasis in plant ecology on trade-offs among plant traits, e.g., Reich et al. (2003), suggests to some that selected materials are extremely unlikely to perform as well as natural-track materials. Selection for one trait is considered to have a probable cost in terms of another trait that may very well be hidden, only to appear during an extreme environmental event (Brown and Amacher 1999). A trade-off among traits may be driven by a physical law, a physiological relationship, a fundamental ecological constraint, genetic linkage on a chromosome, or any combination of these.

Hybridization is the second genetic-manipulation procedure. Its purpose may be to develop a broad-based population for direct use or to increase genetic diversity, the fuel that drives artificial selection, as a prelude to such selection. Second, its purpose may be to generate heterosis, by combining distinct, yet synergistic, genomes into a single genotype. Third, its purpose may be to introduce specific traits from one population or even taxon to another. Plant material developers may choose to implement hybridization for any one or combination of these reasons when planning plant material development.

The third genetic-manipulation procedure is bulking. Bulking is used for self-pollinated or apomictic species and is analogous to the use of hybridization for cross-pollinated species. Multiple populations may be combined into a bulk mixture without hybridization to increase genetic diversity of a single plant material. However, care should be taken that minimal genetic variation is present for traits critical to seed production, e.g., phenological maturity or seed shattering, that would favor some genotypes over others across generations of seed increase. Otherwise, loss of one or more of the bulk components is likely to occur through inadvertent selection, termed "genetic shift" (Ferdinandez et al. 2005). If such variation is desired for a particular application, it can be achieved by combining multiple plant materials of the same species in the seed mix prior to planting.

For example, populations of the self-pollinating Indian ricegrass (*Achnatherum hymenoides* [Roem. & Schult.] Barkworth) commonly feature seed polymorphism associated with seed dormancy (Jones et al. 2007). Seed polymorphism is genetically controlled with a single plant producing seed of a single morph associated with a single germination type. If low-dormancy and high-dormancy morphs are mixed in a plant material, the latter would quickly be lost due to genetic shift through generations of seed multiplication. However, as described above, two morphs could be grown independently and be subsequently combined in the seed mix at planting to generate a polymorphic population for a restoration planting.

The fourth genetic-manipulation procedure is chromosome doubling. This procedure may confer advantages in species or taxonomic groups that possess multiple ploidy levels. For example, seed mass often increases with ploidy level within a polyploid complex (Bretagnolle et al. 1995). These researchers found that, on average, tetraploid orchardgrass (*Dactylis glomerata* L.) lines possessed greater seed mass and seedling growth than diploid lines, but tetraploid and diploid lines with similar seed mass were more comparable for seedling growth. Seed mass of tetraploid perennial ryegrass (*Lolium perenne* L.) cultivars averaged 58% more than diploid cultivars, and seed mass was positively correlated with seedling vigor (Smith et al. 2003). In 2008, we released "Continental" basin wildrye (*Leymus cinereus* [Scribn. & Merr.] A. Löve), whose development involved chromosome doubling to generate an induced octoploid, which was subsequently hybridized with a natural octoploid (Jones et al. 2009).

3 Responses to Seven Common Objections to Genetically Manipulated Plant Materials

Commonly expressed objections to genetically manipulated materials include the following: (1) they do not reflect natural phylogeographic patterns, such as described by Larson et al. (2004), (2) they incur outbreeding depression upon hybridization with remnant indigenous material (Hufford and Mazer 2003), (3) they may themselves be subject to outbreeding depression, (4) they are too well adapted, being competitively exclusive or potentially invasive, (5) they are poorly adapted and lacking in fitness, (6) they possess too much genetic variation as a result of hybridization, and (7) they possess too little genetic variation upon selection. While each of these concerns is legitimate, we argue that genetic manipulation may be performed in a "restoration-appropriate" fashion to respect these concerns. When done properly, genetic manipulation may enhance the probability of restoration success. Below, we respond to each of the seven objections listed above.

3.1 Objection: Manipulated Plant Materials Are Not Genetically Appropriate

This criticism reflects the belief that restoration plant materials should emulate the natural patterns of genetic variation in place prior to disturbance. The intent is to implement plant materials that allow evolutionary processes to proceed along an unaltered trajectory. Nevertheless, both hybridization and selection are essential elements of evolutionary change, the former because it augments genetic variation and the latter because it modifies that variation so that the plant adapts to its environment. Furthermore, both utilize naturally occurring genetic variation. The controversy regarding hybridization and selection, then, does not result from the processes themselves, but rather from the fact that the man is the actor rather than nature. Some argue that man's efforts are futile and deleterious because nature is more complex than man can ever comprehend, much less replicate. However, if artificial selection and hybridization do indeed have the potential to develop plant material with enhanced stress tolerance and competitive ability against invasive plants, their use should not be automatically disqualified when they are most desperately needed and potentially offer great benefit.

Many biologists possess a belief in the near-infallibility of natural selection (Gould and Lewontin 1979), which suggests to them that local material is "best" (Johnson et al. 2004). However, this assumption is beginning to be challenged by restoration ecologists (Broadhurst et al. 2008). The greatest irony of this assertion is that sites with the most compelling need for restoration have been drastically altered by biotic or abiotic disturbances such as overgrazing, invasion by pernicious weeds, soil degradation, and altered fire regime. Furthermore, it should be remembered that it is perfectly possible for two very closely related populations that have been exposed to distinct selection pressures to be profoundly different for critical traits responsive to selection (Antonovics and Bradshaw 1970).

While manipulated material may not strictly reflect natural geographic patterns of genetic variation, the current trend is to accommodate this view by developing plant materials corresponding to geographically defined regions. Various ecoregion, precipitation, hardiness-zone, and heat-zone maps are employed extensively in this regard. While such map boundaries do not necessarily correspond to natural phylogeographic entities such as metapopulations, they do result in a more local approach than has been pursued previously. For species where metapopulation boundaries have been geographically mapped, e.g., bluebunch wheatgrass (*P. spicata* (Pursh) A. Löve) (Larson et al. 2004), these maps are being used as well.

Currently available releases of bluebunch wheatgrass used in the sagebrush steppe, i.e., "Whitmar," "Goldar," Anatone germplasm, and P-7 germplasm, all originate in the general region of southeastern Washington and northeastern Oregon, an area drained by the Columbia River, at least partially because materials from this region possess especially good seed yields. While Larson et al. (2004) identified four metapopulations (E, J, K, and R) to the south that range from east to west across the internally drained Great Basin from western Utah through Nevada, no bluebunch wheatgrass materials have been released from this region. Now that these Great Basin metapopulations have been mapped, work to develop polycrosses from populations collected in each of the four geographical regions may begin. Several researchers have reported that local material performs "well" in its own environment, even if it may not be "best" (Smith et al. 2009); thus, each of the four respective metapopulation polycrosses may offer innate adaptation to its environment. In addition, artificial selection for increased seed yield within each polycross may increase economic appeal of these materials to the seed industry and increase seed availability.

3.2 *Objection: Nonlocal Material May Result in Outbreeding Depression Upon Hybridization with Remnant Indigenous Material*

The second concern of using manipulated plants for restoration is that they generally encompass nonlocal genotypes that may result in "genetic swamping." This occurs when adapted indigenous material is overcome by gene flow of genotypes maladapted to the site originating from an outside source where they are presumably well adapted (Rice and Emery 2003). This, in turn, could lead to outbreeding depression, i.e., reduced fitness in initial or advanced generations of hybrids following on-site hybridization with remnant indigenous germplasm (Rogers and Montalvo 2004).

"Genetic pollution" refers to the loss of purity of local genotypes when introduced material of a contrasting genetic background hybridizes with remnant indigenous material. While contamination with nonlocal material may or may not result in outbreeding depression, it always introduces new genes and modifies gene frequencies. Thus, genetic pollution may be objectionable to the conservation biologist owing to its lack of correspondence to natural genotypic arrays. When the seed for future restoration projects must come from wildland seed harvest because a given species does not lend itself to cost-effective seed production in off-site cultivated fields, introduction of nonindigenous genotypes in restoration projects obviously compromises future restoration efforts that are based on local genetic material. This is probably the best argument against the use of genetically manipulated material, but it applies primarily to species in which cultivated seed production is impractical, e.g., timber species (Millar and Libby 1989).

The potential for genetic pollution has been investigated in switchgrass (*Panicum virgatum* L.), an important native species of the tallgrass prairie of the midwestern U.S. Populations of this species may be categorized as either upland (northern origin) or lowland (southern origin) ecotypes, and these ecotypes can be distinguished genetically (Hultquist et al. 1997). When six local switchgrass populations were grown across 12 locations, the population × location interaction accounted from 10 to 13% of the phenotypic variation for all traits except ground cover (31%) (Casler et al. 2007b). Statistical interaction for ground cover was primarily a result of better performance of upland ecotypes at northern locations and lowland ecotypes at southern locations. Thus, problems with outbreeding depression could likely be avoided by adhering to separate seed-transfer zones for upland (northern) and lowland (southern) ecotypes. While this study did not examine hybrids of the six populations for outbreeding depression, the molecular data suggested that manipulated switchgrass cultivars did not contain unique genotypes relative to natural populations (Casler et al. 2007a); therefore, these authors considered that the materials examined were unlikely to create genetic pollution problems.

The switchgrass example suggests that genetic pollution is more likely when a significant taxonomic boundary is crossed, such as the distinction between lowland and upland ecotypes. Likewise, Montalvo and Ellstrand (2001) encountered outbreeding depression in hybrids between coastal and inland subspecies of *Lotus scoparius* (Torr. & A. Gray) Ottley. Violating such genetic boundaries can usually be avoided with a basic knowledge of the systematics and intraspecific genetic variation of the species, assuming that this has been researched. When outbreeding depression does occur in products of hybridization, Carney et al. (2000) have shown that such individuals are selected against, minimizing the problem.

Outbreeding depression is a much more serious threat to populations of threatened and endangered species, which could easily be swamped. For this reason, close relatives of these species may not be good candidates for restoration plant materials unless they possess a widespread natural distribution (Johnson et al. 2008).

For common species, genetic pollution may be more of a problem when its hybridization products are vigorous and able to self-propagate, rather than being depressed. While this example does not involve manipulated plant materials, interspecific hybrids between Atlantic and Pacific species of *Spartina* have become invasive in San Francisco Bay following the introduction of the Atlantic species into the ecosystem (Sloop et al. 2009).

3.3 Objection: Broad-Based Plant Materials Themselves Are Subject to Outbreeding Depression

Restraint has been urged regarding broad-based materials, which in and of themselves are suspected of harboring outbreeding depression (Rogers and Montalvo 2004). These authors argue that added diversity generated via hybridization may be a "double-edged sword," sometimes beneficial and sometimes detrimental due to genetic load. While these authors agree that the ameliorating effects of natural selection on such hybrid complexes may result in positive outcomes, they argue that it is impossible to know whether the overall impact will be positive or, in fact, negative, thus implying that additional diversity should be avoided altogether. While some impacts may be negative, empirical results suggest that natural selection culls unfavorable genotypic combinations (Leger 2008). It is for precisely this reason that broad-based materials are useful.

Outbreeding depression is more likely to result when wide hybridization across taxonomic boundaries is employed to generate broad-based populations for selection. However, being genetically controlled like any other trait, outbreeding depression may often be ameliorated by artificial selection. However, if outbreeding depression persists despite selection, this becomes increasingly obvious in terms of infertility and/or poor performance as the material is advanced, and such material is typically discarded by the plant material developer. Furthermore, when proposing to "release" a plant material to the seed industry for restoration applications, the developer is usually required to justify the release before an independent committee. The developer presents data comparing the material to comparable previously released materials, this process being intended to circumvent the release of inferior material.

3.4 Objection: Manipulated Plant Materials Are Too Well Adapted

Invasiveness or competitive exclusion is a concern if a plant is too well adapted. However, in modified ecosystems, natural-track plant materials are typically inadequate to compete with and supplant invasive species, otherwise an invasive-species problem would never have materialized. No negative environmental effects of native plant materials have been documented on rangelands in the United States (Booth and Vogel 2006), but protocols have been developed to preclude the release of plant material with risk to the environment. For example, the USDA-Natural Resources Conservation Service has instituted an "environmental evaluation of plant material releases" (copy available from senior author upon request) to determine whether a proposed material may be potentially invasive (USDA Natural Resources Conservation Service 2010). Based on a battery of questions, proposed releases are scored as low, medium, or high for each of four parts: (1) impact on habitats, ecosystems, and land use, (2) ease of management (control), (3) conservation need and plant use, and (4) biological characteristics (ability to self-propagate). A flowchart is then used to integrate the four scores and arrive at a decision to either (1) release, (2) suspend the release process pending further action, or (3) release with qualification such as intended area of use or notification of potential adverse impact. The latter two decisions may require the preparation of an environmental assessment or an environmental impact statement.

3.5 Objection: Manipulated Plant Materials Are Poorly Adapted

Alternatively to the previous objection, concern has been expressed that manipulated material may be poorly adapted. Some contend that nonindigenous materials possess genotypes poorly suited to local conditions, but examples of such problems are admittedly few (Rogers and Montalvo 2004). Certainly, when a plant is placed outside the area in which it is adapted, lack of fitness is to be expected and natural selection is likely to remove the material. A working knowledge of the abiotic environment across the species' distribution, previous experience, and plant-material testing can generally preclude the inappropriate placement of a plant material. When a plant material is found to be maladapted to a certain environment, it is important to acknowledge the error and learn from it.

The question as to whether material is poorly adapted should preferably be answered with data for a particular plant material rather than by generalized conjecture. Plant materials are tested before release to determine how new materials perform relative to previously released materials. Surprisingly, the data collected in support of native plant material releases often seem to be ignored when they are considered for a restoration project. In fact, the primary question nearly always asked when a new native plant material is under discussion is: "where is it from?" Performance is presumed to follow from locality.

We contend that the assumption that local material is best is dubious when the site to be restored has been modified. Instead, we suggest that the primary question for a newly released plant material should be "how does it perform relative to previous releases and to local material?" We also suggest that testing and evaluation data would result in more robust decision-making tools through two mechanisms: (1) compilation and amalgamation of tests across time and space to improve inferences regarding plant-material adaptation and (2) monitoring of restoration projects over time to determine success or failure of various plant materials on different sites and under varied environmental conditions.

The first suggestion arises from the almost-universal testing programs employed by organizations that develop plant materials. These tests are typically conducted at regular intervals, i.e., yearly or biennially, using a common-garden approach. Newly developed experimental materials of a species are compared against previously available materials of that species and often against materials of other species. While the results of these individual evaluations are useful for making these direct comparisons, their overall utility is limited. First, the results are rarely used for more than justification of the release of the experimental material(s). Second, due to various resource constraints, i.e., finances, manpower, and available site locations, the tests necessarily cannot sample all potential environments in which the material might be seeded. Thus, individual tests, while useful, are limited in their ability to provide comprehensive performance data for plant materials.

To remedy these shortcomings, we propose that plant-material testing should be viewed as an ongoing process, by which each new test provides increased information and improved inference regarding the utility of individual plant materials. As a plant material is repeatedly tested at various sites before and after release, the data accumulate and a more robust picture emerges that documents the performance of that plant material across many years and locations. The analysis of the amalgamated data set enhances insight into the plant materials' ecological adaptation, allowing conclusions to be drawn regarding their utility on distinct sites across diverse environments. These results can provide restoration practitioners with a more powerful decision-making tool to infer which plant materials should be included in a restoration project. With sufficient information, models based on site characteristics could be developed that provide users with a list of appropriate plant materials for a particular site. Repeated testing will also provide empirical evidence to determine whether locally-derived native plant materials are truly the best adapted for restoration projects.

The biggest hurdles to accomplishing this objective are (1) successful amalgamation of testing data across years from within individual plant-development programs and (2) the sharing of testing data across programs that work on similar species. An example arises from our own plant-material program based in Logan, Utah in the Intermountain West region of the USA. Over the last 25–30 years, our program has conducted over 30 evaluations across a vast geographic range. The cumulative results would provide much-needed information concerning the overall adaptation of the materials in this geographic area, the importance of genotype × environment interaction, and the magnitude of the plant materials' phenotypic plasticity. To date, these individual tests have never been amalgamated and analyzed as a combined data set. Thus, we have not reaped the full benefits of our extensive and expensive testing program. We are now in the process of completing this analysis, and we look forward to the new information that it will provide.

The second suggestion arises from the need to supplement the information in the release document (the proposal for release and accompanying data) through continued monitoring of the plant materials over time in restoration projects themselves. Unfortunately, monitoring seems to be the forgotten stepchild of restoration. Numerous calls for more data from actual restoration projects have been made (Clewell and Rieger 1997; Rice and Emery 2003). Few data are available to gauge success or to direct future plant material development efforts. However, the information lost to lack of monitoring is critical. Monitoring would allow definitive conclusions to be drawn regarding the success or failure of these projects and provide critical answers concerning the adaptability of various types of plant materials on sites with varying levels of disturbance and abiotic stresses.

Released materials are occasionally withdrawn because they do not perform as anticipated. Much more often, however, materials fail because they are not propagated by the seed or nursery industries and never become commercially available in quantity. This may be because they have propagation problems, making them expensive to produce, or because industry does not perceive that sufficient demand exists for a particular item. Either situation can turn a released plant material into a failed venture.

3.6 Objection: Manipulated Plant Materials Developed via Hybridization Have Too Much Genetic Variation

The Rogers and Montalvo (2004) argument that enhanced genetic variation is a "double-edged sword" (Sect. 3.3) also pertains here. However, an important concern of restoration practitioners is that the materials they use possess a favorable balance between fitness and flexibility (Clewell and Rieger 1997). Concomitantly, plant material developers may seek to augment genetic variation in their populations in order to enhance future success. Rice and Emery (2003) recognized the importance of genetic variation, the raw material upon which natural selection must operate, for adaptation to a changing environment, i.e., global climate change. While a population with an "optimal" genotype and low variance is ideal for a static environment, the optimum changes when the environment changes. Hence, low genetic variation can quickly become a handicap. Because natural populations have become highly fragmented, they often cannot rely on neighboring populations to provide additional genetic variation via gene flow, which limits the potential of natural selection to generate novel adapted genotypes (Davis and Shaw 2001).

Rice and Emery (2003) went so far as to call for "coarse selective tuning," i.e., mixing genotypes from multiple indigenous populations of a similar climate to balance the need for current adaptation and potential future adaptation. The intent of this strategy is to include a broad array of adapted genotypes from the zone of adaptation in order to promote the operation of evolutionary mechanisms in the context of the restoration-site environment, referred to by Rice and Emery (2003) as "fine-tuning." Genetic variation is generally highest in the center of species' distributions because gene flow is intercepted from a wider range of locally adapted populations (Davis and Shaw 2001). In addition, inclusion of genotypes from the edge of a species' distribution may be desirable to incorporate novel adaptation from areas where adaptation to environmental extremes is likely to be found (Rice and Emery 2003).

We applied this approach in the development of P-7 germplasm of bluebunch wheatgrass. As ascertained with AFLP DNA markers, this plant material, a multiple-origin polycross of 25 local populations, aligns with bluebunch wheatgrass meta-population "P," which stretches across southeastern Washington, northeastern Oregon, and western Idaho, and includes Whitmar, Goldar, and Anatone germplasm (Larson et al. 2000, 2004). While most of the constituent populations making up P-7 originate from this core region, P-7 includes smaller amounts of material from distal locations in Utah, Nevada, Montana, and British Columbia (Larson et al. 2000). This strategy may also be construed as a precaution against genetic erosion (Rogers 2004).

3.7 Objection: Cultivars Have Inadequate Levels of Genetic Variation

The agricultural connotation conveyed by the word "cultivar" causes some restoration practitioners to avoid their use, while others avoid them because they purportedly possess too little genetic variation (Jones 2005). This claim seems to be linked to mistaken beliefs that (1) all cultivars have undergone artificial selection and that (2) selection necessarily results in great reduction in genetic variation. Regarding the first belief, cultivars are not defined by the amount of genetic variation they possess, and they may be either natural-track or manipulated-track (Fig. 1). They may vary from near-inbred lines, as in many self-pollinated crop species such as soybeans or wheat, or uniform single-cross hybrids, as for crops such as maize (both of which are probably responsible for this mistaken belief), to highly heterozygous populations generated from single or multiple-source materials.

Regarding the second belief, while a narrow genetic base may be compatible with the objectives of row-crop cultivar development, this would not usually be the desired approach for restoration plant material development. For the purpose of restoration, the goal of the plant material developer should be to modify the level of expression of a trait while maintaining as much genetic diversity as possible over the long term. Retention of genetic diversity is important to ensure continued favorable response to selection. This is accomplished by maintaining a high effective population size, which minimizes the amount of inbreeding from mating among relatives in populations subjected to selection (Falconer 1960). If selection is conducted at too high an intensity level without regard to population size, genetic variation for all traits may quickly erode, resulting in inbreeding depression, particularly undesirable for cross-pollinating species (Rogers and Montalvo 2004).

The goal of artificial selection is to modify the mean of the traits under selection, which in theory reduces genetic variation for this trait. Remarkably, over 100 cycles of selection for high oil and protein concentration in maize (*Zea mays* L.) have not exhausted genetic variation, as theoretically predicted (Dudley 2007). Thus, past assumptions that long-term selection advances cannot be sustained in cross-pollinated species are no longer accepted, though the reasons for this phenomenon are only partially understood. On the other hand, because of the lack of genetic recombination, selection has a much more direct negative impact on genetic variation in self-pollinated species. For these species, then, selection can be recommended only under exceptional circumstances, such as fixing a particular desired simply-inherited trait. It is also important to understand that selection does not result in inbreeding of the population for traits other than the trait under selection, as long as high effective population size is maintained.

4 Conclusion

Genetically manipulated restoration plant materials may be developed in such a way to allay some of the objections made by strict proponents of the evolutionary paradigm. As Gray (2002) has stated, the challenge for restoration plant materials involves selecting genotypes that are both adapted to the environment and that are likely adaptable to long-term environmental change over an evolutionary time scale. To this end, natural and manipulated approaches need not be mutually exclusive, and elements of the two may be combined as desired (Jones 2009; Jones and Monaco 2009). However, our contention is that the decision regarding which of the available materials to use is best made by the restoration practitioner, who is responsible for developing project objectives. By providing a variety of products for different applications, the plant material developer may assist restoration practitioners in their task.

Acknowledgment Appreciation is extended to Andy Meerdo for constructing the image for Fig. 1.

References

Antonovics J, Bradshaw AD (1970) Evolution in closely adjacent plant populations VIII. Clinal patterns at a mine boundary. Heredity 25:349–362

Booth DT, Vogel KP (2006) Revegetation priorities. Rangelands 28(5):24–30

Bretagnolle F, Thompson JD, Lumaret R (1995) The influence of seed size variation on seed germination and seedling vigour in diploid and tetraploid *Dactylis glomerata* L. Ann Bot 76:607–615

Broadhurst LM, Lowe A, Coates DJ, Cunningham SA, McDonald M, Vesk PA, Yates C (2008) Seed supply for broadscale restoration: maximizing evolutionary potential. Evol Appl 1:587–597

Brown RW, Amacher MC (1999) Selecting plant species for ecological restoration: a perspective for land managers. In: Holzworth LK, Brown RW (comps). Revegetation with native species:

proceedings, 1997 Society for Ecological Restoration annual meeting. USDA Forest Service Rocky Mountain Research Station, Ogden, UT, pp 1–16
Burton PJ, Burton CM (2002) Promoting genetic diversity in the production of large quantities of native plant seed. Ecol Restor 20:117–123
Carney SE, Gardner KA, Riesberg LH (2000) Evolutionary changes over the fifty-year history of a hybrid population of sunflowers (*Helianthus*). Evolution 54:462–474
Casler MD, Stendal CA, Kapich L, Vogel KP (2007a) Genetic diversity, plant adaptation regions, and gene pools for switchgrass. Crop Sci 47:2261–2273
Casler MD, Vogel KP, Taliaferro CM, Ehlke NJ, Berdahl JD, Brummer EC, Kallenbach RL, West CP, Mitchell RB (2007b) Latitudinal and longitudinal adaptation of switchgrass populations. Crop Sci 47:2249–2260
Clewell A, Rieger JP (1997) What practitioners need from restoration ecologists. Restor Ecol 5:350–354
Davis MB, Shaw RG (2001) Range shifts and adaptive responses to quaternary climate change. Science 292:673–679
Dudley JW (2007) From means to QTL: the Illinois long-term selection experiment as a case study in quantitative genetics. Crop Sci 47(S3):S20–S31
Falconer DS (1960) Introduction to quantitative genetics. Ronald, New York
Ferdinandez YSN, Coulman BE, Fu YB (2005) Detecting genetic changes over two generations of seed increase in an awned slender wheatgrass population using AFLP markers. Crop Sci 45:1064–1068
Gray AJ (2002) The evolutionary context: a species perspective. In: Pcrrow MR, Davy AJ (eds) Handbook of ecological restoration vol. 1, Principles of restoration. Cambridge University Press, Cambridge, pp 66–80
Gould SJ, Lewontin RC (1979) The spandrels of San Marco and the Panglossian paradigm: a critique of the adaptationist programme. Proc Roy Soc London B 205:581–598
Hobbs RJ, Arica S, Aronson J, Barton JS, Bridgewater P, Cramer VA, Epstein PR, Ewel JJ, Klink CA, Lugo AE, Norton D, Ojima D, Richardson DM, Sanderson EW, Valladares F, Vilà M, Zamora R, Zobel M (2006) Novel ecosystems: theoretical and management aspects of the new ecological world order. Global Ecol Biogeogr 15:1–7
Hufford KM, Mazer SJ (2003) Plant ecotypes: genetic differentiation in the age of ecological restoration. Trends Ecol Evol 18:147–155
Hultquist SJ, Vogel KP, Lee DJ, Arumuyganathan K, Kaeppler S (1997) DNA content and chloroplast DNA polymorphisms among switchgrass from remnant Midwestern prairies. Crop Sci 37:595–598
Johnson DA, Jones TA, Connors KJ, Bhattarai K, Bushman BS, Jensen KB (2008) Notice of release of NBR-1 Germplasm basalt milkvetch. Native Plants J 9:127–132
Johnson GR, Sonensen FC, St. Clair JB, Cronn RC (2004) Pacific Northwest Forest tree seed zones: a template for native plants. Native Plants J 5:131–140
Jones TA (2005) Genetic principles for the use of native seeds. Native Plants J 5:14–24
Jones TA (2009) Conservation biology and plant breeding: special considerations for the development of native plant materials for use in restoration. Ecol Rest 27:8–11
Jones TA, Monaco TA (2009) A role for assisted evolution in designing native plant materials for domesticated landscapes. Front Ecol Environ 7:541–547. doi:10.1890/080028
Jones TA, Monaco TA, James JJ (2010) Launching the counterattack: interdisciplinary deployment of native-plant functional traits for repair of rangelands dominated by invasive annual grasses. Rangelands 32:38–42
Jones TA, Parr SD, Winslow SR, Rosales MA (2009) Notice of release of 'Continental' basin wildrye. Native Plants J 10:57–61
Jones TA, Redinbaugh MG, Zhang Y, Larson SR, Dow BD (2007) Polymorphic Indian ricegrass populations result from overlapping waves of migration. W North Amer Nat 67:338–346
Kareiva P, Watts S, McDonald R, Boucher T (2007) Domesticated nature: shaping landscapes and ecosystems for human welfare. Science 316:1866–1869

Larson SR, Jones TA, Hu ZM, McCracken CL, Palazzo A (2000) Genetic diversity of bluebunch wheatgrass cultivars and a multiple-origin polycross. Crop Sci 40:1142–1147

Larson SR, Jones TA, Jensen KB (2004) Population structure in *Pseudoroegneria spicata* (Poaceae: Triticeae) modeled by Bayesian clustering of AFLP genotypes. Am J Bot 91: 1778–1800

Leger EA (2008) The adaptive value of remnant native plants in invaded communities: an example from the Great Basin. Ecol Appl 18:1226–1235

Meffe GK, Carroll CR (1997) What is conservation biology? In: Meffe GK, Carroll CR (eds) Principles of conservation biology, 2nd edn. Sinauer Associates, Sunderland, MA, pp 3–27

Millar CI, Libby WJ (1989) Disneyland or native ecosystem: genetics and the restorationist. Restor Manage Notes 7:18–24

Milton SJ (2003) 'Emerging ecosystems' – a washing-stone for ecologists, economists and sociologists? S Afr J Sci 99:404–406

Montalvo AM, Ellstrand NC (2001) Nonlocal transplantation and outbreeding depression in the subshrub *Lotus scoparius* (Fabaceae). Am J Bot 88:258–269

Reich PB, Wright IJ, Cavender-Bares J, Craine JM, Oleksyn J, Westoby M, Walters MB (2003) The evolution of plant functional variation: traits, spectra, and strategies. Int J Plant Sci 164: S143–S164

Rice KJ, Emery NC (2003) Managing microevolution: restoration in the face of global change. Front Ecol Environ 1:469–478

Rogers DL (2004) Genetic erosion – no longer just an agricultural issue. Native Plants J 5:112–122

Rogers DL, Montalvo AM (2004) Genetically appropriate choices for plant materials to maintain biological diversity. University of California. Report to the USDA Forest Service, Rocky Mountain Region, Lakewood, Colo., USA. URL: http://www.fs.fed.us/r2/publications/botany/plantgenetics.pdf. Cited 17 June 2010

Sloop CM, Ayres DR, Strong DR (2009) The rapid evolution of self-fertility in *Spartina* hybrids (*Spartina alterniflora* X *foliosa*) invading San Francisco Bay, CA. Biol Invasions 11:1131–1144

Smith BM, Diaz A, Daniels R, Winder L, Holland JM (2009) Regional and ecotypic traits in *Lotus corniculatus* L., with reference to restoration ecology. Restor Ecol 17:12–23

Smith KF, McFarlane NM, Croft VM, Trigg PJ, Kearney GA (2003) The effects of ploidy and seed mass on the emergence and early vigour of perennial ryegrass (*Lolium perenne*) cultivars. Aust J Exp Agr 43:481–486

USDA Natural Resources Conservation Service (2010) Worksheet for documenting an environmental evaluation of NRCS plant releases. National Plant Material: Program Manual, 4th ed., Title 190 Part 540.83

Wilkinson DM (2004) Do we need a process-based approach to nature conservation? Continuing the parable of Green Mountain, Ascension Island. J Biogeogr 31:2041–2042

Young SA, Schrumpf B, Amberson E (2004) The AOSCA native plant connection. URL: http://www.asoca.org/aoscanativeplantbrochure.pdf. Cited 16 June 2010

Photosynthesis and Stomatal Behaviour

Tracy Lawson, Susanne von Caemmerer, and Irene Baroli

Contents

T. Lawson (✉)
Department of Biological Sciences, University of Essex, Wivenhoe Park, Colchester CO4 3SQ, UK
e-mail: tlawson@essex.ac.uk

S. von Caemmerer
Research School of Biology, Australian National University, GPO Box 475, Canberra, ACT2601, Australia
e-mail: Susanne.Caemmerer@anu.edu.au

I. Baroli
Departamento de Biodiversidad y Biología Experimental, Facultad de Ciencias Exactas y Naturales, Universidad de Buenos Aires, Pabellón II, Ciudad Universitaria C1428EHA, Buenos Aires, Argentina
e-mail: baroli@bg.fcen.uba.ar

U. Lüttge et al. (eds.), *Progress in Botany 72*,
DOI 10.1007/978-3-642-13145-5_11,

Abstract In order for plants to use water efficiently, stomata must ensure an appropriate balance between CO_2 demands for photosynthesis and water loss through transpiration. To achieve this, stomatal conductance (*gs*) often correlates with mesophyll photosynthetic rates. However, the underlying mechanisms and signals that promote this relationship are currently unknown. Stomata and photosynthesis respond to a number of environmental cues; however, the dynamics and magnitude of these responses are not identical, with stomatal responses generally an order of magnitude slower than mesophyll photosynthesis. The resulting disconnection between stomatal conductance and photosynthetic rate means that under naturally fluctuating environmental conditions water use efficiency (WUE) can be far from optimal. Manipulation of stomatal behaviour provides an obvious mechanism for producing plants with improved WUE; however, before such an approach is possible we must first understand the hierarchy of stomatal responses to varying environmental parameters, the mechanisms behind these complex signalling pathways, and how stomatal behaviour is tuned to mesophyll photosynthetic rates or capacity.

1 Introduction

Plants require sufficient CO_2 to enter the leaf for photosynthesis while conserving water to avoid dehydration and metabolic disruption. As the leaf is almost impermeable to water and CO_2, almost all of the water transpired as well as the CO_2 absorbed pass through stomatal pores (Cowan and Troughton 1971; Jones 1992) and therefore stomata are essential to plant water and carbon status. Additionally stomata also play key roles in nutrient uptake and evaporative cooling of the leaf tissue (Morison 2003). The transpirational loss of water through stomata is unavoidable because of the plant's need to expose internal cell surfaces to the external air surrounding the leaf for photosynthetic CO_2 uptake. Only about 2% of the water taken up by a plant is used for biochemical reactions, with the remaining 98% being lost by evaporation from the cell surfaces (Morison 2003). The rate of diffusion of gases into or out of the leaf from/to the surrounding environment depends upon the concentration gradient and the resistance of diffusion along the pathway (see Weyers and Meidner 1990). For water loss from the mesophyll cells inside the leaf, the major pathway is therefore from the mesophyll cell wall through the substomatal cavity to the stomatal pore and then out through the layer of air immediately surrounding the leaf (boundary layer) to the mixed air stream. The pathway for CO_2 uptake is essentially the same but in reverse, with an additional resistance component represented by the entry into the mesophyll cell chloroplasts. The resistance of the stomatal pathway depends on the geometry of the pores as well as their density (or frequency). Although the pore area may represent only a small fraction (between 0.5 and 3%) of the total leaf area when fully open, evaporation rates can be equivalent to 50% of that of a wet surface of the same area due to edge effects of diffusion (Willmer and Fricker 1996). Modelling stomatal conductance based on pore geometry, Weyers and Lawson (1997) illustrated that the main determinant of g_s was pore aperture, while stomatal density makes a smaller contribution, although greater than that of depth or length. Stomata and their function therefore, play a central role in determining the amount of carbon gained per unit water lost, known as plant "water use efficiency (WUE)".

Crop breeders and scientists are under growing pressure to produce crop plants with increased yields (or even sustained yield) and greater water and nutrient use efficiency to meet the increasing demand for food and sustainable fuel for the ever-expanding global population. Future plants must also have the ability to maintain production in predicted future climates of increased atmospheric CO_2 concentration, increased temperature and reduced water availability. The regulatory role of stomatal control over water loss and CO_2 uptake for photosynthesis makes these cells an obvious target for manipulation for improving WUE in future crops plants. However, before such an approach is possible, we must first be able to understand the complex signalling and response pathways and mechanisms that enable stomata to respond to environmental stimuli and to mesophyll demands for CO_2 and the interaction and hierarchy of responses that obviously exist in guard cells.

1.1 Stomatal Function, Plant Productivity and Water Use Efficiency

In order for stomata to function effectively (maximising CO_2 uptake while minimising water loss), they respond to changes in external conditions and internal signals (Raschke 1978). Environmental factors can influence stomatal behaviour either directly or indirectly (Wong et al. 1979). Direct factors are those affecting the guard cells themselves, while indirect effects are those affecting stomatal behaviour by influencing the photosynthesis of the mesophyll (Willmer and Fricker 1996). Typically, stomatal pores in C_3 and C_4 plants open with light the extent of which has been shown to be wavelength specific (see Kuiper 1964), low CO_2 concentrations and high humidity (linked with temperature and the driving force of water loss), while closure is induced under conditions of darkness, high CO_2 concentration (see Sect. 2.1), low humidity and high temperatures (see reviews by Assmann 1993; Willmer and Fricker 1996; Outlaw 2003; Vavasseur and Raghavendra 2005; Shimazaki et al. 2007). There are exceptions to these typical stomatal responses dependent upon the photosynthetic mechanisms employed (discussed in further detail in Sect. 3).

Changes in stomatal aperture are driven by changes in turgor pressure of guard cells (Heath 1938) due to the accumulation of ions and/or solutes (Imamura 1943; Fujino 1967; Outlaw 1983), which increases the osmotic potential and lowers the water potential of guard cells (Weyers and Meidner 1990; Willmer and Fricker 1996). Stomatal conductance (g_s) to water and CO_2 is dependent upon stomatal characters such as density (number of stomata per unit leaf area) and stomatal aperture (pore width), both of which are influenced by environmental conditions surrounding the leaf, including gas concentrations, relative humidity and temperature. The number, size and distribution of stomata vary both between and within species (Tichà 1982) and are often dependent upon environmental growth conditions (Weyers et al. 1997; Weyers and Lawson 1997). Anatomical features of stomata and their influence on gas exchange and photosynthesis will be considered in Sect. 4. Considerable heterogeneity in stomatal characters, behaviour and function has also been demonstrated at several levels, from spatial patterns on individual leaves to whole leaves within plants, and also species differences (Lawson 1997; Weyers and Lawson 1997). An understanding of the nature of stomatal heterogeneity and its origins may be important as this could provide plants with some functional advantages or disadvantages with respect to photosynthesis or WUE (e.g. stomatal patchiness; Mott and Peak 2007).

To improve WUE, two approaches are immediately obvious. The first is to improve photosynthetic carbon assimilation/growth rate and yield without the need for increased water loss by the plant. Numerous studies have altered expression of targeted plant enzymes to identify their control on carbon assimilation. Such studies have targeted Calvin cycle enzymes (Harrison et al. 1998; Haake et al. 1999; Henkes et al. 2001; Raines 2003; Lefebvre et al. 2005; Suzuki et al. 2007) and also photorespiratory enzymes (Wingler et al. 1999). Single enzyme transformations have demonstrated that enhanced photosynthetic rates are possible, for example, increased. Sedoheptulose-1,7-bisphosphatase activity resulted in tobacco plants with improvements in carbon assimilation by 6–12% (Lefebvre et al. 2005). The inclusion of thermostable Rubisco activase improved photosynthetic rates in Arabidopsis (Kurek et al. 2007); however, increasing Rubisco content in rice plants showed no enhancement in photosynthetic rate (Suzuki et al. 2009).

The potential outcomes of targeting more than one enzyme have recently been modelled by Zhu et al. (2007). The second approach is to breed plants with an altered stomatal conductance (Jones 1976, 1977). Increased stomatal conductance could remove stomatal limitation on photosynthesis and increase carbon uptake, while reduced stomatal conductance (particularly under water stress conditions) could increase water use but at the expense of carbon gain. However, before contemplating the potential of altering stomatal conductance to increased WUE, the extent to which stomata limit photosynthesis under the plant natural growth environmental must first be determined (Jones 1987). It has been reported for several C_3 species that removal of stomatal limitation increases photosynthetic rates by only 10–20% and that in C_4 species this is even less (Farquhar and Sharkey 1982; Jones 1985). Under drought stress conditions metabolic limitation of photosynthesis may be much greater than that caused by reduced stomatal aperture (Lawlor 2002). A third less-obvious approach for increasing WUE which would include selecting plants with stomata that respond more rapidly to the fluctuating environmental conditions that plants experience in the field environment could be envisaged. However, before such an approach is attempted, we must first understand the mechanisms and response(s) of stomata to varying environmental factors and the combination of such changes, and the mechanisms by which stomatal conductance relates to mesophyll photosynthesis. In the following sections we will examine anatomical and physiological aspects of stomatal function in relation to photosynthetic carbon assimilation. It is worth noting that commercially grown wheat varieties with increased WUE have been identified using an innovative isotope discrimination screen ($\Delta^{13}C$ and $\delta^{18}O$), which provides information on photosynthesis, transpiration and stomatal behaviour (see Rebetzke et al. 2006; Condon et al. 2007; Ripullone et al. 2008). In another case, mutants of the ERECTA gene (which "regulates transpiration efficiency") were isolated in *Arabidopsis* using a similar carbon isotope screening technique (Masle et al. 2005).

2 Stomata Responses to Environmental Parameters

Environmental variables such as light, $[CO_2]$ and temperature are often considered to have the greatest influence on photosynthetic rates, as well as direct and indirect impacts on stomatal behaviour. Despite CO_2 concentration in nature remaining relatively stable, stomata are often assessed relative to $[CO_2]$ because internal CO_2 concentration (C_i), which is ultimately controlled by external CO_2 concentration and photosynthetic carbon assimilation rates, is often believed to provide a key controlling mechanism linking stomatal behaviour with mesophyll photosynthetic rates (see Mott 2009). We have not attempted to review the vast amount of literature currently available on these areas of research but refer readers to two recent comprehensive reviews examining stomatal responses to CO_2 (Vavasseur and Raghavendra 2005) and light (Shimazaki et al. 2007).

2.1 *Stomatal Responses to CO_2 Concentration*

It is well-accepted that stomatal conductance responds to CO_2, a necessary requirement if plants are to optimise water use with carbon demand. In short-term responses, stomata generally reduce aperture with increased CO_2 concentration and open under lowered CO_2 concentrations (in both darkness and light). The signalling and response mechanisms that underlie stomatal responses to CO_2 have proven difficult to identify until recently. Hashimoto et al. (2006) demonstrated that stomata of the Arabidopsis *high leaf temperature 1* mutants (*hrt1-1* and *hrt1-2*) have reduced ability to control stomatal movements in response to altering CO_2 concentrations, indicating that HT1 kinase is an important regulator of stomatal responses to CO_2. Several reports have also postulated an interactive role between ABA and stomatal CO_2 responses (for details see Vavasseur and Raghavendra 2005). Long-term exposure to elevated CO_2 concentrations may result in (1) physiological acclimation of stomata, for example reduced sensitivity to CO_2 concentration compared with those observed under short-term fluctuations (e.g. Berryman et al. 1994; Xu et al. 1994; Lodge et al. 2001; Maherali et al. 2002); (2) changes in anatomy (Woodward 1987; McElwain and Chaloner 1995; McElwain et al. 1995); and (3) attenuated responses to environmental parameters other than $[CO_2]$ (Ainsworth and Rogers 2007). Since the first reports of decreasing aperture with elevated CO_2 concentration (Freudenberger 1940; Heath and Russell 1954), numerous researchers and laboratories have studied the effects of both short- and long-term responses of stomatal aperture to increasing and decreasing CO_2 concentrations (e.g. Raschke 1972) and illustrated a variety of responses that appear to be species dependent (e.g. Maherali et al. 2002) or independent (e.g. Morison 1985) of the plants photosynthetic pathway or biochemistry. Different stomatal responses to CO_2 have been reported for the same species from different laboratories; for example, Tuba et al. (1994) reported little change in *Triticum aestivum* stomatal sensitivity when grown at elevated CO_2 concentration, but a lack of sensitivity could not be confirmed in later studies by Šantrůček and Sage (1996). Stomatal responses to CO_2 do, however, appear to be governed by growth conditions. For example *Vicia faba* plants grown under ambient CO_2 concentrations in a growth chamber had enhanced CO_2 responses compared with plants grown in glass house conditions, which was reported to be an intrinsic property of the guard cells (Frechilla et al. 2002), although combined effects of differing light and humidity could also explain these results. Additionally, plants grown under well watered conditions or in environments with low evaporative demands have been shown to have reduced sensitivity to CO_2 (Mansfield and Atkinson 1990; Mansfield et al. 1990; Mansfield 1994). It is believed that CO_2 sensing occurs within the guard cells themselves, as CO_2 responses have been demonstrated in epidermal peels (Fitzisimons and Weyers 1986) and it is generally accepted that guard cells sense C_i rather than atmospheric CO_2 (C_a) (Mott 1988). However, recent work on transgenic plants may provide evidence that stomata respond to C_a rather than C_i as stomata closed in response to increases in C_a and at ambient C_a stomatal conductance was similar in both wild type (WT) and transgenic plants despite elevated C_i concentrations due to reduced photosynthetic rates (von Caemmerer et al. 2004; Baroli et al. 2008). Stomatal sensitivity to C_a has recently been the focus of an evolutionary study on WUE (see Brodribb et al. 2009) (see below).

2.2 Stomatal Responses to Light

The relationship between light, photosynthesis and stomatal conductance is often discussed in association with the complex interactions with C_i. This is fuelled by the fact that guard cells have two light response components. The first of these light response components is the photosynthesis-independent blue light response, which is associated with rapid opening and is sensed by the guard cells (Zeiger et al. 2002; Kinoshita and Shimazaki 1999; Shimazaki et al. 2007) while the second is thought to be a photosynthetic mediated response (often termed the red light response), which saturates at rates similar to mesophyll photosynthesis (see Sharkey and Raschke 1981). The red light response is often believed to be a C_i response operating through mesophyll consumption of CO_2 (e.g. see Roelfsema et al. 2002). Support for a C_i driven red light response comes from experiments conducted on albino leaves or leaves treated with an inhibitor of carotenoid synthesis, in which stomata responded to blue light, but failed to open under red illumination (Roelfsema et al. 2006). Additional support for a CO_2-induced red light response comes from mutant plants that lack a CO_2 response; these mutants respond to blue light but show no response to red light (Hashimoto et al. 2006; Marten et al. 2008). However, several studies have argued against a direct C_i stomatal-driven response to red light; for example stomata were shown to respond to red light even when C_i was held constant (Messinger et al. 2006; Lawson et al. 2008; Wang et al. 2008). Sharkey and Raschke (1981) confirmed the work of Wong et al. (1979) and reported that stomatal responses to C_i were too low to account for the large changes observed under light. The site of perception for the stomatal red light response is controversial, with some studies suggesting the involvement of guard cell photosynthesis (see Lawson 2009) while others suggest a mesophyll driven signal (Mott 2009). The influence of such mesophyll driven signals on stomatal behaviour has implications for the correlation between stomatal function and photosynthetic capacity (Sect. 3).

The blue light response of stomata has been reported to have faster dynamics than the red light response (although this could be species dependent) and is believed to be involved in rapid dawn opening (Zeiger et al. 1985) and in response to sunflecks (Kirschbaum et al. 1988). However, in tobacco leaves subjected to step increases in red or red and blue light, no difference in rate of stomatal opening was observed (Fig. 1). Only the intensity of the irradiation influenced the rates of opening, and interestingly the final conductance achieved was closely related to the initial rate of opening (Fig. 1). Lawson et al. (2008) reported a similar correlation between red light induced rate of opening and final conductance in plants with differing photosynthetic capacity owing to differing activities of the Calvin cycle enzyme sedoheptulose-1,7-bisphosphatase (SBPase). These data demonstrate the importance of the dynamics of stomatal responses on the final conductance obtained and highlight the potential of manipulating stomatal response in order to control g_s and WUE.

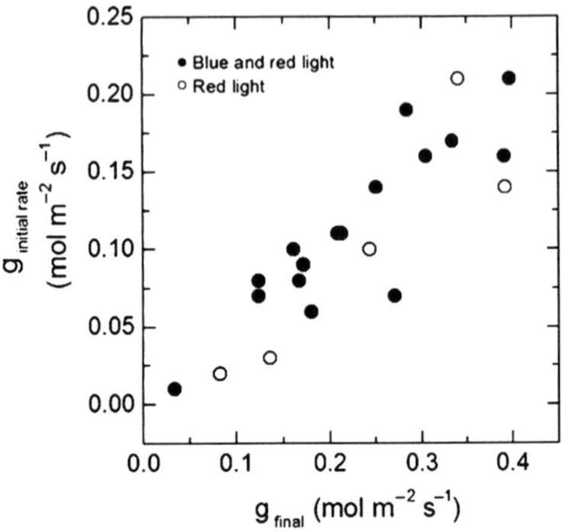

Fig. 1 Relationship between final stomatal conductance and initial rate of opening in tobacco in response to either red or blue light. Leaves were acclimated to 100 μmol m^{-2} s^{-1} followed by step changes in irradiance at different intensities. Open symbol shows stomatal responses to red light, and close symbol blue light

2.3 Temperature Response of Stomata

Variations in leaf and air temperature can alter the rate of transpiration, which will have a direct effect on plant WUE (Šantrůček and Sage 1996), plant productivity (Morison 1993) and crop yield (Lu et al. 2000). In species that lack CO_2 concentrating mechanisms, increasing temperature can cause indirect effects due to metabolic increases in photorespiration and respiration relative to photosynthetic rates, bringing about a rise in internal CO_2 concentration which, if stomata are sensitive to C_i, will bring about stomatal closure (Ball et al. 1987; Willmer and Fricker 1996). Stomatal responses to temperature are variable, depending upon the species and growth conditions (Sage and Sharkey 1987; Aphalo and Jarvis 1991). The effect of temperature on stomata is complicated by the consequential change in leaf–air vapour pressure deficit (VPD) (Hall et al. 1976; Šantrůček and Sage 1996; Willmer and Fricker 1996). At constant VPD, some species show increased g_s with an increasing temperature (Šantrůček and Sage 1996), while others show no change (Aphalo and Jarvis 1991). As temperature increases, g_s will generally decline as a result of increased VPD, although both C_3 (Sage and Sharkey 1987; Kudoyarova et al. 2007) and C_4 (Dwyer et al. 2007) species have been reported to increase g_s with temperature despite increasing VPD. These observations have been attributed to an overriding effect of temperature on g_s that is independent of VPD (Sage and Sharkey 1987; Dwyer et al. 2007) or increased hydraulic conductance in well watered soils (Kudoyarova et al. 2007). Temperature dependent increases in g_s have been attributed to increase metabolic activity in guard cell respiration and increase proton pumping, independent of mesophyll photosynthesis (Lu et al. 2000). Understanding the impact of temperature and the combined effect of temperature, CO_2 and humidity on stomatal behaviour is critical to understanding optimal stomatal conductance and plant WUE. Additionally, long term growth under elevated temperature and/or CO_2 concentration may result in stomatal acclimation, which could counterbalance the expected future increases in WUE as demonstrated and predicted from short-term studies (Šantrůček and Sage 1996).

2.4 Stomatal Responses Under Fluctuating Environmental Conditions

Stomatal responses to environmental signals such as light, CO_2 and VPD have been extensively studied under steady state conditions, but rarely do they vary in isolation under natural conditions (Lawson and Morison 2004). Fewer studies have examined dynamic stomatal behaviour in a natural environment, where multiple signals must be integrated and speed of response can be critical to both daily carbon gain and WUE (Raschke 1970; Kirschbaum et al. 1988; Knapp 1993; Tinoco-Ojanguren and Pearcy 1993; Cardon et al. 1994; Allen and Pearcy 2000; Lawson and Morison 2004; Noe and Giersch 2004). The disconnection in the time response of stomatal opening and photosynthetic response was examined in detail by Pearcy and co-workers, who studied photosynthetic behaviour to sunflecks in plant canopies and demonstrated that stomatal opening in response to a light signal continued long after the signal had ceased (Kirschbaum et al. 1988; Pearcy 1990). An example of stomatal opening in response to light in Fig. 2 shows the difference in the rate of increase in stomatal conductance and CO_2 assimilation rate and highlights the dilemma faced by plants in coordinating the two processes. The duration and intensity of sun (or shade) flecks influence the response of both g_s and A, as illustrated in Fig. 3. After a period of low light, an increase in irradiance does not result in an immediate increase in A, but shows an initial increase followed by a delay before maximum A is achieved (see Fig. 3a). This lag period is due to both mesophyll induction (involving light regulation of enzymes and metabolite pools) as shown in Fig. 3a and, depending on the duration of the sun fleck, changes in stomatal aperture (Fig. 3b). Although the increase of g_s in response to a light increase during sun flecks is faster than the decreasing response to a drop in light, stomatal movements can take up to tens of minutes and can "overshoot" – continuing to open after the fleck has passed (Kirschbaum et al. 1988; Tinoco-Ojanguren and Pearcy 1993). Most work has indicated that the main control of assimilation during the first 10 min of induction is with the biochemical control of assimilation rate and that stomata do not cause a major limitation (Barradas and Jones 1996; Fig. 3a). However, in sun flecks greater than 10 min, stomatal aperture can limit assimilation rate (Fig. 3b).

The effect(s) of shade flecks on g_s and A are less well studied. Decreased light intensity for a period of 5 min or less greatly decreases the A rate and has little impact on stomatal behaviour; therefore, on restoration of the light, identical A rates can be achieved prior to the shade fleck (Fig. 3c). However, during sunflecks longer than about 8 min, g_s will decrease and continue to decrease for a further 5–8 min

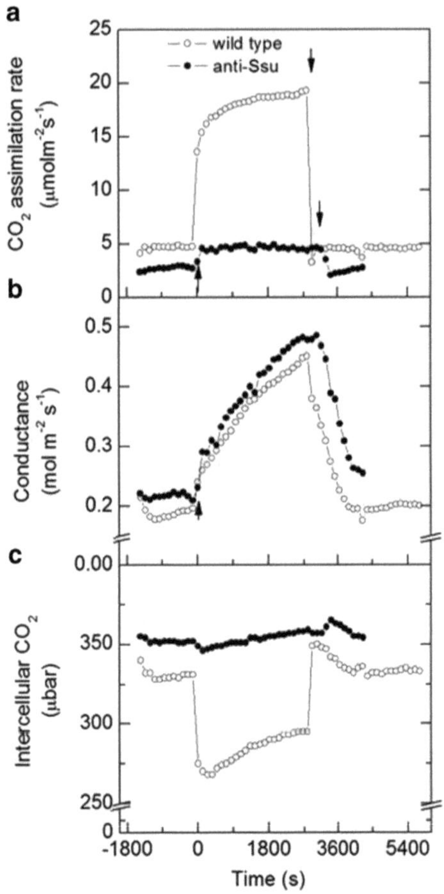

Fig. 2 Examples of changes in (**a**) CO_2 assimilation rate, (**b**) leaf conductance and (**c**) intercellular CO_2 with time after a step change in irradiance from 100 to 1,000 μmol quanta m^{-2} s^{-1}. for wild type (*open circle*) and transgenic tobacco with reduced Rubisco (anti-Ssu, *closed circles*). Ambient CO_2 and water vapour were maintained at 380 μmol mol^{-1} and 23 mmol mol^{-1}. Leaf temperature was maintained at 25°C. Arrows indicate when light was increased from 100 to 1,000 μmol quanta m^{-2} s^{-1} and returned to 100 μmol quanta m^{-2} s^{-1}. (Figure reproduced from von Caemmerer et al. (2004) by permission of Society of Experimental Biology (http://www.sebiology.org) & Oxford University Press)

after the shade fleck has been removed, which restricts restoration of A for about 15 min (Fig. 3d). The examples used here illustrate the potential fluctuating changes in stomatal behaviour relative to A, the responses of which are dependent not only on the duration but also on the intensity of the flecks, proportional change in irradiance relative to the starting irradiance (see, Lawson 1997), as well as the species studied. When scaled up to the plant or crop level, the impact of the disproportional change in g_s and A can result in significant decreases in plant WUE (see Sect. 6). Increasing stomatal response times to fluctuating environmental parameters could potentially decrease lag times and reduce the amount of time stomata restrict carbon assimilation. Additionally, such improvements would reduce the time periods in which plants lose water unnecessarily because of a greater stomatal conductance than is required for the potential carbon gain at that time (e.g. during a shade fleck).

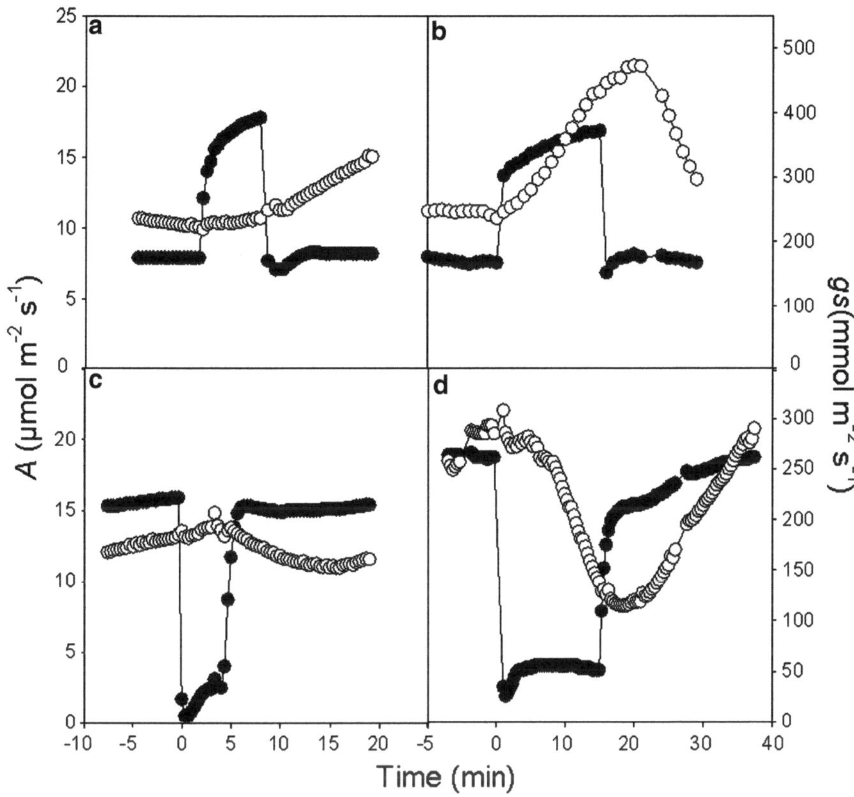

Fig. 3 The effect of sun and shade flecks on CO_2 assimilation rate and stomatal conductance. (**a**) and (**b**) *Sun flecks*: PPFD was increased from 230 to 615 μmol m^{-2} s^{-1} at time zero for (**a**) 5 min, (**b**) 15 min, (**c**) and (**d**) *Shade flecks*: PPFD was decreased from 615 μmol m^{-2} s^{-1} at time zero. Measurements of assimilation rate (*solid circles*) and stomatal conductance (*open circles*) made every 5 s with cuvette CO_2 maintained at 357 μmol mol^{-1}, temperature 25°C and VPD 1.34 kPa

Although under controlled environmental conditions, within-leaf patterns of photosynthesis often correlate with stomatal behaviour (Weyers et al. 1997), in natural fluctuating environments, there is significantly greater within-leaf variation in stomatal conductance and photosynthetic rates, the patterns of which often show no correlation (Weyers and Lawson 1997). The observed variation and lack of correlation between stomatal conductance and assimilation rate over individual leaves of *Phaseolus vulgaris* were attributed to the differences in the induction times for stomata and photosynthesis to respond and keep track of variable conditions (Lawson and Weyers 1999). The rate of stomatal opening and closing reflects the kinetics of multiple processes such as signal transduction, water movement, the establishment of osmotic potential, as well as the influence of the mechanical properties of guard cells and is therefore a complex kinetic function (Kirschbaum et al. 1988; Cardon et al. 1994; Franks and Farquhar 2007). It is clear that the kinetics of stomatal responses (and resulting g_s) to certain environmental parameters are not consistent across species (see Hetherington and Woodward 2003; Sharkey and Raschke 1981) and that direct and indirect responses are significantly influenced by the type of photosynthetic pathway (Huxman and Monson 2003).3

2.5 Night Time Stomatal Conductance

It is generally accepted that stomata close in response to darkness, although significant night time stomatal conductances, which can result in 5–30% of the daily water loss (Snyder et al. 2003), have been observed in a range of species (see Donovan et al. 1999; Daley and Phillips 2006; Caird et al. 2007), including C_3 and C_4 photosynthetic types. Rates of up to 90% of day time conductances have been reported, although rates of water loss depend on VPD, which is generally much lower during the night (Caird et al. 2007). As night time stomatal conductance allows water loss with no carbon gain, and with reduced leaf cooling requirements, the benefit for the plant is at present still unclear (Caird et al. 2007). It has been suggested that night time stomatal conductance could enhance nutrient uptake (Donovan et al. 2001; Caird et al. 2007) or oxygen delivery to respiring cells (Daley and Phillips 2006), repair xylem embolism (Rogiers et al. 2009) and/or could enhance early morning/predawn stomatal opening that has the potential to maximise carbon uptake when temperature and VPD are lower (Caird et al. 2007). However, it appears that rates of night time conductance are correlated with day time stomatal conductance. Night time stomatal conductance could not only have significant impacts on plant water status and WUE, but also has important implications for pollutant uptake (e.g. ozone, Grulke et al. 2004) and isotope signatures for determining environmental impacts on transpiration rates (Barbour et al. 2005).

3 Stomatal Interactions with Photosynthesis

3.1 Photosynthetic Pathways and Stomatal Function

Most reviews on stomatal function are C_3-centric, most likely as a result of the bias toward stomatal research on C_3 plants, as they tend to have larger stomata to work with. In this section, we briefly examine some of the key differences in stomatal behaviour between C_3, C_4 and Crassulation Acid Metabolism (CAM) plants. The global distribution of C_3, C_4 and CAM plants is unequal, driven by global variation in environmental variables. C_3 plants are the most widely distributed plants, dominating the temperate regions of the world and covering an estimated 87.4 million km^2 compared with 18.8 million km^2 for C_4 species (Still et al. 2003).

Plants with the C_4 photosynthetic pathway appeared about 8 million years ago and tend to dominate in warmer, dryer habitats at 21–23° latitudes (Sage et al. 1999), representing less than 4% of all plant species. CAM plants occurred earlier in evolution in the carboniferous and differ greatly in stomata behavioural patterns, opening in darkness and remaining closed during the light period (Osmond 1978; Black and Osmond 2003). In CAM plants, stomata open at night for net CO_2 uptake catalysed by phosphoenolpyruvate carboxylase (PEPC), which is stored in the vacuole as malic acid. During the light period, decarboxylation of vacuole acids releases CO_2 behind closed stomata, enabling high CO_2 concentration to develop for refixation by Rubisco (Cockburn et al. 1979). Stomata remain closed during the light period, except during late afternoon (termed phase IV), when reduced acid concentration and optimal conditions (namely water supply) permit stomata to reopen to perform C_3 photosynthetic CO_2 fixation. Until recently, stomatal behaviour in CAM plant has been associated with changing C_i concentration due to the action of PEPC at night and decarboxylation during the day (Cockburn et al. 1979; Spalding et al. 1979); however, von Caemmerer and Griffiths (2009) demonstrated that stomata of Kalanchoe species did not respond when C_i was manipulated. Stomata of a particular CAM species, *Tillandsia recurvata*, have been shown to be sensitive and directly responsive to changes in ambient air humidity during nocturnal CO_2 fixation. Changes in stomatal aperture were shown to be a direct response to changes in humidity and not to bulk tissue water conditions of the leaves (Lange and Medina 1979). The induction of CAM in the facultative CAM plant *Mesembryanthemum crystallinum* has been shown to abolish the stomatal blue light response, which is apparent when the plant is in C_3 mode (Mawson and Zaugg 1994), a response that has been linked to the a lack of light-induced formation of zeaxanthin (Tallman et al. 1997). Zeaxanthin has previously been linked with blue light responses in guard cells (Zeiger and Zhu 1998; Frechilla et al. 1999). Stomata of *Portulacaria afra* have been shown to have typical blue and red light responses when in C_3 mode, both of which become undetectable in CAM mode (Lee and Assmann 1992). However, red light response was lacking in *M. crystallinum* even in C_3 mode (Mawson and Zaugg 1994). Studies on the rate of opening in response to a decrease in CO_2 concentration in the CAM plants *Kalenchoe daigremontiana* and *Kalanchoe pinnata* showed that the rate of opening in the dark was as rapid as in the light during Phase IV of the CAM cycle, where CO_2 is fixed primarily via C_3 photosynthesis, indicating that energy can be sourced via mitochondrial respiration or stored carbohydrate (von Caemmerer and Griffiths 2009). The fact that stomata and photosynthesis are temporally segregated in CAM plants provides an ideal opportunity to use such systems to examine stomatal responses in isolation from mesophyll C_3 metabolite production or demands (von Caemmerer and Griffiths 2009).

Stomatal function and behaviour in C_4 plants are more similar to those of C_3 plants than of CAM plants, although differences in magnitude and sensitivity to light and CO_2 have been reported between C_3 and C_4 species (Huxman and Monson 2003; Maherali et al. 2002). For example, several studies have demonstrated that stomata of C_4 plants have greater sensitivity to C_i than C_3 species

(Dubbe et al. 1978; Sharkey and Raschke 1981; Ramos and Hall 1982), although Morison and Gifford (1983) reported similar sensitivities. It would, however, be interesting to see if such reported differences in sensitivity hold true if stomatal response was measured as a function of C_a rather than C_i. C_3 species have also been reported to be less prone to closure than C_4 when light was decreased or CO_2 increased (Akita and Moss 1972). Stomatal acclimation to CO_2 responses have been observed in C_3 forbs but not in C_3 or C_4 grasses (Maherali et al. 2002). In response to severe drought, reduced stomatal aperture could limit photosynthesis to a greater degree in C_4 plants than in C_3, due to the fact that C_4 photosynthesis operates closer to the C_i inflection point and there is a steep photosynthetic response with increasing C_i at low C_i (see Ghannoum 2009). Additionally, C_3 and C_4 plants have been reported to differ in their diurnal periodicity of opening, with maximum g_s reported for C_3 plants at 10 am, while for C_4 plants the maximum is at 12 noon (Das and Santakurmari 1977), suggesting that low light levels and low temperature promote stomatal opening in C_3 plants compared with C_4.

Dissimilar sensitivity could be due to different sensory or signalling mechanisms or differences in anatomical features. The more evolutionary advanced dumb-bell shaped guard cells (often found in grasses and C_4 species) are able to open more rapidly compared with kidney shaped guard cells (Hetherington and Woodward 2003; Franks and Farquhar 2007). The speed of stomatal opening in grasses is believed to have evolved to increase photosynthesis and WUE (Grantz and Assmann 1991; Huxman and Monson 2003). The physical explanation for the reported increased speed and efficiency is due the dumb-bell shape of the guard cells enabling a greater magnitude of change in stomatal aperture with relatively small changes in turgor (Raschke 1975) which "maximise the potential of stomata to track changes in environmental conditions" (Hetherington and Woodward 2003). Both opening and closing movements of stomata are active energy consuming steps and according to Assmann and Zeiger (1987) ATP synthesis by either mitochondria or chloroplasts could accommodate the energetic requirements of stomatal opening. The fact that dumb-bell type guard cells require small changes in guard cell turgor would also imply smaller energy requirements (Hetherington and Woodward 2003).

Differences in anatomical features have also been observed between C_4 plants, which have reportedly a high stomatal density on the upper leaf surface (relative to the lower leaf surface) compared with C_3 plants (e.g. Das and Santakurmari 1977). Environmental growth conditions also influence stomatal responses; for example, sun and shade leaves differ in their response to light (Turner 1979) and also stomata on upper and lower surfaces will differ in their response to light, mostly likely due to the fact that their growth microenvironments are different (Pemadasa 1981). Additionally, we often quote that the stomata of C_3 (and C_4) plants are closed in darkness; however, as discussed above, night time stomatal conductance has been observed in a diverse range of both C_3 and C_4 species (see Caird et al. 2007).

3.2 *Correlation Between Stomatal Conductance and Photosynthetic Capacity*

Across species and under a variety of growth conditions, plants regulate their transpiration and photosynthetic rates in parallel, maintaining a balance between stomatal and photosynthetic capacity. This results in the conservation of the ratio of C_i to C_a (Wong et al. 1979, 1985; Hetherington and Woodward 2003). This empirical direct correlation between photosynthesis and stomatal conductance was central to initial models of stomatal control of photosynthesis (Farquhar and Wong 1984; Ball et al. 1987) and has been carried over to more recent models (Dewar 2002; Buckley et al. 2003). However, the underlying regulatory mechanisms are still unclear.

3.2.1 Evidence for and Against a Mesophyll Driven Signal

Guard cells may respond to photosynthetic demand by direct sensing of C_i or C_i/C_a (Ball and Berry 1982; Mott 1988; Roelfsema et al. 2002). Contrary to the predictions of the above mentioned models, transgenic plants with impairments in different steps of the photosynthetic process maintain normal stomatal conductances, resulting in elevated C_is and bringing into question the postulated C_i control of stomatal movement (for review see von Caemmerer et al. 2004). Alternatively, it has been proposed that guard cells sense the metabolic status of the mesophyll via a diffusible factor that is a product of photosynthetic activity in the mesophyll (Wong et al. 1979; Mott et al. 2008) and that stomatal aperture would be inversely proportional to the pool size of such diffusible factor (Farquhar and Wong 1984). Possible metabolites include ATP, NADPH, or ribulose bisphosphate (RuBP), the concentration of which depends strongly on the balance between chloroplast electron transport and the carboxylation reaction catalysed by Rubisco (Messinger et al. 2006).

The first suggestion of a mesophyll driven signal on stomatal behaviour was proposed by Heath and Russell in 1954. These researchers separated an indirect C_i effect from a direct light effect on stomatal behaviour and suggested that there was an indirect effect transmitted from either the epidermal cells or the mesophyll cells by a chemical or an electrical signal (Heath and Russell 1954). The influence of mesophyll on stomatal behaviour was illustrated by Lee and Bowling (1992) who demonstrated different behaviour patterns in isolated epidermis of *Commelina communis* compared with intact leaves. In the same study, the authors reported that epidermis incubated with mesophyll cells showed greater opening responses that those incubated without mesophyll cells or when mesophyll cells were kept in the dark. These results suggested that a product of photosynthesis (which they named stomatin) aided stomatal opening, although the compound was never identified. They ruled out D-glucose, sucrose, malic acid and ATP (Lee and Bowling 1992). Guard cell membrane hyperpolarization by light and CO_2 in intact leaves,

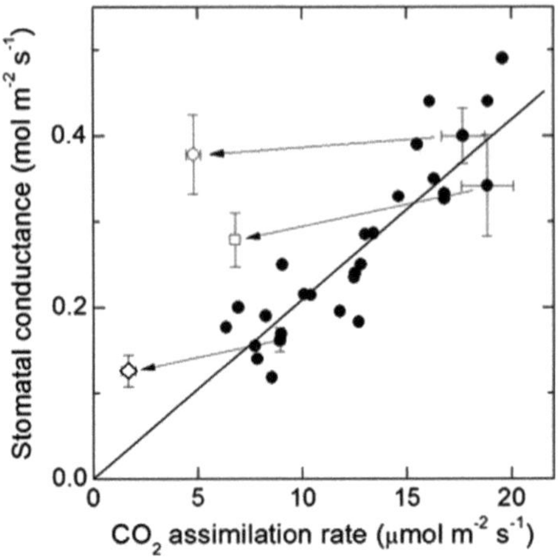

Fig. 4 Relationship between stomatal conductance and CO_2 assimilation rate in wild type and transgenic tobacco plants impaired in photosynthesis by a decrease in Rubisco function (anti-Ssu plants). Data are redrawn from (Baroli et al. 2008; www.plantphysiol.org: Copyright American Society of Plant Biologists) and plants were grown under elevated CO_2 in environmentally controlled chambers and conductance and photosynthesis measurements were performed under ambient CO_2. *Filled circles*, wild type; *open diamond*, mean ± SE ($n = 4$) from low light (LL) grown anti-Ssu plants; *open square*, mean ± SE ($n = 4$) of medium light (ML)-grown anti-Ssu plants, *open circle*, mean ± SE ($n = 5$) from ML-grown anti-Ssu plants assayed in red/blue light (redrawn from von Caemmerer et al. 2004; www.jxb.oxfordjournals.org: Copyright Society of Experimental Biology). *Arrows* link data from anti-Ssu plants with the mean ± SE of 4–5 wild type plants grown and assayed under identical conditions at the same time. The *solid line* represents linear regression fit of all wild type data ($y = 0.0217$ (± 0.00069) $\times x$, $R = 0.90$). Each data point not showing error bars corresponds to an individual plant. Error bars represent SE

but not in isolated epidermis or leaves treated with photosynthetic inhibitors, provided further evidence for a mesophyll driven signal (Lee and Bowling 1993). In a viewpoint paper, Lee and Bowling (1995) collated evidence for mesophyll influence on stomatal opening and concluded that the stomatal mechanism was controlled by command and operations and that "command originates in the mesophyll cells and operation in the guard cells". However, from studies with transgenic plants with impaired photosynthesis, it is clear that the rate of opening is not influenced by the current rate of photosynthesis (Fig. 4) ((von Caemmerer et al. 2004; Baroli et al. 2008).

A recent publication by Mott et al. (2008) has revised the potential for a mesophyll driven signal. Isolated epidermal peels grafted onto mesophyll (from the same species or another leaf) showed rapid reversible responses to light and CO_2. Mott and coworkers put forward two working hypotheses to explain the findings: (1) something produced by the mesophyll sensitises the guard cells to light and CO_2; (2) stomata respond to a signal generated in the mesophyll in response to light and CO_2. Their findings supported the second hypothesis.

3.3 *Involvement of Guard Cell Photosynthesis in Stomatal Responses*

The fact that guard cells contain functional chloroplasts (Willmer and Fricker 1996; Lawson et al. 2003), carry out linear electron transport (Hipkins et al. 1983; Shimazaki and Zeiger 1985; Willmer and Fricker 2006; Lawson et al. 2002), possess many of the main Calvin cycle enzymes (Zemel and Gepstein 1985; Shimazaki and Zeiger 1985; Gotow et al. 1988; Shimazaki et al. 1989) and guard and mesophyll cell photosynthetic efficiency correlate closely (Lawson et al. 2002, 2003) has provided attractive but controversial alternative sensory and/or regulatory mechanisms allowing stomatal behaviour to track mesophyll photosynthetic performance and explain the close correlation between mesophyll photosynthesis and stomatal conductance (Lawson 2009). Guard cell chloroplasts could contribute to stomatal behaviour in several different ways. Electron transport in guard cells, coupled to the production of ATP and/or reductants, is required for osmoregulation (Schwartz and Zeiger 1984; Shimazaki and Zeiger 1985; Tominaga et al. 2001). ATP produced through either linear (Shimazaki and Zeiger 1985; Lawson et al. 2002, 2003) or cyclic electron transport (Lurrie 1977) could provide sufficient ATP to drive ion exchange during stomatal opening. Although the exact mechanisms of ATP export from the chloroplast for use in the cytosol are unclear, specific ATP transporters have recently been identified that may provide an explanation for this mechanism (for detail see Weber and Fischer 2007). Tominaga et al. (2001) conducted experiments on epidermal peels of *Commelina benghalensis* under red light with and without inhibitors of oxidative phosphorylation (oligomycin) and photosystem II (DCMU) and demonstrated that under red light ATP produced during photophosphorylation was utilised by the plasma membrane H^+-ATPase for H^+ pumping. Several independent studies have demonstrated the use of ATP and NADPH from photosynthetic electron transport for the reduction of oxaloacetate (OAA, from starch breakdown or imported) and 3-phosphoglycerate (3-PGA, from guard cell CO_2-fixation or imported from the cytosol) which is subsequently exported to the cytosol via a 3-PGA-triose phosphate shuttle (Shimazaki et al. 1989; Ritte and Raschke 2003).

Guard cell chloroplasts also act as a starch storage facility which provides a second mechanism by which guard cell chloroplasts could contribute to coordinate stomatal movements. Carbon skeletons produced from starch breakdown are used to produce malate via PEPC and CO_2 fixation within the guard cells (Willmer and Dittrich 1974; Raschke and Dittrich 1977; Schnabl et al. 1982; Willmer 1983; Outlaw 1990). As alluded above, OAA (provided from starch breakdown) is reduced to malate (Rao and Anderson 1983; Scheibe et al. 1990) and malate accumulation has been correlated with stomatal aperture (Allaway 1973; Pearson 1973; Pearson and Milthorpe 1974; Vavasseur and Raghavendra 2005). The third and controversial mechanism concerns CO_2 fixation by the Calvin cycle within guard cell chloroplasts, and the use of end products in stomatal movements (see Outlaw 2003; Lawson 2009). Although the presence of Calvin cycle enzymes has been clearly demonstrated (Zemel and Gepstein 1985; Shimazaki and Zeiger 1985; Gotow et al. 1988; Shimazaki 1989), numerous reports have questioned the functional capacity of the Calvin cycle for producing osmotica (Outlaw et al. 1979, 1982; Outlaw 1982, 1987, 1989; Tarczynski et al. 1989). On the other hand, several studies have reported that the Calvin cycle represents a major sink for the products of photosynthetic electron transport (Cardon and Berry 1992; Lawson et al. 2002, 2003) and incorporation of $^{14}CO_2$ into 3-PGA (Gotow et al. 1988). Photosynthetic dependence of sucrose accumulation has also been demonstrated in epidermal peels of *V. faba* exposed to red illumination (Poffenroth et al. 1992). The reported contribution of osmotica produced for carbon reduction within the guard cells for stomatal opening ranges from 2% (Reckmann et al. 1990) suggesting that rates are too low for any functional significance (Outlaw 1989) to 40% (Poffenroth et al. 1992) representing a significant source of osmotica for guard cell function. Guard cell chloroplast function in stomatal movements provides an attractive hypothesis for several reasons: firstly, the guard cell chloroplasts are ideally and conveniently located (as such mechanisms would be required to be in proximity to the stomatal guard cells); secondly, the functional role of these highly conserved guard cell chloroplasts remains elusive; and finally, a mechanism for stomatal movements and behaviour linked to the guard cell chloroplasts would provide an ideal link between stomatal behaviour and the tight correlation observed with mesophyll photosynthetic rates, with both being governed by chloroplast performance.

However, work on photosynthetic transgenic plants with impaired carboxylation (von Caemmerer et al. 2004) or RuBP regeneration (Lawson et al. 2008) has shown that despite similar decreases in photosynthesis in the mesophyll as well as in the guard cells, stomata can achieve equivalent or even greater g_s than wild type plants. These data clearly demonstrate that guard cell photosynthesis is not essential for stomatal function, but may play a subtle role in stomatal opening rate and final conductance under red light conditions (Lawson et al. 2008).

3.4 Sucrose as Signal Between Photosynthesis and Stomatal Behaviour

Reports suggesting relatively low levels of Calvin cycle activity and sucrose production in guard cell chloroplasts led to the alternative suggestion that sucrose produced by mesophyll photosynthesis could provide osmotica for stomatal movements, and thereby link stomatal behaviour with photosynthetic rates (Lu et al. 1995, 1997; Ritte et al. 1999; Outlaw and De Vleighere-He 2001; Kang et al. 2007a). Hite et al. (1993) showed that guard cells could act as carbon sinks, importing sucrose via plasma membrane transporters (Stadler et al. 2003). Sucrose import into the guard cells not only provides an osmoticum for stomatal opening but can also provide a "transpiration-linked, photosynthetic-dependent passive mechanism for modulation of stomatal aperture size" (Kang et al. 2007a). The guard cell apoplastic sucrose can also exert an osmotic effect, which can drive stomatal closure, acting as a possible signal between mesophyll assimilation rate and transpiration (Kang et al. 2007a). Under photosynthetic conditions, sucrose accumulation in the guard cell apoplast driven by transpiration is sufficient to diminish stomatal aperture. The guard cell apoplastic sucrose is a product of recent mesophyll photosynthesis (Lu et al. 1997); Outlaw and co-workers have hypothesised that its concentration is correlated with the rate of photosynthesis as well as transpiration (Outlaw and De Vleighere-He 2001) providing a signal that prevents further stomatal opening when photosynthesis and transpiration are high (Kang et al. 2007a). However, such a mechanism may only be possible in apoplastic phloem loaders (Kang et al. 2007b). It is also difficult to see how this mechanism works in maintaining high photosynthetic rates, as under the hypothesis proposed, limited CO_2 supply as a result of preventing stomatal opening would potentially increase mesophyll demand for CO_2.

3.5 ROS Signalling in Stomata and Relationship with Photosynthesis

Reactive oxygen species (ROS) result from the incomplete reduction of molecular oxygen. They include superoxide, hydrogen peroxide, singlet oxygen and the hydroxyl radical. Besides being unavoidable toxic by-products of metabolism and environmental stress, ROS play a central role as messengers in the signal transduction chain leading to the acclimation response to stress (Van Breusegem et al. 2008; Pfannschmidt et al. 2009). In particular, hydrogen peroxide has been shown to mediate ABA-induced stomatal closure and it is thought that the source of H_2O_2 is the guard cells themselves (reviewed in Wang and Song 2008). MAP kinases have recently been implicated in the down stream function of ROS activity to regulate guard cell ABA signalling positively (Jammes et al. 2009). Chloroplasts are considered to be the primary sources of ROS in plant cells (Asada 2006) and ROS production in the chloroplasts of guard cells has been observed in response to ozone treatment (Joo et al. 2005). However, whether and to what extent chloroplast-generated ROS contribute directly to the signal transduction that leads to stomatal movements has not been defined (Wang and Song 2008). In fact, H_2O_2 generated at the guard cell plasma membrane by the action of NADPH oxidase has emerged as a major player in ABA-mediated signal transduction in guard cells. H_2O_2 is required to initiate stomatal closure (Wang and Song 2008). In Arabidopsis, mutations in two of the ten NADPH oxidase subunits, AtrbohD and AtrbohF, abolish ABA-induced ROS production and stomatal closure and the effect of the mutations is cancelled by the exogenous addition of H_2O_2 (Kwak et al. 2003). The signal transduction cascade involving ABA and ROS has been shown to include cytosolic Ca^{2+} transients, G proteins, protein kinases and phosphatases, phosphatidic acid and transcription factors (Wang and Song 2008; Pham and Desikan 2009; Cho et al. 2009; Zhang et al. 2009). In rice, a zinc-finger transcription factor named DST (for *drought and salt tolerance*), which accumulates H_2O_2, has been recently shown to negatively regulate stomatal closure by direct modulation of the genes related to H_2O_2 homeostasis (including peroxidases), with plants lacking DST exhibiting increased stomatal closure and reduced stomatal density, resulting in enhanced tolerance to drought (Huang et al. 2009).

Scavenging and prevention of excess production of ROS are integral features of cellular metabolism (Niyogi 2000). ROS are detoxified in plant cells by a combination of enzymatic reactions involving superoxide dismutases, ascorbate peroxidases, catalases and glutathione peroxidases, as well as small antioxidant molecules, such as ascorbate and glutathione. Detoxification of ROS is essential for normal stomatal movements. An Arabidopsis mutant lacking a glutathione peroxidase (AtGPX3) exhibits enhanced production of H_2O_2 in guard cells and reduced WUE, whereas plants overexpressing AtGPX3 showed increased WUE and were less sensitive to water stress (Miao et al. 2006). Ascorbate peroxidase (APX1) deficient Arabidopsis plants, which show reduced photosynthetic rates compared to wild type plants, also show abnormal stomatal conductances, in particular with respect to light-to dark transitions, although their response to exogenous ABA applications appears normal (Pnueli et al. 2003).

3.6 Role for Respiration

Although the relationship between g_s and A is key to examining WUE, the effect of respiration (by either guard or mesophyll cells) may play an important role in contributing to stomatal sensory or signalling mechanisms in response to changing environmental parameters (Lawson 2009). Guard cells are known to contain numerous mitochondria (Willmer and Fricker 1996; Vavasseur and Raghavendra 2005) and high metabolic fluxes through the catabolic pathway have been reported (Hampp et al. 1982). Raghavendra and colleagues have suggested that ATP produced through oxidative phosphorylation is important for stomatal movements (Raghavendra and Vani 1989; Parvathi and Raghavendra 1997). The application of inhibitors of oxidative phosphorylation such as KCN has revealed an impaired stomatal response to light induced opening (Schwartz and Zeiger 1984). Transgenic tomato plants with reductions in mitochondrial fumarate hydratase (fumarase) activity (Nunes-Nesi et al. 2007) and malate dehydrogenase (Nunes-Nesi et al. 2005) show reductions in stomatal aperture and CO_2 limitation of photosynthesis (Nunes-Nesi et al. 2007). Additionally, Lu et al. (2000) demonstrated a positive correlation between stomatal conductance and guard cell respiration rates in Pima cotton. A role for respiratory ATP in stomatal opening in CAM plants has also recently been demonstrated (von Caemmerer and Griffiths 2009). It has been suggested that both the photosynthetic and respiratory pathways in guard cells are important in stomatal function (Asai et al. 2000) and that the relative importance of each pathway maybe altered if either one is restricted (see Parvathi and Raghavendra 1997).

4 Environmental Control of Stomatal Development and the Role of Photosynthesis

The developmental control underpinning the coordination between stomatal conductance and photosynthesis is not well understood. Developmental changes during the life span of the leaf have been shown to be independent of the photosynthetic capacity of the plant. For example, Jiang and Rodermel (1995) showed that stomatal conductance followed similar developmental changes with leaf age in antisense small subunit Rubisco (anti-Ssu) plants and wild type tobacco plants despite their different photosynthetic rates. Similarly, it was shown that, despite the very different photosynthetic capacities, the leaf and stomatal developmental response to growth light environment in anti-SSu and wild type plants was similar, resulting in fewer stomata per leaf area in leaves developed under low compared to high light in both genotypes (Baroli et al. 2008).

4.1 The Genetic Pathway of Stomatal Development

Recent work in Arabidopsis has substantially advanced our understanding of the genetics of stomatal differentiation (see Nadeau 2009; Bergmann and Sack 2007 and references therein). The process appears to be highly regulated and involves a series of cell divisions which are asymmetrical and oriented so as to ensure correct stomatal spacing by preventing the formation of adjacent stomata. The stomatal developmental pathway begins when a protodermal cell in the epidermis of the unfolding leaf converts to a meristemoid mother cell, which then undergoes an asymmetric cell division producing a small meristemoid cell and a larger sister cell. The meristemoid cell can undergo several self-renewing asymmetric divisions before differentiating into a guard mother cell, which then divides symmetrically and further differentiates to form the pair of mature guard cells that surround the stoma. The asymmetry of cell divisions in the stomatal lineage was shown recently to be determined by the intracellular distribution of the product of the BREAKING OF ASYMMETRY IN THE STOMATAL LINEAGE (BASL) gene (Dong et al. 2009). Cell–cell signalling in response to positional cues during stomatal development is mediated by the putative cell surface receptors TOO MANY MOUTHS (TMM) and members of the ERECTA family of leucine-rich repeat containing receptor kinases (ER, ERL1 and ERL2). These are proposed to interact with the small secretory peptides EPIDERMAL PATTERNING FACTORS (EPF) 1 and 2, produced by stomatal precursors, which may act as mobile positional signals (Hara et al. 2007; Hunt and Gray 2009). Activation of the receptors stimulates a mitogen-activated protein kinase (MAPK) cascade starting with the MAPKKK YODA (YDA), which in turn activates MKK4/MKK5 and MPK3/MPK6 (Wang et al., 2007a, b). This MAPK cascade negatively regulates stomatal development, and it may target three structurally related transcription factors belonging to the basic helix-loop-helix (bHLH) family SPEECHLESS (SPCH), which acts as a positive regulator of the initial asymmetric cell division of the meristemoid, along with MUTE and FAMA, which control meristemoid differentiation and guard cell morphogenesis, respectively. In addition, MYB transcription factors (MYB124, MYB88 and FOUR LIPS) are also involved in the final fate transitions of the stomatal differentiation pathway.

The importance of identifying the pathways regulating stomatal differentiation in order to develop plants with greater water use efficiencies has recently been reviewed by Wang et al. (2007a, b). However, changes in anatomical stomatal characters do not always result in changes in stomatal conductance and/or increase WUE (Lawson and Morison 2004; Lawson 2009). This has been demonstrated in Arabidopsis plants over-expressing STOMATAL DENSITY AND DISTRUIBTUION 1 (SDD1) gene, which conveyed a 40% reduction in stomatal density, while the sdd1-1 mutants (Berger and Altmann 2000) increased density by 300% relative to the wild type. Yet both plants showed similar stomatal conductance and assimilation rates as WT plants. Decreases in density were compensated for by greater apertures and *vice versa* (Bussis et al. 2006).

4.2 Interaction Between Stomatal Development Genes and Environmental Signals

While considerable knowledge exists on the effect of environmental factors such as light intensity and CO_2 concentration on the signalling mechanisms determining stomatal pore aperture, very little is known about their effects on the modulation of stomatal development. Moreover, it is not clear whether the stomatal lineage cells perceive the environmental stimuli directly (Casson and Gray 2008). However, links are starting to be established between the stomatal developmental genes and other known regulators of gene expression in response to environmental conditions. The bHLH-leucine zipper transcription factor SCREAM1 (SCRM1) has been shown recently to interact directly with and specify the sequential actions of SPCH, MUTE and FAMA (Kanaoka et al. 2008). SCRM1 was previously identified as INDUCER OF CBF EXPRESSION 1 (ICE1), a regulator of the expression of cold-induced genes. Another point of environmental regulation of stomatal patterning may be the MAP kinase cascade (Wang et al. 2007a, b), as its components MKK4, MKK5, MAPK3 and MAPK6 have previously been shown to play a role in the environmental stress response, with MAPK3 and MAPK6 directly involved in osmotic stress (Nakagami et al. 2005). Recent work with photoreceptor mutants in Arabidopsis has placed the YDA gene downstream of a developmental cascade involving the master regulator of photomorphogenesis COP1, the red light photoreceptors phytochrome A and phytochrome B and the blue-light photoreceptor cryptochrome, indicating that the three photoreceptors act additively to promote stomatal development in response to light quality (Kang et al. 2009).

4.3 Systemic Signals and Control of Stomatal Density in Response to the Environment

The stomatal density (the number of pores per unit area) and the stomatal index (number of guard cells relative to total epidermal cells) on the leaf epidermis are regulated during leaf expansion by humidity, temperature, CO_2 concentration and light intensity and once determined they remain unchanged for the lifetime of the leaf (Casson and Gray 2008). Light intensity and CO_2 concentration are the two most studied environmental variables with respect to their effects on stomatal differentiation. In general, both stomatal density and index are higher in plants grown in full sunlight or at high light intensities than in plants grown in shade (Willmer and Fricker 1996; Schoch et al. 1984; Boccalandro et al. 2009; Casson et al. 2009). Although exceptions have been recorded (notably, in free-air CO_2 enrichment (FACE) experiments, see Ainsworth and Rogers 2007) many species acclimate to increases in CO_2 levels by decreasing their stomatal density. This has been observed to occur during the past two centuries in industrial Europe, under laboratory conditions and through plant evolution (Woodward 1987; Hetherington and Woodward 2003), so much so that stomatal density is used as a proxy of paleo-CO_2 levels (Chaloner and McElwain 1997). Because the response to environmental conditions that controls stomatal patterning occurs before leaf expansion, this acclimation process must require the integration of signals at the whole plant level. In fact, experiments in which developing and mature leaves were subjected to different light intensities and CO_2 concentrations have demonstrated that the stomatal pattern of developing leaves is influenced by the conditions experienced by the mature leaves (Schoch et al. 1980; Lake et al. 2001; Thomas et al. 2004; Driscoll et al. 2006; Miyazawa et al. 2006). In Arabidopsis, growth of mature leaves at elevated CO_2 resulted in a 20–30% decrease (relative to control plants grown at ambient CO_2) in stomatal index in developing leaves that were exposed to ambient CO_2 levels. Conversely, in reciprocal experiments where mature leaves were exposed to ambient CO_2 and the developing leaves to elevated CO_2, the stomatal index was increased in developing leaves (Lake et al. 2001). Similar results have been reported for light intensity in *Chenopodium album* (Yano and Terashima 2001).

So far, only the HIGH CARBON DIOXIDE (HIC) gene of Arabidopsis has been identified as having a role in modulating changes in stomatal index in response to elevated CO_2 (Gray et al. 2000). When exposed to elevated levels of CO_2, HIC mutant plants showed a significant increase in stomatal index, whereas the parental ecotype showed a small decrease. HIC is expressed in guard cells and shares high homology with the Arabidopsis KCS1 gene, a 3-ketoacyl CoA synthase involved in the production of very long chain fatty acids found in the cuticular waxes. The mechanism by which HIC affects stomatal patterning in response to CO_2 is unknown. It is possible that a cuticular wax or an intermediate is a signalling compound that influences stomatal development. Alternatively, HIC may indirectly affect stomatal patterning in response to CO_2 by altering the permeability to water, CO_2, or another signalling compound within the epidermis.

The nature of the stomatal differentiation signal that is transported from mature leaves to developing leaves is also not clear. Carbohydrate accumulation and sugar signalling are involved in plant development and cell cycle control (Rolland et al. 2006). However, the fact that increased light intensity and elevated CO_2 both enhance rates of photosynthesis and sugar content of mature leaves, but exert opposite effects on the stomatal density of new leaves, argues against a photosynthetic nature of the signal. Consistent with this is the observation that the levels of sugars increased in mature and developing leaves when they were exposed to high CO_2 and decreased with shade treatment (Coupe et al. 2006). Based on the linear correlation between carbon isotope discrimination values and stomatal density found in cowpeas subjected to different CO_2, water and phosphorus environments, Sekiya and Yano (2008) proposed that the C_i/C_a ratio may be involved in the systemic signalling that determines stomatal density. However, transgenic tobacco plants with decreased Rubisco content show a normal response of stomatal density to light intensity despite maintaining a high C_i/C_a (Baroli et al. 2008), suggesting that, at least when light intensity is the environmental stimulus, the C_i/C_a ratio would not be part of the signalling mechanism. In experiments in which CO_2 concentration, irradiance and water pressure deficit were varied in mature leaves independently of the conditions around the developing leaves, the stomatal index of the developing leaves was positively and highly correlated with the stomatal conductance of the mature leaves and independent of their net photosynthesis (Miyazawa et al. 2006). This result suggests that it is not the carbohydrate production but the transpiration rates of mature leaves that could influence stomatal development, independently of photosynthesis, by controlling the delivery rates of hormones, such as cytokinins and abscisic acid, that are transported through the xylem from the roots to the expanding leaves. Boonman et al. (2007) have recently shown that cytokinin delivery through the xylem is dependent on transpiration rates in Arabidopsis and tobacco and they suggest that cytokinin import rate into the leaf could be a signal for photosynthetic acclimation to environmental variables such as light intensity.

4.4 Hydraulic Conductance Correlates with Stomatal Behaviour

Stomatal density and size are not the only anatomic features of the leaf that can exert control over stomatal behaviour. Hydraulic conductivity, which is a measure of the efficiency of water transport within leaves, has been shown to correlate with maximum stomatal conductance (Brodribb and Holbrook 2004; Brodribb et al. 2005; Nardini and Salleo 2005; Brodribb and Jordan 2008) as well as stomatal sensitivity to perturbations in VPD (Franks and Farquhar 1999). Evidence exists for a hydraulic influence on both long-term adaptation of maximum conductance as well as short-term stomatal responses (Brodribb and Jordan 2008). It has been suggested that species with a high hydraulic conductance might be more buffered and therefore respond less to changes in VPD (Franks and Farquhar 1999). However, there is considerable species–species variation in the physiological influence that plant hydraulic conductance exerts over leaf conductance (Sperry 2000; Meinzer 2002). A strong correlation exists between maximum leaf hydraulic conductance and leaf anatomical characters including vein density, stomatal pore area index and palisade thickness (Aasamaa et al. 2001; Sack and Frole 2006). These observations indicate that hydraulic efficiency is highly adapted (Brodribb and Holbrook 2004), with implications for evolutionary coordination between gas exchange and hydraulic capacity (Franks and Farquhar 1999; Brodribb and Field 2000). The relationship between anatomical characters and hydraulic conductivity could provide a means for paleo-reconstruction of stomatal behaviour/sensitivity in past climatic environments (Brodribb and Jordan 2008).

5 Stomatal Manipulation to Improve Water Use Efficiency

Since altering stomatal anatomical features such as density and size may not necessarily always result in plants with increased WUE, modifications in stomatal behavioural characteristics may provide an alternative strategy (Nilson and Assmann 2007). To exploit such an opportunity, we first need to gain a solid understanding of stomatal metabolism for engineering drought resistance in future crop plants. Several studies have demonstrated the potential of such an approach by examining plants with increased or decreased amounts of a single enzyme and shown an alteration in stomatal conductance. *Zea mays* plants with increased amount of NADP-malic enzyme demonstrated signs of drought resistance with decreased stomatal conductance (Laporte et al. 2002). Increased drought resistance has also been demonstrated in Arabidopsis plants with alterations in guard cell membrane transporters (Klein et al. 2004), calcium dependent protein kinases (Ma and Wu 2007), aquaporin genes (Cui et al. 2008) and ABA biosynthesis or ABA sensitivity (Jakab et al. 2005; Wang et al. 2005; Yang et al. 2005). Hu et al (2006) established drought tolerance in rice over-expressing SNAC1 (STRESS-RESPONSIVE NAC 1, which encodes a NAC transcription factor). NAC genes play important roles in developmental processes, auxin signalling, defense and abiotic stress responses. Recently, a previously unknown zinc finger protein DST (drought and salt tolerance) has been identified as a regulator of stomatal closure through modulation of H_2O_2 signalling pathways in guard cells (Huang et al. 2009), providing possible new avenues of enhanced drought and salt tolerance in rice. Understanding the mechanisms involved in stomatal sensing/signalling pathways in response to changing CO_2 may play a fundamental role in plant WUE as illustrated in a recent study exploring the evolution of stomatal sensitivity to CO_2 (Brodribb et al. 2009). Plant species with reduced CO_2 sensitivity (such as ferns) had a reduced WUE compared to CO_2 sensitive angiosperms, suggesting that atmospheric CO_2

concentration is an evolutionary driving force for optimising WUE (Brodribb et al. 2009). However, in order to increase WUE, a combination of alterations in stomatal function and photosynthetic capacity may be required. For example, increased WUE in the *Erecta* mutants was not only because of modifications to cell expansion and division and stomatal density but also because of alterations in leaf diffusive properties and mesophyll photosynthetic capacity (Masle et al. 2005).

It is also worth bearing in mind that any modifications in stomatal (or photosynthetic) capacity and behaviour would require testing in a field situation, with fluctuating environmental conditions, predation and competition. For example, ABA over-sensitive Arabidopsis mutants with reduced g_s could not compete for water with WT plants (Basco et al. 2008). Although such findings have implications for screening protocols, they may not be so critical for monocultures of crops.

The potential to identify genes involved in stomatal responses to various environmental stresses/factors has recently been explored using various "-omic" approaches (Leonhardt et al. 2004; Coupe et al. 2006). Using microarray technologies, Leonhardt et al (2004) showed reductions in guard cell metabolism when Arabidopsis leaves were sprayed with ABA. Such findings agreed with earlier studies that reported decreased PEPC transcripts under drought conditions (Kopka et al. 1997). A functional proteomic study in Arabidopsis guard cells protoplasts by Zhao et al. (2008) identified new proteins and signalling pathways required for ABA responses in guard cells. Analysis of functional groups of genes revealed only a 1.9% higher representation of photosynthetic genes in mesophyll and guard cells. Transcriptomics analysis has also been used to identify transcriptional factors that are necessary for mediating stomatal movement in response to light (Gray 2005; Casson and Gray 2008) and water deficit (Cominelli et al. 2005). Such information is critical in helping to determine the link between mesophyll photosynthesis and guard cell behaviour. For example, a comparative proteomic study between mesophyll and guard cells in *Brassica napus* revealed functional differentiation between the two cell types. Expression patterns from guard cells were enriched with proteins involved in respiration, transport, transcription, cell structure and signalling, while proteins involved in photosynthesis, starch synthesis and defense mechanisms were more prevalent in mesophyll cells (Zhu et al. 2009). Although such approaches are invaluable in the information they provide on stomatal signalling and response mechanisms, to date most studies have relied upon the use of guard cell protoplasts. There are two main concerns with the use of protoplasts; firstly, the isolation procedure will inevitable result in cellular damage, resulting in the production of ROS, which are known to be involved in signal transduction pathways and alteration of gene expression (Galvez-Valdivieso et al. 2009). Secondly, the isolation of guard cells from the mesophyll will immediately remove the potential of any mesophyll-driven signalling event. A third (most likely less important) point is the use of bulk samples removing possible identification of heterogeneity between guard cells, as it is well established that considerable variation in stomatal behaviour is apparent over individual leaves (Weyers and Lawson 1997).

6 Scaling-Up: From Leaf to Canopy

In this chapter, we have focused on stomatal and photosynthetic responses at the leaf level, reporting results mainly obtained from leaf cuvette gas exchange measurements. An appreciation of the dynamic role of stomata in photosynthetic physiology and WUE at the canopy/crop level (or beyond) is essential for a mechanistic understanding of ecosystem carbon and water fluxes and can be predicted with consideration of models and scaling processes. At the same time, modelling and attempts to scale-up must also pay due regard to the intricacy and variation often found at smaller scales (Weyers et al. 1997). A full review of scaling measurements from the leaf to canopy level is beyond the scope of this chapter and its authors; however, in this section we briefly highlight some of the problems and complexities associated with scaling processes.

Under constant environmental conditions, leaves are tightly coupled with the atmosphere and transpiration is proportional to g_s; however, in the open air the relationship becomes more complex because of additional resistance pathways between stomatal and the bulk atmosphere that feedback on ecosystem evapotranspiration (Bernacchi et al. 2007). Carbon and water fluxes from a vegetation canopy can be predicted using leaf scale parameters by "bottom-up" scaling, using the parallel sum of individual leaf measurements within a canopy structure and taking into account the microenvironment surrounding leaves (Jarvis 1993). A prerequisite for such approaches is knowledge of the variation of parameters throughout the canopy, making such models impractical for regional or global applications (Kruijt et al. 1997). Alternatively, canopy fluxes can be measured directly using eddy covariance techniques and are characterised as "big leaf" models using a "top-down" approach (Kruijt et al. 1997). The degree of influence of leaf or canopy conductance on evaporation to the bulk atmosphere depends on boundary layer conductance and efficiencies of heat and mass transfer between the canopy "surface" and bulk atmosphere (Jarvis and McNaughton 1986; McNaughton and Jarvis 1991). Coupling between stomatal behaviour and the bulk atmosphere depends upon the relative size and structure of the boundary layer, and as boundary layer influences become large, stomatal influences become less important (Jarvis and McNaughton 1986). The influence of stomatal conductance on transpiration gained at one scale can be used to predict what may happen at another scale only if the coupling between saturation deficit at the leaf surface and that of the air outside the leaf boundary layer is known and similar at the two scales (Jarvis and McNaughton 1986). As scale increases, the importance of stomatal behaviour on flux parameters tends to decrease; for example, if the ratio of stomatal conductance to boundary layer conductance is large, only crude models of stomatal responses to environmental variables are required; however, better models are needed if this ratio is small. However, when water stress causes stomatal closure, the importance of leaf conductance increases and more reliable estimates of stomatal responses to changing environmental parameters are required (McNaughton and Jarvis 1991). Vegetation type and canopy structure may also significantly impact on large scale predictive models of transpiration, photosynthesis and WUE. Irrespective of whether the canopy is constructed of a single species or mixed population, significant variation in stomatal behaviour and transpiration will exist between leaves and between plants that make up the canopy, leading to local differences in saturation deficit and coupling between leaf surfaces and the bulk atmosphere. The impact of changing environmental conditions on stomatal behaviour in different plant species will also contribute to local differences. For example, decreases in soil water status will reduce stomatal conductance which will differentially influence photosynthesis and transpiration in C3 and C4 species (Jarvis and McNaughton 1986). Recently, Bernacchi et al. (2007) demonstrated close coupling of stomatal conductance with ecosystem evapotranspiration in a soybean crop grown in the field under elevated CO_2. Experiments conducted over four growing seasons demonstrated an average decrease in g_s by 10% with elevated CO_2 which correlated with an 8.6% decrease in evapotranspiration, clearly demonstrating a close coupling of stomatal conductance with ecosystem evapotranspiration, which was not driven by changes in growth. Such findings demonstrated that despite system feedbacks, changing g_s of upper canopy leaves can modify the transfer of water vapour to the bulk atmosphere, as well as illustrate the importance of stomatal responses to changing climatic conditions on ecosystem evapotranspiration (Bernacchi et al. 2007). The response of g_s to elevated CO_2 is a critical parameter for large scale ecosystem models of photosynthesis and transpiration, as many of the modelling approaches incorporate the Ball et al. (1987) model of stomatal conductance in which stomatal conductance is dependent on the sensitivity of g_s to CO_2 concentration, assimilation rate and relative humidity which would alter with any acclamatory response (Ainsworth and Rogers 2007). However, it is clear from field based FACE experiments that CO_2 effects on stomatal conductance are very much species specific and dependent on the photosynthetic pathways used (Ainsworth and Rogers 2007).

The continual fluctuation in input energy into a canopy via sun and shade flecks also impacts on the accuracy of scaling models, via spatial heterogeneity in stomatal conductance and rate of photosynthesis. The lack of detailed spatial resolution results in many "big-leaf" models overestimating canopy photosynthetic rates. By fractionating the canopy into sun and shade-lit proportions, De Pury and Farquhar (1997) were able to scale photosynthesis from leaves to canopies avoiding the errors associated with big-leaf models.

There are many other considerations and complexities that we have not covered, including the influence of scaling on time and fluctuating dynamics of leaf level measurements and the errors associated with these measurements (Buckley et al. 2003). Temporal and spatial variations and fluctuations in environmental variables will influence the minute-to-minute responses of stomata and photosynthesis with varying degrees of spatial resolution and each of these will have an error associated with them. Such spatial–temporal regulation at the leaf level also has implications for measurement collection protocols and sample sizes.

7 Conclusion

A detailed understanding of stomatal behaviour in relation to photosynthesis is essential to understand the impacts of changing climate on plant performance and WUE. In this chapter, we have examined stomatal behaviour in response to several key environmental parameters, and highlighted the differences observed between different photosynthetic pathways. We have tried to emphasise the role of stomata in determining mesophyll CO_2 assimilation rates and the impact of fluctuating environmental parameters on both photosynthesis and stomatal behaviour. Although we often consider stomatal function to be key to controlling CO_2 and H_2O fluxes, the importance and impact of anatomical features such as stomatal density and size should not be overlooked. Increasing developments and advancements in modern techniques such as proteomics and transcriptomics are providing novel and exciting approaches in the quest to understand the complex and plastic responses that are apparent in stomatal function. By combining modern approaches with traditional plant physiological procedures and employing a holistic approach, we are becoming closer to determining the link that correlates stomata function with photosynthesis which will improve our ability to model and predict ecosystem-level responses to carbon and water fluxes.

References

Aasamaa K, Sober A, Rahi M (2001) Leaf anatomical characteristics associated with shoot hydraulic conductance, stomatal conductance and stomatal sensitivity to changes of leaf water status in temperate deciduous trees. Aust J Plant Physiol 28:765–774

Ainsworth EA, Rogers A (2007) The response of photosynthesis and stomatal conductance to rising [CO_2]: molecular mechanisms and environmental interactions. Plant Cell Environ 30:258–270

Akita S, Moss DN (1972) Differential stomatal response between C_3 and C_4 species to atmospheric CO_2 concentration and light. Crop Sci 12:789–793

Allaway WD (1973) Accumulation of malate in guard cells of *Vicia faba* during stomatal opening. Planta 110:63–70

Allen MT, Pearcy RW (2000) Stomatal behaviour and photosynthetic performance under dynamic light regimes in a seasonally dry tropical rain forest. Oecologia 122:470–478

Aphalo PJ, Jarvis PG (1991) Do stomata respond to relative humidity? Plant Cell Environ 14:127–132

Asada K (2006) Production and scavenging of reactive oxygen species in chloroplasts and their functions. Plant Physiol 141:391–396

Asai N, Nakajima N, Tamaoki M, Kamada H, Kondo N (2000) Role of malate synthesis mediated by phosphoenolpyruvate carboxylase in guard cell in the regulation of stomatal movements. Plant Cell Physiol 41:10–15

Assmann SM, Zeiger E (1987) Guard cell bioenergetics. In: Zeiger E, Farquhar G, Cowan IR (eds) Stomatal function. Stanford University press, Stanford, pp 163–194

Assmann SM (1993) Signal transductions in guard cells. Annu Rev Cell Biol 9:345–375

Ball JT, Berry JA (1982) The C_i/C_a ratio: a basis for predicting stomatal control of photosynthesis, vol 81. Carnegie Institute of Washington Yearbook, Washington, pp 88–92

Ball TJ, Woodrow IE, Berry JA (1987) A model predicting stomatal conductance and its contribution to the control of photosynthesis under different environmental conditions. In: Biggins J (ed) Progress in photosynthesis research. Martinus-Nijhoff, Dordrecht, pp 221–224

Barbour MM, Cernusak LA, Whitehead D, Griffin KL, Turnbull MH, Tissue DT, Farquhar GD (2005) Nocturnal stomatal conductance and implications for modeling $\delta^{18}O$ of leaf-respired CO_2 in temperate tree species. Funct Plant Biol 32:1107–1121

Baroli I, Price GD, Badger MR, von Caemmerer S (2008) The contribution of photosynthesis to the red light response of stomatal conductance. Plant Physiol 146:737–747

Barradas VL, Jones HG (1996) Responses of CO_2 assimilation to changes in irradiance: laboratory and field data and a model for beans (*Phaseolus vulgaris* L.). J Exp Bot 47:639–645

Basco R, Janda T, Galiba G, Papp I (2008) Restricted transpiration may not result in improved drought tolerance in a competitive environment for water. Plant Sci 174:200–204

Berger D, Altmann T (2000) A subtilisin-like serine protease involved in the regulation of stomatal density and distribution in *Arabidopsis thaliana*. Genes Dev 14:1119–1131

Bergmann DC, Sack FD (2007) Stomatal development. Annu Rev Plant Biol 58:163–81

Bernacchi CJ, Kimball BA, Quarles DR, Long SP, Ort DR (2007) Decreases in stomatal conductance of soybean under open-air elevation of [CO_2] are closely coupled with decreases in ecosystem evapotranspiration. Plant Physiol 143:134–144

Berryman CA, Eamus D, Duff GA (1994) Stomatal responses to a range of variables in two tropical tree species grown with CO_2 enrichment. J Exp Bot 45:539–546

Black CC, Osmond CB (2003) Crassulacean acid metabolism photosynthesis: 'Working the night shift'. Photosynth Res 76:329–341

Boccalandro HE, Rugnone ML, Moreno JE, Ploschuk EL, Serna L, Yanovsky MJ, Casal JJ (2009) Phytochrome B enhances photosynthesis at the expense of water-use efficiency in Arabidopsis. Plant Physiol 150:1083–1092

Boonman A, Prinsen E, Gilmer F, Schurr U, Peeters AJ, Voesenek LACJ, Pons TL (2007) Cytokinin import rate as a signal for photosynthetic acclimation to canopy light gradients. Plant Physiol 143:1841–1852

Brodribb TJ, Field TS (2000) Stem hydraulic supply is linked to leaf photosynthetic capacity: evidence from New Caledonian and Tasmanian rainforests. Plant Cell Environ 23:1381–1388

Brodribb T, Holbrook N (2004) Stomatal protection against hydraulic failure: a comparison of coexisting ferns and angiosperms. New Phytol 162:663–670

Brodribb T, Holbrook NM, Zwieniecki MA, Palma B (2005) Leaf hydraulic capacity in ferns, conifers and angiosperms: impacts on photosynthetic maxima. New Phytol 165:839–846

Brodribb T, Jordan G (2008) Internal coordination between hydraulics and stomatal control in leaves. Plant Cell Environ 31:1557–1564

Brodribb TJ, McAdam SMA, Jordan GJ, Field TS (2009) Evolution of stomatal responsiveness to CO_2 and optimization of water-use efficiency among land plants. New Phytol 183:839–847

Buckley TN, Mott KA, Farquhar GD (2003) A hydromechanical and biochemical model of stomatal conductance. Plant Cell Environ 26:1767–1785

Bussis D, von Groll U, Fisahn J, Altmann T (2006) Stomatal aperture can compensate altered stomatal density in *Arabidopsis thaliana* at growth light conditions. Funct Plant Biol 33:1037–1043

Caird MA, Richards JH, Donovan LA (2007) Nighttime stomatal conductance and transpiration in C_3 and C_4 plants. Plant Physiol 143:4–10

Cardon ZG, Berry JA (1992) Effects of O_2 and CO_2 concentration on the steady-state fluorescence yield of single guard cell pairs in intact leaf discs of *Tradescantia albiflora*. Plant Physiol 99:1238–1244

Cardon ZG, Berry JA, Woodrow IE (1994) Dependence of the extent and direction of average stomatal response in *Zea mays* L. and *Phaseolus vulgaris* L. on the frequency of fluctuations in environmental stimuli. Plant Physiol 105:1007–1013

Casson S, Gray JE (2008) Influence of environmental factors on stomatal development. New Phytol 178:9–23

Casson SA, Franklin KS, Gray JE, Grierson CS, Whitelam GC, Hetherington AM (2009) Phytochrome B and PIF4 regulate stomatal development in response to light quantity. Curr Biol 19:229–234

Chaloner WG, McElwain J (1997) The fossil plant record and global climatic change. Rev Palaeobot Palynol 95:73–82

Cho D, Shin D, Jeon BW, Kwak JM (2009) ROS-mediated ABA signalling. J Plant Biol 52:102–113

Cockburn W, Ting IP, Steinberg LV (1979) Relationships between stomatal behaviour and internal carbon dioxide concentration in CAM plants. Plant Physiol 63:1029–1032

Cominelli E, Galbiati M, Vavasseur A, Conti L, Sala T, Vuylsteke M, Leonhardt N, Dellaporta SL, Tonelli C (2005) A guard-cell-specific MYB transcription factor regulates stomatal movements and plant drought tolerance. Curr Biol 15:1196–1200

Condon AG, Reynold MP, Rebetzke GJ, Ginkel M, Richards RR, Farquhar GD (2007) Using stomatal aperture-related traits to select for high yield potential in bread wheat. In: Buck HT, Nisi JE, Salomon N (eds) Developments in plant breeding 11 – wheat production in stressed environments, vol 12. Springer, Berlin, Heidelberg, New York, pp 612–624

Coupe SA, Palmer BG, Lake JA, Overy SA, Oxborough K, Woodward FI, Gray JE, Quick WP (2006) Systemic signaling of environmental cues in *Arabidopsis* leaves. J Exp Bot 57:329–341

Cowan IR, Troughton JH (1971) The relative role of stomata in transpiration and assimilation. Planta 97:325–336

Cui XH, Hai FS, Chen H, Chen J, Wang XC (2008) Expression of the *Vicia faba* VfPIP1 gene in *Arabidopsis thaliana* plants improves their drought resistance. J Plant Res 121:207–214

Daley MJ, Phillips NG (2006) Interspecific variation in nighttime transpiration and stomatal conductance in a mixed New England deciduous forest. Tree Physiol 26:411–419

Das VSR, Santakurmari (1977) Stomatal characteristics of some dicotyledonous plants in relation to the 4 carbon and 3 carbon pathways of photosynthesis. Plant Cell Physiol 18:935–938

De Pury DGG, Farquhar GD (1997) Simple scaling of photosynthesis from leaves to canopies without the errors of big-leaf models. Plant Cell Environ 20:537–557

Dewar RC (2002) The Ball–Berry–Leuning and Tardieu–Davies stomatal models: synthesis and extension within a spatially aggregated picture of guard cell function. Plant Cell Environ 25:1383–1398

Dong J, MacAlister CA, Bergmann DC (2009) BASL controls asymmetric cell division in Arabidopsis. Cell 137:1320–1330

Donovan LA, Grisé DJ, West JB, Pappert RA, Alder NN, Richards JH (1999) Predawn disequilibrium between plant and soil water potentials in two cold-desert shrubs. Oecologia 120:209–217

Donovan LA, Linton MJ, Richards JH (2001) Predawn plant water potential does not necessarily equilibrate with soil water potential under well watered conditions. Oecologia 129:328–335

Driscoll SP, Prins A, Olmos E, Kunert KJ, Foyer CH (2006) Specification of adaxial and abaxial stomata, epidermal structure and photosynthesis to CO_2 enrichment in maize leaves. J Exp Bot 57:381–390

Dubbe DR, Farquhar GD, Raschke K (1978) Effect of abscisic acid on the gain of the feedback loop involving carbon dioxide and stomata. Plant Physiol 62:413–417

Dwyer SA, Ghannoum O, Nicotra A, von Caemmerer S (2007) High temperature acclimation of C_4 photosynthesis is linked to changes in photosynthetic biochemistry. Plant Cell Environ 30:53–66

Farquhar DG, Sharkey TD (1982) Stomatal conductance and photosynthesis. Ann Rev Plant Physiol 33:317–345

Farquhar GD, Wong CS (1984) An empirical model of stomatal conductance. Aust J Plant Physiol 11:191–210

Franks PJ, Farquhar GD (1999) A relationship between humidity response, growth form and photosynthetic operating point I C_3 plants. Plant Cell Environ 22:1337–1349

Franks PJ, Farquhar GD (2007) The mechanical diversity of stomata and its significance in gas-exchange control. Plant Physiol 143:78–87

Fitzisimons PJ, Weyers JDB (1986) Volume changes of *Commelina communis* L. guard cell protoplasts in response to K^+, light and CO_2. Physiol Plant 66:463–468

Frechilla S, Talbott LD, Zeiger E (2002) The CO_2 response of *Vicia faba* guard cells acclimates to growth environment. J Exp Bot 53:545–550

Frechilla S, Zhu J, Talbott LD, Zeiger E (1999) Stomata from *npq1*, a Zeaxanthin-less Arabidopsis mutant, lack a specific response to blue light. Plant Cell Physiol 40:949–954

Freudenberger H (1940) Die Reaktion der Schliesszellen auf Kohlensäure and Sauerstoffentzug. Protoplasma 35:15–54

Fujino M (1967) Role of adenosine triphosphate and adenosine triphosphatase in stomatal movements. Sci Bull Fac Educ Nagasaki Until 18:1–47

Galvez-Valdivieso G, Fryer MJ, Lawson T, Slattery K, Truman W, Smirnoff N, Asami T, Davies WJH, Jones AM, Baker NR, Mullineaux PM (2009) The high light response in Arabidopsis involves ABA signaling between vascular and bundle sheath cells. Plant Cell 21:2143–2162

Ghannoum O (2009) C_4 photosynthesis and water stress. Ann Bot 103:635–644

Grantz DA, Assmann SM (1991) Stomatal response to blue-light – water-use efficiency in sugarcane and soybean. Plant Cell Environ 14:683–690

Gray JE (2005) Guard cells: transcription factors regulate stomatal movements. Curr Biol 15:593–595

Gray JE, Holroyd GH, van der Lee FM, Bahrami AR, Sijmons PC, Woodward FI, Schuch W, Hetherington AM (2000) The HIC signalling pathway links CO_2 perception to stomatal development. Nature 408:713–716

Grulke NE, Alonso R, Nguyen T, Cascio C, Dobrowolski W (2004) Stomata open at night in pole-sized and mature ponderosa pine: implications for O_3 exposure metrics. Tree Physiol 24:1001–1010

Gotow K, Taylor S, Zeiger E (1988) Photosynthetic carbon fixation in guard cell protoplasts of *Vicia faba* L. Evidence from radiolabel experiments. Plant Physiol 86:700–705

Haake V, Geiger M, Walch-Liu P, Engels C, Zrenner R, Stitt M (1999) Changes in aldolase activity in wild-type potato plants are important for acclimation to growth irradiance and carbon dioxide concentration, because plastid aldolase exerts control over the ambient rate of photosynthesis across a range of growth conditions. Plant J 17:479–489

Hall AE, Schulze ED, Lange OL (1976) Current perspectives of steady state stomatal responses to environments. In: Lange OL, Kappen L, Schulze ED (eds) Water and plant life: ecological studies, vol 19. Springer, Berlin, Heidelberg, New York, pp 169–188

Hampp R, Outlaw WH Jr, Tarczynski MC (1982) Profile of basic carbon pathways in guard cells and other leaf cells of *Vicia faba* L. Plant Physiol 70:1582–1585

Hara K, Kajita R, Torii KU, Bergmann DC, Kakimoto T (2007) The secretory peptide gene *EPF1* enforces the stomatal one-cell-spacing rule. Genes Dev 21:1720–1725

Harrison EP, Willingham NM, Lloyd JC, Raines CA (1998) Reduced sedoheptulose-1,7-bisphosphatase levels in transgenic tobacco lead to decreased photosynthetic capacity and altered carbohydrate accumulation. Planta 20:27–36

Hashimoto M, Negi J, Young J, Israelsson M, Schroeder JI, Iba K (2006) *Arabidopsis* HT1 kinase controls stomatal movements in response to CO_2. Nat Cell Biol 8:391–397

Heath OVS (1938) An experimental investigation of the mechanism of stomatal movement, with some preliminary observations upon the response of guard cells to shock. New Phytol 37:385–395

Heath OVS, Russell J (1954) Studies in stomatal behaviour IV. An investigation of the light response of wheat stomata with the attempted elimination of control by the mesophyll. J Exp Bot 5:269–292

Henkes S, Sonnewald U, Badur R, Flachmann R, Stitt M (2001) A small decrease of plastid transketolase activity in antisense tobacco transformants has dramatic effects on photosynthesis and phenylpropanoid metabolism. Plant Cell 13:535–551

Hetherington AM, Woodward FI (2003) The role of stomata in sensing and driving environmental change. Nature 424:901–908

Hipkins ME, Fitzimons PJ, Weyers JDB (1983) The primary processes of photosystem II in purified guard-cell and mesophyll-cell protoplasts from *Commelina communis* L. Planta 159:554–560

Hite DRC, Outlaw WH Jr, Tarczynski MC (1993) Elevated levels of both sucrose-phosphate synthase and sucrose synthase in *Vicia* guard cells indicate cell-specific carbohydrate interconversion. Plant Physiol 101:1217–1221

Hu H, Dai M, YaoJ XB, Li X, Zhang Q, Xiong L (2006) Over-expressing a NAM, ATAF, and CUC (NAC) transcription factor enhances drought resistance and salt tolerance in rice. Proc Natl Acad Sci USA 103:12987–12992

Huang XY, Chao DY, Gao JP, Zhu MZ, Shi M, Lin HX (2009) A previously unknown zinc finger protein, DST, regulates drought and salt tolerance in rice via stomatal aperture control. Genes Dev 23:1805–1817

Hunt L, Gray JE (2009) The signaling peptide EPF2 controls asymmetric cell divisions during stomatal development. Curr Biol 19:864–869

Huxman TE, Monson RK (2003) Stomatal responses of C_3, C_3–C_4, and C_4 *Flaveria* species to light and intercellular CO_2 concentration: implications for the evolution of stomatal behaviour. Plant Cell Environ 26:313–322

Imamura S (1943) Untersuchungen uber den Mechanismus der Turgorschwankung der Spaltöffnungesschließzellen. Jpn J Bot 12:82–88

Jakab G, Ton J, Flors V, Zimmerli L, Metraux J-P, Mauch-Mani B (2005) Enhancing *Arabidopsis* salt and drought stress tolerance by chemical priming for its abscisic acid response. Plant Physiol 139:267–274

Jammes F, Song C, Shin D, Munemasa S, Takeda K, Gu D, Cho D, Lee S, Giordo R, Sritubitim S, Leonhardt N, Elis B, Murata Y, Kwak M (2009) MAP kinases MPK9 and MPK12 are preferentially expressed in guard cells and positively regulate ROS-mediate ABA signaling. Proc Natl Acad Sci USA 106:20520–20525

Jarvis PG (1993) Prospects for bottom-up models. In: Ehleringer J, Field C (eds) Scaling physiological processes; leaf to globe. Academic, New York, pp 115–116

Jarvis PG, McNaughton KG (1986) Stomatal control of transpiration: scaling up from leaf to region. Adv Ecol Res 15:1–49

Jiang CZ, Rodermel SR (1995) Regulation of photosynthesis during leaf development in *RbcS* antisense DNA mutants of tobacco. Plant Physiol 107:215–224

Jones HG (1976) Crop characteristics and the ratio between assimilation and transpiration. J Appl Ecol 13:605–622

Jones HG (1977) Transpiration in barley lines with differing stomatal frequencies. J Exp Bot 23:162–168

Jones HG (1985) Partitioning stomatal and non-stomatal limitation to photosynthesis. Plant Cell Environ 8:95–104

Jones HG (1987) Breeding for stomatal characters. In: Zeiger E, Farquhar GD, Cowan IR (eds) Stomatal function. Stanford University Press, Stanford, pp 431–443

Jones HG (1992) Plants and microclimate: a quantative approach to environmental plant physiology. Cambridge University Press, Cambridge

Joo JH, Wang S, Chen JG, Jones AM, Fedoroff NV (2005) Different signaling and cell death roles of heterotrimeric G protein alpha and beta subunits in the Arabidopsis oxidative stress response to ozone. Plant Cell 17:957–970

Kanaoka MM, Pillitteri LJ, Fujii H, Yoshida Y, Bogenschutz NL, Takabayashi J, Zhu J-K, Torii K (2008) SCREAM/ICE1 and SCREAM2 specify three cell-state transitional steps leading to Arabidopsis stomatal differentiation. Plant Cell 20:1775–1785

Kang Y, Lian HL, Wang FF, Huang JR, Yang HQ (2009) Cryptochromes, phytochromes, and COP1 regulate light-controlled stomatal development in Arabidopsis. Plant Cell 21: 2624–2641

Kang Y, Outlaw WH Jr, Anderson PC, Fiore GB (2007a) Guard cell apoplastic sucrose concentration – a link between leaf photosynthesis and stomatal aperture size in apoplastic phloem loader *Vicia faba* L. Plant Cell Environ 30:551–558

Kang Y, Outlaw WH Jr, Fiore GB, Riddle KA (2007b) Guard cell apoplastic photosynthate accumulation corresponds to a phloem-loading mechanisms. J Exp Bot 58:4061–4070

Klein M, Geiser M, Suh SJ, Kolukisaoglu HU, Azevedo L, Plaza S, Curtis MD, Richter A, Weder B, Schulz B, Martinoia E (2004) Disruption of AtMRP4, a guard cell plasma membrane ABCC-type ABC transporter, leads to deregulation of stomatal opening and increased drought susceptibility. Plant J 39:219–236

Kirschbaum MUF, Gross LJ, Pearcy RW (1988) Observed and modelled stomatal responses to dynamic light environments in the shade plant *Alocasia macrorrhiza*. Plant Cell Environ 11:111–121

Kinoshita T, Shimazaki K (1999) Blue light activates the plasma membrane H+-ATPase by phosphorylation of the C-terminus in stomatal guard cells. EMBO J 18:5548–5558

Knapp AK (1993) Gas-exchange dynamics in C_3 and C_4 grasses – consequences of differences in stomatal conductance. Ecology 74:113–123

Kopka J, Provart NJ, Muller-Rober B (1997) Potato guard cells respond to drying soil by a complex change in the expression of genes related to carbon metabolism and turgor regulation. Plant J 11:871–882

Kruijt B, Ongeri S, Jarvis PG (1997) Scaling of PAR absorption, photosynthesis and transpiration from leaves to canopy. In: van Gardingen P, Foody G, Curran P (eds) Scaling up, society for experimental biology seminar series. Cambridge University Press, Cambridge, pp 79–104

Kudoyarova GR, Veselov DS, Faizov RG, Veselova SV, Ivanov EA, Farkhutdinov RG (2007) Stomata response to changes in temperature and humidity in wheat cultivars grown under contrasting climatic conditions. Russ J Plant Physiol 54:46–49

Kuiper PJC (1964) Dependence upon wavelength of stomatal movement in epidermal tissue of *Senecio odoris*. Plant Physiol 39:952–955

Kurek I, Chang TK, Bertain SM, Madrigal A, Liu L, Lassner MW, Zhu GH (2007) Enhanced thermostability of Arabidopsis Rubisco activase improves photosynthesis and growth rates under moderate heat stress. Plant Cell 19:3230–3241

Kwak JM, Mori IC, Pei ZM, Leonhardt N, Torres MA, Dangl JL, Bloom RE, Bodde S, Jones JDG, Schroeder JI (2003) NADPH oxidase AtrbohD and AtrbohF genes function in ROS-dependent ABA signaling in Arabidopsis. EMBO J 22:2623–2633

Lake JA, Quick WP, Beerling DJ, Woodward FI (2001) Plant development: signals from mature to new leaves. Nature 411:154

Lange OL, Medina E (1979) Stomata of the CAM plant *Tillandsia recurvata* respond directly to humidity. Oecologia 40:357–363

Laporte MM, Shen B, Tarczynski C (2002) Engineering for drought avoidance: expression of maize NADP-malic enzyme in tobacco results in altered stomatal function. J Exp Bot 53:699–705

Lawlor DW (2002) Limitation to photosynthesis in water-stressed leaves: stomata vs. metabolism and the role of ATP. Ann Bot 89:871–885

Lawson T (1997) Heterogeneity in stomatal characteristics, PhD thesis, Dundee

Lawson T (2009) Guard cell photosynthesis and stomatal function. New Phytol 181:13–34

Lawson T, Lefebvre S, Baker NR, Morison JIL, Raines C (2008) Reductions in mesophyll and guard cell photosynthesis impact on the control of stomatal responses to light and CO_2. J Exp Bot 59:3609–3619

Lawson T, Morison JIL (2004) Stomatal function and physiology. In: Hemsley AR, Poole I (eds) The evolution of plant physiology; from whole plants to ecosystem. Elsevier Academic, Cambridge, pp 217–242

Lawson T, Oxborough K, Morison JIL, Baker NR (2002) Responses of photosynthetic electron transport in stomatal guard cells and mesophyll cells in intact leaves to light, CO_2, and humidity. Plant Physiol 128:52–62

Lawson T, Oxborough K, Morison JIL, Baker NR (2003) The response of guard cell photosynthesis to CO_2, O_2, light and water stress in a range of species are similar. J Exp Bot 54:1734–1752
Lawson T, Weyers JDB (1999) Spatial and temporal variation in gas exchange over the lower surface of *Phaseolus vulgaris* primary leaves. J Exp Bot 50:1381–1391
Lee DM, Assmann SM (1992) Stomatal responses to light in the facultative Crassulacean acid metabolism species, *Portulacaria afra*. Plant Physiol 85:35–42
Lee J-S, Bowling DJF (1992) Effect of the mesophyll on stomatal opening in *Commelina communis*. J Exp Bot 43:951–957
Lee J-S, Bowling DJF (1993) The effect of a mesophyll factor on the swelling of guard cell protoplasts of *Commelina communis* L. J Plant Physiol 142:203–207
Lee J-S, Bowling DJF (1995) Influence of the mesophyll on stomatal opening. Aust J Plant Physiol 22:357–363
Lefebvre S, Lawson T, Zakhleniuk OV, Lloyd JC, Raines CA (2005) Increased sedoheptulose-1,7-bisphosphatase activity in transgenic tobacco plants stimulates photosynthesis and growth from an early stage in development. Plant Physiol 138:451–460
Leonhardt N, Kwak JM, Robert N, Waner D, Leonhardt G, Schroeder J (2004) Microarray expression analyses of *Arabidopsis* guard cells and isolation of a recessive abscisic acid hypersensitive protein phosphatase 2C mutant. Plant Cell 16:596–615
Lodge RJ, Dijkstra P, Drake BG, Morison JIL (2001) Stomatal acclimation to increased CO_2 in a Florida scrub oak species *Quercus myrtifolia* Willd. Plant Cell Environ 24:77–88
Lu P, Outlaw WH Jr, Smith BD, Freed GA (1997) A new mechanism for the regulation of stomatal aperture size in intact leaves: accumulation of mesophyll-derived sucrose in guard cell walls of *Vicia faba*. Plant Physiol 114:109–118
Lu P, Zhang SQ, Outlaw WH Jr, Riddle KA (1995) Sucrose: a solute that accumulates in the guard-cell- apoplast and guard cell symplast of open stomata. FEBS Lett 326:180–184
Lu Z, Quiñones MA, Zeiger E (2000) Temperature dependence of guard cell respiration and stomatal conductance co-segregate in an F_2 population of Pima cotton. Aust J Plant Physiol 27:457–462
Lurrie S (1977) Photochemical properties of guard cell chloroplasts. Plant Sci Let 10:219–223
Ma SY, Wu WH (2007) AtcPK23 functions in Arabidopsis responses to drought and salt stresses. Plant Mol Biol 65:511–518
Maherali H, Reid CD, Polley HW, Johnson HB, Jackson RB (2002) Stomatal acclimation over a sub-ambient to elevated CO_2 gradient in a C_3/C_4 grassland. Plant Cell Environ 25:557–566
Mansfield TA (1994) Some aspects of stomatal physiology relevant to plants cultured *in vitro*. In: Lumsden PJ, Nicholas JR, Davies WJ (eds) Physiology, growth and development of plants in culture. Kluwer, Dordrecht, pp 120–131
Mansfield TA, Hetherington AM, Atkinson CJ (1990) Some current aspects of stomatal physiology. Annu Rev Plant Physiol Plant Mol Biol 41:55–75
Mansfield TA, Atkinson CJ (1990) Stomatal behaviour in water stressed plants. In: Alscher RG, Cumming JR (eds) Stress responses in plants: adaptation and acclimation mechanisms. Wiley-Liss, New York, pp 241–264
Marten H, Hyun T, Gomi K, Seo S, Hedrich R, Roelfsema MRG (2008) Silencing of *Nt*MPK4 impairs CO_2-induced stomatal closure, activation of anion channels and cytosolic Ca^{2+} signals in *Nicotiana tabacum* guard cells. Plant J 55:698–708
Masle J, Gilmour SR, Farquhar GD (2005) The ERECTA gene regulates plant transpiration efficiency in *Arabidopsis*. Nature 436:866–870
Mawson BT, Zaugg MW (1994) Modulation of light-dependent stomatal opening in isolated epidermis following induction of crassulacean acid metabolism in *Mesembryanthemum crystallinum* L. Plant Physiol 144:740–746
McElwain JC, Chaloner WG (1995) Stomatal density and index of fossil plants track atmospheric carbon dioxide in the palaeozoic. Ann Bot 76:389–395
McElwain JC, Mitchell FJG, Jones MB (1995) Relationship of stomatal density and index of *Salix cinerea* to atmospheric carbon dioxide concentration of the Holocene. Holocene 5:216–219

Meinzer FC (2002) Co-ordination of vapour and liquid phase water transport properties in plants. Plant Cell Environ 25:265–274
Messinger SM, Buckley TN, Mott KA (2006) Evidence for involvement of photosynthetic processes in the stomatal response to CO_2. Plant Physiol 140:771–778
Miao Y, Lv D, Wang P, Wang X-C, Chen J, Wang XC, Chen J, Miao C, Song CP (2006) An Arabidopsis glutathione peroxidase functions as both a redox transducer and a scavenger in abscisic acid and drought stress responses. Plant Cell 18:2749–2766
Miyazawa S-I, Livingston NJ, Turpin DH (2006) Stomatal development in new leaves is related to the stomatal conductance of mature leaves in poplar (*Populus trichocarpa* x *P. deltoides*). J Exp Bot 57:373–380
McNaughton KG, Jarvis PG (1991) Effects of spatial scale on stomatal control of transpiration. Agricult Forest Meteorol 54:279–301
Morison JIL (1985) Intercellular CO_2 concentration and stomatal response to CO_2. In: Zeiger E, Cowan IR, Farquhar GD (eds) Stomatal function. Stanford University Press, Stanford
Morison JIL (1993) Response of plants to CO_2 under water limited conditions. Vegetation 105:193–209
Morison JIL (2003) Plant water use, stomatal control. In: Stewart BA, Howell TA (eds) Encyclopedia of water science. Marcel Dekker, New York, pp 680–685
Morison JIL, Gifford RM (1983) Stomatal sensitivity to carbon dioxide and humidity: a comparison of two C_3 and two C_4 grass species. Plant Physiol 71:789–796
Mott KA (1988) Do stomata respond to CO_2 concentrations other than intercellular? Plant Physiol 86:200–203
Mott KA (2009) Opinion: stomatal responses to light and CO_2 depend on the mesophyll. Plant Cell Environ 32:1479–1486
Mott KA, Peak D (2007) Stomatal patchiness and task-performing networks. Ann Bot 99:219–226
Mott KA, Sibbernsen ED, Shope JC (2008) The role of the mesophyll in stomatal responses to light and CO_2. Plant Cell Environ 31:1299–1306
Nadeau JA (2009) Stomatal development: new signals and fate determinants. Curr Opin Plant Biol 12:29–35
Nakagami H, Pitzschke A, Hirt H (2005) Emerging MAP kinase pathways in plant stress signalling. Trends Plant Sci 10:339–346
Nardini A, Salleo S (2005) Water stress-induced modifications on leaf hydraulic architecture in sunflower: co-ordination with gas exchange. J Exp Bot 56:3093–3101
Nilson SE, Assmann SM (2007) The control of transpiration. Insights from Arabidopsis. Plant Physiol 143:19–27
Niyogi KK (2000) Safety valves for photosynthesis. Curr Opin Plant Biol 3:455–460
Noe SM, Giersch C (2004) A simple dynamic model of photosynthesis in oak leaves: coupling leaf conductance and photosynthetic carbon fixation by a variable intracellular CO_2 pool. Funct Plant Biol 31:1195–1204
Nunes-Nesi A, Carrari F, Gibon Y, Sulpice R, Lytovchenko A, Fisahn J, Graham J, Ratcliffe RG, Sweetlove LJ, Fernie AR (2007) Deficiency of mitochondrial fumarase activity in tomato plants impairs photosynthesis via an effect on stomatal function. Plant J 50:1093–1106
Nunes-Nesi A, Carrari F, Lytovchenko A, Smith AM, Loureiro ME, Ratcliffe RG, Sweetlove LJ, Fernie AR (2005) Enhanced photosynthetic performance and growth as a consequence of decreasing mitochondrial malate dehydrogenase activity in transgenic tomato plants. Plant Physiol 137:611–622
Osmond CB (1978) Crassulacean acid metabolism: a curiosity in context. Annu Rev Plant Physiol 29:397–414
Outlaw WH Jr (1982) Carbon metabolism in guard cells. Recent Adv Phytochem 17:185–222
Outlaw WH Jr (1983) Current concepts on the role of potassium in stomatal movements. Physiol Plant 59:302–311
Outlaw WH Jr (1987) A minireview: comparative biochemistry of photosynthesis in palisade cells, spongy cells and guard cells of C_3 leaves. In: Biggins J (ed) Progress in photosynthesis research, vol IV, 5. Martinus Nijhoff, Dordrecht, pp 265–272

Outlaw WH Jr (1989) Critical examination of the quantitative evidence for and against photosynthetic CO_2 fixation in guard cells. Physiol Plant 77:275–281
Outlaw WH Jr (1990) Kinetic properties of guard-cell phosphoenolpyruvate carboxylase. Biochem Physiol Pflanz 186:317–325
Outlaw WH Jr (2003) Integration of cellular and physiological functions of guard cells. Crit Rev Plant Sci 22:503–529
Outlaw WH Jr, De Vleighere-He X (2001) Transpiration rate: an importance factor in controlling sucrose content of the guard cell apoplast of broad bean. Plant Physiol 126:1717–1724
Outlaw WH Jr, Manchester J, Di Camelli CA, Randall DD, Rapp B, Veither GM (1979) Photosynthetic carbon reduction pathway is absent in chloroplasts of *Vicia faba* guard cells. Proc Natl Acad Sci USA 76:6371–6375
Outlaw WH Jr, Tarczynski MC, Anderson LC (1982) Taxonomic survey for the presence of ribulose-1,5-bisphosphate carboxylase activity in guard cells. Plant Physiol 70:1218–1220
Parvathi K, Raghavendra AS (1997) Both Rubisco and phosphoenolpyruvate carboxylase are beneficial for stomatal function in epidermal strips of *Commelina benghalensis*. Plant Sci 124:153–157
Pearcy RW (1990) Sunflecks and photosynthesis in plant canopies. Annu Rev Plant Physiol Plant Mol Biol 41:421–453
Pearson CJ (1973) Daily changes in stomatal aperture and in carbohydrates and malate within epidermis and mesophyll of leaves of *Commelina cyanea* and *Vicia faba*. Aust J Biol Sci 26:1035–1044
Pearson CJ, Milthorpe FL (1974) Structure, carbon dioxide fixation and metabolism of stomata. Aust J Plant Physiol 1:221–236
Pemadasa MA (1981) Photocontrol of stomatal movements. Biol Rev Camb Phil Soc 56:551–559
Pfannschmidt T, Bräutigam K, Wagner R, Dietzel L, Schröter Y, Steiner S, Nykytenko A (2009) Potential regulation of gene expression in photosynthetic cells by redox and energy state: approaches towards better understanding. Ann Bot 103:599–607
Pham J, Desikan R (2009) ROS signalling in stomata. In: Rio LA, Puppo A (eds) Reactive oxygen species in plant signaling. Springer, Berlin, Heidelberg, New York, pp 55–71
Pnueli L, Liang H, Rozenberg M, Mittler R (2003) Growth suppression, altered stomatal responses, and augmented induction of heat shock proteins in cytosolic ascorbate peroxidase (Apx1)-deficient Arabidopsis plants. Plant J 34:187–203
Poffenroth M, Green DB, Tallman G (1992) Sugar concentrations in guard cells of *Vicia faba* illuminated with red or blue light: analysis by high performance liquid chromatography. Plant Physiol 98:1460–1471
Raghavendra AS, Vani T (1989) Respiration in guard cells: pattern and possible role in stomatal function. J Plant Physiol 135:3–8
Rao M, Anderson LE (1983) Light and stomatal metabolism. 1. Possible involvement of light-modulation of enzymes in stomatal movement. Plant Physiol 71:456–459
Raines CA (2003) The Calvin cycle revisited. Photosynth Res 75:1–10
Ramos C, Hall AE (1982) Relationships between leaf conductance, intercellular CO_2 partial pressure and CO_2 uptake rate in two C_3 and two C_4 plant species. Photosynthetica 16:343–355
Raschke K (1970) Temperature dependencies and apparent activation energies of stomatal opening and closing. Planta 95:1–17
Raschke K (1978) Movements using turgor mechanisms. In: Haupt W, Feinleib ME (eds) Encyclopedia of plant physiology, physiology of movements. Springer, Berlin, Heidelberg, New York, pp 387–441
Raschke K (1972) Saturation kinetics of the velocity of stomatal closing to CO_2. Plant Physiol 49:229–234
Raschke K (1975) Stomatal action. Annu Rev Plant Physiol 26:309–340
Raschke K, Dittrich P (1977) [^{14}C] carbon dioxide fixation by isolated leaf epidermis with stomata closed or open. Planta 134:69–75
Rebetzke GJ, Richards RA, Condon AG, Farquhar GD (2006) Inheritance of carbon isotope discrimination in bread wheat (*Triticum aestivum* L.). Euphytica 150:97–106

Reckmann U, Scheibe R, Raschke K (1990) Rubisco activity in guard cell compared with the solute requirement for stomatal opening. Plant Physiol 92:246–253
Ripullone F, Matsuo N, Stuart-Williams H, Wong SC, Borghetti M, Tani M, Farquhar GD (2008) Environmental effects on oxygen isotope enrichment of leaf water in cotton leaves. Plant Physiol 146:729–736
Ritte G, Raschke K (2003) Metabolite export of isolated guard cell chloroplasts of *Vicia faba*. New Phytol 159:195–202
Ritte G, Rosenfeld J, Rohrig K, Raschke K (1999) Rates of sugar uptake by guard cell protoplasts of *Pisum satvum* L. related to the solute requirements for stomatal opening. Plant Physiol 121:647–655
Roelfsema MRG, Hanstein S, Fell HH, Hedrich R (2002) CO_2 provides an intermediate link in the red light response of guard cells. Plant J 32:65–75
Roelfsema MRG, Konrad KR, Marten H, Psaras GK, Hartung W, Hedrich H (2006) Guard cells in albino leaf patches do not respond to photosynthetically active radiation, but are sensitive to blue light, CO_2 and abscisic acid. Plant Cell Environ 29:1595–1605
Rogiers SY, Greer DH, Hutton RJ, Landsberg JJ (2009) Does night time transpiration contribute to anisohydric behaviour in a *Vitus vinifera* cultivar? J Exp Bot 60:3763–3763
Rolland F, Baena-Gonzalez E, Sheen J (2006) Sugar sensing and signaling in plants: conserved and novel mechanisms. Annu Rev Plant Biol 57:675–709
Sack L, Frole K (2006) Leaf structural diversity is related to hydraulic capacity in tropical rainforest trees. Ecology 87:483–491
Sage RF, Sharkey TD (1987) The effect of temperature on the occurrence of O_2 and CO_2 insensitive photosynthesis in field grown plants. Plant Physiol 84:658–664
Sage RF, Wedin DA, Li M (1999) The biogeography of C_4 photosynthesis: patterns and controlling factors. In: Sage RF, Monson RK (eds) C_4 biology. Academic, San Diego, pp 313–374
Šantrůček J, Sage RF (1996) Acclimation of stomatal conductance to a CO_2-enriched atmosphere and elevated temperature in *Chenopodium album*. Aust J Plant Physiol 23:467–478
Scheibe R, Reckmann U, Hedrich R, Raschke K (1990) Malate dehydrogenases in guard cells of *Pitsum sativum*. Plant Physiol 93:1358–1364
Schnabl H, Elbert C, Krammer G (1982) The regulation of the starch-malate balances during volume changes of guard cell protoplasts. J Exp Bot 33:996–1003
Schoch PG, Zinsou C, Sibi M (1980) Dependence of the stomatal index on environmental factors during stomatal differentiation in leaves of *Vigna sinensis* L. I Effect of light intensity. J Exp Bot 31:1211–1216
Schoch PG, Jacques R, Lecharny A, Sibi M (1984) Dependence of the stomatal index on environmental factors during stomatal differentiation in leaves of *Vigna sinensis* L.II Effect of different light quality. J Exp Bot 35:1405–1409
Schwartz A, Zeiger E (1984) Metabolic energy for stomatal opening. Role of photophosphorylation and oxidative phosphorylation. Planta 161:129–136
Sekiya N, Yano K (2008) Stomatal density of cowpea correlates with carbon isotope discrimination in different phosphorus, water and CO_2 environments. New Phytol 179:799–807
Sharkey TD, Raschke K (1981) Separation and measurement of direct and indirect effects of light on stomata. Plant Physiol 68:33–40
Shimazaki K (1989) Ribulose bisphosphate carboxylase activity and photosynthetic O_2 evolution rate in *Vicia* guard cell protoplasts. Plant Physiol 91:459–463
Shimazaki K-I, Doi M, Assmann SM, Kinoshita T (2007) Light regulation of stomatal movements. Annu Rev Plant Biol 58:219–247
Shimazaki K, Zeiger E (1985) Cyclic and non-cyclic phosphorylation in isolated guard cell protoplasts from *Vicia faba* L. Plant Physiol 78:211–214
Shimazaki K, Terada J, Tanaka K, Kondo N (1989) Calvin–Benson cycle enzymes in guard-cell protoplasts from *Vicia faba* L. Plant Physiol 90:1057–1064
Snyder KA, Richards JH, Donovan LA (2003) Night-time conductance in C_3 and C_4 species: do plants lose water at night? J. Exp Bot 54:861–865

Spalding MH, Stumpf DK, Ku MSB, Burris RH, Edwards GE (1979) Crassulacean acid metabolism and diurnal variations of internal CO_2 and O_2 concentrations in *Sedum praealtum*. Aust J Plant Physiol 8:557–567

Sperry JS (2000) Hydraulic constraints on plant gas exchange. Agr Forest Meteorol 104:13–23

Stadler R, Buttner M, Ache P, Hedrich R, Ivashikina N, Melzer M, Shearson SM, Smith SM, Sauer N (2003) Diurnal and light-regulated expression of AtsSTP1 in guard cells of Arabidopsis. Plant Physiol 133:528–537

Still CJ, Berry JA, Collatz GJ, DeFries RS (2003) Global distribution of C_3 and C_4 vegetation: carbon cycle implications. Global Biogeochem Cycles 17:1006

Suzuki Y, Miyamoto T, Yoshizawa R, Mae T, Makino A (2009) Rubisco content and photosynthesis of leaves at different positions in transgenic rice with an over-expression of RBCS. Plant Cell Environ 32:417–427

Tallman G, Zhu J, Mawson BT, Amodeo G, Nouhi Z, Levy K, Zeiger E (1997) Induction of CAM in *Mesembryanthemum crystallinum* abolishes the stomatal response to blue light and light-dependent Zeaxanthin formation in guard cell chloroplasts. Plant Cell Physiol 38:236–242

Tarczynski MC, Outlaw WH Jr, Arold N, Neuhoff V, Hampp R (1989) Electrophoretic assay for ribulose 1,5-bisphosphate carboxylase/oxygenase in guard cells and other leaf cells of *Vicia faba* L. Plant Physiol 89:1088–1093

Thomas PW, Woodward FI, Quick WP (2004) Systemic irradiance signalling in tobacco. New Phytol 161:193–198

Tichà I (1982) Photosynthetic characteristics during ontogenesis of leaves. 7. Stomatal density and sizes. Photosynthetica 16:375–471

Tinoco-Ojanguren C, Pearcy RW (1993) Stomatal dynamics and its importance to carbon gain in 2 rain forest Piper species.1. VPD Effects on the transient stomatal response to light flecks. Oecologia 94:388–394

Tominaga M, Kinoshita T, Shimazaki K (2001) Guard-cell chloroplasts provide ATP required for H^+ pumping in the plasma membrane and stomatal opening. Plant Cell Physiol 42:795–802

Tuba Z, Szente K, Koch J (1994) Response of photosynthesis, stomatal conductance, water use efficiency and production to long-term elevated CO_2 in winter wheat. J Plant Physiol 144:661–668

Turner NC (1979) Differences in response of adaxial and abaxial stomata to environmental variables. In: Sen DS, Chawan DD, Bansal RP (eds) Structure function and ecology of the stomata. Bishen Singh Mahendra Pal Singh, Dehra Dun, pp 320–329

Van Breusegem F, Bailey-Serres J, Mittler R (2008) Unraveling the tapestry of networks involving reactive oxygen species in plants. Plant Physiol 147:978–984

Vavasseur A, Raghavendra AS (2005) Guard cell metabolism and CO_2 sensing. New Phytol 165:665–682

von Caemmerer S, Griffiths H (2009) Stomatal responses to CO_2 during a diel Crassulacean acid metabolism cycle in *Kalanchoe daigremontiana* and *Kalanchoe pinnata*. Plant Cell Environ 32:567–576

von Caemmerer S, Lawson T, Oxborough K, Baker NR, Andrews TJ, Raines CA (2004) Stomatal conductance does not correlate with photosynthetic capacity in transgenic tobacco with reduced amounts of Rubisco. J Exp Bot 55:1157–1166

Wang Y, Chen X, Xiang C-B (2007a) Stomatal density and bio-water saving. J Integr Plant Biol 49:1435–1444

Wang H, Ngwenyama N, Liu Y, Walker JC, Zhang S (2007b) Stomatal development and patterning are regulated by environmentally responsive mitogen-activated protein kinases in Arabidopsis. Plant Cell 19:63–73

Wang P, Song CP (2008) Guard-cell signalling for hydrogen peroxide and abscisic acid. New Phytol 178:703–718

Wang Y, Ying JF, Kuzma M, Chalifoux M, Sample A, McArthur C, Uchacz T, Sarvas C, Wan JX, Dennis DT, McCourt P, Huang YF (2005) Molecular tailoring of farnesylation for plant drought tolerance and yield protection. Plant J 43:413–424

Wang Y, Noguchi KO, Terashima I (2008) Distinct light responses of adaxial and abaxial stomata in leaves of *Helianthus annuus* L. Plant Cell Environ 31:1307–1315

Weber PM, Fischer K (2007) Making the connections – the crucial role of metabolite transporters at the interface between chloroplast and cytosol. FEBS Lett 581:2215–2222

Weyers JDB, Lawson T (1997) Heterogeneity in stomatal characteristics. Adv Bot Res 26:317–352

Weyers JDB, Lawson T, Peng Z (1997) Variation in stomatal characteristics at the whole leaf level. In: van Gardingen P, Foody G, Curran P (eds) Scaling up. Society for experimental biology seminar series. Cambridge University Press, Cambridge, pp 129–149

Weyers JDB, Meidner H (1990) Methods in stomatal research. Longman Scientific and Technical, Harlow

Willmer CM (1983) Phosphoenolpyruvate carboxylase activity and stomatal operation. Physiol Veg 21:943–953

Willmer CM, Dittrich P (1974) Carbon dioxide fixation by epidermal and mesophyll tissues of *Tulipa* and *Commelina*. Planta 117:123–132

Willmer CM, Fricker M (1996) Stomata, 2nd edn. Chapman and Hall, London

Wingler A, Quick WP, Bungard RA, Bailey KJ, Lea PJ, Leegood RC (1999) The role of photorespiration during drought stress: an analysis utilizing barley mutants with reduced activities of photorespiratory enzymes. Plant Cell Environ 22:361–373

Wong SC, Cowan IR, Farquhar GD (1979) Stomatal conductance correlates with photosynthetic capacity. Nature 282:424–426

Wong SC, Cowan IR, Farquhar GD (1985) Leaf conductance in relation to rate of CO_2 assimilation II. Effects of short-term exposures to different photon flux densities. Plant Physiol 78:826–829

Woodward FI (1987) Stomatal numbers are sensitive to increases in CO_2 from pre-industrial levels. Nature 327:617–618

Xu DQ, Terashima K, Crang RFE, Chen XM, Hesketh JD (1994) Stomatal and non-stomatal acclimation to a CO_2-enriched atmosphere. Biotronics 23:1–9

Yang CY, Chen YC, Jauh GY, Wang CS (2005) A lily ASR protein involves abscisic acid signaling and confers drought and salt resistance in *Arabidopsis*. Plant Physiol 139:836–846

Yano S, Terashima I (2001) Separate localization of light signal perception for sun or shade type chloroplast and palisade tissue differentiation in *Chenopodium album*. Plant Cell Physiol 42:1303–1310

Zeiger E, Iino M, Ogawa T (1985) The blue light response of stomata: pulse kinetics and some mechanistic implications. Photochem Photobiol 42:759–763

Zeiger E, Talbott LD, Frechilla S, Srivastava A, Zhu J (2002) The guard cell chloroplast: a perspective for the twenty-first century. New Phytol 153:415–424

Zeiger E, Zhu JX (1998) Role of zeaxanthin in blue light photoreception and the modulation of light-CO_2 interactions in guard cells. J Exp Bot 49:433–442

Zemel E, Gepstein S (1985) Immunological evidence for the presence of ribulose bisphosphate carboxylase in guard cell chloroplasts. Plant Physiol 78:586–590

Zhao Z, Zhang W, Stanley BA, Assmann SM (2008) Functional proteomics of *Arabidopsis thaliana* guard cells uncovers new stomatal signaling pathways. Plant Cell 20:3210–3226

Zhang Y, Zhu H, Zhang Q, Li M, Yan M, Wang R, Wang L, Welti R, Zhang W, Wang X (2009) Phospholipase Dal1 and phosphatidic acid regulate NADPH oxidase activity and production of reactive oxygen species in ABA-mediated stomatal closure in Arabidopsis. Plant Cell 21:2357–2377

Zhu M, Dai S, McClung S, Yan X, Chen S (2009) Functional differentiation of *Brassica napus* guard cells and mesophyll cells revealed by comparative proteomics. Mol Cell Proteom 8:752–766

Zhu XG, de Sturler E, Long SP (2007) Optimizing the distribution of resources between enzymes of carbon metabolism can dramatically increase photosynthetic rate: a numerical simulation using an evolutionary algorithm. Plant Physiol 145:513–526

Impacts of Ultraviolet Radiation on Interactions Between Plants and Herbivorous Insects: A Chemo-Ecological Perspective

Franziska Kuhlmann and Caroline Müller

Contents

Abstract Plant–insect interactions are mainly triggered by plant metabolites that serve as nutrients and influence acceptance behaviour. For generalist herbivores, several metabolites act as feeding deterrents or toxins, whereas they can be attractive and stimulatory for specialists that are adapted to the chemical compounds. However, the qualitative and quantitative composition of plant natural products can be strongly modified by environmental factors. One important factor is the ultraviolet (UV) radiation that affects plant morphology and physiology and consequently also chemistry. In this review, general plant responses to UV are shortly summarised.

F. Kuhlmann and C. Müller (✉)
Department of Chemical Ecology, Bielefeld University, Universitätsstr. 25, D-33615 Bielefeld, Germany
e-mail: caroline.mueller@uni-bielefeld.de

U. Lüttge et al. (eds.), *Progress in Botany 72*,
DOI 10.1007/978-3-642-13145-5_12, © Springer-Verlag Berlin Heidelberg 2011

Then investigation methods of UV effects on plant–herbivore interactions are presented. In the main part, various compound classes are discussed with regard to their chemical nature, their inducibility by UV and their role in plant–insect interactions. Furthermore, direct effects of UV on herbivores as well as visual responses of insects to UV are delineated and tritrophic interactions are highlighted. The most pronounced metabolic changes caused by UV occur in phenolic substances, whereas effects on alkaloids and terpenoids are less consistent. In more than half of the studies, it has been shown that UV-exposed plants are better protected against herbivores than plants grown under low UV conditions, but this pattern cannot be generalised and it cannot be directly related to the increased concentrations of phenolic substances and UV-exposed plants. Only few studies proved a clear link between an UV-induced compound and its effect on an herbivore. Overall, insect responses to differently UV-challenged plants are highly species-specific and depend, furthermore, on the developmental stage of the plant and the insect. Although many studies have investigated UV-mediated plant–insect interactions, there is still a lack of knowledge on the mechanisms underlying changes in herbivore behaviour and performance caused by UV irradiation on plants.

1 Introduction

Solar radiation is the essential source of energy for life on earth, driving photosynthesis of plants. The radiation relevant for plants is composed of ultraviolet B (UV-B; 280–315 nm), UV-A (315–400 nm), photosynthetically active radiation (PAR, 400–700 nm) and near-infrared radiation (700 nm–1 mm). UV-B is absorbed by stratospheric ozone to a large extent, whereas UV-A is not absorbed. The UV-B radiation at a given location is not only influenced by ozone, but also by the angle of sun rays, altitude and latitude, cloud cover, season, aerosols, surface reflectance, shading and plant canopy structure (Madronich et al. 1998; McKenzie et al. 2003; Paul and Gwynn-Jones 2003; Jenkins and Brown 2007). Therefore, the UV-B radiation reaching a plant can vary tremendously from one place to the other. The specific radiation a plant receives influences its morphology, physiology and chemical composition, which has consequences on all organisms feeding or parasitising this plant. Substantial insight on the impact of UV-B radiation on plants and higher trophic levels was obtained in the context of the still ongoing stratospheric ozone depletion and research on its consequences (Caldwell et al. 2003).

This review focuses on the chemo-ecological impacts of UV-A and UV-B radiation on terrestrial plant–insect interactions. After a short introduction on the UV perception of plants, knowledge on various stress-responses to UV and photomorphogenic changes in plants is summarised. We then delineate the methods that have been used to study UV impacts on plant–insect interactions. The major section of the review deals with UV-induced changes of individual chemical plant traits, including epicuticular waxes, phytohormones, proteinase inhibitors as well as various secondary plant metabolites, and how these affect plant–insect interactions. We hereby focus on the role of insects as herbivores and exclude pollinator aspects. Additionally, direct impacts of UV on insect herbivores and their natural enemies are summarised. Finally, the most important findings are discussed considering evolutionary aspects, and conclusions for applied crop protection and for essential future research are drawn.

2 UV Perception and Responses of Plants

Plants need to be able to cope with unsteady environmental radiation conditions. Sunburns of plant tissue are rare in nature; this indicates that the light signal perception and transduction in plants as well as repair and protection mechanisms must be highly fine-tuned. Plants have a high light sensitivity to the UV-B, UV-A/ blue, red and far-red part of the solar spectrum. Photoreceptors for UV-A and blue light are well characterised. They are phototropins, cryptochromes and members of the Zeitlupe family, whereas phytochromes respond predominantly to red/far-red light (Chen et al. 2004; Favory et al. 2009). In contrast, the molecular identity of a UV-B photoreceptor is still unknown, although its existence is not questioned (Frohnmeyer and Staiger 2003; Jenkins 2009). Furthermore, the basic UV-B light perception and transduction mechanisms in plants are until now only partly understood (Jenkins and Brown 2007; Wargent et al. 2009). For a recent review on signal transduction in response to UV-B radiation, we refer to Jenkins (2009). Two independent signal transduction pathways are proposed, a non-specific (stress) and a UV-B specific (photomorphogenic) pathway (Frohnmeyer and Staiger 2003); however, responses of both pathways overlap partially.

2.1 UV-B Stress Responses of Plants

Detrimental harms may occur, if plants are not acclimatised and suddenly exposed to high fluence rates of UV-B. Hereby, UV-B radiation can generate reactive oxygen species (ROS) by more than one mechanism (Jenkins 2009) and induce the formation of cyclobutane pyrimidine dimers and pyrimidine pyrimidinone dimers. The ROS can cause oxidative damage to DNA, lipids, proteins and membranes (Jenkins and Brown 2007) and further unspecific effects on most cellular processes. Plants respond rapidly with various repair and protection mechanisms. DNA repair is achieved by photoreactivation mediated by photolyase, base excision and recombination. Additionally, the production of anti-oxidants and the activities of anti-oxidant enzymes are enhanced (Jenkins and Brown 2007). Furthermore, high UV-B radiation leads to an alteration of hormone concentrations, e.g. jasmonic acid, salicylic acid and ethylene, which are also involved in defence and wound signalling pathways (Rozema et al. 1997; Jansen et al. 1998; Mackerness et al. 1999; see Sect. 4.2). Finally, *pathogenesis-related 1* (*PR-1*), *PR-2*, *PR-5*, *PDF1.2* (defencin gene) and proteinase inhibitor genes can be expressed under UV-B stress (Frohnmeyer and Staiger 2003; Jenkins and Brown 2007). Whereas high fluence rates can induce stress responses, low fluence rates can initiate regulatory, photomorphogenic responses (Jenkins 2009).

2.2 *Photomorphogenic Plant Responses to UV*

Photomorphogenic plant responses induced by low UV-B fluence rates are essential for UV-B acclimatisation and tolerance (Ulm and Nagy 2005; Oravecz et al. 2006; Favory et al. 2009; Wargent et al. 2009) and are likely regulated by a UV-B photoreceptor (see above; Jenkins 2009). They comprise changes in plant texture and metabolite composition (see Sect. 4), and therefore also affect other organisms interacting with a plant. The most characteristic acclimatisation responses of plant morphology to UV are an increase in leaf thickness (see Sect. 4.1) and an inhibition of plant growth (Jansen et al. 1998; Zavala and Ravetta 2002; Caldwell et al. 2003). In UV-B-exposed plants, hypocotyl elongation is inhibited, internodes are shorter and leaf area is reduced, resulting in an overall reduced above-ground biomass (Caldwell et al. 2003). It is unlikely that growth is inhibited as a result of changed photosynthesis per leaf area or a damaged photosynthetic apparatus (Jansen et al. 2010), as gas exchange and net assimilation rates in different plant species were not affected by UV-B (Beyschlag et al. 1988; Zavala and Ravetta 2002). Instead, growth inhibition as response to UV-B exposure may be due to decreased concentrations of the phytohormone indol-3-acetic acid (IAA), which go along with increases in peroxidase and IAA oxidase activity (Huang et al. 1997; Jansen et al. 2004; Yang et al. 2004) (see Sect. 4.2). Flavonoids like quercetin and kaempferol are induced by UV-B (see Sect. 4.4.2) and are known to inhibit IAA transport. Therefore, they influence IAA distribution in plants (Brown et al. 2001; Jansen et al. 2001; Jansen 2002). The detailed mechanistic interplay between peroxidase activity, IAA, flavonoid concentrations and growth responses upon UV-B irradiation needs further investigation.

Notable growth responses to UV are mainly found during early development. Broccoli plants (*Brassica oleracea* L. convar. *botrytis*, Brassicaceae) germinating and growing under different UV-conditions differed significantly in growth, with plants under high UV irradiation being smaller. In contrast, when plants germinated under low UV conditions and were transferred at a later life stage to different ambient UV regimes in filter tents, shoot length, leaf area, and overall biomass did not differ between these plants (Kuhlmann and Müller 2009a). Thus, trade-offs may be especially high during early growth. Costs involved in production of phenolics or other protective metabolic processes may reduce the resources that otherwise could be used for growth (Herms and Mattson 1992), as indicated in the negative correlation between shoot biomass and flavonoid concentration (see Fig. 1; Kuhlmann and Müller 2009b).

Changes were also found in the trichome coverage of the plant surface in response to UV. *Arabidopsis thaliana* Heynh. (Brassicaceae) showed an increased

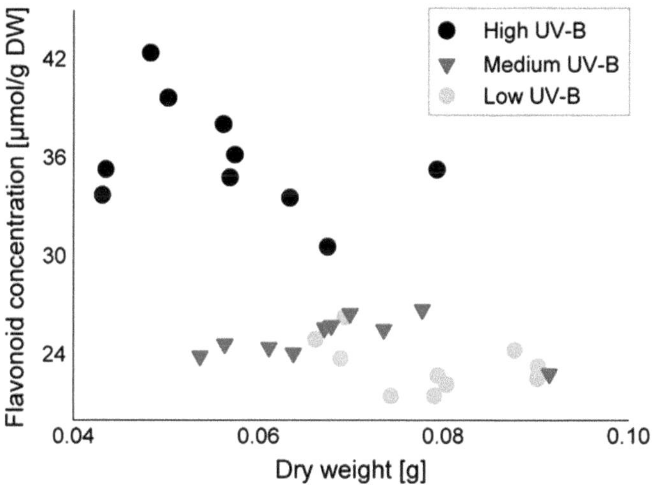

Fig. 1 Relationship between shoot biomass and flavonoid concentration in 17 days old broccoli plants, grown in greenhouses covered with material of different UV-B-transmittance (*black circles*: 80%; *triangles*: 23%; *grey circles*: 4% UV-B-transmittance) (for details see Kuhlmann and Müller 2009b). Spearman rank correlation $R = -0.663$, $P < 0.001$; $n = 30$ (ten per treatment)

trichome density due to UV irradiation (Lake et al. 2009) and trichomes can act as efficient defence against herbivorous insects. This demonstrates that not only chemical defences (see Sect. 4) but also mechanical defences can be induced by this abiotic factor.

3 Investigation Methods of UV Impacts on Plant–Insect Interactions

Investigations of UV impacts on plants started already in the 1920s (Schanz 1920), whereas research of UV effects on plant–insect interactions was initiated much later (e.g. Berenbaum 1978; Hatcher and Paul 1994). Experimental approaches range from artificial supplementation of UV with lamps in climate chambers or in the field to attenuation of UV using different filters or specific glass qualities. Most studies investigated short-term effects of UV-B or UV-A and UV-B on plant responses.

3.1 Supplemental UV

UV lamps (often in combination with respective filter materials) have been used in climate chambers (Hatcher and Paul 1994; Grant-Petersson and Renwick 1996; Lindroth et al. 2000; Caasi-Lit 2005; Foggo et al. 2007) or greenhouses

Fig. 2 Experimental designs to investigate effects of UV on plant biology and plant–insect interactions. (**a**) Sun simulator (8 m^2 effective area, 2.5 m height) of the phytotron facility at Helmholtz Centre of Munich (Germany) (photo kindly provided by A. Albert) with natural diurnal fluctuations of simulated UV-B. (**b**) Tents (1.20 m × 1 m ground area, 1.3 m maximum height) covered with filters of different quality to attenuate UV-B or UV-A and UV-B. (**c**) Greenhouses (4.2 m × 3 m ground area, 3.9 m maximum height) covered with innovative materials, transmitting either high (80%), medium (23%) or low (4%) levels of UV-B (**b**, **c**, Würzburg University, Germany)

(McCloud and Berenbaum 1994, 1999; Caasi-Lit 1998; Lavola et al. 1998; Warren et al. 2002; Izaguirre et al. 2003) to investigate effects of UV-B free and UV-B increased environmental conditions on plants and plant–insect interactions. Such lamps were also applied under field conditions (Salt et al. 1998; Buck and Callaghan 1999; Gwynn-Jones 1999; Veteli et al. 2003). The artificial supplementation of UV was mainly used to simulate predictable scenarios of increased UV-B due to depletion of the ozone layer (e.g. Newsham et al. 1996; Björn et al. 1997). A major drawback of these experiments is the fact that plants do not respond exclusively to UV-B. Instead, the relative composition of UV-B, UV-A and PAR influences their physiological responses (Caldwell 1971; Mirecki and Teramura 1984; Krizek 2004). To expose plants to the natural composition of UV and PAR radiation under controlled conditions, highly sophisticated sun simulators mimicking natural diurnal fluctuations of UV-B were installed at the Helmholtz Centre in Munich (Germany) (Fig. 2a; Seckmeyer and Payer 1993; Lütz et al. 2005; Favory et al. 2009). Until now only UV effects on plants but not on plant–insect interactions were investigated in these sun simulators.

3.2 Selective UV Exclusion

In contrast to an artificial increase of UV-B, another approach is to filter specific wavelengths of the ambient radiation to diminish selectively UV-B or UV-B and UV-A (e.g. Bergvinson et al. 1994; Ballaré et al. 1996; Rousseaux et al. 1998; Mazza et al. 1999b; Zavala et al. 2001; Izaguirre et al. 2003; Kuhlmann and Müller 2009b; Reifenrath and Müller 2009). Therefore, plants are grown, for example, in tents, which are covered with materials of different filter properties (Fig. 2b). Usually, the Northern wall is kept open for air circulation, which also allows access of naturally occurring herbivores. The growth conditions are not necessarily standardised under these tents, as temperature, humidity and light intensity cannot be regulated. Therefore, filters should be chosen that result in similar microclimatic conditions to allow comparisons between plants placed in tents covered by foils with different filter qualities (Kuhlmann and Müller 2009a). In order to highly standardise the growth conditions under selective UV-B exclusion regimes, temperature- and humidity-controlled greenhouses entirely covered with material of different filter qualities were built at three localities in Germany, Würzburg, Jülich and Bonn (Fig. 2c; for construction see Kuhlmann and Müller 2009b). These houses are not accessible for insects from the outside. However, insects can be released in a controlled way to study the effects of different UV-B conditions on plant–insect interactions (Kuhlmann and Müller 2010).

One draw-back of all these experiments is the number of true replicates. Whereas some studies used only two replicate constructions per filter type (Reifenrath and Müller 2009), six tents per filter type were used in a recent study to diminish pseudo-replication (Kuhlmann and Müller 2009a). However, highly expensive covering material and elaborate, space-, and cost-consuming constructions allow often only for one construction per type (e.g. the greenhouses in Kuhlmann and Müller 2010). In these cases, scientists must at least repeat their experiment to ensure that the described effects are reproducible.

Overall, expectable changes of UV-B irradiation, not arbitrarily high increases, should be mimicked if we want to study possible UV-B-impacts on plant–insect interactions. Furthermore, care must be taken to separate direct from plant-mediated UV effects on insect herbivores (see also Sect. 5).

4 Effects of UV on Plant Chemistry and Plant–Insect Interactions

Changes in plant chemistry in response to UV exposure are numerous. UV can affect the cuticle composition, phytohormone levels, proteinase inhibitor activities, as well as numerous secondary plant metabolites to varying extent. In the following, a short description of the chemical nature of the various compound classes and their biosynthesis is given. Then effects of UV on these metabolites and their role in plant–insect interactions are discussed.

4.1 Epicuticular Waxes

The surface of above-ground plant tissue is covered by the cuticle, which is composed of cuticular waxes and cutin. The cuticular waxes are esters of long-chain fatty-acids with long-chain primary alcohols and cyclic compounds

(Riederer and Müller 2006), whereas the polyester-type biopolymer cutin forms the matrix of the cuticle. Epicuticular waxes can reflect up to 30% of the incident UV-B radiation (Holmes 1997), indicating that increased wax layers are involved in UV-protection (Kakani et al. 2003). Indeed, enhanced UV-B radiation can cause a 28% increase in leaf wax layers of different crop plant species (Steinmüller and Tevini 1985). Similarly, cuticle thickness increased on the upper and especially on the lower leaf side due to UV-B exposure in two evergreen sclerophyll species (Grammatikopoulos et al. 1998). Next to thickness, the chemical composition of waxes has been shown to be influenced by UV-B radiation (Tevini and Steinmüller 1987; Barnes et al. 1996). However, a direct link between changes in cuticular wax composition and increased UV-B protection has been discussed controversially (Kakani et al. 2003). Broccoli grown in greenhouses with 80% UV-B transmission (see Fig. 2c) for about 3 weeks revealed a significantly lower total leaf wax coverage compared with plants grown under attenuated UV-B conditions, with lower concentrations of aldehydes, alkanes and one ketone but higher alkene concentrations in high-UV-B- compared to low-UV-B-exposed plants (Kuhlmann and Müller 2010). Similarly, in seedlings of *Brassica napus* L., decreased total wax contents were found in plants kept at enhanced UV-B radiation (Sangtarash et al. 2009). In a long-term study of UV-B treatment for 7 years, reduced adaxial trichome densities under UV-B were found in one *Vaccinium* species (Ericaceae), but waxes of several subarctic heathland species were not affected (Semerdjieva et al. 2003). There is a clear need to disentangle the role of the epicuticular waxes in UV-protection in more detail and many more plant species should be investigated to be able to draw a general conclusion.

Epicuticular waxes are also involved in host plant recognition by herbivorous insects, where they can either lead to acceptance or rejection (Müller and Riederer 2005; Müller 2008). For aphids, the texture, the wax coverage and the chemical composition of a plant influence host suitability and whether aphids will settle or not (Powell et al. 2006). UV-B-mediated changes in the composition of cuticular waxes on broccoli may therefore result in a modified acceptance and performance of aphids prolifering on differently UV-exposed plants (Kuhlmann and Müller 2010). Other herbivores may be likewise affected by UV impacts on the cuticular waxes but further studies in this field are lacking. Next to plant acceptance, the wax structure of a given plant affects the attachment of insects on this plant (Eigenbrode 2004; Müller 2006). However, no studies have been carried out to our knowledge until now to investigate whether UV-mediated changes of plant surfaces impact the attachment of herbivores and their predators. Finally, the specific optical properties of a given plant surface depend on the composition of epicuticular waxes (Pfündel et al. 2006). It is entirely unclear whether UV affects these properties and thereby interacts with the host finding behaviour of herbivorous insects, which is not only chemically but also visually mediated (Müller 2008). It is surprising that so little is known on the role of epicuticular waxes in UV-affected plant–insect interactions, although the cuticle is the first contact zone between the plant and an approaching insect.

4.2 Phytohormones

Plant development, growth and stress responses are regulated by phytohormones. The major growth-regulating hormones are auxin, brassinosteroids, cytokinin and gibberellin, whereas stress responses are mediated mainly by abscisic acid, ethylene, jasmonic acid and salicylic acid (Vlot et al. 2009; Wolters and Jürgens 2009). Signalling pathways of various hormones are interlinked and can repress or enforce each other. For example, insect attack induces the synthesis of ethylene and jasmonic acid, whereas salicylic acid is mainly involved in defence against biotrophic pathogens and inhibits the synthesis of jasmonic acid and subsequent ethylene production. Ethylene and jasmonic acid are positively controlled by each other (Kessler and Baldwin 2002; Wang et al. 2002). Furthermore, ROS, including singlet oxygen, superoxide ions and peroxides, are generated in plant metabolism in response to biotic and abiotic stresses (see Sect. 2.1). Plants need to control these oxidants, because they can cause detrimental effects. On the other hand, ROS act as signalling molecules affecting phytohormones like salicylic acid, jasmonic acid and ethylene (Mackerness 2000; Navrot et al. 2007).

Indole-3-acetic acid (IAA) is the most abundant auxin. It is synthesised either from tryptophan, or from a tryptophan precursor in a tryptophan-independent pathway, or by hydrolysis of IAA conjugates (Bartel 1997; Normanly and Bartel 1999). This plant growth regulating hormone decreases in response to UV-B exposure, whereas peroxidase and IAA oxidase activities increase (Huang et al. 1997; Jansen et al. 2004; Yang et al. 2004). Auxin reduction may be a general stress response, because herbivore damage can lead to decreased auxin concentrations as well (Baldwin et al. 1997; Cheong et al. 2002; Schmelz et al. 2003). A decrease of auxin enhances the resistance of plants against pathogens, herbivores and UV-B radiation, which may reveal a trade-off between growth and defence (Herms and Mattson 1992; Jansen 2002).

The alkene ethylene is a gaseous phytohormone, which is synthesised from the precursors methionine, *S*-adenosylmethionine and 1-aminocyclopropane-1-carboxylic acid in the Yang cycle (Wang et al. 2002). Ethylene induction in response to UV-B irradiation has been demonstrated in members of several plant families, including Brassicaceae, Rosaceae, Solanaceae and Poaceae (Predieri et al. 1995; Mackerness et al. 1999; An et al. 2006; Wang et al. 2006; Rakitin et al. 2008; Sangtarash et al. 2009). Its production depends on the developmental stage, being significantly induced in the vegetative but not in the reproductive stage of *Silene noctiflora* L. (Caryophyllaceae) (Qaderi et al. 2008). In response to UV-B exposure of *A. thaliana*, ethylene led to an up-regulation of *PR* and pathogen defence-associated *PDF1.2* genes (Mackerness et al. 1999). Oviposition of *Plutella xylostella* was similar on wildtype *A. thaliana* and the ethylene-constitutive mutant *eto-2* irrespective of the UV-B treatment of the plants (Caputo et al. 2006). Therefore, ethylene probably plays no role in UV-B affected *A. thaliana*-insect interactions, in contrast to jasmonic acid.

The cyclopentanone derivative jasmonic acid and its methyl ester are synthesised from the precursor linolenic acid via the octadecanoid pathway (Sembdner and Parthier 1993; Creelman and Mullet 1997). Jasmonic acid signalling mediates protection against necrotrophic pathogens and insects, for example, by activation of proteinase inhibitor synthesis (see Sect. 4.3) and synthesis of defence-related secondary metabolites (Farmer and Ryan 1992; McConn et al. 1997; Vlot et al. 2009; Wu and Baldwin 2009). Responses of jasmonic acid to UV-B are highly species-specific. For example, UV-B exposure increased jasmonic acid concentrations in *A. thaliana* (Mackerness et al. 1999) but not in tomato (*Lycopersicon esculentum* Mill., Solanaceae) (Stratmann et al. 2000). However, in both species jasmonic acid is induced due to wounding or herbivory (Stratmann et al. 2000; Textor and Gershenzon 2009). Involvement of UV-B-induced jasmonic acid signalling in plant–insect interactions has been nicely demonstrated by use of mutants. The oviposition rate of female *P. xylostella* L. (Lepidoptera) was reduced on *A. thaliana* wildtype (Col-0) irradiated with solar UV-B compared with plants kept under attenuated conditions. Plants of the mutant *jar1-1* with impaired jasmonic acid sensitivity were preferred over the wildtype only when plants were grown under solar UV-B, whereas moths did not discriminate between wildtype and *jar1-1* plants grown under attenuated UV-B. The mutants had reduced levels of UV-absorbing compounds, which may explain the behavioural answer of the herbivores (Caputo et al. 2006). Overall, mutants with reduced jasmonic acid sensitivity are generally more susceptible to insect herbivory (Walling 2000; Kessler and Baldwin 2002).

The phenolic phytohormone salicylic acid (*o*-hydroxybenzoic acid) is biosynthesised via two enzymatic pathways with chorismate or chorismate-derived L-phenylalanine as precursor (Vlot et al. 2009). Salicylic acid has many functions; it influences responses to abiotic stress and developmental growth processes and plays a major role in resistance against biotrophic pathogens and viruses in plants. Salicylic acid concentrations increase in response to UV-B irradiation resulting in an induction of *PR* gene expression in *A. thaliana*, which is similar to pathogen responses (Surplus et al. 1998). Similarly, insect feeding by the generalist *Spodoptera exigua* (Hübn.) has been recently shown to lead to a burst of salicylic acid (Diezel et al. 2009), demonstrating that salicylic acid also plays a role in plant–insect interactions. Finally, salicylic acid can act synergistically with UV-B; phenol oxidase activities were induced in *Capsicum annuum* L. (Solanaceae) stronger when plants were treated with salicylic acid and were UV-B exposed than when plants were treated with the phytohormone but were not UV-B exposed (Mahdavian et al. 2008).

It is surprising that the actual concentrations of the different signalling molecules have been rarely directly measured in plants after different UV exposure although these phytohormones are often discussed to be involved in response to several stress factors including UV-B radiation and to mediate changes of other metabolic pathways. However, recent evidence suggests that jasmonate levels are not necessarily increased but instead UV-B-induced changes of secondary metabolite concentrations may be due to changes in jasmonate sensitivity

(Ballaré et al. 2009). Studies with mutants deficient in certain pathway genes or transgenic lines (e.g. Jansen et al. 2004; Caputo et al. 2006) are highly useful tools to further shed light on the role of individual signalling molecules.

4.3 Proteinase Inhibitors

One of the essential dietary components for herbivores is nitrogen, which is mainly provided by plant proteins. Proteinase inhibitors can inhibit dietary protein hydrolysis of insects by tightly binding proteolytic enzymes (Kehr 2006) such as serine, cysteine, aspartic- and metallo-proteinases (Ryan 1990; Fan and Wu 2005). The proteinase inhibitor-related reduction in amino acid assimilation causes starvation and interferes with development and growth of herbivorous insects (Ryan 1990; Fan and Wu 2005; Zhu-Salzman and Zeng 2008). Oxylipins, such as jasmonic acid (see Sect. 4.2), act as signalling molecules that lead amongst others to the production of proteinase inhibitors (Ryan 2000).

Indications for a direct effect of UV-B radiation on proteinase inhibitor activities are sparse, but activities may be induced stronger in UV-B-exposed plants that are challenged by additional stresses. Proteinase inhibitor activities were not influenced by different UV exposure conditions in broccoli (Brassicaceae) and in *Salvia pratensis* L. (Lamiaceae) (Ulmann 2007; Kuhlmann and Müller 2009a). Similarly, they were not affected in tomato plants (Solanaceae) irradiated with UV-A/UV-B (Stratmann et al. 2000). However, when another stress was added by wounding, systemic proteinase inhibitor activities significantly increased in tomato exposed to UV-B prior wounding significantly more than in unexposed plants (Stratmann et al. 2000). Members of the Solanaceae respond differently to UV-B. In *Nicotiana longiflora* Cav., an insect responsive proteinase inhibitor gene was down-regulated, whereas in *Nicotiana attenuata* Torr. Ex W., proteinase inhibitor gene expression was up-regulated upon UV-B irradiation (Izaguirre et al. 2003). In *N. attenuata*, proteinase inhibitor activity was, however, strongly enhanced in response to simulated herbivory when plants faced UV-B (Izaguirre et al. 2003), as found for tomato (see above). Therefore, UV may have an impact on plant–insect interactions via modification of the proteinase inhibitor activity, at least in some species. Effects of UV on other digestibility-reducing proteins have not been investigated in relation to plant–insect interactions. Thus, further research is needed in this field.

4.4 Phenolic Compounds

Phenylpropanoids represent the largest group of secondary metabolites, comprising almost 20% of total carbon in the terrestrial biosphere (Yu and Jez 2008). Their core structure is formed by one or more aromatic benzene rings with one or several hydroxyl groups and/or sugars. These metabolites are derived from the shikimic acid pathway, sharing phenylalanine and tyrosine as common precursors (Bassman 2004). The major classes of phenolic compounds are phenolic and hydroxycinnamic acids, flavonoids, tannins, lignins, suberin, styrylpyrones, stilbenes, coumarins and furanocoumarins (Buchanan et al. 2000). These different classes can have various functions. For example, they are required for signalling, plant stabilisation and flower colouration. Furthermore, they play an important role in UV-protection and have anti-microbial as well as anti-herbivore properties (Heldt 1999; Bassman 2004; Vlot et al. 2009).

4.4.1 Hydroxycinnamic Acid Derivatives

Hydroxycinnamic acids are hydroxyl derivatives of cinnamic acid (Bassman 2004). Next to flavonoids (see Sect. 4.4.2), hydroxycinnamic acids are considered to be the most important UV-screening pigments due to their strong absorbance in this wavelength region (Caldwell et al. 1983; Burchard et al. 2000). Especially, the ester sinapoyl malate contributes highly to UV-B protection of *A. thaliana* as demonstrated by enhanced UV-B sensitivity of a mutant defective in the ability to synthesise sinapic acid (Landry et al. 1995). Hydroxycinnamic acid derivatives seem to be predominantly involved in UV-protection of young, developing rye plants (*Secale cereale* L., Poaceae), whereas they are replaced by flavonoids during development (Burchard et al. 2000). A similar shift was also observed in *A. thaliana* (Lake et al. 2009). In rye and different species of Brassicaceae, hydroxycinnamic acid derivatives are constitutively present (Burchard et al. 2000; Reifenrath and Müller 2007), whereas in other plant species a significant induction of these metabolites by UV was shown (e.g. *Vitis vinifera* L., Vitaceae, Kolb et al. 2001). Thus, the inducibility of these acids and their derivatives by UV is highly species-specific and development-dependent. However, it is not clear whether all hydroxycinnamic acids that are induced by UV in turn also act as UV protection.

Hydroxycinnamic acid esters are not only involved in UV-protection but can also play a role in herbivore defence. For example, they cause mortality and developmental inhibition of larvae of *Cylas puncticollis* Boheman (Coleoptera: Apionidae) (Stevenson et al. 2009). Furthermore, phenolic acids have been shown to increase oxidative stress in herbivores (Summers and Felton 1994). An increase of chlorogenic acid and dicaffeoylspermidine isomers was caused not only by UV-B irradiation but also by simulated herbivory in tobacco (Solanaceae) (Izaguirre et al. 2007). Thus, an induction of these phenolic compounds due to UV-B may also result in an improved herbivore protection, but to our knowledge this relationship has never been directly investigated.

4.4.2 Flavonoids

Flavonoids are heterocyclic compounds based on a C_{15} skeleton, which consist of two aromatic rings and one γ-pyrone ring with an ether-linked oxygen. Modifications of the basic structure can lead to approximately 4,500 metabolites (Bassman 2004). In plants, most of the flavonoids occur as flavonoid glycosides with one or more sugars substituted to the phenolic hydroxyl groups (Harborne 1991). The sugars protect the flavonoids from enzymatic degradation and keep the otherwise toxic compounds in an inactive stage (Harborne 1991). Furthermore, the glycosylation renders the flavonoids water soluble for storage in the vacuole. Moreover, acyl groups can be attached to one or more hydroxyl sugar groups. Flavonoids are biosynthesised from three malonate units condensing with a phenylalanine-derived precursor (Harborne 1991). Chalcone synthase and chalcone isomerase are further key enzymes involved in the condensation and formation of the flavonoid skeleton (Bassman 2004). Flavonoids have essential functions in growth, development, fertility, disease resistance and UV protection of plants but are also involved in flower colouration (Harborne and Williams 2000; Peer and Murphy 2007). UV exposure of plants induces flavonoid accumulation in leaf epidermal cells, in leaf wax and in trichomes (Harborne and Williams 2000). Here, flavonoids absorb UV-B efficiently due to their conjugated double bonds. In contrast, flavonoid-deficient mutants are highly sensitive to UV radiation (Lois and Buchanan 1994; Schmitz-Hoerner and Weissenböck 2003). The pattern of flavonoid accumulation is species-specific and varies within a plant in dependence of its developmental stage and in an irradiation-dependent manner (Burchard et al. 2000; Greenham et al. 2007; Kuhlmann and Müller 2009a). In almost all investigated systems, a significant induction by UV radiation could be detected (Table 1), but individual flavonoids are differently regulated, which is mirrored, for example, in changed ratios of kaempferol and quercetin glycosides due to UV exposure (Markham et al. 1998; Olsson et al. 1998; Reifenrath and Müller 2007; Winter and Rostás 2008; Kuhlmann and Müller 2009b). Quercetins are stronger induced, as they likely offer a more efficient UV-protection than kaempferols due to their additional *ortho*-dihydroxyl group in the B-ring, leading to an improved anti-oxidative capacity (Olsson et al. 1998).

Flavonoids have also been described to function as feeding and oviposition deterrents against herbivorous insects (Harborne and Williams 2000; Renwick et al. 2001; Ho et al. 2003; Caasi-Lit 2005; Caasi-Lit et al. 2007) and can inhibit the digestibility of dietary proteins (Pourcel et al. 2007). Therefore, they are potentially useful as biocontrol agents (Rajkumar and Jebanesan 2008). Flavonoids added to artificial diets reduced larval consumption and survival of several Lepidopteran species (Harborne and Williams 2000). In contrast, some specialist insects are attracted to these metabolites and use these compounds as phagostimulants or oviposition stimulants (Harborne and Williams 2000). For a few herbivore specialists, flavonoids only act as costimulants, increasing the stimulatory activity of other metabolites such as sugars or glucosinolates (van Loon et al. 2002; Kim and Mullin 2007; Reifenrath and Müller 2008). The flavonoids myricitrin and quercitrin added to an artificial diet did not affect feeding of *Operophtera brumata* L. (Lepidoptera: Geometridae) but the larvae showed a preference for UV-B-exposed plant material (Lavola et al. 1998). Thus, changes in insect behaviour in response to UV exposure of host plants may not necessarily be related to increases in flavonoid

Table 1 Influences of UV radiation on plant metabolites and responses of insect herbivores (and parasitoids)

Type of experiment[a]	Plant species	Herbivore species	Parasitoid species	UV effects on plant chemistry (compared with lower UV-cond.)	UV effects on herbivore (compared with lower UV cond.)	UV effects on parasitoid (compared with lower UV cond.)	Potential mechanism explaining UV effect on herbivores[b]	Reference
Artificial diet								
C_S UV	Artificial diet with xanthotoxin (occurring in Apiaceae and Rutaceae)	*Spodoptera eridania* (Cramer) (Lep., Noctuidae)	–		Enhanced mortality	–	Phototoxicity of xanthotoxin (P)	(Berenbaum 1978)
C_S UV	Linear furano-coumarins (psoralen, bergapten, xanthotoxin) at four conc. (occurring in Apiaceae and Rutaceae)	*Trichoplusia ni* Hbn. (Lep., Noctuidae)	*Copidosoma floridanum* (Ashmead) (Hym., Encyrtidae)		Increasing furanocoumarin conc. prolonged larval development, increased pupal mass, supplemental UV slowed development, no effects on survival	Increased mortality at high concentration of furanocoumarins, survival not affected by supplemental UV	Direct effect of furanocoumarins (P)	(Reitz and Trumble 1996)
C_S UV	Linear furanocoumarins, (psoralen, bergapten, xanthotoxin) at four conc. (occurring in Apiaceae and Rutaceae)	*Spodoptera exigua* (Hbn.) (Lep., Noctuidae)	*Archytas marmoratus* (Townsend) (Dipt., Tachinidae)		Increasing mortality with higher furanocoumarin conc., prolonged larval development, no effect on pupal-developmental time and pupal mass	Increasing mortality with increasing furanocoumarin conc., developmental time and size not affected	Direct effect of furanocoumarins (P)	(Reitz and Trumble 1997)
Betulaceae								
G_S UV-B	*Betula pendula* Roth	*Operophtera brumata* L.(Lep., Geometridae)	–	Higher flavonoid glycoside conc., phenolic acid conc. slightly increased, condensed tannins unaffected	More consumption, no feeding preferences for artificial diet containing flavonoids	–	Differences in plant quality, but not flavonoids (P)	(Lavola et al. 1998)

F_S, C_S UV-B	*Betula pubescens* Ehrh.	*Epirrita autumnata* (Lep., Geometridae)	–	–	More herbivory, direct preference for UV-B in the laboratory	–	UV-B preference (P)	(Buck and Callaghan 1999)
Brassicaceae								
C_S UV-B	*Arabidopsis thaliana* (L.) Heynh. Col-0 and camalexin-deficient mutant *pad3-1*	*Brevicoryne brassicae* (L.) (Ster., Aphididae)	–	Camalexin induced	Less aphid progeny on camalexin-accumulating Col-0 plants compared with *pad3-1* mutants	–	Camalexin or camalexin-related metabolites potentially as defence	(Kuśnierczyk et al. 2008)
C_S UV-B	*Arabidopsis thaliana* L*er*	*Pieris rapae* L. (Lep., Pieridae) (specialist), *Trichoplusia ni* (Hbn.) (Lep., Noctuidae) (generalist)	–	Higher flavonoid conc.	Less herbivory and less insect weight gain of *P. rapae*, no effects on *T. ni*	–	Differences in plant chemistry	(Grant-Petersson and Renwick 1996)
F_E, G_S UV-B	*Arabidopsis thaliana* Col-0, JA-insensitive *jar1-1*, ET-insensitive *ein-4*, ET-constitutive-response *eto-2*	*Plutella xylostella* L. (Lep., Plutellidae)	–	Higher conc. of phenolics	Reduced natural herbivory in the field, feeding preference and performance not affected in the laboratory, less oviposition by adults (but not in *jar1-1*)	–	Alteration in plant traits used by females for host choice	(Caputo et al. 2006)
F_E UV	*Brassica oleracea* L. convar. *botrytis*	Aleyrodidae (*Aleyrodes proletella* L.), Aphididae, Thripidae	–	No effects on C/N, glucosinolates and proteinase inhibitors, higher conc. of phenolic compounds	Aleyrodidae and Aphididae preferred +UV cond., Thripidae preferred –UV cond.	–	Direct visual orientation to UV (P for *A. proletella*); chemistry probably not relevant	(Kuhlmann and Müller 2009a)
G_E UV-B → F	*Brassica oleracea* convar. *botrytis*	Aleyrodidae, Aphididae, Thripidae	–	Lower C/N ratio, higher flavonoid conc., minor effect on aliphatic glucosinolates, indolyl glucosinolates unaffected	No infestation pattern in field	–	Differences in plant chemistry probably not relevant	(Kuhlmann and Müller 2009b)

(*continued*)

Table 1 (continued)

Type of experiment[a]	Plant species	Herbivore species	Parasitoid species	UV effects on plant chemistry (compared with lower UV-cond.)	UV effects on herbivore (compared with lower UV cond.)	UV effects on parasitoid (compared with lower UV cond.)	Potential mechanism explaining UV effect on herbivores[b]	Reference
G_E UV-B	*Brassica oleracea* convar. *botrytis*	*Brevicoryne brassicae* (L.) (specialist), *Myzus persicae* (Sulzer) (generalist) (Ster., Aphididae)	–	Reduced wax coverage, amino acids not affected except proline (reduced conc.), higher flavonoid conc., no effects on glucosinolates	Lower reproduction of *B. brassicae*, no effect on *M. persicae*	–	Differences in plant chemistry or tissue structure	(Kuhlmann and Müller 2010)
C_S UV-B	*Brassica oleracea* L. var. *capitata*	*Plutella xylostella* (Lep., Plutellidae)	*Cotesia plutellae* Kurdjurnov (Hym., Braconidae)	–	Lower meal sizes by larvae, lower pupal masses, less oviposition by adults	Females preferred herbivore larvae and plants	Differences in plant chemistry	(Foggo et al. 2007)
F_E UV	*Sinapis alba* L.	*Phaedon cochleariae* F: (Col., Chrysomelidae)	–	Fourfold increase in flavonoid conc., decreases in minor aromatic glucosinolates	No effects on feeding preferences	–	High plasticity of herbivore	(Reifenrath and Müller 2008)
F_E UV	*Sinapis alba*, *Nasturtium officinale* L.	*Phaedon cochleariae* (Col., Chrysomelidae)	–	C/N, soluble protein, glucose and flavonoid conc. significantly affected by UV, glucosinolates mostly not affected, myrosinases not affected	Growth rates and developmental times unaffected	–	High plasticity of herbivore	(Reifenrath and Müller 2009)
Ericaceae								
F_S UV-B	*Calluna vulgaris* (L.) Hull	*Strophingia ericae* (Curtis) (Ster., Aphalaridae)	–	Amino acids not affected except isoleucine (reduced conc.), N, C/N and water soluble phenolics not affected	Reduced population numbers	–	Unknown	(Salt et al. 1998)

F_S UV-B	*Vaccinium myrtillus* L., *V. uliginosum* L., *V. vitis-idaea* L.	Insects not identified	–	–	More herbivory on *V. myrtillus*, less on *V. uliginosum*, no effect on *V. vitis-idaea*	–	Unknown	(Gwynn-Jones et al. 1997)
Fabaceae								
$F_{E,S}$ UV-B	*Glycine max* L.	*Caliothrips phaseoli* Hood (Thys., Thripidae)	–	–	Less herbivory	–	Differences in plant quality, UV-B perception and avoidance	(Mazza et al. 1999b)
F_E UV	*Glycine max* L.	*Spodoptera frugiperda* (Smith) (Lep., Noctuidae) (generalist)	*Cotesia marginiventris* Cresson (Hym., Braconidae)	Higher total flavonoid conc., emission of VOCs not influenced	No effect on performance	No effect on preference behaviour	Generalist tolerates differences in plant chemistry	(Winter and Rostás 2008)
F_E UV-B	*Glycine max* L.	*Diabrotica speciosa* (Germar.) (Col., Chrysomelidae), Lep., Orth.	–	Nitrogen and hemicellulose unaffected, increased phenolic compound conc., reduced lignin conc.	Less herbivory	–	Plant-mediated or direct UV-B effect	(Zavala et al. 2001)
F_E UV-B	*Glycine max* L.	*Anticarsia gemmatalis* Hbn. (Lep., Noctuidae)	–	Nitrogen and hemicellulose unaffected, higher conc. of phenolic compounds, reduced lignin conc.	Less herbivory, slower growth, higher mortality	–	Differences in plant quality	(Zavala et al. 2001)
C_S UV-B	*Pisum sativum* L.	*Autographa gamma* L. (Lep., Noctuidae)	–	Higher nitrogen and total phenolics conc.	Increased larval growth rate and food utilisation efficiency, reduction in consumption	–	Higher nitrogen (R)	(Hatcher and Paul 1994)
C_S UV-B	*Trifolium repens* L	*Spodoptera litura* F., *Graphania mutans* (Walker) (Lep., Noctuidae)	–	Slight increases of nitrogen, decreases of carbohydrates, increases in cyanogenic activity, no changes in UV-absorbing compounds	Only marginal effects on long- and short-term feeding and performance	–	Differences in plant chemistry	(Lindroth et al. 2000)

(*continued*)

Table 1 (continued)

Type of experiment[a]	Plant species	Herbivore species	Parasitoid species	UV effects on plant chemistry (compared with lower UV-cond.)	UV effects on herbivore (compared with lower UV cond.)	UV effects on parasitoid (compared with lower UV cond.)	Potential mechanism explaining UV effect on herbivores[b]	Reference
Fagaceae								
F_S UV-B/UV-A	*Quercus robur* L.	Various Lep.	–	–	Increased herbivory	–	Differences in plant quality	(Newsham et al. 1999a)
F_S UV-B/UV-A	*Quercus robur* L.	Various Lep.	–	–	Increased herbivory	–	Plant-mediated or direct UV effect	(Newsham et al. 1996)
Gunneraceae								
F_E UV-B	*Gunnera magellanica* Lam.	Lep., Noctuidae	–	Nitrogen increased, hemicellulose and phenolic compounds unaffected	Less herbivory	–	Plant-mediated UV-B effect (R)	(Rousseaux et al. 1998)
F_E UV-B	*Gunnera magellanica*	Various chewing insects (not identified)	–	Phenolic compounds unaffected	Less herbivory	–	Unknown	(Rousseaux et al. 2001)
Lamiaceae								
F_EUV	*Salvia pratensis* L.	*Cassida canaliculata* Laich. (Col., Chrysomelidae)	–	No effects on C/N, hydroxycinnamic acids, and flavonoids, reduced terpene conc., no effects on soluble proteins and proteinase inhibitors	No effects on oviposition and feeding preferences of adults, higher body masses of pupae and adults	–	Differences in plant chemistry	(Ulmann 2007)
Nothofagaceae								
F_E UV-B	*Nothofagus antarctica* (G. Forst.) Oerst.	Insect larvae (Lep., Geometridae; Col., Chrysomelidae), adults (Col., Curculionidae)	–	Pentagalloylglucose unaffected, lower gallic acid, digalloylglucose, trigalloylglucose, and tetragalloylglucose conc., higher flavonoid conc.	Less herbivory	–	Differences in plant chemistry	(Rousseaux et al. 2004)
Plantaginaceae								
G_S UV-B	*Plantago lanceolata* L.	*Precis coenia* Hbn. (Lep., Nymphalidae), *Trichoplusia ni* (Hbn.) (Lep., Noctuidae)		Nitrogen increased in old leaves, no effect on iridoid glycosides, verbascosides increased in young leaves	No effect on growth and survivorship of *P. coenia*, growth of *T. ni* accelerated on excised leaves, depressed on potted plants		Inhibitory effect of UV-B (P) and differences in plant chemistry	(McCloud and Berenbaum 1999)

Poaceae								
F_S UV-B/ UV-A	*Lolium perenne* L., *Festuca rubra* L., *F. arundinacea* Schreb., *F. pratensis* Huds	*Schistocerca gregaria* (Forsskal) (Orth., Acrididae)	–	*F. pratensis*: loline (alkaloid) conc. unaffected	No effects on consumption, preferences in *F. pratensis* modified by endophyte × UV interaction	–	Complex interaction of UV and endophyte	(McLeod et al. 2001)
G_S UV-B	*Oryza sativa* L.	*Marasmia patnalis* Bradley (Lep., Pyralidae)	–	Age-dependent, higher conc. of phenolics	Higher larval survival and growth index but cultivar-dependent; less oviposition	–	Low phenolics conc. involved in larval feeding stimulation, act as sterilants for adults	(Caasi-Lit 1998)
C_S UV-B	*Oryza sativa*, crude and partially purified extracts	*Helicoverpa armigera* (Hbn.) (Lep., Noctuidae)	–	Higher conc. of phenolics (proven in a previous study)	Reduced growth, development and reproduction	–	Phenolic compounds (P)	(Caasi-Lit 2005)
C_S UV-B	*Oryza sativa*, purified extracts	*Helicoverpa armigera* (Lep., Noctuidae)	–	Increased flavonoid conc.	Isovitexin-2′-*O*-β-[6-*O*-E-*p*-coumaroyl glucopyranoside] reduced male fecundity and egg fertility, no effect of closely related flavonoids	–	Specific flavonoids (P)	(Caasi-Lit et al. 2007)
F_E UV	*Zea mays* L.	*Ostrinia nubilalis* (Hbn.) (Lep., Pyralidae)	–	In tendency higher nitrogen, reduced hydroxamic acid (DIMBOA) and hydroxycinnamic acid conc., higher total cell wall bound cyclobutane dimer conc.	Less herbivory	–	Differences in plant chemistry	(Bergvinson et al. 1994)
Rutaceae								
G_S UV-B	*Citrus jambhiri* Lush. + artificial diet	*Trichoplusia ni* (Lep., Noctuidae)		No effects on soluble protein, higher furanocoumarin conc., specifically in young leaves, 2.5 fold increase in psoralen, 1.8 fold increase in bergapten conc.	Decrease in survivorship and growth		Furanocoumarins (P)	(McCloud and Berenbaum 1994)

(*continued*)

Table 1 (continued)

Type of experiment[a]	Plant species	Herbivore species	Parasitoid species	UV effects on plant chemistry (compared with lower UV-cond.)	UV effects on herbivore (compared with lower UV cond.)	UV effects on parasitoid (compared with lower UV cond.)	Potential mechanism explaining UV effect on herbivores[b]	Reference
Salicaceae								
G_S, F_S UV-B	*Populus trichocarpa* Torr. & Gray	*Chrysomela scripta* Fab.(Col., Chrysomelidae)	–	Nitrogen and C/N unaffected, sulphur conc. unaffected or increased, UV-absorbing compounds increased, tannins reduced	Higher feeding preference, reduced consumption efficiency	–	Preference for high conc. of salicylate phenolic glycosides (use for defence)	(Warren et al. 2002)
F_S UV-B/UV-A	*Salix myrsinifolia* (Salisb.), *S. phylicifolia* L.	*Phratora vitellinae* L. (Col., Chrysomelidae), Hym., Col., Lep.	–	Nitrogen highest in UV-A treatment, tannins, salicylates, flavonoids, acids not affected	In field: more *P. vitellinae* adults and oviposition on *S. myrsinifolia,* but not higher damage by naturally occurring herbivores; in laboratory: growth of *P. vitellinae* not affected on *S. myrsinifolia* but impaired on *S. phylicifolia*	–	Plant quality or direct UV effect	(Veteli et al. 2003)
Solanaceae								
F_E UV-B	*Datura ferox* L.	Various insects, mostly Col., Chrysomelidae	–	Increased cyclobutane-pyrimidine dimers per unit DNA	Less herbivory	–	Plant quality (P)	(Ballaré et al. 1996)
G_S, F_E UV-B	*Nicotiana longiflora* Cav., *N. attenuata* Torr. Ex W.	*Manduca sexta* (L.) (Lep., Sphingidae)	–	Proteinase inhibitor conc. increased in *N. attenuata,* unaffected in *N. longiflora*	Slower growth of first-instar larvae	–	Plant quality	(Izaguirre et al. 2003)

[a]Plants of different growth conditions were exposed for different durations to the UV treatments; *C* climate chamber; *F* field; *G* greenhouse (per se usually without UV-B); subscript S: UV supplementation with lamps or control; subscript E: UV exclusion (mostly of ambient UV) or control

[b](*P*) proven with synthetic compounds or bioassay; (*R*) correlation found; *Col.* Coleoptera; *Dipt.* Diptera; *Hym.* Hymenoptera; *Lep.* Lepidoptera; *Orth.* Orthoptera; *Ster.* Sternorrhyncha; *Thys.* Thysanoptera; *ET* ethylene; *JA* jasmonic acid; *conc.* concentration; *cond.* condition; *VOC* volatile organic compound

concentrations, although this has been often proposed (see Table 1). More research is needed to identify the underlying principles of changed insect performance on, and behaviour towards UV-exposed plants using artificial diets supplemented with flavonoids or mutants with flavonoid-deficiencies.

Whether insect feeding can also induce flavonoids is discussed controversially and may depend on the plant tissue. Damage of soybean seeds (*Glycine max* (L) Merr., Fabaceae) by the stink bug *Nezara viridula* (Heteroptera: Pentatomidae) led to an induction of flavonoids (Piubelli et al. 2003), whereas damage to leaf tissue of different Brassicaceae species by aphids had almost no effect on flavonoid content (Kuhlmann and Müller unpublished). Under natural radiation conditions, flavonoid concentrations are usually already rather high and may be near to the upscale threshold of a certain plant species. Therefore, plants may not have evolved the ability to induce flavonoids even more in response to herbivore attack. The crucial role of flavonoids in plant protection against UV-irradiation is unquestioned. In contrast, the low performance of some herbivorous species on UV-exposed plant material is often attributed to flavonoids, however, without testing this explicitly and without analysing other plant metabolites, which may have been induced as well by UV and may have a higher impact on the herbivores than the flavonoids.

4.4.3 Tannins

Tannins occur either as condensed tannins, which are not susceptible to hydrolysis, or as hydrolysable tannins (Buchanan et al. 2000). Condensed tannins are polymers of flavonoid units linked by carbon–carbon bonds, whose base unit is flavone. Hydrolysable tannins are bound by carboxylic ester linkages, and their base unit is gallic acid (Hagerman and Butler 1991). Enhanced UV-B radiation caused an increase in condensed tannin concentrations only in few plant species, such as two *Betula* species (Betulaceae) (Lavola 1998) and *Laurus nobilis* L. (Lauraceae) (Grammatikopoulos et al. 1998), whereas in various other plant species minor or no effects of UV-B radiation on tannins were found (Tegelberg and Julkunen-Tiitto 2001; Bassman 2004). Thus, a tannin-induction by UV-B is obviously highly species-specific.

Tannins act astringently and therefore as defence against herbivores. Their ability to precipitate proteins renders plant material with high tannin concentrations indigestible (Hagerman and Butler 1991), but they can also act directly toxic to insects (Bernays et al. 1989). Hatcher and Paul (1994) determined phenolic contents as tannic acid equivalents; in *Pisum sativum* L. (Fabaceae) exposed to UV-B phenolics increased, but nitrogen levels increased even more. Larvae of *Autographa gamma* L. (Lepidoptera: Noctuidae) responded with a significantly higher efficiency of food conversion on these plants and a reduced consumption (Hatcher and Paul 1994). Adjustment of the consumption is probably mainly driven by nitrogen and less by phenolics in this species. These results demonstrate that herbivores can compensate for changes in plant tissue quality caused by increased UV-B radiation. Tannins of *Populus trichocarpa* Torr. & Gray (Salicaceae) responded differently in a greenhouse and a field experiment using artificial UV-B supplementation (Warren et al. 2002). In the former, tannin concentrations slightly decreased in young leaves grown under enhanced UV-B, whereas in the latter, no effects on tannin levels by UV-B irradiation were found. Thus, other factors than UV-B supplementation might have affected the tannin levels. Adults of *Chrysomela scripta* Fab. (Coleoptera: Chrysomelidae) showed no significant feeding preferences on differently UV-B-exposed leaves. A reduced consumption efficiency of larvae on plants grown under high UV-B conditions compared with plants of low-UV-B conditions was likely related to phenolic compounds other than tannins (Warren et al. 2002). Thus, clear indications for a significant role of UV-B-affected tannin levels in plant–insect interactions are lacking.

4.4.4 Lignins

Lignins are highly methoxylated polymers, which are biosynthesised by condensation of monolignols such as paracoumaryl alcohol, coniferyl alcohol and sinapyl alcohol (Buchanan et al. 2000). Lignin is the second most abundant organic natural product and forms an important reinforcement of tissues and cell walls (Hagerman and Butler 1991). UV-B seems to down-regulate lignin production in several species such as soybean and *Quercus robur* L. (Fagaceae) (Newsham et al. 1999b; Zavala et al. 2001), resulting in lower lignin:nitrogen ratios, but also the opposite could be found in one of two populations of *Pinus taeda* L. (Pinaceae) (Cybulski et al. 2000).

Due to the attribute of lignins to increase the toughness of tissues by incrustation of cell wall carbohydrates, lignins can reduce the tissue digestibility for herbivorous insects (Hagerman and Butler 1991). Larvae of *Anticarsia gemmatalis* Hbn. (Lepidoptera: Noctuidae) had a slightly higher mortality and reduced growth rates when fed on plants exposed to solar UV-B compared with plants grown under attenuated UV-B radiation. This pattern was not related to the lignin content of the host plant tissue, because lignins were actually higher in plants grown under UV-attenuated conditions (Zavala et al. 2001). Thus, similarly to tannins, lignins do not seem to be of importance in plant-mediated UV-B-influences on feeding and performance, although they can impair herbivory in principle.

4.4.5 Furanocoumarins

Furanocoumarins are phototoxic compounds, which absorb wavelengths at 330 nm in the UV-A region and are mainly found in Apiaceae and Rutaceae (Larson and Berenbaum 1988), growing in open habitats (Berenbaum 1981). They are biosynthesised via the phenylpropanoid pathway and the mevalonic acid pathway. In linear furanocoumarins, the furan ring is in line with the benz-2-pyrone nucleus, whereas in the angular furanocoumarins, the furan ring is oriented at an angle (Larson and Berenbaum 1988). In presence of UV radiation, furanocoumarins can crosslink DNA and thus interfere with transcription, causing toxicity (Berenbaum and Zangerl 1998). UV irradiation affects furanocoumarins differently. In *Pastinaca sativa* L. (Apiaceae), most but not all furanocoumarins increased in concentration upon UV exposure; therefore, their relative composition changed as well (Zangerl and Berenbaum 1987). However, these metabolites likely do not play a photoprotective role for the plant (Zangerl and Berenbaum 1987), but are important in plant–insect interactions.

Various furanocoumarins are known to be toxic and deterrent to insects and can even increase resistance against specialists (Zangerl and Berenbaum 1987), although adapted insects can detoxify furanocoumarins by cytochrome P450 monooxygenases to some degree (Zangerl and Berenbaum 2003). The activity of the furanocoumarins depends on their photoactivation by UV. Xanthotoxin toxicity to a polyphagous Lepidopteran species was shown to be related to UV radiation (Berenbaum 1978). Specifically young caterpillars of *Trichoplusia ni* Hbn. (Lepidoptera: Noctuidae) developed slower, when they were reared on *Citrus* plants (Rutaceae) exposed to increased UV-B-levels (McCloud and Berenbaum 1994). Feeding experiments with specific furanocoumarins added to an artificial diet proved that these metabolites are responsible for the negative effect on larval performance, with psoralen being more toxic than bergapten. The combination of dietary furanocoumarins and UV-B exposure of caterpillars had a more deleterious effect than furanocoumarins alone, potentially due to the phototoxicity of UV-excited metabolites (McCloud and Berenbaum 1994) or due to direct negative UV-B-effects on the insects (see Sect. 5).

Increasing concentrations of linear furanocoumarins increased the larval developmental time and mortality of *T. ni* and *Spodoptera exigua* (Hübner) (Lepidoptera: Noctuidae) (Reitz and Trumble 1996, 1997). This also affected natural enemies of these herbivores; high furanocoumarin levels in the food chain caused increased mortality of the parasitoids *Copidosoma floridanum* (Ashmead) (Hymenoptera) and *Archytas marmoratus* (Townsend) (Diptera) by exerting toxicity on the parasitoid larvae (Reitz and Trumble 1996, 1997). The results demonstrate that UV effects on plant metabolites can extent up to the third trophic level.

4.5 Alkaloids

Alkaloids are cyclic compounds containing nitrogen in negative oxidation stage and are usually biosynthesised from amino acids (Hartmann 1991). Although alkaloid occurrence is restricted to certain taxa, more than 12,000 compounds are known (Bassman 2004). Here we only refer to alkaloid classes, which have been discussed in the context of UV protection and interaction with herbivores, namely glucosinolates, camalexin, nicotine, and polyamines. A few other alkaloids (e.g. terpenoid indole alkaloids, cannabinoid alkaloids) were also investigated for their response to UV-B radiation in plants (Bassman 2004) but are not further discussed here, as they were not studied for effects on herbivores.

4.5.1 Glucosinolates

Glucosinolates are amino-acid derived alkaloids, whose basic skeleton consists of a β-D-glucose residue linked to (*Z*)-*N*-hydroximinosulfate ester by sulphur (Halkier and Gershenzon 2006). The 120 known glucosinolates differ in their side chain, with indolyl glucosinolates originating from tryptophan, aliphatic glucosinolates originating from alanine, methionine, isoleucine, leucine or valine and aromatic glucosinolates originating from phenylalanine or tyrosine. These metabolites are primarily characteristic for the Brassicaceae but occur in a few other plant families as well (Fahey et al. 2001; Wittstock and Halkier 2002). Glucosinolates and their hydrolysis products, which are mainly produced by the activity of the β-thioglucosidase myrosinase after tissue-rupture (Halkier and Gershenzon 2006), have anti-fungal, anti-microbial and anti-herbivore functions at least against non-adapted organisms (Bednarek et al. 2009; Clay et al. 2009; Hopkins et al. 2009).

Brassicaceae have been intensively studied with regard to UV effects on plant–insect interactions (see Table 1), because this plant family contains important crop plants as well as the model plant *A. thaliana*. Glucosinolates are not involved in UV-protection. However, effects on their concentrations mediated by UV radiation are conceivable, because glucosinolates are inducible by jasmonic acid (Doughty et al. 1995; Textor and Gershenzon 2009), which in turn has been shown to be induced by UV-B radiation at least in *A. thaliana* (Mackerness et al. 1999; Mackerness 2000; see Sect. 4.2). In contrast to expectations, glucosinolate concentrations were only marginally influenced by UV radiation in the studied Brassicaceae species. Glucosinolate concentrations of broccoli grown under filter tents including or excluding ambient UV-A and UV-B radiation did not differ significantly (Kuhlmann and Müller 2009a). Broccoli grown in three greenhouses transmitting different levels of UV-B radiation (4%, 23% and 80%; see Fig. 2c) showed significant differences in minor aliphatic glucosinolates of shoot tissue. However, the concentrations were not directly correlated with UV-B radiation fluence rates, as lowest aliphatic glucosinolate concentrations were found in plants grown under medium UV-B fluence rates (Kuhlmann and Müller 2009b). In *Sinapis alba* L. and *Nasturtium officinale* L., a few minor glucosinolates were reduced in plants that were exposed for 2 days in filter tents transmitting UV compared with plants exposed in tents under UV-exclusion. Activities of soluble myrosinase remained unaffected by UV in *S. alba* and *N. officinale*, whereas insoluble myrosinase activities were significantly lower in UV-exposed compared with UV-excluded *S. alba* plants (Reifenrath and Müller 2007). Overall, glucosinolate concentrations vary much more between leaves of different age or between different plant species than due to different UV exposure conditions (Reifenrath and Müller 2007, 2009).

Insects are highly sensitive to the glucosinolate concentrations of their host plants. Generalists are deterred by these compounds, whereas specialists can be stimulated for feeding and oviposition by glucosinolates (Halkier and Gershenzon 2006). Even though the above described effects of UV radiation on the glucosinolate–myrosinase system of different Brassicaceae species were not very pronounced, changes of both substrate concentrations and enzyme activities may impact both herbivorous specialists and generalists. Next to glucosinolate concentrations and myrosinase activities, other leaf compounds differed significantly in mature *S. alba* and *N. officinale* plants exposed for 2 days to two UV treatments in filter tents (Table 1). Nevertheless, the performance of the specialist *Phaedon cochleariae* F. (Coleoptera: Chrysomelidae) reared on these differently pretreated plant materials was not affected by the food quality modified due to UV impacts on plants (Reifenrath and Müller 2009). Therefore, these specialised beetles must have a high plasticity and must be able to compensate for nutritional deficits. In accordance, the feeding response of *P. cochleariae* was unaffected by the significantly different relative composition of glucosinolates and flavonoids in *S. alba* leaf tissue of greenhouse-kept (without UV-B) and outdoor-exposed plants (Reifenrath and Müller 2008). In other studies on Brassicaceae–insect interactions, different UV exposure of host plants strongly affected the performance of Lepidopteran herbivores (Grant-Petersson and Renwick 1996; Caputo et al. 2006) and their parasitoids (Foggo et al. 2007). These studies were conducted without investigating glucosinolate concentrations or myrosinase activities. It remains to be analysed whether modifications of the glucosinolate–myrosinase system occurred in these plant species due to UV radiation and whether these mediated some of the observed insect responses or whether other nutritional factors were of higher importance.

If plants are challenged by herbivores, effects on glucosinolate concentrations are much more pronounced than due to UV exposure, and glucosinolate concentrations can increase up to threefold or more (Kuhlmann and Müller 2009b; Textor and Gershenzon 2009). Moreover, a significant interaction effect of UV-B treatment and aphid infestation occurred on indolyl glucosinolate concentrations in broccoli shoot tissue exposed in greenhouses with different UV-B transmission (Kuhlmann and Müller 2010). This demonstrates that plants respond to their environmental challenges with high specificity. Protection against UV-B and against herbivores is regulated partially in separate ways. Overall, glucosinolate concentrations differ only little in plants exposed to varying UV-B conditions and are not involved in UV-protection of Brassicaceae. However, plants grown under low fluence rates of UV-B can be more susceptible to herbivores and may be, therefore, earlier forced to activate specialised defences, such as glucosinolates, against feeding insects.

4.5.2 Camalexin

The indole alkaloid camalexin (3-thiazol-2′-yl-indole) is a phytoalexin occurring in members of the Brassicaceae and is biosynthesised from the amino acid tryptophan. Camalexin is mainly induced by pathogens and has anti-microbial properties (Glawischnig 2007). Furthermore, artificial UV-B radiation (supplied by lamps) as well as feeding by aphids led to camalexin accumulation in *A. thaliana* (Mert-Turk et al. 2003; Kuśnierczyk et al. 2008). Increased camalexin concentrations induced by the aphid *Brevicoryne brassicae* (L.) (Sternorrhyncha: Aphididae) reduced in turn the fitness of these aphids (Kuśnierczyk et al. 2008). It needs further investigation to what extent natural ambient UV-B radiation modifies camalexin levels in plants and thereby affects other herbivorous species.

4.5.3 Nicotine

The alkaloid nicotine is derived from the amino acid nicotinic acid (vitamin B_3). It has a high molar extinction coefficient in the UV range. Nicotine concentrations increased after UV exposure in *Datura stramonium* L. (Solanaceae) cultured in a nicotine-containing hydroponic solution within 48 h but began to level off thereafter, whereas nicotine levels in unexposed plants remained constant (Baldwin and Huh 1994). Although a potent quenching of free radical species by nicotine through UV absorption was shown in vitro, a protective role of nicotine against UV-damage could not be proven in vivo. Plants with high nicotine levels were not more resistant to UV-induced damage than nicotine-free controls but had stronger decreases in photosynthetic capacity (Baldwin and Huh 1994). In contrast, nicotine concentrations are highly inducible by herbivory and are known to play an important role in herbivore defence (Baldwin 1989, 1999). It remains to be tested whether UV-mediated changes of nicotine concentrations occur under natural conditions and whether they may have an impact on herbivore responses.

4.5.4 Polyamines

Polyamines are small aliphatic molecules derived from arginine and methionine, which are omnipresent in all plant cells, particularly in the cytoplasm, in vacuoles, chloroplasts and mitochondria (Bouchereau et al. 1999; Groppa and Benavides 2008; Kusano et al. 2008). Polyamines bind to macromolecules like DNA, RNA, chromatin and proteins, because they occur positively charged in cells (Kusano et al. 2008). They are involved in gene regulation, cell signalling, membrane stabilisation and protection against oxidative stress (Groppa and Benavides 2008; Kusano et al. 2008). Furthermore, morphogenesis and developmental processes of plants are regulated by polyamines (Alcázar et al. 2006; Liu et al. 2007; Groppa and Benavides 2008). Polyamines accumulate in response to biotic and abiotic stresses and contribute to plant stress tolerance (Bouchereau et al. 1999; Alcázar et al. 2006; Liu et al. 2007). UV-B exposure led to an increase in polyamines such as putrescine, spermidine and spermine in various herbaceous plant species (An et al. 2004; Lütz et al. 2005; Rakitin et al. 2008). In contrast, polyamine concentrations were not influenced by enhanced UV-B conditions in tree species such as willows and silver birch seedlings (Tegelberg et al. 2006, 2008). Extensive research is needed to elucidate a possible connection between UV-B tolerance and herbivore resistance of plants via polyamines.

4.6 *Terpenoids*

Terpenoids (or isoprenoids) are the largest and most diverse class of organic plant metabolites, whose carbon skeletons are composed of characteristic C_5 units

(Gershenzon and Croteau 1991). More than 23,000 terpenoids have been identified (Cheng et al. 2007). Monoterpenes (C_{10}) and sesquiterpenes (C_{15}) are volatile, whereas diterpenes (C_{20}) and terpenoids with higher numbers of C_5-units are not volatile. The C_5 precursor isopentenyl diphosphate is either derived from the mevalonic acid pathway localised in the cytosol or formed from pyruvate and glyceraldehyde-3-phosphate (2-*C*-methyl-D-erythritol 4-phosphate pathway) in plastids. Prenyltransferases and terpene synthases are responsible for final formation of terpenes (Cheng et al. 2007). Terpenes often accumulate in oil glands or resin ducts (Gershenzon and Croteau 1991). The terpene concentrations change to varying degrees in different plant species in response to UV exposure. UV-B exposure resulted in accumulation of resin, consisting mainly of diterpenes, in *Grindelia chiloensis* (Corn.) Cabr. (Asteraceae). The resin may protect the plant from UV radiation, because it has high absorbance in the UV range (280–400 nm) (Zavala and Ravetta 2002). Similarly, increased concentrations of terpenes were detected in fruits of *Cucumis melo* L. (Cucurbitaceae) and in leaves of *Ocimum basilicum* L. (Lamiaceae) exposed to UV light (Johnson et al. 1999; Lamikanra et al. 2002). In contrast, in *Salvia pratensis* L. (Lamiaceae), a significantly reduced concentration of several monoterpenes and of the sesquiterpene germacrene D was found in plants grown under nearly ambient UV radiation compared with plants grown under attenuated UV-A and UV-B conditions (Ulmann 2007). In wood and needles of several tree seedlings, terpenoids were unaffected by UV-B (Tegelberg et al. 2002; Turtola et al. 2006). It remains to be studied whether these different findings are due to the different UV-qualities tested or due to species-specific differences in plant responses of terpenes.

Mono-, sesqui-, di- and triterpenes are known to act as feeding deterrents, oviposition deterrents and partly even as toxins against various generalist insects. But similarly to other plant secondary metabolites, several specialist insects are highly attracted and stimulated by terpenes (Gershenzon and Croteau 1991). The monophagous leaf beetle *Cassida canaliculata* Laich. (Coleoptera: Chrysomelidae) reached higher body masses when reared on *S. pratensis* plants exposed to almost ambient UV-A and UV-B compared with plants grown under attenuated conditions. However, it is unclear whether this was caused by reduced terpene levels in these plants (see above) or other changes in leaf chemistry (Ulmann 2007). Further studies are needed to elucidate the role of UV-induced changes in terpene-profiles on insect performance and behaviour. This might be specifically promising in insects feeding on aromatic plants rich in terpenes.

4.6.1 Iridoid Glycosides

Iridoid glycosides are cyclopentanoid monoterpene-derived compounds, which are produced by plants presumably via the mevalonic acid pathway (Bowers 1991). The glucoside usually occurs as an *O*-linked glycoside at C-1. At least 500 iridoids are known (Bowers 1991). Iridoids occur in plants as aglycon as well as in the gylcosidic form. The latter can be hydrolysed by β-glycosidases present in the guts of herbivores and form unstable aglyca with differing biological activity (Bowers 1991; Marak et al. 2002). Iridoid concentrations were not affected by elevated UV-B radiation in *Plantago lanceolata* L. (Plantaginaceae) (McCloud and Berenbaum 1999). This is in accordance with the expectation, because iridoid glycosides do not have UV-protective properties.

In interaction with herbivores, iridoid glycosides have pronounced deterrent activities on generalists but act stimulatory on specialist insects (Bowers 1991). UV-B-exposed plant material of *P. lanceolata* did not cause changes in growth and survivorship of a specialist herbivore but affected survivorship of a generalist caterpillar species. However, effects were probably due to direct negative influences of UV-B on this herbivore species and thus unrelated to iridoid glycosides (McCloud and Berenbaum 1999). Therefore, iridoid glycosides are important herbivore-defences but do not play any role as UV-protection, demonstrating again that plants have different strategies for protection against biotic vs. abiotic harms.

4.6.2 Volatile Organic Compounds

Plants release various volatile organic compounds (VOCs) into their surrounding atmosphere from reproductive and vegetative tissues. The main biogenic VOCs are isoprenes and monoterpenes, but also alkanes, alkenes, carbonyls, alcohols, esters, ethers and acids belong to this group (Kesselmeier and Staudt 1999). The volatiles can function as defence against abiotic and biotic stress (Dudareva et al. 2006; Unsicker et al. 2009). Ten years ago it was suggested that isoprene emissions are not sensitive to long-term influences of elevated UV-B (Kesselmeier and Staudt 1999). However, abiotic stress often leads to an increase of ROS (see Sect. 2.1), and volatile isoprenoids are discussed to reduce the oxidising load of stressed plants either directly or indirectly via defence signalling responses of the plants (Vickers et al. 2009). Thus, an induction of VOCs by UV could be expected. Enhanced UV-B radiation led indeed to a significant increase of isoprene emission in a subarctic peatland community (Tiiva et al. 2007), whereas isoprene emission per leaf mass was not significantly influenced by enhanced UV-B conditions in various tree species (Harley et al. 1996). Obviously, VOCs are therefore not that efficient in UV-B-protection.

Whereas responses of VOCs to UV-B are indifferent, these metabolites are highly induced by insect feeding in the vegetative plant parts. On herbivores, these VOCs can act either as attractants or repellents, depending on VOC concentration and species (Arimura et al. 2009). Moreover, VOCs are involved in indirect defence, because they can lead predators or parasitoids to their hosts (Arimura et al. 2009). But until now little evidence exists that UV-B modifies the profiles of herbivore-induced VOCs. Enhanced UV-B radiation did not change constitutive and herbivore-induced volatiles in *Picea abies* L. Karst. (Pinaceae) (Blande et al. 2009). Similarly, emission of VOCs by herbivore-damaged plants of *G. max* cv. London (Fabaceae) did not differ between plants grown under almost ambient or attenuated UV-A and UV-B radiation. In accordance, the parasitoid *Cotesia marginiventris* (Cresson) (Hymenoptera) did not discriminate between odours of these plants (Winter and Rostás 2008). In contrast, the parasitoid *Cotesia plutellae* Kurdjumov (Hymenoptera) preferred odours of the plant–herbivore complex (*Brassica oleracea* L. – *Plutella xylostella* L.) exposed to supplemental UV-B over odours of the complex grown under UV-B-depleted conditions, indicating that VOC profiles were modified by UV-B radiation in this system (Foggo et al. 2007). These findings suggest that some but not all tritrophic interactions, which are guided by VOCs, remain stable under different radiation conditions, but the inducibility of VOCs as well as the response of parasitoids depends highly on the species that are involved in the tritrophic system. More research is needed to elucidate the indirect influence of UV radiation on parasitoid species.

5 Direct Effects of UV on Herbivores and Their Natural Enemies

Next to plant-mediated effects of UV on herbivores, UV radiation can directly harm insects. UV-B-irradiated larvae of a chrysomelid species had increased mortality and reached lower pupal body masses compared with larvae kept under an UV-B-absorbing filter (Bacher and Luder 2005). Survivorship of a generalist caterpillar species decreased, when insects were reared under UV-B exposure on artificial diet, whereas a specialist Lepidopteran species was not affected by UV-B exposure (McCloud and Berenbaum 1999). UV-B radiation also increased the mortality of two-spotted spider mites (*Tetranychus urticae* Koch, Acari: Tetranychidae) and reduced the egg-laying capacity of the females, which was reflected in an avoidance of UV-B radiation by these arthropods (Barcelo 1981; Ohtsuka and Osakabe 2009). In the field, specifically the delicate, soft larvae of herbivores may avoid exposure to high UV-B irradiation by feeding on the lower leaf side, which additionally allows them to hide from predators. This avoidance of UV by insects may not have been accounted for and may have been a confounding variable in some studies, in which insects were kept directly on UV-exposed plants. A clear distinction between direct and plant-mediated effects of UV-B on insect herbivores is not always possible (e.g. Newsham et al. 1996; Zavala et al. 2001; Veteli et al. 2003) and thus data should be interpreted with care.

Avoidance behaviour may be evoked because UV radiation influences the visual orientation of herbivores in host finding (Antignus 2000). Thrips are able to perceive and avoid UV-B radiation, whereas UV-A radiation is attractive to them (Mazza et al. 1999b, 2002; Kuhlmann and Müller 2009a). However, contrasting results were found regarding the visual orientation of different thrips species tested with UV-absorbing and UV-transmitting filter materials (Antignus et al. 1996; Costa and Robb 1999; Costa et al. 2002; Díaz et al. 2006). The aphid species tested until now prefer UV-inclusion conditions (Antignus et al. 1996; Costa et al. 2002;

Chyzik et al. 2003; Díaz et al. 2006; Doukas and Payne 2007; Kuhlmann and Müller 2009a). Similarly, whiteflies are attracted to UV-inclusion conditions and avoid UV-deficient environments (Antignus et al. 1996; Costa and Robb 1999; Antignus et al. 2001; Kuhlmann and Müller 2009a). The species-specific and life-history trait-specific responses of herbivores to visual cues are probably driven by adaptations to different habitats. Under applied aspects, UV-absorbing filters can be used in integrated pest management programs at least against certain herbivorous insects (Antignus et al. 1996; Raviv and Antignus 2004).

UV radiation may not only influence survival and behaviour of herbivores but also of their enemies. However, evidence is scarce. The fecundity and host finding activity of the parasitoid *Aphidius matricariae* (Haliday) (Hymenoptera) was not influenced by UV radiation (Chyzik et al. 2003). Three other hymenopteran parasitoid species showed a clear preference for ambient light over UV-depleted light, when their orientation behaviour was tested in a Y-tube with two different filters. However, their parasitism rates did not differ in greenhouses covered either with regular plastic or UV-absorbing plastic sheet (Chiel et al. 2006). The host location ability of parasitoids in different UV environments has high implications for their use as biocontrol agents introduced to field or greenhouse crops.

6 Conclusions and Outlook

UV radiation influences the production and induction of plant metabolites to differing extent, depending on the compound class. Within the phytohormones, auxin probably plays a central role as it also regulates the photomorphogenic changes that occur specifically in young, developing plants. Within the phenolic compounds, mainly flavonoids are consistently highly induced by UV radiation and involved in UV-protection. Some compound classes, such as furanocoumarins, seem to be regularly induced by UV, but likely do not act photoprotective. In contrast, other compound classes, such as proteinase inhibitors, tannins, polyamines, terpenoids, and alkaloids, are only induced in certain taxonomic groups, but are also not involved in protection against UV. Little or no effects of UV were found on glucosinolates and iridoid glycosides, whereas lignin seems to be actually down-regulated by UV. In most studies, only one or a few target metabolites were analysed in plant tissues that were obtained from contrasting UV irradiation treatments. Metabolic profiling may be a powerful tool to find other metabolites that are induced but have not been considered until now. It would further allow disentangling the interplay of different up- and down-regulated metabolites to understand underlying signalling pathways that are induced by UV. Mutants differing in these signalling pathways or transgenic plants affected in biosynthesis of selected metabolites could then be used to confirm these findings. The function of secondary metabolites induced by UV radiation is not fully understood, especially, when these metabolites do not act as UV-protectants. Strikingly, several metabolites that are induced by UV play an important role as herbivore defences. Therefore, it could be argued that such defence metabolites have evolved as general plant defences against environmental impacts and that their induction by UV is highly adaptive. Among the secondary metabolites, the development of the phenolic compound metabolism must have played a very important role in plant evolution (Rozema et al. 1997). Selection may favour UV induction of those phenolics that are also toxic to insects.

Several UV-inducible compounds are likewise induced by insect feeding, such as hydroxycinnamic acids, camalexin and to some extent flavonoids. Plant responses to abiotic harms (UV) and biotic harms (insect feeding) result partly in overlapping gene expression and signalling (Izaguirre et al. 2003), but there is also a clear distinction between UV- and herbivore-induced patterns (Pandey and Baldwin 2008; Kuhlmann and Müller 2009b). Compounds, such as flavonoids, are consistently induced by UV, whereas others, such as glucosinolates and nicotine, are typically induced by herbivory. In several studies, it is difficult to disentangle the effects caused by UV on plant metabolites from the effects caused by herbivory (e.g. Kuhlmann and Müller 2009a, b). The signalling molecules jasmonic acid, salicylic acid and ethylene likely play a central role in these induction responses (Thompson and Goggin 2006). Their specific cross-talk and induction pattern leads to a fine-tuned response pattern in each plant. Plant responses are therefore highly species- and elicitor-specific (Thompson and Goggin 2006; Textor and Gershenzon 2009).

In most studies, it is intricate to relate a negative or positive effect of UV-treated plant material on a given herbivore species to one particular compound (Table 1). Only experiments, in which these compounds were incorporated in diets and were tested with a certain insect species, gave clear evidence (e.g. Berenbaum 1978; McCloud and Berenbaum 1994; Reitz and Trumble 1996, 1997). Mutants or transgenic plants may help to disentangle the role of specific plant metabolites in UV-mediated plant–insect interactions and to understand the underlying mechanisms.

The reviewed effects of UV exposure on plant–insect interactions are highly system-specific. In more than half of the studies, positive effects of UV exposure on plants could be determined, meaning that insects fed less or performed poorer on plants grown under high UV conditions compared with their development on plants grown under low UV conditions. In 13% of all studies, no differences were found, whereas in more than a quarter of the studies negative effects on the plants occurred, with insects developing better on plants grown under high UV conditions (Fig. 3). Even within one insect species, different developmental stages can respond differently to UV-exposed tissue (Caputo et al. 2006). The contrasting findings for various plant–insect interactions could be due to two factors, which may either act alone or in concert: (1) plants as well as insects respond highly species-specifically to UV and (2) different methods were applied to test the UV effects. Experiments ranged from short exposure of plants for 2 days (Reifenrath and Müller 2009) up to 7-year treatments (Semerdjieva et al. 2003). Short exposure is a typical test for stress response, whereas experiments carried out over longer durations or in which plants are grown from the germination onwards under different UV conditions give

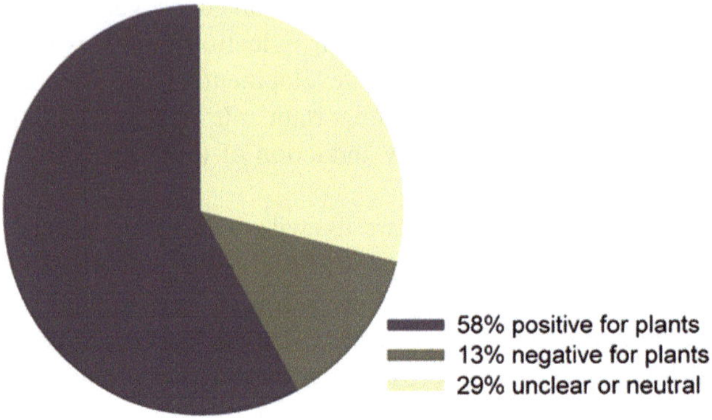

Fig. 3 Literature survey on studies of chemically mediated UV impacts on plant–insect interactions. Results of 38 studies were reviewed (see Table 1). Positive effects on plants (*less herbivory, poorer insect performance*), negative (*more herbivory, improved insect performance*) or unclear/neutral effects on plants exposed to high UV radiation conditions compared with low conditions are given in percent

information about adaptations to environmental conditions that are not stress-related. Furthermore, many studies were carried out under rather artificial conditions, adding high fluence rates of UV-B by use of UV-lamps. More than a decade ago, Rozema et al. (1997) noted that there is a need for more outdoor studies under natural light conditions. Since then less than half of the following studies were conducted under more natural situations (see Table 1), excluding or including ambient UV-A and UV-B radiation. Very few studies investigated long-term effects (e.g. Semerdjieva et al. 2003; Turtola et al. 2006), but only with regard to UV-B effects on plants. Long-term studies of UV impacts on plant–insect relationships are missing.

Finally, little research has been conducted on the effects of UV on tritrophic interactions. These studies focussed on UV effects on indirect defences, with the hypothesis that UV exposure may change the profile of herbivore-induced VOC-emission (e.g. Foggo et al. 2007; Winter and Rostás 2008; see Sect. 4.6.2). No general conclusions can be drawn for the effects of UV radiation on these relationships, as one study found an UV effect on parasitoid attraction, the other did not. Another important indirect defence of plants against herbivores can be mediated by extrafloral nectaries, which excrete higher amounts of nutrients when attacked by herbivores, thereby attracting carnivores (Heil 2008). There is a complete lack of knowledge of UV-B effects on extrafloral nectar production (Davis 2003) and much more needs to be done to understand the influence of UV-B on tritrophic interactions.

Increasing UV-B radiation is only one of the scenarios of global climate change. The interaction between different players involved in climate change, such as CO_2, temperature, ozone and various other gases, needs to be considered to evaluate the final outcome on species interactions (Caldwell et al. 2003; Bidart-Bouzat and Imeh-Nathaniel 2008; Tegelberg et al. 2008). From an applied aspect, it is important to understand how UV-B radiation can influence plant quality and plant suitability to herbivores. Specifically in greenhouse cultures with otherwise controlled conditions, the cover material of the greenhouse constructions should be carefully chosen. Growth conditions for cultivated plant species should be adapted to optimise plant chemistry with regard to human dietary aspects (Harborne and Williams 2000; Gomes et al. 2008; Jansen et al. 2008; Zhang and Björn 2009). In addition, herbivore and parasitoid behaviour of potential biocontrol agents under different UV-B light should be taken into account, including their visual orientation. Finally, costs for UV-B acclimatisation of plants need to be considered. A reduction in biomass, seed biomass and grain yield was found in several crop species in response to UV-B exposure (Teramura 1983; Mazza et al. 1999a; Kakani et al. 2003).

In summary, there is plenty of evidence that UV-B radiation can influence plant–insect interactions, whereby direct UV-B and plant-mediated effects may act in concert on the herbivorous insects. However, the interactions depend on species-specific and adaptive behavioural predispositions. Detailed research is needed to disentangle which plant compound(s), other than the best examined and very important UV absorbing phenolics, influence plant–insect relationships under UV-B irradiation. In addition, morphological plant changes induced by UV-B that may modify the digestibility or access to food sources must be examined and should be linked to the overall changes in plant chemistry. Furthermore, herbivore behaviour in response to different wavebands should be studied in more detail to clarify the role of visual cues involved in UV-B-mediated plant–insect interactions, in addition to chemical cues.

References

Alcázar R, Marco F, Cuevas JC, Patron M, Ferrando A, Carrasco P, Tiburcio AF, Altabella T (2006) Involvement of polyamines in plant response to abiotic stress. Biotechnol Lett 28:1867–1876

An LG, Liu GX, Zhang MX, Chen T, Liu YH, Feng HY, Xu SJ, Qiang WY, Wang XL (2004) Effect of enhanced UV-B radiation on polyamine content and membrane permeability in cucumber leaves. Russ J Plant Physiol 51:658–662

An LZ, Xu XF, Tang HG, Zhang MX, Hou ZD, Liu YH, Zhao ZG, Feng HY, Xu SJ, Wang XL (2006) Ethylene production and 1-aminocyclopropane-1-carboxylate (ACC) synthase gene expression in tomato (*Lycopersicon esculentum* Mill.) leaves under enhanced UV-B radiation. J Integr Plant Biol 48:1190–1196

Antignus Y (2000) Manipulation of wavelength-dependent behaviour of insects: an IPM tool to impede insects and restrict epidemics of insect-borne viruses. Virus Res 71:213–220

Antignus Y, Mor N, Joseph RB, Lapidot M, Cohen S (1996) Ultraviolet-absorbing plastic sheets protect crops from insect pests and from virus diseases vectored by insects. Environ Entomol 25:919–924

Antignus Y, Nestel D, Cohen S, Lapidot M (2001) Ultraviolet-deficient greenhouse environment affects whitefly attraction and flight-behavior. Environ Entomol 30:394–399

Arimura G, Matsui K, Takabayashi J (2009) Chemical and molecular ecology of herbivore-induced plant volatiles: proximate factors and their ultimate functions. Plant Cell Physiol 50:911–923

Bacher S, Luder S (2005) Picky predators and the function of the faecal shield of a cassidine larva. Funct Ecol 19:263–272

Baldwin IT (1989) Mechanism of damage-induced alkaloid production in wild tobacco. J Chem Ecol 15:1661–1680

Baldwin IT (1999) Inducible nicotine production in native *Nicotiana* as an example of adaptive phenotypic plasticity. J Chem Ecol 25:3–30

Baldwin I, Huh S (1994) Primary function for a chemical defense? Nicotine does not protect *Datura stramonium* L from UV damage. Oecologia 97:243–247

Baldwin IT, Zhang ZP, Diab N, Ohnmeiss TE, McCloud ES, Lynds GY, Schmelz EA (1997) Quantification, correlations and manipulations of wound-induced changes in jasmonic acid and nicotine in *Nicotiana sylvestris*. Planta 201:397–404

Ballaré CL, Demkura PV, Abdala G, Baldwin IT (2009) Specific effects of solar UV-B radiation on plant defense mediated by the jasmonate signaling pathway. Comp Biochem Physiol A-Mol Integr Physiol 153A:S200–S200

Ballaré CL, Scopel AL, Stapleton AE, Yanovsky MJ (1996) Solar ultraviolet-B radiation affects seedling emergence, DNA integrity, plant morphology, growth rate, and attractiveness to herbivore insects in *Datura ferox*. Plant Physiol 112:161–170

Barcelo JA (1981) Photoeffects of visible and ultraviolet radiation on the two-spotted spider mite, *Tetranychus urticae*. Photochem Photobiol 33:703–706

Barnes PW, Ballaré CL, Caldwell MM (1996) Photomorphogenic effects of UV-B radiation on plants: consequences for light competition. J Plant Physiol 148:15–20

Bartel B (1997) Auxin biosynthesis. Annu Rev Plant Physiol Plant Mol Biol 48:49–64

Bassman JH (2004) Ecosystem consequences of enhanced solar ultraviolet radiation: secondary plant metabolites as mediators of multiple trophic interactions in terrestrial plant communities. Photochem Photobiol 79:382–398

Bednarek P, Pislewska-Bednarek M, Svatoš A, Schneider B, Doubský J, Mansurova M, Humphry M, Consonni C, Panstruga R, Sanchez-Vallet A, Molina A, Schulze-Lefert P (2009) A glucosinolate metabolism pathway in living plant cells mediates broad-spectrum antifungal defense. Science 323:101–106

Berenbaum M (1978) Toxicity of a furanocoumarin to armyworms – case of biosynthetic escape from insect herbivores. Science 201:532–534

Berenbaum MR (1981) Patterns of furanocoumarin distribution and insect herbivory in the Umbelliferae: plant chemistry and community structure. Ecology 62:1254–1266

Berenbaum MR, Zangerl AR (1998) Chemical phenotype matching between a plant and its insect herbivore. Proc Natl Acad Sci USA 95:13743–13748

Bergvinson D, Arnason J, Hamilton R, Tachibana S, Towers G (1994) Putative role of photodimerized phenolic acids in maize resistance to *Ostrinia nubilalis* (Lepidoptera: Pyralidae). Environ Entomol 23:1516–1523

Bernays EA, Driver GC, Bilgener M (1989) Herbivores and plant tannins. Adv Ecol Res 19:263–302

Beyschlag W, Barnes PW, Flint SD, Caldwell MM (1988) Enhanced UV-B irradiation has no effect on photosynthetic characteristics of wheat (*Triticum aestivum* L) and wild oat (*Avena fatua* L) under greenhouse and field conditions. Photosynthetica 22:516–525

Bidart-Bouzat MG, Imeh-Nathaniel A (2008) Global change effects on plant chemical defenses against insect herbivores. J Integr Plant Biol 50:1339–1354

Björn LO, Callaghan TV, Johnsen I, Lee JA, Manetas Y, Paul ND, Sonesson M, Wellburn AR, Coop D, Heide-Jorgensen HS, Gehrke C, Gwynn-Jones D, Johanson U, Kyparissis A, Levizou E, Nikolopoulos D, Petropoulou Y, Stephanou M (1997) The effects of UV-B radiation on European heathland species. Plant Ecol 128:252–264

Blande JD, Turunen K, Holopainen JK (2009) Pine weevil feeding on Norway spruce bark has a stronger impact on needle VOC emissions than enhanced ultraviolet-B radiation. Environ Pollut 157:174–180

Bouchereau A, Aziz A, Larher F, Martin-Tanguy J (1999) Polyamines and environmental challenges: recent development. Plant Sci 140:103–125

Bowers MD (1991) Iridoid glycosides. In: Rosenthal GA, Berenbaum MR (eds) Their interactions with secondary plant metabolites, vol 1: chemical participants, vol 1. Academic Press, Harcourt Brace Jovanovich, San Diego, pp 297–325

Brown DE, Rashotte AM, Murphy AS, Normanly J, Tague BW, Peer WA, Taiz L, Muday GK (2001) Flavonoids act as negative regulators of auxin transport in vivo in *Arabidopsis*. Plant Physiol 126:524–535

Buchanan BB, Gruissem W, Jones RL (2000) Biochemistry and molecular biology of plants. American Society of Plant Physiologists, Rockville, Maryland

Buck N, Callaghan T (1999) The direct and indirect effects of enhanced UV-B on the moth caterpillar *Epirrita autumnata*. Ecol Bull 47:68–76

Burchard P, Bilger W, Weissenböck G (2000) Contribution of hydroxycinnamates and flavonoids to epidermal shielding of UV-A and UV-B radiation in developing rye primary leaves as assessed by ultraviolet-induced chlorophyll fluorescence measurements. Plant Cell Environ 23:1373–1380

Caasi-Lit M (1998) Effects of ultraviolet-B irradiated rice plants on the growth and development of the rice leaffolder, *Marasmia patnalis* Bradley. Philipp Ent 12:179–193

Caasi-Lit MT (2005) Effects of crude and partially purified extracts from UV-B-irradiated rice leaves on *Helicoverpa armigera* (Hübner). Photochem Photobiol 81:1101–1106

Caasi-Lit MT, Tanner GJ, Nayudu M, Whitecross MI (2007) Isovitexin-2′-*O*-beta- [6-*O*-E-*p*-coumaroylglucopyranoside] from UV-B irradiated leaves of rice. *Oryza sativa* L. inhibits fertility of *Helicoverpa armigera*. Photochem Photobiol 83:1167–1173

Caldwell MM (1971) Effects of solar UV radiation on the growth and development of higher plants. In: Giese AC (ed) Photophysiology VI, vol 1. Academic Press, New York, pp 131–177

Caldwell MM, Ballaré CL, Bornman JF, Flint SD, Björn LO, Teramura AH, Kulandaivelu G, Tevini M (2003) Terrestrial ecosystems, increased solar ultraviolet radiation and interactions with other climatic change factors. Photochem Photobiol Sci 2:29–38

Caldwell MM, Robberecht R, Flint SD (1983) Internal filters: prospects for UV-acclimation in higher plants. Physiol Plant 58:445–450

Caputo C, Rutitzky M, Ballaré CL (2006) Solar ultraviolet-B radiation alters the attractiveness of Arabidopsis plants to diamondback moths (*Plutella xylostella* L.): impacts on oviposition and involvement of the jasmonic acid pathway. Oecologia 149:81–90

Chen M, Chory J, Fankhauser C (2004) Light signal transduction in higher plants. Annu Rev Genet 38:87–117

Cheng AX, Lou YG, Mao YB, Lu S, Wang LJ, Chen XY (2007) Plant terpenoids: biosynthesis and ecological functions. J Integr Plant Biol 49:179–186

Cheong YH, Chang HS, Gupta R, Wang X, Zhu T, Luan S (2002) Transcriptional profiling reveals novel interactions between wounding, pathogen, abiotic stress, and hormonal responses in *Arabidopsis*. Plant Physiol 129:661–677

Chiel E, Messika Y, Steinberg S, Antignus Y (2006) The effect of UV-absorbing plastic sheet on the attraction and host location ability of three parasitoids: *Aphidius colemani*, *Diglyphus isaea* and *Eretmocerus mundus*. Biocontrol 51:65–78

Chyzik R, Dobrinin S, Antignus Y (2003) Effect of a UV-deficient environment on the biology and flight activity of *Myzus persicae* and its hymenopterous parasite *Aphidius matricariae*. Phytoparasitica 31:467–477

Clay NK, Adio AM, Denoux C, Jander G, Ausubel FM (2009) Glucosinolate metabolites required for an *Arabidopsis* innate immune response. Science 323:95–101

Costa HS, Robb KL (1999) Effects of ultraviolet-absorbing greenhouse plastic films on flight behavior of *Bemisia argentifolii* (Homoptera: Aleyrodidae) and *Frankliniella occidentalis* (Thysanoptera: Thripidae). J Econ Entomol 92:557–562

Costa HS, Robb KL, Wilen CA (2002) Field trials measuring the effects of ultraviolet-absorbing greenhouse plastic films on insect populations. J Econ Entomol 95:113–120
Creelman RA, Mullet JE (1997) Biosynthesis and action of jasmonates in plants. Annu Rev Plant Physiol Plant Mol Biol 48:355–381
Cybulski WJ, Peterjohn WT, Sullivan JH (2000) The influence of elevated ultraviolet-B radiation (UV-B) on tissue quality and decomposition of loblolly pine (*Pinus taeda* L.) needles. Environ Exp Bot 44:231–241
Davis AR (2003) Influence of elevated CO_2 and ultraviolet-B radiation levels on floral nectar production: a nectary-morphological perspective. Plant Syst Evol 238:169–181
Díaz BM, Biurrún R, Moreno A, Nebreda M, Fereres A (2006) Impact of ultraviolet-blocking plastic films on insect vectors of virus diseases infesting crisp lettuce. HortScience 41:711–716
Diezel C, von Dahl CC, Gaquerel E, Baldwin IT (2009) Different lepidopteran elicitors account for cross-talk in herbivory-induced phytohormone signaling. Plant Physiol 150:1576–1586
Doughty KJ, Kiddle GA, Pye BJ, Wallsgrove RM, Picket JA (1995) Selective induction of glucosinolates in oilseed rape leaves by methyl jasmonate. Phytochemistry 38:347–350
Doukas D, Payne CC (2007) The use of ultraviolet-blocking films in insect pest management in the UK; effects on naturally occurring arthropod pest and natural enemy populations in a protected cucumber crop. Ann Appl Biol 151:221–231
Dudareva N, Negre F, Nagegowda DA, Orlova I (2006) Plant volatiles: recent advances and future perspectives. Crit Rev Plant Sci 25:417–440
Eigenbrode SD (2004) The effects of plant epicuticular waxy blooms on attachment and effectiveness of predatory insects. Arthropod Struct Dev 33:91–102
Fahey JW, Zalcmann AT, Talalay P (2001) The chemical diversity and distribution of glucosinolates and isothiocyanates among plants. Phytochemistry 56:5–51
Fan SG, Wu GH (2005) Characteristics of plant proteinase inhibitors and their applications in combating phytophagous insects. Bot Bul Acad Sin 46:273–292
Farmer EE, Ryan CA (1992) Octadecanoid precursors of jasmonic acid activate the synthesis of wound-inducible proteinase-inhibitors. Plant Cell 4:129–134
Favory JJ, Stec A, Gruber H, Rizzini L, Oravecz A, Funk M, Albert A, Cloix C, Jenkins GI, Oakeley EJ, Seidlitz HK, Nagy F, Ulm R (2009) Interaction of COP1 and UVR8 regulates UV-B-induced photomorphogenesis and stress acclimation in *Arabidopsis*. EMBO J 28: 591–601
Foggo A, Higgins S, Wargent JJ, Coleman RA (2007) Tri-trophic consequences of UV-B exposure: plants, herbivores and parasitoids. Oecologia 154:505–512
Frohnmeyer H, Staiger D (2003) Ultraviolet-B radiation-mediated responses in plants. Balancing damage and protection. Plant Physiol 133:1420–1428
Gershenzon J, Croteau R (1991) Terpenoids. In: Rosenthal GA, Berenbaum MR (eds) Herbivores. Their interactions with secondary plant metabolites, vol 1. Academic Press, San Diego, New York, Boston, London, Sydney, Tokyo, Toronto, pp 165–219
Glawischnig E (2007) Camalexin. Phytochemistry 68:401–406
Gomes A, Fernandes E, Lima JLFC, Mira L, Corvo ML (2008) Molecular mechanisms of anti-inflammatory activity mediated by flavonoids. Curr Med Chem 15:1586–1605
Grammatikopoulos G, Kyparissis A, Drilias P, Petropoulou Y, Manetas Y (1998) Effects of UV-B radiation on cuticle thickness and nutritional value of leaves in two mediterranean evergreen sclerophylls. J Plant Physiol 153:506–512
Grant-Petersson J, Renwick JAA (1996) Effects of ultraviolet-B exposure of *Arabidopsis thaliana* on herbivory by two crucifer-feeding insects (Lepidoptera). Environ Entomol 25:135–142
Greenham JR, Grayer RJ, Harborne JB, Reynolds V (2007) Intra- and interspecific variations in vacuolar flavonoids among *Ficus* species from the Budongo Forest, Uganda. Biochem Syst Ecol 35:81–90
Groppa MD, Benavides MP (2008) Polyamines and abiotic stress: recent advances. Amino Acids 34:35–45
Gwynn-Jones D (1999) Enhanced UV-B radiation and herbivory. Ecol Bull 47:77–83

Gwynn-Jones D, Lee J, Callaghan T (1997) Effects of enhanced UV-B radiation and elevated carbon dioxide concentrations on a sub-arctic forest heath ecosystem. Plant Ecol 128:242–249

Hagerman AE, Butler LG (1991) Tannins and lignins. In: Rosenthal GA, Berenbaum MR (eds) Herbivores: their interactions with secondary plant metabolites, the chemical participants, vol 1. Academic Press, San Diego, pp 355–388

Halkier BA, Gershenzon J (2006) Biology and biochemistry of glucosinolates. Annu Rev Plant Biol 57:303–333

Harborne JB (1991) Flavonoid pigments. In: Rosenthal GA, Berenbaum MR (eds) Herbivores: their interactions with secondary plant metabolites. The chemical participants, vol 1. Academic Press, San Diego, pp 389–430

Harborne JB, Williams CA (2000) Advances in flavonoid research since 1992. Phytochemistry 55:481–504

Harley P, Deem G, Flint S, Caldwell M (1996) Effects of growth under elevated UV-B on photosynthesis and isoprene emission in *Quercus gambelii* and *Mucuna pruriens*. Glob Chang Biol 2:149–154

Hartmann T (1991) Alkaloids. In: Rosenthal GA, Berenbaum MR (eds) Herbivores: their interactions with secondary plant metabolites. The chemical participants, vol 1. Academic Press, San Diego, pp 79–121

Hatcher PE, Paul ND (1994) The effect of elevated UV-B radiation on herbivory of pea by *Autographa gamma*. Entomol Exp Appl 71:227–233

Heil M (2008) Indirect defence via tritrophic interactions. New Phytol 178:41–61

Heldt H (1999) Pflanzenbiochemie. Spektrum Akademischer Verlag, Heidelberg, Berlin

Herms DA, Mattson WJ (1992) The dilemma of plants: to grow or defend. Q Rev Biol 67:283–335

Ho SH, Wang J, Sim KY, Ee GCL, Imiyabir Z, Yap KF, Shaari K, Goh SH (2003) Meliternatin: a feeding deterrent and larvicidal polyoxygenated flavone from *Melicope subunifoliolata*. Phytochemistry 62:1121–1124

Holmes MG (1997) Action spectra for UV-B effects on plants: monochromatic approaches for analyzing plant responses. In: Lumsden PJ (eds) Plants and UV-B: responses to environmental change, vol. 1, Society for Experimental Biology, Seminar Series 64. Cambridge University Press, Cambridge, pp 31–50

Hopkins RJ, van Dam NM, van Loon JJA (2009) Role of glucosinolates in insect-plant relationships and multitrophic interactions. Annu Rev Entomol 54:57–83

Huang SB, Dai QJ, Peng SB, Chavez AQ, Miranda LL, Visperas RM, Vergara BS (1997) Influence of supplemental ultraviolet-B on indoleacetic acid and calmodulin in the leaves of rice (*Oryza sativa* L). Plant Growth Regul 21:59–64

Izaguirre MM, Mazza CA, Svatoš A, Baldwin IT, Ballaré CL (2007) Solar ultraviolet-B radiation and insect herbivory trigger partially overlapping phenolic responses in *Nicotiana attenuata* and *Nicotiana longiflora*. Ann Bot 99:103–109

Izaguirre MM, Scopel AL, Baldwin IT, Ballaré CL (2003) Convergent responses to stress. Solar ultraviolet-B radiation and Manduca sexta herbivory elicit overlapping transcriptional responses in field-grown plants of *Nicotiana longiflora*. Plant Physiol 132:1755–1767

Jansen MAK (2002) Ultraviolet-B radiation effects on plants: induction of morphogenic responses. Physiol Plant 116:423–429

Jansen MAK, Elfstrand M, Heggie L, Sitbon F, Dix PJ, Thorneley RNF (2004) Over-expression of phenol-oxidising peroxidases alters the UV-susceptibility of transgenic *Nicotiana tabacum*. New Phytol 163:585–594

Jansen MAK, Gaba V, Greenberg BM (1998) Higher plants and UV-B radiation: balancing damage, repair and acclimation. Trends Plant Sci 3:131–135

Jansen MAK, Hectors K, O'Brien NM, Guisez Y, Potters G (2008) Plant stress and human health: do human consumers benefit from UV-B acclimated crops? Plant Sci 175:449–458

Jansen MAK, Le Martret B, Koornneef M (2010) Variations in constitutive and inducible UV-B tolerance; dissecting photosystem II protection in *Arabidopsis thaliana* accessions. Physiol Plant 138:22–34

Jansen MAK, van den Noort RE, Tan MYA, Prinsen E, Lagrimini LM, Thorneley RNF (2001) Phenol-oxidizing peroxidases contribute to the protection of plants from ultraviolet radiation stress. Plant Physiol 126:1012–1023

Jenkins GI (2009) Signal transduction in responses to UV-B radiation. Annu Rev Plant Biol 60:407–431

Jenkins GI, Brown BA (2007) UV-B perception and signal transduction. In: Whitelam GC, Halliday KJ (eds) Light and plant development, vol 30. Blackwell Publishing, Oxford, pp 155–182

Johnson CB, Kirby J, Naxakis G, Pearson S (1999) Substantial UV-B-mediated induction of essential oils in sweet basil (*Ocimum basilicum* L.). Phytochemistry 51:507–510

Kakani VG, Reddy KR, Zhao D, Sailaja K (2003) Field crop responses to ultraviolet-B radiation: a review. Agric For Meteorol 120:191–218

Kehr J (2006) Phloem sap proteins: their identities and potential roles in the interaction between plants and phloem-feeding insects. J Exp Bot 57:767–774

Kesselmeier J, Staudt M (1999) Biogenic volatile organic compounds (VOC): an overview on emission, physiology and ecology. J Atmos Chem 33:23–88

Kessler A, Baldwin IT (2002) Plant responses to insect herbivory: the emerging molecular analysis. Annu Rev Plant Biol 53:299–328

Kim JH, Mullin CA (2007) An isorhamnetin rhamnoglycoside serves as a costimulant for sugars and amino acids in feeding responses of adult western corn rootworms (*Diabrotica virgifera virgifera*) to corn (*Zea mays*) pollen. J Chem Ecol 33:501–512

Kolb CA, Käser MA, Kopecký J, Zotz G, Riederer M, Pfündel EE (2001) Effects of natural intensities of visible and ultraviolet radiation on epidermal ultraviolet screening and photosynthesis in grape leaves. Plant Physiol 127:863–875

Krizek DT (2004) Influence of PAR and UV-A in determining plant sensitivity and photomorphogenic responses to UV-B radiation. Photochem Photobiol 79:307–315

Kuhlmann F, Müller C (2009a) Development-dependent effects of UV radiation exposure on broccoli plants and interactions with herbivorous insects. Environ Exp Bot 66:61–68

Kuhlmann F, Müller C (2009b) Independent responses to ultraviolet radiation and herbivore attack in broccoli. J Exp Bot 60:3467–3475

Kuhlmann F, Müller C (2010) UV-B impact on aphid performance mediated by plant quality and plant changes induced by aphids. Plant Biol 12:676–684

Kusano T, Berberich T, Tateda C, Takahashi Y (2008) Polyamines: essential factors for growth and survival. Planta 228:367–381

Kuśnierczyk A, Winge P, Jorstad TS, Troczynska J, Rossiter JT, Bones AM (2008) Towards global understanding of plant defence against aphids – timing and dynamics of early *Arabidopsis* defence responses to cabbage aphid (*Brevicoryne brassicae*) attack. Plant Cell Environ 31:1097–1115

Lake JA, Field KJ, Davey MP, Beerling DJ, Lomax BH (2009) Metabolomic and physiological responses reveal multi-phasic acclimation of *Arabidopsis thaliana* to chronic UV radiation. Plant Cell Environ 32:1377–1389

Lamikanra O, Richard OA, Parker A (2002) Ultraviolet induced stress response in fresh cut cantaloupe. Phytochemistry 60:27–32

Landry LG, Chapple CCS, Last RL (1995) *Arabidopsis* mutants lacking phenolic sunscreens exhibit enhanced ultraviolet-B injury and oxidative damage. Plant Physiol 109:1159–1166

Larson RA, Berenbaum MR (1988) Environmental phototoxicity – solar ultraviolet-radiation affects the toxicity of natural and man-made chemicals. Environ Sci Technol 22:354–360

Lavola A (1998) Accumulation of flavonoids and related compounds in birch induced by UV-B irradiance. Tree Physiol 18:53–58

Lavola A, Julkunen-Tiitto R, Roininen H, Aphalo P (1998) Host–plant preference of an insect herbivore mediated by UV-B and CO_2 in relation to plant secondary metabolites. Biochem Syst Ecol 26:1–12

Lindroth RL, Hofman RW, Campbell BD, McNabb WC, Hunt DY (2000) Population differences in *Trifolium repens* L. response to ultraviolet-B radiation: foliar chemistry and consequences for two lepidopteran herbivores. Oecologia 122:20–28

Liu JH, Kitashiba H, Wang J, Ban Y, Moriguchi T (2007) Polyamines and their ability to provide environmental stress tolerance to plants. Plant Biotechnol 24:117–126

Lois R, Buchanan BB (1994) Severe sensitivity to ultraviolet-radiation in an *Arabidopsis* mutant deficient in flavonoid accumulation. 2. Mechanisms of UV-resistance in *Arabidopsis*. Planta 194:504–509

Lütz C, Navakoudis E, Seidlitz HK, Kotzabasis K (2005) Simulated solar irradiation with enhanced UV-B adjust plastid- and thylakoid-associated polyamine changes for UV-B protection. Biochim Biophys Acta Bioenerg 1710:24–33

Mackerness SA-H (2000) Plant responses to ultraviolet-B (UV-B: 280-320 nm) stress: what are the key regulators? Plant Growth Regul 32:27–39

Mackerness SA-H, Surplus SL, Blake P, John CF, Buchanan-Wollaston V, Jordan BR, Thomas B (1999) Ultraviolet-B-induced stress and changes in gene expression in *Arabidopsis thaliana*: role of signalling pathways controlled by jasmonic acid, ethylene and reactive oxygen species. Plant Cell Environ 22:1413–1423

Madronich S, McKenzie RL, Björn LO, Caldwell MM (1998) Changes in biologically active ultraviolet radiation reaching the Earth's surface. J Photochem Photobiol B 46:5–19

Mahdavian K, Ghorbanli M, Kalantari KM (2008) Role of salicylic acid in regulating ultraviolet radiation-induced oxidative stress in pepper leaves. Russ J Plant Physiol 55:560–563

Marak HB, Biere A, van Damme JMM (2002) Two herbivore-deterrent iridoid glycosides reduce the in-vitro growth of a specialist but not of a generalist pathogenic fungus of *Plantago lanceolata* L. Chemoecology 12:185–192

Markham KR, Tanner GJ, Caasi-Lit M, Whitecross MI, Nayudu M, Mitchell KA (1998) Possible protective role for 3′,4′-dihydroxyflavones induced by enhanced UV-B in a UV-tolerant rice cultivar. Phytochemistry 49:1913–1919

Mazza CA, Battista D, Zima AM, Szwarcberg-Bracchitta M, Giordano CV, Acevedo A, Scopel AL, Ballaré CL (1999a) The effects of solar ultraviolet-B radiation on the growth and yield of barley are accompanied by increased DNA damage and antioxidant responses. Plant Cell Environ 22:61–70

Mazza CA, Izaguirre MM, Zavala J, Scopel AL, Ballaré CL (2002) Insect perception of ambient ultraviolet-B radiation. Ecol Lett 5:722–726

Mazza CA, Zavala J, Scopel AL, Ballaré CL (1999b) Perception of solar UVB radiation by phytophagous insects: behavioral responses and ecosystem implications. Proc Natl Acad Sci USA 96:980–985

McCloud ES, Berenbaum M (1994) Stratospheric ozone depletion and plant-insect interactions: effects of UVB radiation on foliage quality of *Citrus jambhiri* for *Trichoplusia ni*. J Chem Ecol 20:525–539

McCloud ES, Berenbaum M (1999) Effects of enhanced UV-B radiation on a weedy forb (*Plantago lanceolata*) and its interactions with a generalist and specialist herbivore. Entomol Exp Appl 93:233–246

McConn M, Creelman RA, Bell E, Mullet JE, Browse J (1997) Jasmonate is essential for insect defense *Arabidopsis*. Proc Natl Acad Sci USA 94:5473–5477

McKenzie RL, Björn LO, Bais A, Ilyas M (2003) Changes in biologically active ultraviolet radiation reaching the Earth's surface. Photochem Photobiol Sci 2:354–354

McLeod A, Rey A, Newsham K, Lewis G, Wolferstan P (2001) Effects of elevated ultraviolet radiation and endophytic fungi on plant growth and insect feeding in *Lolium perenne*, *Festuca rubra*, *F. arundinacea* and *F. pratensis*. J Photochem Photobiol B 62:97–107

Mert-Turk F, Bennett MH, Mansfield JW, Holub EB (2003) Quantification of camalexin in several accessions of *Arabidopsis thaliana* following inductions with *Peronospora parasitica* and UV-B irradiation. Phytoparasitica 31:81–89

Mirecki RM, Teramura AH (1984) Effects of ultraviolet-B irradiance on soybean.5. The dependence of plant-sensitivity on the photosynthetic photon flux-density during and after leaf expansion. Plant Physiol 74:475–480

Müller C (2006) Plant–insect interactions on cuticular surfaces. In: Riederer M, Müller C (eds) Biology of the plant cuticle, vol 1. Blackwell Publishing, Oxford, pp 398–422

Müller C (2008) Resistance at the plant cuticle. In: Schaller A (ed) Induced plant resistance to herbivory, vol. 1, Springer, Berlin, pp 107–129

Müller C, Riederer M (2005) Plant surface properties in chemical ecology. J Chem Ecol 31:2621–2651

Navrot N, Rouhier N, Gelhaye E, Jacquot JP (2007) Reactive oxygen species generation and antioxidant systems in plant mitochondria. Physiol Plant 129:185–195

Newsham K, Greenslade P, McLeod A (1999a) Effects of elevated ultraviolet radiation on Quercus robur and its insect and ectomycorrhizal associates. Glob Chang Biol 5:881–890

Newsham K, McLeod AR, Greenslade P, Emmett B (1996) Appropriate controls in outdoor UV-B supplementation experiments. Glob Chang Biol 2:319–324

Newsham KK, Greenslade PD, Kennedy VH, McLeod AR (1999b) Elevated UV-B radiation incident on *Quercus robur* leaf canopies enhances decomposition of resulting leaf litter in soil. Glob Chang Biol 5:403–409

Normanly J, Bartel B (1999) Redundancy as a way of life – IAA metabolism. Curr Opin Plant Biol 2:207–213

Ohtsuka K, Osakabe M (2009) Deleterious effects of UV-B radiation on herbivorous spider mites: they can avoid it by remaining on lower leaf surfaces. Environ Entomol 38:920–929

Olsson LC, Veit M, Weissenböck G, Bornman JF (1998) Differential flavonoid response to enhanced UV-B radiation in *Brassica napus*. Phytochemistry 49:1021–1028

Oravecz A, Baumann A, Mate Z, Brzezinska A, Molinier J, Oakeley EJ, Adam E, Schäfer E, Nagy F, Ulm R (2006) CONSTITUTIVELY PHOTOMORPHOGENIC1 is required for the UV-B response in *Arabidopsis*. Plant Cell 18:1975–1990

Pandey SP, Baldwin IT (2008) Silencing RNA-directed RNA polymerase 2 increases the susceptibility of *Nicotiana attenuata* to UV in the field and in the glasshouse. Plant J 54:845–862

Paul ND, Gwynn-Jones D (2003) Ecological roles of solar UV radiation: towards an integrated approach. Trends Ecol Evol 18:48–55

Peer WA, Murphy AS (2007) Flavonoids and auxin transport: modulators or regulators? Trends Plant Sci 12:556–563

Pfündel EE, Agati G, Cerovic ZG (2006) Optical properties of plant surfaces. In: Riederer M, Müller C (eds) Biology of the plant cuticle, vol 1. Blackwell Publishing, London, pp 216–249

Piubelli GC, Hoffmann-Campo CB, De Arruda IC, Franchini JC, Lara FM (2003) Flavonoid increase in soybean as a response to *Nezara viridula* injury and its effect on insect-feeding preference. J Chem Ecol 29:1223–1233

Pourcel L, Routaboul JM, Cheynier V, Lepiniec L, Debeaujon I (2007) Flavonoid oxidation in plants: from biochemical properties to physiological functions. Trends Plant Sci 12:29–36

Powell G, Tosh CR, Hardie J (2006) Host plant selection by aphids: behavioral, evolutionary, and applied perspectives. Annu Rev Entomol 51:309–330

Predieri S, Norman HA, Krizek DT, Pillai P, Mirecki RM, Zimmerman RH (1995) Influence of UV-B radiation on membrane lipid-composition and ethylene evolution in 'Doyenne-d'Hiver' pear shoots grown in-vitro under different photosynthetic photon fluxes. Environ Exp Bot 35:151–160

Qaderi MM, Yeung EC, Reid DM (2008) Growth and physiological responses of an invasive alien species, *Silene noctiflora*, during two developmental stages to four levels of ultraviolet-B radiation. Ecoscience 15:150–159

Rajkumar S, Jebanesan A (2008) Bioactivity of flavonoid compounds from *Poncirus trifoliata* L. (Family: Rutaceae) against the dengue vector, *Aedes aegypti* L. (Diptera: Culicidae). Parasitol Res 104:19–25

Rakitin VY, Prudnikova ON, Karyagin VV, Rakitina TY, Vlasov PV, Borisova TA, Novikova GV, Moshkov IE (2008) Ethylene evolution and ABA and polyamine contents in *Arabidopsis thaliana* during UV-B stress. Russ J Plant Physiol 55:321–327
Raviv M, Antignus Y (2004) Invited Review: UV radiation effects on pathogens and insect pests of greenhouse-grown crops. Photochem Photobiol 79:219–226
Reifenrath K, Müller C (2007) Species-specific and leaf-age dependent effects of ultraviolet radiation on two Brassicaceae. Phytochemistry 68:875–885
Reifenrath K, Müller C (2008) Multiple feeding stimulants in *Sinapis alba* for the oligophagous leaf beetle *Phaedon cochleariae*. Chemoecology 18:19–27
Reifenrath K, Müller C (2009) Larval performance of the mustard leaf beetle (*Phaedon cochleariae*, Coleoptera, Chrysomelidae) on white mustard (*Sinapis alba*) and watercress (*Nasturtium officinale*) leaves in dependence of plant exposure to ultraviolet radiation. Environ Pollut 157:2053–2060
Reitz SR, Trumble JT (1996) Tritrophic interactions among linear furanocoumarins, the herbivore *Trichoplusia ni* (Lepidoptera: Noctuidae), and the polyembryonic parasitoid *Copidosoma floridanum* (Hymenoptera: Encyrtidae). Environ Entomol 25:1391–1397
Reitz SR, Trumble JT (1997) Effects of linear furanocoumarins on the herbivore *Spodoptera exigua* and the parasitoid *Archytas marmoratus*: host quality and parasitoid success. Entomol Exp Appl 84:9–16
Renwick JAA, Zhang WQ, Haribal M, Attygalle AB, Lopez KD (2001) Dual chemical barriers protect a plant against different larval stages of an insect. J Chem Ecol 27:1575–1583
Riederer M, Müller C (2006) Biology of the plant cuticle. Annual reviews of plant biology. Blackwell Publishing, Oxford
Rousseaux MC, Ballaré CL, Scopel AL, Searles PS, Caldwell MM (1998) Solar ultraviolet-B radiation affects plant–insect interactions in a natural ecosystem of Tierra del Fuego (southern Argentina). Oecologia 116:528–535
Rousseaux MC, Julkunen-Tiitto R, Searles PS, Scopel AL, Aphalo PJ, Ballaré CL (2004) Solar UV-B radiation affects leaf quality and insect herbivory in the southern beech tree *Nothofagus antarctica*. Oecologia 138:505–512
Rousseaux MC, Scopel AL, Searles PS, Caldwell MM, Sala OE, Ballaré CL (2001) Responses to solar ultraviolet-B radiation in a shrub-dominated natural ecosystem of Tierra del Fuego (southern Argentina). Glob Chang Biol 7:467–478
Rozema J, van de Staaij J, Björn LO, Caldwell M (1997) UV-B as an environmental factor in plant life: stress and regulation. Trends Ecol Evol 12:22–28
Ryan CA (1990) Protease inhibitors in plants – genes for improving defenses against insects and pathogens. Annu Rev Phytopathol 28:425–449
Ryan CA (2000) The systemin signaling pathway: differential activation of plant defensive genes. Biochim Biophys Acta-Protein Struct Mol Enzymol 1477:112–121
Salt DT, Moody SA, Whittaker JB, Paul ND (1998) Effects of enhanced UVB on populations of the phloem feeding insect *Strophingia ericae* (Homoptera: Psylloidea) on heather (*Calluna vulgaris*). Glob Chang Biol 4:91–96
Sangtarash MH, Qaderi MM, Chinnappa CC, Reid DM (2009) Differential sensitivity of canola (*Brassica napus*) seedlings to ultraviolet-B radiation, water stress and abscisic acid. Environ Exp Bot 66:212–219
Schanz F (1920) Versuche über die Wirkung ultravioletter Strahlen des Tageslichtes auf die Vegetation. Pfluger's Arch Gesamte Physiol Mensch Tiere 181:229–284
Schmelz EA, Engelberth J, Alborn HT, O'Donnell P, Sammons M, Toshima H, Tumlinson JH (2003) Simultaneous analysis of phytohormones, phytotoxins, and volatile organic compounds in plants. Proc Natl Acad Sci USA 100:10552–10557
Schmitz-Hoerner R, Weissenböck G (2003) Contribution of phenolic compounds to the UV-B screening capacity of developing barley primary leaves in relation to DNA damage and repair under elevated UV-B levels. Phytochemistry 64:243–255

Seckmeyer G, Payer HD (1993) A new sunlight simulator for ecological research on plants. J Photochem Photobiol B-Biol 21:175–181

Sembdner G, Parthier B (1993) The biochemistry and the physiological and molecular actions of jasmonates. Annu Rev Plant Physiol Plant Mol Biol 44:569–589

Semerdjieva SI, Phoenix GK, Hares D, Gwynn-Jones D, Callaghan TV, Sheffield E (2003) Surface morphology, leaf and cuticle thickness of four dwarf shrubs from a sub-Arctic heath following long-term exposure to enhanced levels of UV-B. Physiol Plant 117:289–294

Steinmüller D, Tevini M (1985) Action of ultraviolet radiation (UV-B) upon cuticular waxes in some crop plants. Planta 164:557–564

Stevenson PC, Muyinza H, Hall DR, Porter EA, Farman DI, Talwana H, Mwanga ROM (2009) Chemical basis for resistance in sweetpotato Ipomoea batatas to the sweetpotato weevil *Cylas puncticollis*. Pure Appl Chem 81:141–151

Stratmann JW, Stelmach BA, Weller EW, Ryan CA (2000) UVB/UVA radiation activates a 48 kDa myelin basic protein kinase and potentiates wound signaling in tomato leaves. Photochem Photobiol 71:116–123

Summers CB, Felton GW (1994) Prooxidant effects of phenolic acids on the generalist herbivore *Helicoverpa zea* (Lepidoptera, Noctuidae) – potential mode of action for phenolic compounds in plant anti-herbivore chemistry. Insect Biochem Mol Biol 24:943–953

Surplus SL, Jordan BR, Murphy AM, Carr JP, Thomas B, Mackerness SA-H (1998) Ultraviolet-B-induced responses in *Arabidopsis thaliana*: role of salicylic acid and reactive oxygen species in the regulation of transcripts encoding photosynthetic and acidic pathogenesis-related proteins. Plant Cell Environ 21:685–694

Tegelberg R, Aphalo PJ, Julkunen-Tiitto R (2002) Effects of long-term, elevated ultraviolet-B radiation on phytochemicals in the bark of silver birch (*Betula pendula*). Tree Physiol 22:1257–1263

Tegelberg R, Julkunen-Tiitto R (2001) Quantitative changes in secondary metabolites of dark-leaved willow (*Salix myrsinifolia*) exposed to enhanced ultraviolet-B radiation. Physiol Plant 113:541–547

Tegelberg R, Julkunen-Tiitto R, Vartiainen M, Paunonen R, Rousi M, Kellomaki S (2008) Exposures to elevated CO_2, elevated temperature and enhanced UV-B radiation modify activities of polyphenol oxidase and guaiacol peroxidase and concentrations of chlorophylls, polyamines and soluble proteins in the leaves of *Betula pendula* seedlings. Environ Exp Bot 62:308–315

Tegelberg R, Turtola S, Rousi M, Meier B, Julkunen-Tiitto R (2006) Soluble polyamines in *Salix myrsinifolia* and *S. myrsinites x S. myrsinifolia* plantlets exposed to increased UV-B irradiation and decreased watering. Trees Struct Funct 20:299–303

Teramura AH (1983) Effects of ultraviolet-B radiation on the growth and yield of crop plants. Physiol Plant 58:415–427

Tevini M, Steinmüller D (1987) Influence of light, UV-B radiation, and herbicides on wax biosynthesis of cucumber seedlings. J Plant Physiol 131:111–121

Textor S, Gershenzon J (2009) Herbivore induction of the glucosinolates-myrosinase defense system: major trends, biochemical basis and ecological significance. Phytochem Rev 8:149–170

Thompson GA, Goggin FL (2006) Transcriptomics and functional genomics of plant defence induction by phloem-feeding insects. J Exp Bot 57:755–766

Tiiva P, Rinnan R, Faubert P, Rasanen J, Holopainen T, Kyro E, Holopainen JK (2007) Isoprene emission from a subarctic peatland under enhanced UV-B radiation. New Phytol 176:346–355

Turtola S, SallaS L, Holopainen JK, Julkunen-Tiitto R, Kainulainen P (2006) Long-term exposure to enhanced UV-B radiation has no significant effects on growth or secondary compounds of outdoor-grown Scots pine and Norway spruce seedlings. Environ Pollut 144:166–171

Ulm R, Nagy F (2005) Signalling and gene regulation in response to ultraviolet light. Curr Opin Plant Biol 8:477–482

Ulmann S (2007). Einfluss von UV-B auf den monophagen Blattkäfer *Cassida canaliculata* und seine Wirtspflanze *Salvia pratensis*. Diploma Thesis. Universität Würzburg, pp 86

Unsicker SB, Kunert G, Gershenzon J (2009) Protective perfumes: the role of vegetative volatiles in plant defense against herbivores. Curr Opin Plant Biol 12:479–485

van Loon JJA, Wang CZ, Nielsen JK, Gols R, Qiu YT (2002) Flavonoids from cabbage are feeding stimulants for diamondback moth larvae additional to glucosinolates: chemoreception and behaviour. Entomol Exp Appl 104:27–34

Veteli TO, Tegelberg R, Pusenius J, Sipura M, Julkunen-Tiitto R, Aphalo PJ, Tahvanainen J (2003) Interactions between willows and insect herbivores under enhanced ultraviolet-B radiation. Oecologia 137:312–320

Vickers CE, Gershenzon J, Lerdau MT, Loreto F (2009) A unified mechanism of action for volatile isoprenoids in plant abiotic stress. Nat Chem Biol 5:283–291

Vlot AC, Dempsey DMA, Klessig DF (2009) Salicylic acid, a multifaceted hormone to combat disease. Annu Rev Phytopathol 47:177–206

Walling LL (2000) The myriad plant responses to herbivores. J Plant Growth Regul 19:195–216

Wang KLC, Li H, Ecker JR (2002) Ethylene biosynthesis and signaling networks. Plant Cell 14: S131–S151

Wang YB, Feng HY, Qu Y, Cheng JQ, Zhao ZG, Zhang MX, Wang XL, An LZ (2006) The relationship between reactive oxygen species and nitric oxide in ultraviolet-B-induced ethylene production in leaves of maize seedlings. Environ Exp Bot 57:51–61

Wargent JJ, Gegas VC, Jenkins GI, Doonan JH, Paul ND (2009) UVR8 in *Arabidopsis thaliana* regulates multiple aspects of cellular differentiation during leaf development in response to ultraviolet B radiation. New Phytol 183:315–326

Warren JM, Bassman JH, Eigenbrode S (2002) Leaf chemical changes induced in *Populus trichocarpa* by enhanced UV-B radiation and concomitant effects on herbivory by *Chrysomela scripta* (Coleoptera: Chrysomelidae). Tree Physiol 22:1137–1146

Winter TR, Rostás M (2008) Ambient ultraviolet radiation induces protective responses in soybean but does not attenuate indirect defense. Environ Pollut 155:290–297

Wittstock U, Halkier BA (2002) Glucosinolate research in the *Arabidopsis* era. Trends Plant Sci 7:263–270

Wolters H, Jürgens G (2009) Survival of the flexible: hormonal growth control and adaptation in plant development. Nat Rev Genet 10:305–317

Wu JQ, Baldwin IT (2009) Herbivory-induced signalling in plants: perception and action. Plant Cell Environ 32:1161–1174

Yang H, Zhao ZG, Qiang WY, An LZ, Xu SJ, Wang XL (2004) Effects of enhanced UV-B radiation on the hormonal content of vegetative and reproductive tissues of two tomato cultivars and their relationships with reproductive characteristics. Plant Growth Regul 43:251–258

Yu O, Jez JM (2008) Nature's assembly line: biosynthesis of simple phenylpropanoids and polyketides. Plant J 54:750–762

Zangerl AR, Berenbaum MR (1987) Furanocoumarins in wild parsnip – effects of photosynthetically active radiation, ultraviolet-light, and nutrients. Ecology 68:516–520

Zangerl AR, Berenbaum MR (2003) Phenotype matching in wild parsnip and parsnip webworms: causes and consequences. Evolution 57:806–815

Zavala JA, Ravetta DA (2002) The effect of solar UV-B radiation on terpenes and biomass production in *Grindelia chiloensis* (Asteraceae), a woody perennial of Patagonia, Argentina. Plant Ecol 161:185–191

Zavala JA, Scopel AL, Ballaré CL (2001) Effects of ambient UV-B radiation on soybean crops: impact on leaf herbivory by *Anticarsia gemmatalis*. Plant Ecol 156:121–130

Zhang WJ, Björn LO (2009) The effect of ultraviolet radiation on the accumulation of medicinal compounds in plants. Fitoterapia 80:207–218

Zhu-Salzman K, Zeng RS (2008) Molecular mechanisms of insect adaptation to plant defense: lessons learned from a Bruchid beetle. Insect Sci 15:477–481

Space as a Resource

Thorsten E.E. Grams and Ulrich Lüttge

Contents

Abstract In this assay, we aim at defending the hypothesis that physically spoken sheer "empty" space in itself without any other commodities has the function of resource for plant life. Definitions of space, niche, and resource are examined. We consider competition for resources and space occupation and exploitation above ground, where light is often a decisive limiting factor. Steady-state and dynamic situations are assessed with respect to cost–benefit relations and using the examples of epiphytic niches and gaps of tropical rain forests. In a similar vein, belowground

T.E.E. Grams (✉)
Ecophysiology of Plants, Department of Ecology and Ecosystem Management, Technische Universität München, Am Hochanger 13, 85354 Freising, Germany
e-mail: grams@tum.de

U. Lüttge
Institut für Botanik, Technical University of Darmstadt, Schnittspahnstrasse 3-5, 64287 Darmstadt, Germany
e-mail: luettge@bio.tu-darmstadt.de

U. Lüttge et al. (eds.), *Progress in Botany 72*,
DOI 10.1007/978-3-642-13145-5_13,

relations are evaluated with the morphological and physiological responses of roots, the role of different types of mycorrhiza characterized by fungal occupation of soil–host rhizospheres, and the dynamic recognition of self and foreign by roots in the soil. Sharing of space is exemplified by describing vegetation islands on sand plains and epiphyte nests in tree canopies, where facilitation is distinguished from competition. Another example for natural facilitation is hydraulic redistribution. Applied facilitation with anthropogenic manipulation is illustrated by nurse effects in tree plantations. The new concept of "empty" space that can become a limiting resource, i.e., the central hypothesis of the essay, is supported by considering the four dimensions of space, namely the three dimensions of extension in length plus the dimension of time. However, in the simplest case of the one dimension of space provided for atmospheric bromeliads by a telephone-line wire devoid of any other commodities, we may recognize the most straightforward compelling argument for the conclusion we want to promote in this assay, namely, that sheer space in itself has the function of resource.

1 Introduction: Space–Niche–Resource

1.1 "Empty" Space

Defining space, the Oxford Encyclopedic Dictionary (1991) has: "*a continuous unlimited area or expanse which may or may not contain objects*" and also "*an empty area.*" We would rather take it as a volume than an area, but the major question arising is what we may understand as "empty." In the purist's view, it would mean a complete vacuum. Physically this may not really exist at all, not even in outer space of our universe if we consider its matter and irradiance and the particle/wave dualism. Thus, here we also arrive at metaphysical dimensions. Nevertheless, in fact, asking in this essay the question if space as such and in itself, i.e., without any objects that it may contain, can be considered as resource for life we theoretically mean space in the purist way. However, as we see, practically this is not feasible, and indeed, dealing with life it is appropriate to return to more earthly dimensions. Even the air is not an empty space. It contains objects, gasses in this case, where carbon dioxide certainly is a basic material substrate and resource for the life of plants. This implies that when talking about "empty," we must use it in parentheses.

Hence, what do we mean by space in the context of our following considerations? The air as space is not available for plants which cannot fly. However, it is a prerequisite for terrestrial life to get established in it. Something needs to provide access to it. Metaphorically, we may recall that one speaks of the works of architects as creating space. Thus, the wall of a concrete building or the sand of a desert on which biofilms can develop or a telephone wire on which atmospheric bromeliads can get established but which does not provide anything else but access, metaphorically can be considered as space and as such also as a resource.

1.2 Space and Niche

While considering space in a rather abstract way as something that can be occupied but does not provide material resources that can be utilized, it must be clear that we do not refer to ecological niche here. Traditionally, niche is not considered purely as space but implies functional aspects. Extended concepts of species niches have two main components: (1) species requirements for survival in a given environment, and (2) the impacts of species on their environment (Chase and Leibold 2003). Then, evidently the "empty" environment where a species may get established is not a niche. Clearly, all life is always affecting its own environment. It will create niches after arrival in "empty" space. Hence, the use of a niche is not what we mean here by the use of "empty" space as a resource. Our new concept of space is more abstract than the much broader and concrete concept of ecological niches. On the other hand, space is, or becomes, part of a niche, and therefore, it is unavoidable in the following to touch the niche concept when we exemplify, but it would distract from our intention if we were merging with it.

1.3 Space and Resource

For resource, the Oxford Encyclopedic Dictionary (1991) as the first entry has: "*an expedient or device*," and it gives an example: "*escape was their only resource*." We may then ask the question of escape where to? This immediately generates the connection with the idea of space. For space, the same dictionary has: "*a continuous unlimited area or expanse which may or may not contain objects*." Hence, evidently space is a resource where organisms can escape to or which they can conquer, which they can occupy, which they must defend, from which they can set out on exploration or new escape. When space contains objects, these may function as "means" or "supply," which are other definitions the dictionary has for resource. Thus, these objects, which space may contain, make up the quality or value of space. Regarding plant life on earth, this elicits a plethora of associations, such as propagation, competition, nurse effects, or their opposite defense. In terms of Darwinian evolution, escape as a resource implies adaptive radiation and this directly connects to the ecological niche concept that embraces function in space, and hence also has temporal aspects. This means that like the theoretical physicists, we need to consider space with four dimensions: the three dimensions of extension in lengths plus the dimension of time.

Actual observation teaches that on earth, life can occupy almost any type of space adapting to even quite extreme conditions. Microbiotic crusts, i.e., biofilms and soil crusts, consisting of bacteria, fungi, algae, lichens, and bryophytes occupy almost any surface available (Belnap and Lange 2003). Use of the resource space by such crusts has important global implications. Global annual net uptake of carbon by microbiotic crusts is ca. 3.6 Pg a^{-1}, which corresponds to ca. 6% of the estimated global net carbon uptake by terrestrial vegetation, and the estimated rate of nitrogen fixation by microbiotic crusts is ca. 45 Tg a^{-1} or 40% of the global estimate of biological nitrogen fixation (Elbert et al. 2009). Examples of extreme surfaces colonized by biotic crusts are those of sand dunes, and the bare ground of deserts (Belnap and Lange 2003) and biofilms often dominated by cyanobacteria are found on the highly sun-exposed rocks of tropical inselbergs (Lüttge 1997, 2008) as well as on concrete buildings (Lüttge 1997). This certainly nourishes the expectation that there may be also extraterrestrial life somewhere in the universe, i.e., in outer "space" (Morris 2003).

1.4 Is Space in Itself a Resource?

Thus, in this review, we ask the question "is space in itself a resource." We shall explore this not so much from a theoretical perspective but from mainly empirical observations with eukaryotic organisms, i.e., higher plants and fungi involved in mycorrhizas. The dynamics of occupation of space can be seen almost everywhere in nature if one looks at it from the perspective of whether space is a resource. Life first evolved in the water, but then it conquered the land with remarkable evolutionary selection and adaptations driven by the extreme stress of changing the aquatic for the terrestrial environment. Much more recent famous examples are the invasion of newly formed maritime islands after volcanic events, such as Krakatau or the Galápagos-Islands. Hence, any choice of examples for the role and function of space may appear arbitrary. Here we ask if space as such is a resource. When we discuss this, we must bear in mind that it may only become a resource by objects it contains. This comprises the question addressed above (Sect. 1.1) if space is ever empty? Examples we chose from our own observations and research experience to examine this shall focus on space above and below ground. This comprises facilitation and competition in canopy-dominated habitats including epiphytes and nurse forests, dynamics of vegetation and vegetation islands in occupation of open space and in space belowground, the rhizosphere with allelopathy and mycorrhiza.

2 Competition for Resources: Space Occupation and Exploitation

2.1 Aboveground Competition for Light

2.1.1 Steady-State Situation

Cost–Benefit Relations

Light is often the limiting resource that plants are competing for, and successful competition for light determines their growth success. This is most obvious in the vertical structure of strata of forests especially in the rich vegetation of rain forests, where crowns compete for space and light (see below: Küppers (1989). Besides gradients of temperature and air humidity, carbon dioxide and mineral nutrients in the vertical space of such forests' gradients of irradiance constitute a dominating factor (Lüttge 2008). The competitive territoriality of tree crowns demonstrates that space can be a limiting resource. This can be quantified by modeling of allometric distribution of material. In allometry, during growth the linear dimensions, such as crown diameter and length, are changing nonproportionally to each other as compared to isometry, where all linear dimensions change proportionally (Pretzsch 2010). In a most evocative way, the results of such competition for occupying space are illustrated by vertical fish-eye camera views from the forest floor to the canopy (Fig. 3.27 in Lüttge (2008), frontispiece of current issues of the journal *Trees: Structure and Function*). An understanding of space filling and area preservation based on the general structural principles of growth and form is achieved (Enquist and Niklas 2001; West et al. 2001, 2009). It appears that in competitive interactions, allometric plasticity keeps the spatial scaling of growing volume constant at least under moderate competitive pressure (H. Pretzsch, Forest Yield Science, Technische Universität München, personal communication). An appealing example of such occupation of space is also given by phyllotaxis where both empirical (Pearcy and Yang 1996, 1998) and purely mathematically analytical approaches (King et al. 2004) result in the conclusion that the golden angle is the best solution for optimal capture of light.

As compared to multidirectional resources in vertical space, light is a mostly unidirectional resource because the proportion of diffuse irradiance is typically low. Unidirectional resources are subjected to size-asymmetrical competition, where resource capture is disproportionate to size, i.e., the biomass involved in resource capturing (Schwinning and Weiner 1998). In other words, unidirectional resources are "pre-emptable" and allow for shading. Therefore, even small biomass such as a small number of leaves may have a large shading effect on leaves of a competitor. Hence, the amount of resources captured becomes disproportionate to the involved biomass. In this situation, the spatial arrangement of biomass (e.g., leaves) appears to be of higher importance than its physiological performance (e.g., rate of photosynthesis) (Anten and Hirose 2001; Gayler et al. 2006; Grams and Andersen 2007), stressing the importance of space occupation versus maximizing the resource capture by space exploitation. Accordingly, in a seedling experiment with five tropical tree species, Reekie and Bazzaz (1989) concluded that canopy height and architecture were the major factors determining their competitive ability. Likewise, Küppers and List (1997) demonstrated in a simulation study that even a ten times higher photosynthetic activity does not counterbalance the architectural disadvantage of blackthorn (*Prunus spinosa*) when growing in competition with hedge maple (*Acer campestre*).

As suggested by Küppers (1989), analysis of aboveground competition should focus on "*an integrated view of carbon gain, increment of biomass and its architectural arrangement in space* (...), *especially in situations where crowns compete for space and light.*" In other words, integration of the spatial arrangement of biomass with the flux of resources along this structure (e.g., C gain) may facilitate mechanistic understanding of plant competitiveness (Schulze et al. 1986; Tremmel and Bazzaz 1995). This concept was tested in a series of growth chamber experiments with juvenile trees of European beech (*Fagus sylvatica*) and Norway spruce (*Picea abies*) grown in dense canopies as either mono-or mixed cultures (Grams et al. 2002; Kozovits et al. 2005a, b; Luedemann et al. 2005, 2009). The competitive success of the juvenile trees was determined by the competition for canopy space. In particular, occupation of canopy volume at low biomass investment (i.e., efficiency of space occupation) was of crucial importance and related to the higher biomass increment of spruce over beech in mixed cultures (Grams et al. 2002, 2007; Kozovits et al. 2005a, b; Matyssek et al. 2005). The authors concluded that this type of cost/benefit analysis is adequate for quantitatively comparing competitiveness across species and growth conditions. More recently, it was suggested to integrate this space-related concept of competitiveness with the time-integrative power of stable isotopes, such as ^{13}C, ^{15}N, and ^{18}O, to increase the potential for clarifying mechanisms of competition (Grams and Matyssek 2010).

Optimization of structural biomass investment for space occupation appears to be of general importance in aboveground competition for light as filling of canopy space is reached by plants with surprisingly high efficiency, i.e., low biomass investment per crown volume. Across ecosystems and life forms, canopy biomass densities (e.g., biomass produced per canopy volume) is surprisingly low as less than 3% of the canopy space is occupied by plant material, such as branches and leaves (Chiarucci et al. 2002; Kozovits et al. 2005b; Reiter et al. 2005 and citations therein).

Cost-Efficient Strategies in Competition for Space: Epiphytes, Lianas, and Stranglers

In Sect. 1.3, we have alluded to the ubiquitous occupation of any surface available by biotic crusts and biofilms exemplifying the resource nature of space. Perhaps, epiphytism is one of the best examples supporting our hypothesis that space is a resource. Epiphytism is primarily the conquest of space (Lüttge 1989b). Support is needed on which plants can grow and within vegetation this is given by other plants, i.e., phorophytes as we name plants carrying other plants as epiphytes. Epiphytism is seen in fresh water and marine aquatic habitats, where microscopic as well as macroscopic algae grow on each other. Vascular plants as epiphytes constitute a particularly rich biodiversity in tropical forests (Lüttge 1989a; Benzing 1990). The dense occupation of some phorophyte habitats by epiphytes vividly illustrates the nature of space as a limiting resource.

Observing epiphytes and among them particularly bromeliads, Schimper (1888, 1898) has proposed the idea that the evolution of epiphytism occurred in a struggle for light. However, specifically to the Bromeliaceae, Schimper's hypothesis does not apply. In this family, epiphytism evolved several times polyphyletically and separately from an ancestral group. The putative ancestor must have been already adapted to exposed habitats (Lüttge 1985; Lüttge et al. 1986; Smith 1989), so that in the Bromeliaceae, the evolution of epiphytism appears as a struggle for space rather than for light, although in other groups of plants, Schimper's hypothesis may well apply (Lüttge 1985; Lüttge et al. 1986; Smith 1989). The most extremely adapted epiphytic bromeliads even occupy the wires of fences or telephone lines (Lüttge 1989b; Benzing 1989, 1990). They are called atmospheric bromeliads as they receive all their resources, such as CO_2 and water with nutrients dissolved or in dust, from the air, i.e., the "objects" in the space conquered. These atmospheric bromeliads demonstrate that a vegetation environment with living phorophytes is not exclusively essential for "epiphytism."

This has raised the question of whether epiphytes are parasites. There are no direct metabolic interactions between epiphytes and phorophytes, which are normally characteristic of parasitism. So, these are not a basic requirement of epiphytism. However, epiphytism can be considered as parasitism for space. Epiphytes can severely injure their phorophytes when tree branches break down under their heavy load. Epiphytes may overgrow their phorophytes to an extent that the latter are deprived of light for photosynthesis. Some hemi-epiphytes, e.g., species of the genera *Ficus* and *Clusia*, may first establish themselves epiphytically on host trees and then develop positive gravitropic roots to get contact with the ground and nongravitropic roots that strangle the host. The nongravitropic roots are girdling the bark of the host, so that eventually it is killed, and the original hemi-epiphyte on the pseudo-stems of its aerial roots develops to an independent tree. One would better call these plants murderers rather than parasites (Lüttge 1989b).

It is most drastically seen with the atmospheric bromeliads but pertinent to epiphytism in general that the occupation of space offered by phorophytes brings about a number of important problems of the acquisition of resources necessary for plant life (Lüttge 1989a, 2008; Benzing 1990, 2000). Evolution of epiphytism selected several solutions to these problems, i.e., adaptations evidently essential for the conquest of phorophyte space. One of the problems is light when the epiphytes do not occupy the outer canopy of trees, which is often the case (Johansson 1975). However, the major problem of epiphytes and hemi-epiphytes, which have no root–soil contact, is acquisition of water and the mineral nutrients dissolved (Zotz et al. 2001). Epiphytes have evolved morphotypic traits providing adaptation, such as phytothelmata especially the tanks of the bromeliads with the absorptive trichromes on the leaves (Smith 1989; Benzing 2000) or the absorptive *velamen radicum* of the aerial roots of epiphytic orchids (Goh and Kluge 1989), leaves and pseudo-bulbs often form water storage tissues (hydrenchymas). Formation of epiphyte nests with ants, and mycorrhiza (Sect. 3.2.3) is supported by morphological adaptations. A physiotypic adaptation is that very many epiphytes have polyphyletically evolved crassulacean acid metabolism (CAM) as an effective water-conserving mode of photosynthesis (Griffiths 1989). Fifty-seven percent of all vascular epiphyte species are CAM plants (Lüttge 2008). These observations underline the specific space–resource relations of the epiphytic habitat.

2.1.2 Dynamic Situation: The Dimension of Time

The dynamic situation is a spatiotemporal phenomenon, where the fourth dimension, time, is entering the game. Let us consider the simplest case of two competitors. If they compete primarily for space where they can get established and not for other types of resources, we may expect time to be the most important dimension. An example is given by the clonal growth of two species of bromeliads, *Nidularium innocentii* and *Nidularium procerum*, in a Brazilian swamp forest (Scarano et al. 1997, 1999; Freitas et al. 2003). The two species occupy distinct, monospecific stands in directly adjacent patches of the forest. *N. innocentii* is a C_3-plant and dominates in deep shade in temporarily flooded sections of the forest. *N. procerum* is a CAM-plant dominating permanently flooded areas under open canopies. However, studies of their ecophysiological performance show that in principle both species can adapt to the flooding and light regimes of both of the different patches in the forest. The essential observation then is that where the clones of the two species meet in a narrow boundary zone between their patches each species can occupy the seemingly preferred habitat of the other species. Thus, it appears that ecophysiological performance does not explain the segregation of the two species in the swamp forest but that logically the timing of arrival and establishment in space and subsequent exclusion of the competitor from the limiting resource of space were decisive.

In systems with many more species involved, it is much more complex when we follow the regrowth of vegetation in gaps of tropical rain forests. Clearings may be created in tropical rainforests naturally by falling trees, hurricanes, earthquakes, volcanic eruptions, fires, landslides, or anthropogenic in relation to farming. Giant falling trees create two adjacent gaps, i.e., the space where the crown once was established and the space where it was falling down to which is also named epicenter. By the formation of clearings, often also called chablis, the space is modified with respect to its properties.

Reinvasion of the space after destruction occurs in different phases (Jacobs 1988). First, regrowth is started by pioneer plants that are characterized by having a higher light demand than species of the closed forest. Second, a fierce competition is arising between these plants mainly for light and nutrients. Third, a homeostatic phase is attained. However, this is a dynamic equilibrium not stable continuously, and fourth, another phase of fierce competition follows where light-demanding pioneer species die, and then fifth, a new homeostatic phase is attained with a closing rain forest canopy. With the high floristic diversity given in tropical forests, the new homeostatic forest may be quite different from the one occupying the space before creation of the chablis. In reinvasion, "the selection is of unpredictable irregularity" (Jacobs 1988). Factors determining reinvasion are the diversity and age of the adjacent vegetation, the light climate, the availability and viability of propagules, the seed and seedling bank composition, factors of dispersal, differences in growth and mortality rates among species with different patterns of allocation of carbon and nutrients, edaphic specialization, and soil resources (Torquebiau 1988; Newell et al. 1993; Dalling et al. 1998; Phillips et al. 2003; Palmitto et al. 2004; Valencia et al. 2004; Massey et al. 2006; Kuptz et al. 2010).

This led Remmert (1985, 1991) to challenging the climax theory, where climax associations are thought to be homeostatic ecological equilibria governed by the natural environmental factors determined by the zonal climate. In contrast, he considered tropical forests with their many chablis in different stages of formation and reinvasion as cyclically changing mosaic patterns.

These dynamics of gaps in tropical rainforests (Orians 1982; van der Meer and Bongers 1996) give an excellent example of the role of time as the fourth dimension of the four-dimensional space. With this consideration of space as a dynamic resource, we advance to considering the nonlinear dynamics in ecology. This is often misunderstood as being stochastic. However, it is governed by rules, such as deterministic chaos, which has early been recognized to be a fundamental attribute of population ecology (May 1976; Hastings et al. 1993), self-organization, self-similarity, and fractals (Solé et al. 1994; Solé and Manrubia 1995a, b; Manrubia and Solé 1996; Levin and Muller-Landau 2001; Lüttge 2008).

2.2 Belowground Competition for Water and Nutrients

In contrast to the resource light, supply with belowground resources such as water and nitrate are often multidirectional as they move through the soil by mass flow or diffusion. Accordingly, competition for belowground resources was generally believed to be size-symmetric, i.e., resource gain was expected to be proportional to the biomass involved, because shading effects are less frequent as resources are not "pre-emptable" (Casper and Jackson 1997; Cahill and Casper 2000; Bartelheimer et al. 2008). In this situation, morphological plasticity in space occupation would be less important than physiological plasticity in resource uptake (i.e., space exploitation, Grams and Andersen 2007). However, belowground competition for patchy resources, as often found in soils, and for immobile nutrients such as phosphate are discussed to be size-asymmetric (Fransen et al. 2001; Hodge 2006). Accordingly, efficient space occupation may be an advantageous strategy in competition for these resources similar to aboveground competition for light. In fact, spatial segregation of roots has frequently been observed for both individual roots and whole root systems, and there is evidence that allelopathy (i.e., chemical inhibition) and nontoxic signals contribute to this segregation. Thus, the question arises if roots are defending the occupied belowground space or as asked by Schenk et al. (1999): "*Are roots territorial*?"

Roots display a plethora of morphological and physiological responses to water and nutrient availabilities, and their behavior appears to be more sophisticated than previously thought (Schenk 2006; Hodge 2009; Novoplansky 2009). For example, novel genes, *miz1* and *miz2*, have been discovered in *Arabidopsis thaliana* that are involved in hydrotropismic responses of roots to soil moisture (Kobayashi et al. 2007; Miyazawa et al. 2009). Effects of nutrient concentrations and presence of competitors on root development have been studied in *Pisum sativum* (pea) plants grown with their roots split between two pots. The so-called "fence-sitter" plants allocated their root biomass in response to nutrient availabilities in the two pots. Introduction of a competing plant in one of the pots resulted in competition avoidance by the fence-sitter plants as they moved their roots away from the pots with competitors and resource depletion (Gersani et al. 1998). When the nutrient concentrations in pots with one competitor were doubled, fence-sitters again distributed their roots evenly between the two pots, stressing the importance of nutrient availability for root growth responses.

Another interesting feature of roots is the circumventing of obstacles in the soils, which is known for a long time and represents an important feature of roots for effective occupation of the soil space (Darwin and Darwin 1896; Schenk 2006). Before they even make contact with an inert object, roots are able to adjust their growth in response to it. For example, Falik et al. (2005) grew *P. sativum* plants in the presence of a nylon string on one side of the root system. While the total number of roots remained unchanged, they were shortened by about 30% in response to the nylon string, even before roots made physical contact with it. The authors hypothesized that this behavior is based on the sensitivity of roots to an unknown compound in their own exudates that accumulates in the vicinity of the physical obstacle. Accordingly, root growth recovered by the addition of activated carbon powder that decomposed root-released compounds (Falik et al. 2005).

Roots not only respond to inert objects in the soil and resource-depletion zones but also behave differently when encountering their own versus foreign roots. Thus, plants are capable to distinguish their own roots from those of a competitor, even among genetically identical individuals. For example, roots of buffalo grass (*Buchloe dactyloides*) grow less in the presence of their own roots. However, genetically identical plant cuttings from the same individual became increasingly alienated, i.e., they did not recognize each other, and with time, they eventually acted similar to genetically different individuals. This suggested a physiological recognition mechanism (Gruntman and Novoplansky 2004). Such self/non-self-recognition of roots intuitively makes sense for successful space occupation of soils as it reduces crowding of own roots and increases the plant's competitiveness at lower cost-benefit ratios for soil occupation (Grams et al. 2002; Kozovits et al. 2005b; Schenk 2006). However, the precise mechanisms for the self/non-self-root discrimination are still obscure (Novoplansky 2009).

If soil space in itself is a resource for plants, it appears likely that roots exudate allelopathic chemicals to generate territoriality in the soil (Schenk et al. 1999). Evidence for direct competitive interference by allelopathy seems convincing (Schenk et al. 1999; Hierro and Callaway 2003). For example, allelochemicals exuded by roots of the desert shrub *Larrea tridentata* have been reported to inhibit not only growth of its own roots but also those of a competing shrub species (Mahall and Callaway 1992). More recently, an allelochemical (i.e., (-)-catechin) exudated by the invasive exotic species *Centaurea maculosa* has been shown to inhibit the growth of native species under field conditions (Bais et al. 2003, 2004). Thus, allelochemicals may be involved in the invasive success of exotic species because the native vegetation did not have the time to evolve defensive mechanisms against those chemicals (Schenk 2006).

Following the dictum of Begon et al. (1990) that "*most higher plants do not have roots, they have mycorrhizae*," mycorrhizal associations must not be ignored when discussing belowground space occupation and exploitation. One of the most important and obvious effects of mycorrhizae in belowground competition for space is their ability to increase the total length of the nutrient and water-absorbing structure. In the case of vesicular-arbuscular mycorrhizae (VAM), the length of the absorbing structure (i.e., sum of roots and extraradical hyphae) is increased by 2–3 orders of magnitude compared to the root length alone (Smith and Gianinazzi-Pearson 1988). Accordingly, the capacity for exploring the soil space is considerably enhanced. A similar effect on soil occupation is observed for root associations with ectomycorrhizal fungi (ECM). Agerer (2001) described a suit of distinct types of ECM, classified by their morphology and tendency to develop different amounts of emanating hyphae and by the presence of rhizomorphs. The so-called "exploration types" of ECM vary widely in their spatial arrangement in the soil, space occupation potential, and ecological importance (Agerer 2001, 2007, 2009). For example, ECM fungi with exploration types of low biomass, such as the contact type, will have only small amounts of extraradical hyphae occupying on average only 100 mm^2 of soil space. Conversely, the medium and long distance types of ECM occupy on average 210–1,710 mm^2 (Agerer and Raidl 2004; R. Agerer and S. Raidl, Botanical Institute of the University of Munich, personal communication). Interestingly, the exploration types of ECM fungi are reflected in their nitrogen isotope composition ($\delta^{15}N$). Fungi with exploration types of high biomass (i.e., high proportion of extraradical hyphae) are about 5‰ enriched in ^{15}N compared to fungi with exploration types of low biomass, indicating $\delta^{15}N$ to provide insights into belowground functioning and biomass of ECM fungi (Hobbie and Agerer 2010).

3 Sharing the Space: Facilitation

3.1 Importance of Facilitation Relative to Competition

Facilitation is a spatiotemporal process where our concept of "empty" space tends to combine with the niche concept (Sect. 1.2). A good example for facilitation is the endomycorrhiza of orchid seeds. These seeds are so small with a mass of only 0.3–15 μg per seed and carry so little resources that they already need mycorrhiza for germination as a prerequisite for the occupation of space. When plants facilitate other plants, we call this a "nurse effect." Such nursing is actually quite frequent. Most pertinent to the present review is the nursing of tree seedlings by mature trees via mycorrhiza bridges established in the soil (Simard and Durall 2004; Teste et al. 2009). However, not only nutritional aspects as involved in mycorrhizas are important. The nurse-plant syndrome always takes place when species shelter seedlings, young and/or adult individuals of other species through their ontogeny

(Franco and Nobel 1989; Dias and Scarano 2007). The nurse-plant enhances fitness, survival, and/or growth of the associated species (Callaway et al. 2002; Bruno et al. 2003). However, the nurse-plant syndrome is not free of competition. Competitive interactions build up with time (Callaway and Walker 1997). There is no "gratitude" and the nursed plants compete with their nurse plants and may even outcompete them as they develop and occupy the space. Is nursing then just delaying competition? A few examples should shed light on this problem.

3.2 *Examples for Facilitation*

3.2.1 Hydraulic Redistribution

A well-documented example for trees having a positive influence (facilitation) on their neighbors is the redistribution of water by plant root systems in the soil, a phenomenon called "hydraulic redistribution," which typically moves water from deep moist layers to the dryer soil surface (Caldwell et al. 1998; Hodge 2010). Hydraulic redistribution is a passive water movement in a root system along a gradient of water potential in the soil. It was first observed in arid and semiarid environments (Richards and Caldwell 1987) but is not limited to those species and ecosystems (Brooks et al. 2002; Warren et al. 2008) and has more recently been demonstrated for trees in the Amazon rainforest (Oliveira et al. 2005). The redistribution of water within a plant can be seen as a mechanism to extend root longevity in dry or temporarily drying soils (Bauerle et al. 2008). At the same time, it allows plants with shallow root systems to occupy dry soil space and also facilitates the access to nutrients in the upper soil horizon (Caldwell et al. 1998). Moreover, hydraulic redistribution may represent an effective mechanism for water conservation at depth, in particular in semiarid environments. Root systems have been shown to effectively recharge deeper soil layers after rain events and by this increasing water storage at depth and prolonging water availability during periods of drought (Ryel et al. 2003, 2004; Burgess et al. 2001).

3.2.2 Vegetation Islands

Ecosystems, where the dynamics of new occupation of apparently "empty" space with facilitation based on the nurse plant syndrome can be well studied, are those of sandy plains near coast lines. One example is the Atlantic restinga of Brazil. The sandy coastal plains of the restinga are quaternary terrains (Scarano 2002; Scarano et al. 2005). Nurse plants that are pioneers on the bare sand comprise some small palms (e.g., *Allagoptera arenaria*) and bromeliads. It is often observed in the neotropics that other plants germinate and develop to seedlings and even maturing plants within the phytothelmata of the bromeliad tanks. Such nursing by bromeliads includes that of species of *Clusia*, e.g., *Clusia rosea* in the tanks of *Aechmea lingulata* (Ball et al. 1991), and in the restigas most conspicuously *Clusia hilariana* mainly in the bromeliad *Aechmea nudicaulis*. *C. hilariana*, which is the dominating *Clusia* species in the restingas (Sampaio et al. 2005), then becomes an important nurse plant itself and is effectively involved in the establishment of vegetation islands on the sand plain. Underneath its canopy, higher species richness is found than under any other woody species (Zaluar 1997; Sampaio et al. 2005; Dias et al. 2005; Dias and Scarano 2007).

The view that in progressive successions such vegetation islands develop to denser vegetation with grass land, patchy formation of forests and closed dry forest as proposed by Walter (1973) when he saw the sand plains near Chichiriviche at the Caribbean coast of Venezuela is not always correct. Competition arises on the vegetation islands and with additional stress given by climatic fluctuations islands may disappear again. Indeed, dying islands were particularly observed on the sand plains of Chichiriviche (Lüttge 2008). The vegetation islands are metastable states. Like the chablis of tropical rainforests (Sect. 2.1.2), the vegetation islands on the sand plains are an illustrative example of oscillating mosaics of the spatiotemporal organization and occupation of space.

3.2.3 Epiphyte Nests

In Sect. 2.1.1.2, we have seen that epiphytes compete with phorophytes for space. Epiphytes are also a good example for facilitation involved in such competition. Epiphytes can form so-called epiphyte nests where different species support each other. A nest may start with a nurse plant, e.g., a bromeliad offering its tanks, a fern covering a phorophyte with mantle leaves and forming baskets, a *Clusia* germinating in a small accumulation of humus between branching shoots. Humus will collect in growing nests often supported by ants importing small soil particles and mycorrhizas develop. Mutual support particularly facilitates acquisition of nutrients, so that eventually real epiphyte gardens can grow on the phorophytes (Benzing 1989, 1990).

3.3 Applied Facilitation: Exotic Forest Plantations and Regeneration of Native Vegetation

Plantations of exotic trees are frequently used in reforestation of degraded pasture and range lands especially in the tropics for the purpose of obtaining timber. Forest plantations have increased at a high rate (Cohn 1995). For instance, in Brazil in the years of 1965–1995, plantations of *Eucalyptus* have increased from ca. 0.5 to 4 million ha (Da Silva et al. 1995; Feyera et al. 2002). Exotic trees most frequently used are species of the genera *Eucalyptus*, *Acacia*, *Casuarina*,

Cupressus, and *Pinus* (Table 1 in Feyera et al. 2002). However, occupation of land by exotic, i.e., nonnative trees supported by forestry management is not really rehabilitation of degraded land. Exotic tree monocultures have many disadvantages for the land, e.g., given by the enormous water demand of *Eucalyptus*. In general, the disadvantages are (1) in terms of management, detrimental effects on physical, chemical, and biological soil properties, (2) in terms of productivity and diversity displacement of local vegetation, and (3) in terms of community relations increased susceptibility to epidemic diseases and pests (Feyera et al. 2002). Conversely, there are also advantages of exotic forest plantations, which are (1) in terms of management high silvicultural experience, (2) in terms of productivity initial fast growth and wood production, and (3) in terms of community relations nurse effects, such as improvement of microclimate, reduction of erosion, and enhancement of litter and humus production (Feyera et al. 2002). The nurse effects may lead to regrowth of native woody species within exotic tree plantations of between about 15 up to more than 175 species in various plantations (Da Silva et al. 1995; Parrotta 1993, 1995; Geldenhuys 1997; Keenan et al. 1997; Feyera et al. 2002).

Based on the nurse effect, a remarkable example of applied facilitation using exotic forest plantation for regeneration of native vegetation is the reestablishment of a forest of the native tree *Podocarpus falcatus* out of a plantation of *Eucalyptus saligna* in the Munessa-Shashemene Forest at Degaga in the eastern escarpment of the Great Rift Valley of Ethiopia. Studies of the ecophysiological performance of the trees in this plantation (Lüttge et al. 2003; Fetene and Beck 2004) show that the photosynthetic capacity of *E. saligna* is not much superior to that of *P. falcatus*. The ratio of *E. saligna*/*P. falctus* for apparent photosynthetic electron transport rate (ETR) is 1.3 and for maximum ETR it is 1.1. However, a competitive advantage of *P. falcatus* is that *E. saligna* uses much more water. It is amphistomatous, i.e., carries stomata on both leaf surfaces. The ratio *E. saligna*/*P. falctus* for leaf conductance for water vapor is 1.8 for the upper leaf surface of *E. saligna* and 5.4 for the lower surface and the ratio for loss of water vapor by evapotranspiration from the leaves is 2.0 and 6.0, upper and lower surface of *E. saligna* leaves, respectively.

An important factor for the regeneration of undergrowth in forest plantations is the penetration of light (Geldenhuys 1997). Thus, in various forest plantations in Ethiopia, it was observed that the ground cover with forbs and graminoids may range from 0 to 100% as penetrating irradiance ranges from 0.2 to 55% of full sun light (Michelsen et al. 1996; Feyera et al. 2002). Hence, careful thinning is considered to be an important management option in assisting natural processes of succession (van Wyk et al. 1995). Forestry management in the Munessa-Shashemene Forest of Ethiopia uses regular coppicing of the *Eucalyptus* in about 7-year rhythms. This supplies the local population of farmers with timber and with logs to build their huts, and thus also takes care of essential socioeconomic aspects; it ascertains light penetration and gives the *Eucalyptus* a certain handicap versus the *Podocarpus*.

The nurse effect of *E. saligna* is evident. Seeds of *P. falcatus* are propagated by birds from a nearby solitary large female tree. However, nowhere in the whole area except inside the *E. saligna* plantation germination and establishment of *P. falcatus* seedlings is observed. It is likely that partial shading is the decisive factor. Nursing by *E. saligna* cannot readily develop to competition because *E. saligna* is not physiologically much superior to *P. falcatus* (see above) and due to the coppicing. The example shows that in a complex way of managing interactions between (1) facilitation by applying nurse effects via the establishment of the *Eucalyptus* forest and (2) manipulating competition giving the *Eucalyptus* a handicap versus *Podocarpus*, forestry can affect regrowth of a *P. falcatus* forest, which although being a secondary forest is akin to natural indigenous *P. falcatus* vegetation. The applied facilitation allows *P. falcatus* to regain its resource of space.

4 Synthesis

Is space in itself a resource for plants? This may be a question of dimension. For soil crusts or plants growing on the water surface (e.g., *Lemna* species), growth is obviously constrained by two-dimensional space. Similarly, unidirectional resources (e.g., light) are subjected to shading effects, which reduce the third dimension of resource distribution in space, and plants may encounter similar growth limitations. The situation may be less clear for three-dimensional space. Occupation of space by branches or roots can occur independently of its present resource availability, but this may change with time, i.e., the fourth dimension. In the following, we summarize the evidence to support our conclusion that "empty" space itself is a resource to plants.

When biomass development of organisms primarily occurs in two dimensions, for example in the case of biological crusts or biofilms colonizing surfaces such as soils, rocks, or stems, the resource nature of space is apparent (Armstrong 1982; Hestmark et al. 1997). Thus, in a two-dimensional view, space may limit growth and has to be considered as a resource by itself. In a similar way, this may be the case in competition for unidirectional resources such as light. As elaborated in Sect. 2.1.1.1, shading effects may eliminate or at least strongly diminish the third dimension of resource availability. Thus, in closed and homogeneous canopies, competition for light is reduced to a two-dimensional problem (Chiarucci et al. 2002), and space itself may become a limiting resource plants are competing for as we have exemplified for the aboveground competition in juvenile beech and spruce trees (Grams et al. 2002; Kozovits et al. 2005b). Conversely, in heterogeneous stands such as low land tropical forests, emerging trees reduce light availability for the intermediate canopy layer only by about 50%, whereas the lower canopy receives less than 10% of irradiance (Jacobs 1988). Here at least to a certain extent, light is available in a three-dimensional manner. This situation is similar to competition for multidirectional resources such as water or nitrate when shading effects are absent (Sect. 2.2). Under those conditions, plant growth has been proposed to be not limited by the sheer space (Chiarucci et al. 2002). Nevertheless, Schenk et al.

(1999) suggest belowground space in itself to be a resource, in particular in soils with low nutrient and water supply. In fact, McConnaughay and Bazzaz (1991) gave evidence for the resource nature of soil space irrespective of nutrient availability. They grew plants in different sized pots at equal amounts of nutrients and found that biomass production and reproductive output were increased with pot size.

Most interestingly, the importance of below- and aboveground space occupation is stressed when considering the fourth dimension, i.e., time. As reported by Nyanumba (2007) and Novoplansky (2009), plants respond to spatial and temporal gradients in resource supply as such gradients often exhibit predictable trajectories. For example, growth in increasing soil volumes produced more biomass and seeds as growth in the largest yet constant rooting volume. The temporal aspect in aboveground space occupation was also emphasized by Reiter et al. (2005) who studied C balances of branches of adult beech and spruce trees. Shaded branches with negative C balances were kept by the trees for at least 5 years. This was interpreted as a "sit-and -wait" or "gambling strategy" (cf. de Kroon and Hutchings 1995; Falster and Westoby 2003) as the space occupied by these currently unproductive branches may be exposed to increased irradiance in the future after gap formation (Sect. 2.1.2). Hence, it may be advantageous to occupy and compete for this aboveground canopy space although resource supply is currently low. The importance of time in developing competition for light is nicely illustrated by the performance of two grasses typical of roadside plant communities in Central Europe, namely *Elymus repens* and *Puccinellia distans*. Both species have the same ecological amplitude and in monocultures they grow similarly well under given identical conditions. However, morphologically *E. repens* has the capacity to grow taller than *P. distans*. Hence, in mixed cultures with time in the course of the growing season due to its structural advantage, *E. repens* overgrows and shades *P. distans* that eventually dies (Beyschlag et al. 1992).

Finally and at the end, let us return to the simplest case, i.e., the one dimension. Is the wire of a telephone line not just one dimension? And does it have anything else to offer to plant life than space? Indeed, it offers nothing else than sheer "empty" space. Yet, we see that this space is occupied by plants, namely the atmospheric bromeliads (Sect. 2.1.1.2). We also observe competition between epiphytes for such space showing that it is a limiting resource. This occupation of and competition for space devoid of any other commodities we may take as the most simple because very straightforward compelling argument for the conclusion we want to promote in this assay, namely that sheer space in itself has the function of resource.

Acknowledgments We thank Fabio R. Scarano and an anonymous reviewer for valuable suggestions and thought-provoking comments. The new concept of "empty" space as a resource emerges from the work of the Sonderforschungsbereich (SFB) 607 "Growth and parasite defense" of the Deutsche Forschungsgemeinschaft (DFG). Both authors thank the members of the SFB for discussions and continuous stimulating exchange and T.E.E.G. thanks the DFG for the financial support.

References

Agerer R (2001) Exploration types of ectomycorrhizae – a proposal to classify ectomycorrhizal mycelial systems according to their patterns of differentiation and putative ecological importance. Mycorrhiza 11:107–114

Agerer R (2007) Diversität der Ektomykorrhizen im unter- und oberirdischen Vergleich: die Explorationstypen. Z Mykol 73:61–88

Agerer R (2009) Bedeutung der Ektomykorrhiza für Waldökosysteme. Rundgespräche der Kommission für Ökologie, vol 37, Ökologische Rolle von Pilzen. Verlag Dr. Friedrich Pfeil, München, pp 111–121

Agerer R, Raidl S (2004) Distance-related semi-quantitative estimation of the extramatrical ectomycorrhizal mycelia of *Cortinarius obtusus* and *Tylospora asterophora*. Mycol Prog 3:57–64

Anten NPR, Hirose T (2001) Limitations on photosynthesis of competing individuals in stands and the consequences for canopy structure. Oecologia 129:186–196

Armstrong RA (1982) Competition between 3 saxicolous species of *Parmelia* (lichens). New Phytol 90:67–72

Bais HP, Vepachedu R, Gilroy S, Callaway RM, Vivanco JM (2003) Allelopathy and exotic plant invasion: from molecules and genes to species interactions. Science 301:1377–1380

Bais HP, Park SW, Weir TL, Callaway RM, Vivanco JM (2004) How plants communicate using the underground information superhighway. Trends Plant Sci 9:26–32

Ball E, Hann J, Kluge M, Lee HSJ, Lüttge U, Orthen B, Popp M, Schmittt A, Ting IP (1991) Ecophysiological comportment of the tropical CAM-tree *Clusia* in the field. I growth of *Clusia rosea* Jacq. On St. John, US Virgin Islands, Lesser Antilles. New Phytol 117:473–481

Bartelheimer M, Steinlein T, Beyschlag W (2008) ^{15}N-nitrate-labelling demonstrates a size symmetric competitive effect on belowground resource uptake. Plant Ecol 199:243–253

Bauerle TL, Richards JH, Smart DR, Eissenstat DM (2008) Importance of internal hydraulic redistribution for prolonging the lifespan of roots in dry soil. Plant Cell Environ 31:177–186

Begon ME, Harper JL, Townsend CR (1990) Ecology. Blackwell Science, London

Belnap J, Lange OL (eds) (2003) Biological soil crusts: structure, function and management, vol 150, 2nd edn, Ecological studies. Springer, Berlin

Benzing DH (1989) The mineral nutrition of epiphytes. In: Lüttge U (ed) Vascular plants as epiphytes. Evolution and ecophysiology. Ecologial studies, vol 76. Springer, Berlin

Benzing DH (1990) Vascular epiphytes. Cambridge University Press, Cambridge

Benzing DH (2000) Bromeliaceae: profile and adaptive radiation. Cambridge University Press, Cambridge

Beyschlag W, Ryel RJ, Ullmann I (1992) Experimental and modelling studies of competition for light in roadside grasses. Bot Acta 105:285–291

Brooks JR, Meinzer FC, Coulombe R, Gregg J (2002) Hydraulic redistribution of soil water during summer drought in two contrasting Pacific Northwest coniferous forests. Tree Physiol 22:1107–1117

Bruno JF, Stachowicz JJ, Bertness MD (2003) Inclusion of facilitation into ecological theory. Trend Ecol Evol 18:119–125

Burgess SSO, Adams MA, Turner NC, White DA, Ong CK (2001) Tree roots: conduits for deep recharge of soil water. Oecologia 126:158–165

Cahill JF, Casper BB (2000) Investigating the relationship between neighbor root biomass and belowground competition: field evidence for symmetric competition belowground. Oikos 90:311–320

Caldwell MM, Dawson TE, Richards JH (1998) Hydraulic lift: consequences of water efflux from the roots of plants. Oecologia 113:151–161

Callaway RM, Walker LR (1997) Competition and facilitation: a synthetic approach to interactions in plant community. Ecology 78:1958–1965

Callaway RM, Brooker RW, Choler P, Kikvidze Z, Lortie CJ, Michalet R, Paolini L, Pugnaire FI, Newingham B, Aschehoug ET, Armas C, Kikodze D, Cook BJ (2002) Positive interactions among alpine plants increase with stress. Nature 417:844–848

Casper BB, Jackson RB (1997) Plant competition underground. Annu Rev Ecol Syst 28:545–570

Chase JM, Leibold MA (2003) Ecological niches: linking classical and contemporary approaches. The University of Chicago Press, Chicago

Chiarucci A, Mistral M, Bonini I, Anderson BJ, Wilson JB (2002) Canopy occupancy: how much of the space in plant communities is filled? Folia Geobot 37:333–338

Cohn I (1995) The myths and realities of industrial timber plantations. For Perspect 2:5–8

Da Silva Junior MC, Scarano FR, De Souza CF (1995) Regeneration of an Atlantic forest formation in the understory of a *Eucalyptus grandis* plantation in south-eastern Brazil. J Trop Ecol 11:147–152

Dalling JW, Hubbell SP, Silvera K (1998) Seed dispersal, seedling establishment and gap partitioning among tropical pioneer trees. J Ecol 86:674–689

Darwin C, Darwin F (1896) The power of movement in plants. D. Appleton, New York

de Freitas CA, Scarano FR, Biesboer DD (2003) Morphological variation in two facultative epiphytic bromeliads growing on the floor of a swamp forest. Biotropica 35:546–550

de Kroon H, Hutchings MJ (1995) Morphological plasticity in clonal plants: the foraging concept reconsidered. J Ecol 83:143–152

Dias ATC, Scarano FR (2007) *Clusia* as nurse plant. In: Lüttge U (ed) *Clusia* a woody neotropical genus of remarkable plasticity and diversity. Ecological studies, vol 194. Springer, Berlin

Dias ATC, Zaluar HLT, Ganade G, Scarano FR (2005) Canopy composition influencing plant patch dynamics in a Brazilian sandy coastal plain. J Trop Ecol 21:343–347

Elbert W, Weber B, Büdel B, Andreae MO, Pöschl U (2009) Microbiotic crusts on soil, rock and plants: neglected major players in the global cycles of carbon and nitrogen. Biogeosci Discuss 6:1–33

Enquist BJ, Niklas KJ (2001) Invariant scaling relations across tree-dominated communities. Nature 410:655–660

Falik O, Reides P, Gersani M, Novoplansky A (2005) Root navigation by self inhibition. Plant Cell Environ 28:562–569

Falster DS, Westoby M (2003) Plant height and evolutionary games. Trends Ecol Evol 18:337–343

Fetene M, Beck E (2004) Water relations of indigenous versus exotic tree species, growing at the same site in a tropical montane forest in southern Ethiopia. Trees 18:428–435

Feyera S, Beck E, Lüttge U (2002) Exotic trees as nurse-trees for the regeneration of natural tropical forests. Trees 16:245–249

Franco AC, Nobel PS (1989) Effect of nurse plants on the microhabitat and growth of cacti. J Ecol 77:870–886

Fransen B, de Kroon H, Berendse F (2001) Soil nutrient heterogeneity alters competition between two perennial grass species. Ecology 82:2534–2546

Gayler S, Grams TEE, Kozovits AR, Winkler JB, Luedemann G, Priesack E (2006) Analysis of competition effects in mono- and mixed cultures of juvenile beech and spruce by means of the plant growth simulation model PLATHO. Plant Biol 8:503–514

Geldenhuys CJ (1997) Native forest regeneration in pine and eucalypt plantations in Northern Province, South Africa. For Ecol Manage 99:101–115

Gersani M, Abramsky Z, Falik O (1998) Density-dependent habitat selection in plants. Evol Ecol 12:223–234

Goh CJ, Kluge M (1989) Gas exchange and water relations in epiphytic orchids. In: Lüttge U (ed) Vascular plants as epiphytes. Evolution and ecophysiology. Ecological studies, vol 76. Springer, Berlin, pp 139–166

Grams TEE, Andersen CP (2007) Competition for resources in trees: physiological versus morphological plasticity. In: Esser K, Lüttge U, Beyschlag W, Murata J (eds) Progress in botany, vol 68. Springer, Berlin, pp 356–381

Grams TEE, Kozovits AR, Reiter IM, Winkler JB, Sommerkorn M, Blaschke H, Häberle KH, Matyssek R (2002) Quantifying competitiveness in woody plants. Plant Biol 4:153–158

Grams TEE, Kozovits AR, Häberle K-H, Matyssek R, Dawson TE (2007) Combining $\delta^{13}C$ and $\delta^{18}O$ analyses to unravel competition, CO_2 and O_3 effects on the physiological performance of different-aged trees. Plant Cell Environ 30:1023–1034

Grams TEE, Matyssek R (2010) Stable isotope signatures reflect competitiveness between trees under changed CO_2/O_3 regimes. Environmental Pollution 158:1036–1042

Griffiths H (1989) Carbon dioxide concentrating mechanisms and the evolution of CAM in vascular epiphytes. In: Lüttge U (ed) Vascular plants as epiphytes. Evolution and ecophysiology. Ecological studies, vol 76. Springer, Berlin, pp 42–86

Gruntman M, Novoplansky A (2004) Physiologically mediated self/non-self discrimination in roots. Proc Natl Acad Sci USA 101:3863–3867

Hastings A, Hom CL, Ellner S, Turchin P, Godfray HCJ (1993) Chaos in ecology: is mother nature a strange attractor? Annu Rev Ecol Syst 24:1–33

Hestmark G, Schroeter B, Kappen L (1997) Intrathalline and size-dependent patterns of activity in *Lasallia pustulata* and their possible consequences for competitive interactions. Funct Ecol 11:318–322

Hierro JL, Callaway RM (2003) Allelopathy and exotic plant invasion. Plant Soil 256:29–39

Hobbie EA, Agerer R (2010) Nitrogen isotopes in ectomycorrhizal sporocarps correspond to belowground exploration types. Plant and Soil 327:71–83

Hodge A (2006) Plastic plants and patchy soils. J Exp Bot 57:401–411

Hodge A (2009) Root decisions. Plant Cell Environ 32:628–640

Hodge A (2010) Roots: the acquisition of water and nutrients from the heterogeneous soil environment. Prog Bot 71:307–337

Jacobs M (1988) The tropical rain forest. Springer, Berlin

Johansson DR (1975) Ecology of epiphytic orchids in West African rain forests. Am Orchid Soc Bull 46:703–707

Keenan R, Lamb D, Woldring O, Irvine T, Jensen R (1997) Restoration of plant biodiversity beneath tropical tree plantations in Northern Australia. For Ecol Manage 99:117–131

King S, Beck F, Lüttge U (2004) On the mystery of the golden angle in phyllotaxis. Plant Cell Environ 27:685–695

Kobayashi A, Takahashi A, Kakimoto Y, Miyazawa Y, Fujii N, Higashitani A, Takahashi H (2007) A gene essential for hydrotropism in roots. Proc Natl Acad Sci USA 104: 4724–4729

Kozovits AR, Matyssek R, Blaschke H, Göttlein A, Grams TEE (2005a) Competition increasingly dominates the responsiveness of juvenile beech and spruce to elevated CO_2 and/or O_3 concentrations throughout two subsequent growing seasons. Glob Change Biol 11:1387–1401

Kozovits AR, Matyssek R, Winkler JB, Göttlein A, Blaschke H, Grams TEE (2005b) Aboveground space sequestration determines competitive success in juvenile beech and spruce trees. New Phytol 167:181–196

Küppers M (1989) Ecological significance of above-ground architectural patterns in woody plants – a question of cost–benefit relationships. Trends Ecol Evol 4:375–379

Küppers M, List R (1997) MADEIRA – a simulation of carbon gain, allocation, canopy architecture in competing woody plants. In: Jeremonidis G, Vincent JFV (eds) Plant Biomechanics 1997. The University of Reading, Reading, pp 321–329

Kuptz D, Grams TEE, Günther S (2010) Light acclimation of four native tree species in felling gaps within a tropical mountain rain forest. Trees-Structure and Function 24:117–127

Levin J, Muller-Landau HC (2001) The emergence of diversity in plant communities. CR Acad Sci Paris Sci de la Vie 323:129–139

Luedemann G, Matyssek R, Fleischmann F, Grams TEE (2005) Acclimation to ozone affects host/pathogen interaction and competitiveness for nitrogen in juvenile *Fagus sylvatica* and *Picea abies* trees infected with *Phytophthora citricola*. Plant Biol 7:640–649

Luedemann G, Matyssek R, Winkler JB, Grams TEE (2009) Contrasting ozone × pathogen interaction as mediated through competition between juvenile European beech (*Fagus sylvatica*) and Norway spruce (*Picea abies*). Plant Soil 323:47–60

Lüttge U (1985) Epiphyten: evolution und ökophysiologe. Naturwissenschaften 72:557–566
Lüttge U (ed) (1989a) Vascular plants as epiphytes. Evolution and ecophysiology. Ecological studies, vol 76. Springer, Berlin
Lüttge U (1989b) Vascular epiphytes: setting the scene. In: Lüttge U (ed) Vascular plants as epiphytes. Evolution and ecophysiology. Ecological studies, vol 76. Springer, Berlin, pp 1–14
Lüttge U (1997) Cyanobacterial Tintenstrich communities and their ecology. Naturwissenschaften 84:526–534
Lüttge U (2008) Physiological ecology of tropical plants, 2nd edn. Springer, Berlin
Lüttge U, Ball E, Kluge M, Ong BL (1986) Photosynthetic light requirements of various tropical vascular epiphytes. Physiol Vég 24:315–331
Lüttge U, Berg A, Fetene M, Nauke P, Peter D, Beck E (2003) Comparative characterization of photosynthetic performance and water relations of native trees and exotic plantation trees in an Ethiopian forest. Trees 17:40–50
Mahall BE, Callaway RM (1992) Root communication mechanism and intracommunity distributions of two Mojave desert shrubs. Ecology 73:2145–2151
Manrubia SC, Solé RV (1996) Self-organized criticality in rainforest dynamics. Chaos Solitons Fractals 7:523–541
Massey FP, Massey K, Press MC, Hartley SE (2006) Neighbourhood composition determines growth, architecture and herbivory in tropical rain forest seedling. J Ecol 94:646–655
Matyssek R, Schnyder H, Munch JC, Oßwald W, Pretzsch H, Treutter D (2005) Resource allocation in plants – the balance between resource sequestration and retention. Plant Biol 7:557–559
May RM (1976) Simple mathematical models with very complicated dynamics. Nature 261:459–467
McConnaughay KDM, Bazzaz FA (1991) Is physical space a soil resource? Ecology 72:94–103
Michelsen A, Lisanework N, Friis I, Holst N (1996) Comparisons of understory vegetation and soil fertility in plantations and adjacent natural forests in the Ethiopian highlands. J Appl Ecol 33:627–642
Miyazawa Y, Ito Y, Moriwaki T, Kobayashi A, Fujii N, Takahashi H (2009) A molecular mechanism unique to hydrotropism in roots. Plant Sci 177:297–301
Morris SC (2003) Life's solution. Inevitable humans in a lonely universe. Cambridge University Press, New York
Newell EA, McDonald EP, Strain BR, Denslow JS (1993) Photosynthetic responses of *Miconia* species to canopy openings in a lowland tropical rainforest. Oecologia 94:49–56
Novoplansky A (2009) Picking battles wisely: plant behaviour under competition. Plant Cell Environ 32:726–741
Nyanumba SM (2007) Developmental strategies of plants under temporal variation in growth conditions. MSc thesis, Ben-Gurion University of the Negev, Israel
Oliveira RS, Dawson TE, Burgess SSO, Nepstad DC (2005) Hydraulic redistribution in three Amazonian trees. Oecologia 145:354–363
Orians GH (1982) The influence of tree-fall in tropical forest on tree species richness. Trop Ecol 23:255–279
Palmitto PA, Davies SJ, Vogt KA, Ashton MS, Vogt DJ, Ashton PS (2004) Soil related habitat specialization in dipterocarp rain forest tree species in Borneo. J Ecol 92:609–623
Parrotta JA (1993) Secondary forest regeneration on degraded tropical lands. The role of plantations as "foster ecosystems". In: Lieth H, Lohmann M (eds) Restoration of tropical forest ecosystems. Kluwer, Dordrecht, pp 66–73
Parrotta JA (1995) Influence of overstory composition on understory colonization by native species in plantations on a degraded tropical site. J Veg Sci 6:627–636
Pearcy RW, Yang W (1996) A three-dimensional crown architecture model for assessment of light capture and carbon gain by understory plants. Oecologia 86:1–12
Pearcy RW, Yang W (1998) The functional morphology of light capture and carbon gain in the redwood forest understory plant *Adenocaulon bicolor* Hook. Funct Ecol 12:543–552

Phillips OL, Vargas PN, Monteagudo AL, Cruz AP, Zans MEC, Sánchez WG, Yli-Halla M, Rose S (2003) Habitat association among Amazonian tree species: a landscape-scale approach. J Ecol 91:757–775
Pretzsch H (2010) Re-evaluation of allometry: state-of-the-art and perspective regarding individuals and stands of woody plants. Prog Bot 71:339–369
Reekie EG, Bazzaz FA (1989) Competition and patterns of resource use among seedlings of five tropical trees grown at ambient and elevated CO_2. Oecologia 79:212–222
Reiter IM, Häberle KH, Nunn AJ, Heerdt C, Reitmayer H, Grote R, Matyssek R (2005) Competitive strategies in adult beech and spruce: space-related foliar carbon investment versus carbon gain. Oecologia 146:337–349
Remmert H (1985) Was geschieht im Klimax-Stadium? Ökologisches Gleichgewicht durch Mosaik aus desynchronen Zyklen. Naturwissenschaften 72:505–512
Remmert H (1991) The mosaic cycle of ecosystems. Ecological studies, vol 85. Springer, Berlin
Richards JH, Caldwell MM (1987) Hydraulic lift – substantial nocturnal water transport between soil layers by *Artemisia tridentata* roots. Oecologia 73:486–489
Ryel RJ, Caldwell MM, Leffler AJ, Yoder CK (2003) Rapid soil moisture recharge to depth by roots in a stand of *Artemisia tridentata*. Ecology 84:757–764
Ryel RJ, Leffler AJ, Peek MS, Ivans CY, Caldwell MM (2004) Water conservation in Artemisia tridentata through redistribution of precipitation. Oecologia 141:335–345
Sampaio MC, Picó FX, Scarano FR (2005) Ramet demography of a nurse bromeliad in Brazilian restingas. Am J Bot 92:674–681
Scarano FR (2002) Structure, function and floristic relationships of plant communities in stressful habitats marginal to the Brazilian Atlantic rain forest. Ann Bot 90:517–524
Scarano FR, Ribeiro KT, Moraes LFD, Lima HC (1997) Plant establishment on flooded and non-flooded patches of a swamp forest in southeastern Brazil. Brazil J Trop Ecol 14:793–803
Scarano FR, de Mattos EA, Franco AC, Herzog B, Ball E, Grams TEE, Mantovani A, Barreto S, Haag-Kerwer A, Lüttge U (1999) Habitat segregation of C_3 and CAM *Nidularium* (Bromeliaceae) in response to different light regimes in the understory of a swamp forest in southeastern Brazil. Flora 194:281–288
Scarano FR, Duarte HM, Franco AC, Geßler A, de Mattos EA, Rennenberg H, Lüttge U (2005) Physiological synecology of tree species in relation to geographic distribution and ecophysiological parameters at the Atlantic forest periphery in Brazil: an overview. Trees 19:493–496
Schenk HJ (2006) Root competition: beyond resource depletion. J Ecol 94:725–739
Schenk HJ, Callaway RM, Mahall BE (1999) Spatial root segregation: Are plants territorial? Advances in Ecological Research, vol 28, pp 145–180
Schimper AFW (1888) Botanische Mitteilungen aus den Tropen. II. Epiphytische Vegetation Amerikas. G Fischer, Jena
Schimper AFW (1898) Pflanzengeographie auf physiologischer Grundlage. G Fischer, Jena
Schulze E-D, Küppers M, Matyssek R (1986) The roles of carbon balance and branching pattern in the growth of woody species. In: Givnish TJ (ed) On the economy of plant form and function. Cambridge University Press, Cambridge, pp 585–602
Schwinning S, Weiner J (1998) Mechanisms determining the degree of size asymmetry in competition among plants. Oecologia 113:447–455
Simard SW, Durall DM (2004) Mycorrhizal networks: a review of their extent, function, and importance. Can J Bot 82:1140–1165
Smith JAC (1989) Epiphytic bromeliads. In: Lüttge U (ed) Vascular plants as epiphytes. Evolution and ecophysiology. Ecological studies, vol 76. Springer, Berlin, pp 109–138
Smith SE, Gianinazzi-Pearson V (1988) Physiological interactions between symbionts in vesicular-arbiscular mycorrhizal plants. Annu Rev Plant Physiol Plant Mol Biol 39:221–244
Solé RV, Manrubia SC (1995a) Self-similarity in rain forests: evidence for a critical state. Phys Rev E 51:6250–6253
Solé RV, Manrubia SC (1995b) Are rainforests self-organized in a critical state? J Theor Biol 173:31–40

Solé RV, Manrubia SC, Luque B (1994) Multifractals in rainforest ecosystems: modeling and simulations. In: Novak NM (ed) Fractals in the natural and applied sciences. Elsevier, Amsterdam, pp 397–407

Teste FP, Simard SW, Durall DM, Guy RD, Jones MD, Schoonmaker AL (2009) Access to mycorrhizal networks and roots of trees: importance for seedling survival and resource transfer. Ecology 90:2808–2822

Torquebiau EF (1988) Photosynthetically active radiation environment, patch dynamics and architecture in a tropical rainforest in Sumatra. Aust J Plant Physiol 15:327–342

Tremmel DC, Bazzaz FA (1995) Plant architecture and allocation in different neighbourhoods: implications for competitive success. Ecology 76:262–271

Valencia R, Foster RB, Villa G, Condit R, Svenning JC, Hernández C, Romoleroux K, Losos E, Magård E, Balsev H (2004) Tree species distributions and local habitat variation in the Amazon: large forest plot in eastern Ecuador. J Ecol 92:214–229

van der Meer PJ, Bongers F (1996) Patterns of tree-fall and branch-fall in a tropical rain forest in French Guiana. J Ecol 84:19–29

Van Wyk GF, Everard DA, Geldenhuys CJ (1995) Forest ecotone development and succession: experimental results and guidelines for forest rehabilitation and protection. Report FOR DEA 867. Division of Forest Science and Technology. CSIR, Pretoria

Walter H (1973) Die Vegetation der Erde in ökologischer Betrachtung, Die tropischen und subtropischen Zonen, vol 1. G Fischer, Jena

Warren JM, Brooks JR, Meinzer FC, Eberhart JL (2008) Hydraulic redistribution of water from *Pinus ponderosa* trees to seedlings: evidence for an ectomycorrhizal pathway. New Phytol 178:382–394

West GB, Brown JH, Enquist BJ (2001) A general model for ontogenetic growth. Nature 413:628–631

West GB, Enquist BJ, Brown JH (2009) A general quantitative theory of forest structure and dynamics. Proc Natl Acad Sci USA 106:7040–7045

Zaluar HLT (1997) Espécies focais e a formação de moitas na restinga abertga de Clusia, Carpebus, RJ. M.Sc. Dissertation, Universidade Federal do Rio de Janeiro, Rio de Janeiro

Zotz G, Hietz P, Schmidt G (2001) Small plants, large plants: the importance of plant size for the physiological ecology of vascular epiphytes. J Exp Bot 52:2051–2056

Photorespiration in Phase III of Crassulacean Acid Metabolism: Evolutionary and Ecophysiological Implications

Ulrich Lüttge

Contents

Abstract Ribulose-bis-phosphate carboxylase/oxygenase (RubisCO), the enzyme of primary CO_2-fixation in photosynthesis, from its early evolution in a high CO_2 and low O_2 atmosphere has inherited a dual affinity for both CO_2 and O_2. The reaction with O_2 makes photorespiration an unavoidable affix of photosynthesis. The relative CO_2/O_2 specificity of RubisCO has improved during evolution. However, O_2 affinity has been conserved to date and CO_2 affinity has remained rather low. Hence, various inorganic carbon concentrating mechanisms have evolved. Among them, the CO_2-concentrating mechanism of plants with crassulacean acid metabolism (CAM) achieves the strongest increase in CO_2 concentration within the photosynthesizing organs. In the so-called Phase III of CAM, CO_2 fixed via

U. Lüttge
Institut für Botanik, Fachbereich Biologie, Technische Universität Darmstadt, D-64 287 Darmstadt, Germany
e-mail: luettge@bio.tu-darmstadt.de

U. Lüttge et al. (eds.), *Progress in Botany 72*,
DOI 10.1007/978-3-642-13145-5_14,

phospho*enol*pyruvate carboxylase in the dark period (Phase I of CAM) and stored nocturnally in the form of malate in the vacuoles is remobilized behind closed stomata, and this leads to a 2-fold up to 60-fold increase of CO_2 partial pressures in the leaf air spaces as compared with atmospheric partial pressure. However, this does not eliminate photorespiration because with vigorous CO_2 assimilation at high concentration behind closed stomata, photosynthetic oxygen evolution simultaneously also leads to O_2 concentrating of up to 40% within the leaves. Hence, photorespiration in CAM plants is not only active in Phase IV of CAM when the store of malate is exhausted and stomata open for CO_2 uptake and standard C_3-photosynthesis, but also in Phase III where CO_2 concentrating is counterbalanced by O_2 concentrating. The question arises whether in Phase III photorespiration is quantitatively similar to that in C_3-photosynthesis or suppressed at least to some extent. The particular question asked in this essay is if the ratio of the reaction rates of RubisCO with O_2 and CO_2, v^{O_2}/v^{CO_2}, can be calculated using enzyme kinetics formalism with data given in the literature for partial pressures of O_2 and CO_2 in the leaves during Phase III of CAM. Depending on assumptions inherent in kinetic formalisms, the calculations yield different conclusions. According to one approach photorespiration is strongly suppressed in Phase III of CAM. Another assessment suggests that under physiological conditions of C_3-photosynthesis and Phase III of CAM, oxygenase activity in relation to carboxylase activity of RubisCO is not fundamentally different. At the very most in Phase III of CAM, v^{O_2}/v^{CO_2} is about half of that in C_3 photosynthesis but mostly it is much less reduced than that. The result of the second approach currently appears to be more likely in comparison with experimental findings of gas exchange measurements. However, more experimental data are needed for comparison with the theoretical evaluations.

1 Introduction: Conserved Early Properties of Ribulose-Bis-Phosphate Carboxylase/Oxygenase and the Evolution of Photorespiration

Photosynthesis and its central carboxylating enzyme RubisCO evolved with the earliest stages of life on earth. In evolution, RubisCO most likely originally developed 3.5 billion years ago for CO_2 fixation in the early chemo-autotrophic bacteria. At that time, the earth had a very different atmosphere from now. One assumes that the CO_2-pressure was 5–10 bar in contrast to a current $3.8–4.0 \times 10^{-4}$ bar and a total atmospheric pressure of only 1 bar. Thus, a low affinity of RubisCO to the substrate CO_2 could be afforded. The mechanism of carboxylation at the enzyme involves binding of ribulose-bis-phosphate (RubP) which is converted to the 2,3-enediol(ate) form (enol-RubP) to which CO_2 can then be bound. The consequence of this chemical configuration is, however, that O_2 can also be bound. As the atmospheric oxygen concentrations at the origin of photosynthesis and RubisCO were lower than 0.2%, this oxygen affinity could also be afforded. It is noteworthy, however, that the low CO_2 affinity of RubisCO has been retained in extant plants to date, with a Michaelis constant for CO_2, $K_M^{CO_2}$, of about 9 μmol l^{-1}, and a remaining O_2 affinity is also still present given by a Michaelis constant for O_2 of about 535 μmol l^{-1} (Heldt and Piechulla 2008). The reaction with CO_2, i.e., binding of CO_2 to RubP yields two molecules of 3′-phosphoglyceric acid which feed into the Calvin cycle of assimilatory carbon reduction. The reaction with O_2 yields only one molecule of 3′-phosphoglyceric acid for the Calvin cycle and one molecule of phospho-glycolate which is the starting metabolite of the metabolic pathway called photorespiration.

2 Evolution of RubisCO Specificity and Carbon Concentrating Mechanisms

The relative specificity of RubisCO for its two substrates is characterized by the so-called specificity factor, Σ:

$$\sum = \frac{V_{max}^{CO_2}}{V_{max}^{O_2}} \cdot \frac{K_M^{O_2}}{K_M^{CO_2}}, \tag{1}$$

where K_M are the Michaelis constants and V_{max} the maximal reaction rates with CO_2 and O_2, respectively, as indicated by the superscripts. The specificity factor is a constant for each individual RubisCO enzyme (Spreitzer and Salvucci 2002). When we compare Σ values obtained for extant photosynthetic organisms of different levels of organization or different evolutionary distances from early photosynthesis, we may note a certain improvement with increasing bias toward CO_2, because as shown in Table 1 Σ increased as higher organisms evolved. Evidently, Σ is quite low in some prokaryotes and dinoflagellates. The thermophilic red alga *Galdieria partita* is a special case. At 25°C it has the highest Σ ever recorded for any RubisCO, but Σ is much lower at its natural growth conditions of 45°C (Uemura et al. 1997). From the prokaryotic cyanobacterial subgroup of the eubacteria to the eukaryotic photosynthesising organisms, there is a 1.5-fold increase to the green algae and a more than twofold jump to the Σ of angiosperms.

Thus, the evolutionary trend of increasing the relative CO_2 specificity of RubisCO is evident. Yet the effect was not large enough to have attained satisfactory operation at low atmospheric CO_2. As an effective alternative, photosynthesizing organisms have evolved inorganic carbon concentrating mechanisms. Although there are several potential molecular targets for altering the structure of RubisCO for improved CO_2-fixation as discussed by Spreitzer and Salvucci (2002), CO_2-concentrating mechanisms may well be the more easy and promising path. Cyanobacteria and many eukaryotic algae operate with HCO_3^- transporting and trapping mechanisms which concentrate CO_2 around RubisCO localized in carboxysomes and pyrenoïds, respectively. Vascular plants use primary fixation of CO_2 by the carboxylating enzyme phospho*enol*pyruvarte carboxylase (PEPC), which has a 60-fold higher affinity to its substrate HCO_3^- than RubisCO has to

Table 1 Specificity factors, Σ (1) of extant photosynthetic organisms of increasing evolutionary level (major general reference: Spreitzer and Salvucci 2002)

Organisms	Σ	References
Prokaryotes Dinoflagellates	15	Jordan and Ogren 1981, Whitney and Andrews 1998
Cyanobacteria	40	Jordan and Ogren 1981, Spreitzer 1999
Galdieria partita, thermophilic red alga	240 at 25°C 80 at 45°C	Uemura et al. 1997
Green algae	60	Jordan and Ogren 1981, Spreitzer 1999
Land plants, in general	80–100 80–130	Jordan and Ogren 1981, Spreitzer 1999, Buchanan et al. 2000
Spinacea oleracea, a C_3-plant	95	Griffiths et al. 2008
Two species of *Kalanchoë*, terrestrial CAM plants	88	Griffiths et al. 2008
Isoëtes lacustris, submerged freshwater CAM plant	88	Griffiths, personal communication

its substrate CO_2 and also has no O_2 binding capacity. In the so-called C_4 plants, the C_4-compound malate produced via PEPC is transported from the location of its synthesis (so-called mesophyll tissue) to the site of its decarboxylation (so-called bundle sheath tissue) for regeneration of CO_2 and assimilation via RubisCO in the Calvin cycle. This may lead to a 6–10-fold increase of CO_2 concentration at the RubisCO enzyme. In plants with CAM, malate formation via PEPC occurs in the dark period, and malate is stored in the vacuole. Malate remobilization from the vacuole and its decarboxylation with regeneration of CO_2 and assimilation via RubisCO in the Calvin cycle is performed behind closed stomata in the light period. This achieves the strongest increase of internal CO_2 partial pressures known of CO_2-concentrating mechanisms. In the light period, CO_2 levels in the intercellular air spaces of photosynthesizing organs of CAM plants, $p_i^{CO_2}$, have been shown to range between 0.08 and 2.50%, i.e., from 2 to over 60-fold higher than the atmospheric partial pressures of CO_2 (Lüttge 2002).

It is noteworthy inspecting Table 1 that the specificity factor, Σ, of RubisCO did not change in the evolution of CAM. It is identical in the CAM-fern ally *Isoëtes* (Pteridophytina) and the CAM performing angiosperms of the genus *Kalanchoë*. That Σ of the CAM species is at the lower end of the range given for vascular plants may raise the idea of occurrence of a retro-evolution under the influence of internal CO_2 concentrating. However, the data set available is much too small to support that at this stage.

3 Oxidative Stress, Antioxidative Reactions, and Photorespiration in CAM Plants

For a long while in the history of CAM research under the impression of this dramatic increase of $p_i^{CO_2}$, it was assumed that photorespiration would not occur in CAM plants and that they were largely exempt from oxidative stress. However, it was then discovered that CAM plants do have the entire enzymatic complement of the photorespiratory pathway, that they do produce reactive oxygen species (ROS), and that they do employ antioxidative reactions (AOR) for protection. In a seminal review in Progress in Botany, Niewiadomska and Borland (2007) have recently surveyed the phenomena and consequences of oxidative stress in CAM plants. This also comprises an overview of the evidence for the occurrence of photorespiration including the presence of the enzyme complement of the respiratory carbon flow (glyoxylate cycle). The present attempt of a quantitative estimation may be considered as a follow up and based on the literature assessed by Niewiadomska and Borland (2007).

Photorespiration is an oxygen-consuming and energy-dissipating pathway. In a progressing escalation of possible protective mechanisms it is effective under increasing stress as an early AOR which serves prevent over energization of the photosynthetic light harvesting and electron transport systems. CAM has a phase of CO_2 uptake in the light period which occurs in the afternoon after the nocturnally stored vacuolar malate is consumed. Under favorable conditions – especially with respect to the availability of water – the CAM plants may then open their stomata and take up CO_2 just like C_3 plants. Then, of course, the CO_2-concentrating mechanism is not effective any more. This is called phase IV of CAM (Osmond 1978). When the photorespiratory capacity of CAM plants was first discovered, it therefore was concluded that this was a particular attribute of phase IV which in no way differs from C_3-photosynthesis with all its implications of ROS and the consequences of the RubisCO-affinity for O_2. The CO_2-concentrating mechanism depends on malate decarboxylation and is therefore effective in the earlier part of the light period which we call Phase III. (Note that Phase I is nocturnal malate synthesis and accumulation, and Phase II is a transition phase in the very early light period; Osmond 1978.) Thus, it was assumed that photorespiration would not occur in Phase III due to very high $p_i^{CO_2}$ and that its photoprotective benefits were also not needed in this phase.

4 O_2 Concentrating and Photorespiration in Phase III of CAM

The latter assumption ignored a very important facet of CAM although it had been discovered in principle as early as in 1800 by Alexander von Humboldt and was later underlined by detailed analyses of Spalding et al. (1979). This is the fact that during vigorous photosynthetic CO_2 assimilation at high $p_i^{CO_2}$ in Phase III naturally, water splitting, electron transport, and oxygen evolution are also accelerated and oxygen is also concentrated in the leaf air spaces behind closed stomata. In February 1800, Alexander von Humboldt worked in Venezuela with *Clusia rosea*, a plant of which we now know that it is a CAM species. He observed that in day time it had no gas exchange with the atmosphere, which as we also know now is characteristic of Phase III of CAM, and he measured an internal oxygen concentration of 35% (Faak 2000, see Lüttge 2002). Some 179 years later, analyses

Table 2 Partial pressures, p_i, of O_2 and CO_2 in the gas phase for ambient air, within the leaves of C_3 plants and as measured within the leaves of various named CAM species in Phase III of CAM and the derived concentrations, c, for equilibrium with an aqueous solution at 25°C

	$p_i^{O_2}/p_i^{CO_2}$	$p_i^{CO_2}$ (%)	$p_i^{O_2}$ (%)	c^{CO_2} (μM)	c^{O_2} (μM)	c^{O_2}/c^{CO_2}
Ambient air	525	0.04	21	13.6	273	20
C_3 photosynthesis	840	0.025	21	8.5	273	32
CAM-species						
Opuntia monacantha	205	0.12	24.6	40.8	319	8
Ananas comosus	161	0.13	20.9	44.2	272	6
		0.50	80.5	170.0	1047	6
Hoya carnosa	285	0.08	22.8	27.2	296	11
Huernia sp.	174	0.14	24.4	47.6	317	7
Kalanchoë gastonis-bonnieri	155	0.27	41.9	91.8	545	6
Kalanchoë tomentosa	88	0.35	30.8	119.0	400	3
Sedum praealtum	81	0.29	23.5	98.6	306	3

Superscripts O_2 and CO_2 refer to the respective gasses. After data of Spalding et al. (1979)

by Spalding et al. (1979) greatly confirmed this using gas chromatography. They showed that in the light period in different CAM species, internal O_2 concentration varied between 21 and 42% with an extreme case even at 80% (Table 2).

Thus, with respect to the dual substrate affinity of RubisCO, we must realize that the effect of CO_2 concentrating in Phase III is counterbalanced by simultaneous O_2 concentrating. Clearly, the high internal $p_i^{O_2}$ in Phase III increases oxidative load and supports photorespiration. There is no doubt that photorespiration is going on in Phase III of CAM. Also for the C_3/CAM intermediate plant *Clusia minor*, which can reversibly switch between the two modes of photosynthesis, it was shown that photorespiration is active in the light period (Duarte and Lüttge 2007a). It even shows circadian oscillations in continuous light in *C. minor* in both modes of photosynthesis (Duarte and Lüttge 2007b). However, while Phase III photorespiration is now generally recognized, an open question remaining is if we can relate it quantitatively to what is going on in C_3 plants. The following sections will attempt to assess if notwithstanding O_2-concentrating the CO_2-concentrating in Phase III of CAM will still suppresses photorespiration and to which extent this may occur quantitatively.

5 Calculation of v^{O_2}/v^{CO_2} of RubisCO for C_3-Photosynthesis and Phase III of CAM

5.1 *Calculation of the Relevant Concentrations of CO_2 and O_2 in Aqueous Solution*

Spalding et al. (1979, see also Lüttge 2002) present internal leaf partial pressures of CO_2 in the gas phase and the corresponding O_2/CO_2 ratios during phase III in various CAM plants. From these data, we can calculate the ratio of the reaction rates of RubisCO with O_2 and CO_2, v^{O_2}/v^{CO_2}. The original data from Spalding et al. (1979) provide the partial pressures of O_2 and CO_2 in the leaf gas phase, $p_i^{O_2}$ and $p_i^{CO_2}$, in percent (Table 2). However, as RubisCO is not operating in the gas phase for assessing the reaction rates, we need the concentrations in aqueous solution (Buchanan et al. 2000: page 719). Due to the very different solubility of CO_2 and O_2 in water, the ratios of partial pressures and of concentrations of the two gasses are very different (Table 2). The concentrations, c, in water in equilibrium with the partial pressures in the gas phase are obtained from the Henry–Dalton law of absorption, i.e.,

$$C_H = \frac{c}{p_i}, \tag{2}$$

where C_H is the Henry constant, which at 25°C has a value of 1.3 mmol l^{-1} atm^{-1} for O_2 and a value of 34 mmol l^{-1} atm^{-1} for CO_2. The concentrations calculated from the partial pressures in phase III of CAM in Spalding et al. (1979) are also presented in Table 2. For *Ananas comosus*, these authors have two different entries. The partial CO_2-pressure in ambient air is taken as 0.04%. For the intracellular space in the leaves of C_3 plants, a partial pressure of CO_2 of 0.025% is assumed, because by ongoing photosynthesis the partial pressure is normally reduced to such a level.

There are some principle restrictions using (2) for the calculation of c^{CO_2} because unlike O_2 which is only physically dissolved in water CO_2 also is subjected to a chemical reaction with water, i.e.,

$$CO_2 + H_2O \leftrightarrow H_2CO_3. \tag{3}$$

However, in pure water at pH 7 the equilibrium is far on the left side with only 0.2% H_2CO_3. This is relevant for our calculations because the substrate of RubisCO is CO_2 and not HCO_3^- as it would be generated from

$$CO_2 + 2H_2O \leftrightarrow H_3O^+ + HCO_3^-. \tag{4}$$

In vivo, this reaction is catalyzed by carbonic anhydrase. At slightly alkaline pH as it is given in the cytosol and also in the stroma within the chloroplasts in the presence of carbonic anhydrase, the equilibrium of (4) is shifted to the right side. At pH 8 and 25°C, the ratio of CO_2/HCO_3^- in the equilibrium is 1:50. This is relevant for maintaining a gradient for inward diffusion of CO_2 into the chloroplasts, which is also facilitated by aquaporins (Uehlein et al. 2008). It is also important for carboxylation reactions using HCO_3^- as substrate, e.g., phospho*enol*pyruvate carboxylase (PEPC). However, the substrate of RubisCO is CO_2 and not HCO_3^-. Hence, notwithstanding the restrictions mentioned for assessment of the reaction rates of RubisCO with CO_2 with a reasonable approximation, we may still use the concentrations in aqueous solution in equilibrium with the partial pressures in the gas phase as given by (2).

Another point worth mentioning is that the equilibrium of (3) and (4) is pH-dependent. In CAM plants at the beginning of Phase III when malic acid is remobilized from the vacuoles into the cytoplasm, the cytoplasm is slightly acidified as the cytoplasmic pH is reduced from about 7.5 to 7.1 (Hafke et al. 2001). This would tend to shift the equilibrium vs. CO_2 and in comparison to C_3 plants, it would tend to increase the CO_2 concentration available to RubisCO. However, with the small change of pH in this range, the effect would be small. Moreover, it is not the cytoplasmic pH but the pH of the chloroplast stroma that is relevant. Although we do not have any data on this, it appears rather unlikely that the small cytoplasmic pH shift at the beginning of Phase III in CAM plants is directly influencing stroma pH. Hence, we may assume that all the reservations for application of (2) discussed above apply similarly to C_3 plants and CAM plants. Thus, they may somewhat affect assessment of absolute reaction rates but not evaluation of relative rates of C_3 and CAM species related to each other as relevant for the comparison of photorespiration in the two modes of photosynthesis attempted here. With acceptable approximation for this purpose we may then use the concentrations, c, calculated by (2) in Table 2 in the following assessments.

5.2 Assessment of v^{O_2}/v^{CO_2} Using Straightforward Michaelis–Menten Formalism with RubisCO-Kinetics Separately for CO_2 and O_2

If we take the Michaelis–Menten equation

$$v = \frac{V_{\max} \cdot [c]}{(K_{\mathrm{M}} + [c])}, \tag{5}$$

we can derive for the ratio of the reaction rates with O_2 and CO_2

$$\frac{v^{O_2}}{v^{CO_2}} = \frac{V_{\max}^{O_2} \cdot c^{O_2}}{\left(K_{\mathrm{M}}^{O_2} + c^{O_2}\right)} \cdot \frac{\left(K_{\mathrm{M}}^{CO_2} + c^{CO_2}\right)}{V_{\max}^{CO_2} \cdot c^{CO_2}} \tag{6}$$

or

$$\frac{v^{O_2}}{v^{CO_2}} = \frac{V_{\max}^{O_2}}{V_{\max}^{CO_2}} \cdot \frac{c^{O_2}}{c^{CO_2}} \cdot \frac{\left(K_{\mathrm{M}}^{CO_2} + c^{CO_2}\right)}{\left(K_{\mathrm{M}}^{O_2} + c^{O_2}\right)}. \tag{7}$$

Table 3 Kinetic constants of RubisCO for C_3 and CAM plants from Badger et al. (1974)

		K_M(μM)	V_{max} (μmol min^{-1} mg^{-1} chlorophyll)
Carboxylase	CAM	18	3.5
	C_3	16–21	5.6–9.7
Oxygenase	CAM	310	0.83
	C_3	180–280	1.1–2.6
$V_{max}^{O_2}/V_{max}^{CO_2}$	CAM	0.24	
	C_3	0.11–0.46	

Table 4 Ratios of the reaction rates of RubisCO with O_2 and CO_2, v^{O_2}/v^{CO_2}, for C_3 plants and for the O_2 and CO_2 concentrations measured in the leaves of various named CAM species in Phase III of CAM

	Equation (7)		Equation (9)	
	1	2	1	2
Ambient air	0.34	0.12–0.70	0.20	0.13–1.07
C_3 photosynthesis	0.41	0.16–0.96	0.32	0.20–1.72
CAM-species				
Opuntia monacantha	0.27	0.18	0.08	0.11
Ananas comosus	0.24	0.15	0.06	0.08
	0.42	0.20	0.06	0.08
Hoya carnosa	0.28	0.20	0.11	0.15
Huernia sp.	0.26	0.18	0.07	0.10
Kalanchoë gastonis-bonnieri	0.33	0.18	0.06	0.08
Kalanchoë tomentosa	0.27	0.14	0.03	0.04
Sedum praealtum	0.23	0.14	0.03	0.04

Calculations were done by (7) and (9), respectively. Calculations using constants from the textbook of Heldt and Piechulla (2008) and Buchanan et al. (2000) in columns 1 and from Badger et al. (1974, Table 2) in columns 2

In these equations, v is the actual reaction rate, V_{max} is maximum reaction rate, K_M is Michaelis constant, and c is the concentration, while superscripts indicate the gasses oxygen (O_2) and carbon dioxide (CO_2), respectively.

For the calculation of v^{O_2}/v^{CO_2} in addition to the concentrations, c^{O_2} and c^{CO_2}, from Table 2 we also need the Michaelis constants and we need to know the ratio of $V_{max}^{O_2}/V_{max}^{CO_2}$. We can take the values from textbooks, i.e., $K_M^{O_2} = 535\,\mu mol\,l^{-1}$ and $K_M^{CO_2} = 9\,\mu mol\,l^{-1}$ (Heldt and Piechulla 2008), and we may derive $V_{max}^{O_2}/V_{max}^{CO_2}$ from the specificity factor Σ of RubisCO (1). For extant vascular plants, Σ values reported are ranging between 80 and 130 (Table 1). For the calculations, we use a medium value of 100. Alternatively, we may use the values given by Badger et al. (1974) as listed in Table 3. With (7), we then arrive at the respective v^{O_2}/v^{CO_2} ratios presented in Table 4, i.e., the two left columns, where column 1 uses the Michaelis constants of Heldt and Piechulla (2008) and $\Sigma = 100$, and column 2 the values of Badger et al. (1974, Table 3). The latter give ranges of values for C_3 plants, and in the calculations of Table 4, we have taken combinations of values from these ranges yielding the lowest and the highest possible v^{O_2}/v^{CO_2} ratio, respectively.

5.3 Assessment of v^{O_2}/v^{CO_2} Using Michaelis–Menten Formalism with Kinetics of Mutual Competitive Inhibition of RubisCO by CO_2 and O_2

An alternative of using (7) as derived from (5) is taking into account that the two substrates, CO_2 and O_2, are actually mutually competitive inhibitors of the enzyme RubisCO. For considering competitive inhibition, we need to multiply K_M in the Michaelis–Menten equation (5) by $(1 + c^i/K_M^i)$, where i stands for inhibitor. The reaction rates are then

$$v^{CO_2} = \frac{V_{max}^{CO_2} \cdot c^{CO_2}}{\left(1 + \frac{c^{O_2}}{K_M^{O_2}}\right) \cdot K_M^{CO_2} + c^{CO_2}} \quad \text{and} \quad v^{O_2} = \frac{V_{max}^{O_2} \cdot c^{O_2}}{\left(1 + \frac{c^{CO_2}}{K_M^{CO_2}}\right) \cdot K_M^{O_2} + c^{O_2}}. \tag{8}$$

From this, we receive for the ratio of the reaction rates:

$$v^{O_2}/v^{CO_2} = \frac{1}{\sum} \cdot \frac{c^{O_2}}{c^{CO_2}} = \frac{V_{max}^{O_2}}{K_M^{O_2}} \cdot \frac{K_M^{CO_2}}{V_{max}^{CO_2}} \cdot \frac{c^{O_2}}{c^{CO_2}}. \tag{9}$$

An important assumption in this approach is that c^{CO_2} and c^{O_2} are smaller than the respective K_M values. This is given for O_2. It is just fulfilled for CO_2 in C_3-photosynthesis; however, with the high $p_i^{CO_2}$, it would not apply in Phase III of CAM. With (9) and assuming $p_i^{CO_2} = 0.036\%$, $c^{CO_2} = 8\,\mu M$ and $p_i^{O_2} = 21\%$, $c^{O_2} = 250\,\mu M$, Buchanan et al. (2000) arrive at a v^{O_2}/v^{CO_2} value of 0.3, which is very close to the value calculated for C_3-photosynthesis on the basis of the concentrations given in Table 2 and using the constants of Heldt and Piechulla (2008) and Buchanan et al. (2000). With the latter concentrations and constants (7) gave similar results for C_3-photosynthesis. Using the constants of Badger et al. (1974), results obtained from (7) and (9) for C_3-photosynthesis are also in the same range.

However, for the CO_2 and O_2 concentrations in Phase III of CAM (Table 2), the two approaches give contrasting results. Using (7) with the Michaelis constants of Heldt and Piechulla (2008) (column 1 in Table 4), the range of v^{O_2}/v^{CO_2} ratios obtained for the CAM plants in Phase III is from 0.23 to 0.33 if we do not consider the rather high values of c^{CO_2} and c^{O_2} of one of the data sets given for *Ananas comosus* by Spalding et al. (1979). With the $p_i^{CO_2}$ value of ambient air, C_3 plants would operate at the upper end of this range, and with an actual value of $p_i^{CO_2} = 0.025\%$ under photosynthesising conditions, they would show a relative rate of oxygenase activity to carboxylation activity higher by a factor of about 1.2–1.8 than given by the range obtained for Phase III in CAM plants. With the constants of Badger et al. (1974), v^{O_2}/v^{CO_2} ratios of CAM plants are at the lower end of the range given for C_3 plants (column 2 in Table 4). Using (9), relative rates of oxygenase activity to carboxylation activity in C_3-photosynthesis are higher by factors of about 2.8–10.3 [constants of Heldt and Piechulla (2008)] or 1.3–5.0 [constants of Badger et al. (1974)] than given by the range obtained for Phase III in CAM plants. Hence, (7) suggests a rather modest reduction of photorespiration in Phase III of CAM while (9) suggests a very strong suppression of photorespiration in Phase III.

6 Concomitant Measurements of CO_2 and O_2 Gas Exchange

In *Clusia minor*, Duarte and Lüttge (2007a, b) have simultaneously studied photorespiratory oxygen uptake, J_{O_2}, and photosynthetic CO_2 uptake, J_{CO_2}. They used 20 min gas pulses with only 1% O_2 to create nonphotorespiratory conditions. This allowed measuring J_{O_2}, in direct comparison to J_{CO_2}. For the C_3/CAM intermediate species *Clusia minor* in the C_3 state (Duarte and Lüttge 2007a) and for the peaks of gas exchange during endogenous rhythmicity under continuous light in the C_3 mode and in the CAM mode (Duarte and Lüttge 2007b), they obtained the values extracted from their gas exchange curves in Table 5.

The related ratios of J_{O_2}/J_{CO_2} are in very good agreement with the v^{O_2}/v^{CO_2} ratios calculated with (7)/column 1 of Table 4. The v^{O_2}/v^{CO_2} ratios obtained with the constants of Badger et al. (1974) and (7)/column 2 of Table 4 are somewhat lower than that, which is mainly due to the fact that these authors give lower $K_M^{O_2}$ and higher $K_M^{CO_2}$ than the values listed in the most recent textbook of Heldt and Piechulla (2008). The $K_M^{CO_2}$ given by Heldt and Piechulla (2008) is 9 μmol l^{-1} (see above) and quite close to the values of ca. 14 and ca. 11 μmol l^{-1} reported for the C_3 plant *Spinacia oleracea* and the CAM *Kalanchoës*, respectively, by Griffiths et al. (2004, Table 1). For CAM, the J_{O_2}/J_{CO_2} ratios of Duarte and Lüttge (2007a, b) are much higher than the v^{O_2}/v^{CO_2} ratios calculated by (9) (Table 4).

7 Evaluation of Assumptions

The calculations presented required assumptions. At the outset, it was necessary to calculate the concentrations of CO_2 and O_2 in the aqueous phase from the partial pressures in the gas phase given in the literature. Evidently, this was essential because RubisCO is not operating in the gas phase. Using the partial pressures in the gas phase for C_3-photosynthesis, one would obtain v^{O_2}/v^{CO_2} ratios of 8 using both (7) and (9). This is a totally unrealistic value, which underlines the need of

Table 5 Rates of photorespiratory oxygen uptake, J_{O_2}, and photosynthetic CO_2 uptake, J_{CO_2}, extracted from the curves of Duarte and Lüttge (2007a, b) and the corresponding ratios of J_{O_2}/J_{CO_2}

	C_3 mode	Peaks of endogenous oscillations	
		C_3 mode	CAM mode
J_{O_2} (μmol m^{-2} s^{-1})	2.4	1.5–2.3	1.7
J_{CO_2} (μmol m^{-2} s^{-1})	4.6	5.6	5.4
J_{O_2}/J_{CO_2}	0.52	0.27–0.41	0.32

calculating the concentrations. These calculations are fraught with assumptions which, however, appear to be reasonable approximations as detailed in Sect. 5.1.

Equation (7) is based on a separate consideration of enzyme reactions of RubisCO with either CO_2 or O_2 and ignores the formalism of the kinetics of competitive inhibition. Conversely, (9) incorporates mutual competitive inhibition of CO_2 and O_2 at the RubisCO enzyme. However, this requires the assumption that the concentrations are lower than the respective K_M values (Buchanan et al. 2000). This is largely given for O_2 ($c^{O_2} = 272$ to 545 μM, Table 2; $K_M^{O_2} = 535$ μM) and for CO_2 in C_3-photosynthesis ($c^{CO_2} = 8.5$ μM, $K_M^{CO_2} = 9$ μM), but is by far not fulfilled for the very high concentrations of CO_2 in Phase III of CAM ($c^{CO_2} = 27$ to 119 μM, Table 2; $K_M^{CO_2} = 9$ μM).

Hence, it is notable that (7) and (9) give similar results on v^{O_2}/v^{CO_2} ratios for C_3-photosynthesis. Evidently, the assumption inherent in considering the Michaelis–Menton kinetics for v^{CO_2} and v^{O_2} separately, (7) is acceptable because the implementation of inhibitor kinetics (9) does not greatly affect the result while the assumption of CO_2-concentration being equal to $K_M^{CO_2}$ required by (9) is fulfilled. For Phase III of CAM, where the latter assumption is not fulfilled, (9) indicates a much larger suppression of photorespiration than (7). As compared with C_3-photosynthesis in Phase III of CAM, v^{O_2}/v^{CO_2} is suppressed by a factor of 1.2–1.8 according to (7) and by a factor of 2.8–10.3 according to (9) (Sect. 5.3).

Of course, since there are different assumptions inherent in (7) and (9), it is hard to decide at this stage which one to follow. For this purpose, however, it may be helpful to consider the actual measurements presented in Table 5. Here we see that measurements of gas exchange give a factor for J_{O_2}/J_{CO_2} in C_3-photosynthesis to Phase III of CAM of 0.8–1.6 (depending on which combination of values we use from the bottom line of Table 5). This matches with the results for v^{O_2}/v^{CO_2} obtained from (7) but not (9). It may well be then that for Phase III of CAM, (7) is a better approach than (9).

8 Conclusions and Outlook

If we then dwell on (7), the two comparisons given in (7)/column 1 and 2, respectively, of Table 4 show that v^{O_2}/v^{CO_2} values of Phase III tend to be lower than those obtained for C_3 photosynthesis. However, they are not completely different and well within the same order of magnitude. With the assumptions inherent in the calculations of (7)/column 1, the v^{O_2}/v^{CO_2} values would show a relative rate of oxygenase activity to carboxylation activity about 25–80% higher for C_3 photosynthesis than for Phase III in CAM plants. The ratios of J_{O_2}/J_{CO_2} in Table 5 suggest that in *C. minor*, oxygenase activity in the C_3 mode may be identical to that in the CAM mode or up to 1.6-fold higher. The assessments of (7)/column 2 in Table 4 show that v^{O_2}/v^{CO_2} ratios of C_3-photosynthesis and Phase III of CAM overlap at the lower end of the range of C_3-photosynthesis but the highest v^{O_2}/v^{CO_2} ratios given for C_3 photosynthesis may be 6.5-fold larger than those of Phase III of CAM. This would allow concluding that under physiological conditions of C_3-photosynthesis and Phase III of CAM oxygenase activity in relation to carboxylase activity of RubisCO is not fundamentally different. At the very most in Phase III of CAM, it would be about half of that in C_3 photosynthesis, but mostly it would be much less reduced than that. Notwithstanding its CO_2-concentrating mechanism, all implications of photorespiration would appear fully relevant to CAM not only in its Phase IV but also in Phase III. Conversely, taking the results obtained with (9), photorespiration would be very strongly suppressed in Phase III. At the moment, I consider this as less likely as argued in Sect. 7. It remains to be seen though. More measurements such as those described in Table 5 and concurrent measurements of carboxylase and oxygenase activities (Cousins et al. 2010) should shed more light on this and tell us, to which extent we can draw conclusions based on gas concentrations and using enzyme kinetic formalisms.

Another current restriction is that data sets which are available for comparative evaluations like the ones presented here are limited. Particularly, we have no information about physiological and eco-physiological implications. We do not know to which extent different conditions of measurements would differentially affect the comparisons between C_3-photosynthesis and CAM-Phase III. With the exception of the measurements of Table 5, we are completely lacking quantitative comparative measurements of photorespiration during C_3-photosynthesis and CAM-Phase III under identical environmental and eco-physiological conditions.

For an advancement of our understanding the evolutionary and ecophysiological role of photorespiration in the performance of CAM plants, it also would be important to assess the relative contributions of Phases III and IV as compared with C_3-photosynthesis. Such a broader approach in future work may emerge from the experiments of Maxwell et al. (1998). Occurrence of photorespiration in Phase IV has been early documented by Heber et al. (1996). As noted above (Sect. 3), it was the particular aim of the present essay to restrict itself to Phase III because the other aspects have been well covered recently in Progress in Botany by Niewiadomska and Borland (2007) and a detailed treatment of Phase IV would have been repetitive here. The relative expression of Phases III and IV is modulated by environmental conditions. An extended Phase IV is only expressed when supply of water is sufficient for stomata to open and allow CO_2 and O_2 gas exchange with the atmosphere (Smith and Lüttge 1985) and its modulation of v^{O_2}/v^{CO_2} of RubisCO.

Acknowledgment I thank Susanne von Caemmerer and C. Barry Osmond, Canberra, for valuable comments and suggestions.

References

Badger MR, Andrews TJ, Osmond CB (1974) Detection in C_3, C_4 and CAM plant leaves of a low K_m (CO_2) form of RuDP carboxylase having high RuDP oxygenase activity at physiological pH. In: Avron M (ed) Proc 3rd Int Congr Photosnythesis, Elsevier, Amsterdam, 1421–1429

Buchanan BB, Gruissem W, Jones RL (2000) Biochemistry and molecular biology of plants. American Society of Plant Physiologists, Rockville, Maryland

Cousins AB, Ghannoum O, von Caemmerer S, Badger MR (2010) Simultaneous determination of Rubisco carboxylase and oxygenase kinetic parameters in *Triticum aestivum* and *Zea mays* using membrane inlet mass spectrometry. Plant Cell Environ 33:444–452

Duarte HM, Lüttge U (2007a) Correlation between photorespiration, CO_2 assimilation and spatiotemporal dynamics of photosynthesis in leaves of the C_3-photosynthesis/crassulacean acid metabolism-intermediate species *Clusia minor* L. (Clusiaceae). Trees 21:531–540

Duarte HM, Lüttge U (2007b) *Clusia*'s clock. The background of endogenous rhythmicity of C_3- and C_4-photosynthesis and crassulacean acid metabolism (CAM). In: Lüttge U (ed) *Clusia*. A woody neotropical genus of remarkable plasticity and diversity. Ecological studies, vol 194. Springer-Verlag, Berlin, Heidelberg, pp 245–256

Faak M (ed) (2000) Alexander von Humboldt. Reise durch Venezuela. Auswahl aus den amerikanischen Reisetagebüchern. Beiträge zur Alexander von Humboldt-Forschung 12. Akademie-Verlag, Berlin, p 193

Griffiths H, Robe WE, Girnus J, Maxwell K (2008) Leaf succulence determines the interplay between carboxylase systems and light use during Crassulacean acid metabolism in *Kalanchoë* species. J Exp Bot 59:1851–1861

Hafke JB, Neff R, Hütt MT, Lüttge U, Thiel G (2001) Day-to-night variations of cytoplasmic pH in a crassulacean acid metabolism plant. Protoplasma 216:164–170

Heber U, Neimanis S, Kaiser WM (1996) Regulation of crassulacean acid metabolism in *Kalanchoë pinnata* as studied by gas exchange and measurements of chlorophyll fluorescence. In: Winter K, Smith JAC (eds) Crassulacean acid metabolism: biochemistry, ecophysiology and evolution. Ecological studies, vol 114. Springer, Berlin, Heidelberg, New York, pp 78–96

Heldt HW, Piechulla B (2008) Pflanzenbiochemie, 4th edn. Spektrum Akademischer Verlag, Heidelberg

Jordan DB, Ogren WL (1981) Species variation in the specificity of ribulosebisphosphate carboxylase/oxygenase. Nature 291:513–515

Lüttge U (2002) CO_2-concentrating: consequences in crassulacean acid metabolism. J Exp Bot 53:2131–2142

Maxwell K, Badger MR, Osmond CB (1998) A comparison of CO_2 and O_2 exchange patterns and the relationship with chlorophyll fluorescence during photosynthesis in C_3 and CAM plants. Aust J Plant Physiol 25:45–52

Niewiadomska E, Borland AM (2007) Crassulacean acid metabolism: a cause or consequence of oxidative stress in planta? Prog Bot 69:247–266

Osmond CB (1978) Crassulacean acid metabolism: a curiosity in context. Annu Rev Plant Physiol 29:379–414

Smith JAC, Lüttge U (1985) Day-night changes in leaf water relations associated with the rhythm of crassulacean acid metabolism in *Kalanchoë daigremontiana*. Planta 163:272–282

Spalding MH, Stumpf DK, Ku MSB, Burris RH, Edwards GE (1979) Crassulacean acid metabolism and diurnal variation of internal CO_2 and O_2-concentrations in *Sedum praealtum* DC. Aust J Plant Physiol 6:557–567

Spreitzer RJ (1999) Questions about the complexity of chloroplast ribulose-1,5-bisphosphate carboxylase/oxygenase. Photosynth Res 60:29–42

Spreitzer RJ, Salvucci ME (2002) RUBISCO: structure, regulatory interactions, and possibilities for a better enzyme. Annu Rev Plant Biol 53:449–475

Uehlein N, Otto B, Hanson DT, Fischer M, McDowell N, Kaldenhoff R (2008) Function of *Nicotiana tabacum* aquaporins as chloroplast gas pores challenges the concept of membrane CO_2 permeability. Plant Cell 20:648–657

Uemura K, Anwaruzzaman M, Miyachi S, Yokota A (1997) Ribulose-1,5-bisphosphate carboxylase/oxygenase from thermophilic red algae with a strong specificity for CO_2 fixation. Biochem Biophys Res Commun 233:568–571

Whitney SM, Andrews TJ (1998) The CO_2/O_2 specificity of single-subunit ribulose-bisphosphate carboxylase from the dinoflagellate, *Amphidinium carterae*. Aust J Plant Physiol 25:131–138

Index

MIX
Papier aus verantwortungsvollen Quellen
Paper from responsible sources
FSC® C105338

If you have any concerns about our products,
you can contact us on
ProductSafety@springernature.com

In case Publisher is established outside the EU,
the EU authorized representative is:
Springer Nature Customer Service Center GmbH
Europaplatz 3, 69115 Heidelberg, Germany

Printed by Libri Plureos GmbH
in Hamburg, Germany